METHODS OF ENGINEERING MATHEMATICS

METHODS OF ENGINEERING MATHEMATICS

Edward Haug
Kyung K. Choi

The University of Iowa

PRENTICE HALL, Englewood Cliffs, NJ 07632

Library of Congress Cataloging-in-Publication Data

Haug, Edward J.
Methods of engineering mathematics / Edward J. Haug, Kyung Choi.
p. cm.
Includes bibliographical references and index.
ISBN 0-13-579061-1
1. Engineering Mathematics. I. Choi, Kyung K. II. Title.
TA330.H39 1993
620'.001'515--dc20
92-40569
CIP

Acquisitions editor: **PETER JANZOW**
Production editor: **JOE SCORDATO**
Cover design: **JOE DiDOMENICO**
Prepress buyer: **LINDA BEHRENS**
Manufacturing buyer: **DAVID DICKEY**
Editorial assistant: **PHYLLIS MORGAN**
Supplements editor: **ALICE DWORKIN**

Printed in the United States of America

10 9 8 7 6 5 4 3 2 1

ISBN 0-13-579061-1

Prentice-Hall International (UK) Limited, London
Prentice-Hall of Australia Pty. Limited, Sydney
Prentice-Hall Canada Inc., Toronto
Prentice-Hall Hispanoamericana, S.A., Mexico
Prentice-Hall of India Private Limited, New Delhi
Prentice-Hall of Japan, Inc., Tokyo
Simon & Schuster Asia Pte. Ltd., Singapore
Editora Prentice-Hall do Brasil, Ltda., Rio de Janeiro

ISBN 0-13-579061-1

90000
9 780135 790618

Dedicated to

ALEXANDER J. SCHWARZKOPF

Inspired Research Manager, Mentor, and Friend

CONTENTS

PREFACE

Purpose and Uses of Text

Emergence of the digital computer as the dominant tool in engineering and applied sciences has profoundly influenced the mathematical approach to formulation, analysis, and solution of equations that characterize the behavior of physical systems. Normed linear spaces, function spaces, linear operators, and variational methods that were once the province of abstract mathematics are now the foundation for systematic methods of engineering mathematics that underlie scientific and engineering computation. The purpose of this text is to develop and illustrate the use of basic mathematical methods that are essential for advanced study in a number of engineering disciplines. The fundamentals of linear algebra, matrices, infinite series, function spaces, ordinary differential equations, and partial differential equations are developed at an intermediate level of mathematical sophistication. Statements of definitions and presentations of theorems and their proofs are at the level of mathematical rigor that is required for advanced applications and research in Mechanical, Civil, Industrial, Aeronautical, Chemical, and Biomedical Engineering and related disciplines that employ modern methods of mechanics. Model engineering applications are employed throughout the text to illustrate the use of methods presented and to provide the engineer with a concrete framework in which mathematical methods may be used and understood.

This text has evolved from a two-course sequence that has been taught at the advanced undergraduate and first-year graduate level in the College of Engineering at The University of Iowa over the past twenty years. The first course, which constitutes Chapters 1 through 8 of this text, is a corequisite for intermediate-level courses in mechanics of materials, structural mechanics, finite element methods, dynamics, fluid mechanics, and thermal sciences. This material and the contents of Chapters 9 through 12 are prerequisites for advanced courses in continuum mechanics, elasticity, structural mechanics, dynamics of machines, finite element methods, fluid mechanics, and design optimization. The level of rigor of the text is the minimum essential to support modern engineering applications.

Topical coverage in this text has been selected to support basic graduate-level courses in engineering. Every attempt has been made to avoid making the text a cookbook that covers a plethora of methods, but without an adequate level of rigor. As a result, coverage has been limited to methods that are essential in numerous engineering courses and in engineering applications. Breadth of coverage has been sacrificed in favor of a reasonably thorough and rigorous treatment of central topics and mathematical methods that support the engineering sciences.

Organization of Text

Linear algebraic equations occur in all branches of applied science and engineering. Chapter 1 develops basic tools of matrix algebra that play an important role in the solution of linear equations and in high-speed digital computation. For most engineers, the material in Chapter 1 should be a review of techniques learned in their undergraduate matrix algebra course.

Chapter 2 presents concepts of linear algebra and matrix theory in a vector space setting, by first introducing the algebraic representation of geometric vectors. Vector algebra in finite dimensional vector spaces is represented in terms of matrix operations, which reduce vector operations to a form that can be easily implemented on a digital computer. Vector space concepts are also used to determine existence and uniqueness of solutions of linear equations, without necessarily solving them. The concept of a norm is introduced as an extension of the idea of distance between vectors and scalar product as an extension of the concept of the scalar product of geometric vectors. These generalizations of conventional vector properties provide the engineer with a geometric viewpoint that takes advantage of basic ideas of three-dimensional vector analysis, even in analysis of partial differential equations.

Eigenvalue problems that arise in engineering applications such as vibration and stability and in mathematical techniques such as separation of variables and eigenvector expansion methods are introduced in Chapter 3. Another concept introduced in Chapter 3 is quadratic forms that are associated with energy in mechanical systems. Under physically meaningful conditions, it is shown that the solution of equilibrium equations in mechanics is equivalent to minimization of a quadratic form, which may be the total potential energy of the system, setting the stage for variational methods of analysis.

To support series solution techniques for ordinary and partial differential equations, infinite series whose terms are constants and functions are studied in Chapter 4. The important concept of uniform convergence of infinite series whose terms are functions is introduced and used to establish the convergence of power series solutions of ordinary differential equations. The Frobenius method is presented and special functions, such as Bessel functions, are introduced.

Many engineering applications deal with continuum behavior of a solid or fluid medium, which may be described by a function $u(\mathbf{x}, t)$ of a spatial variable $\mathbf{x}$ and possibly time t. To help the engineer gain insight in continuum applications, the idea of function spaces of candidate solutions is introduced in Chapter 5. The concept of a collection of functions that are candidate solutions is shown to be a natural extension of the finite dimensional vector space ideas developed in Chapters 1 through 3. Much as in Chapter 2, function space algebra is defined and the concepts of scalar product and norm are used to establish a natural algebra and geometry of function spaces. With a norm and scalar product, concepts of closeness of approximation and orthogonality of functions are developed. One of the most important function spaces introduced in this chapter is the space $L_2(\Omega)$ of square integrable functions on a domain Ω in the physical Euclidean space R^n. Similarities in basic properties and concepts of vector analysis in the finite dimensional space R^n and in the infinite dimensional function space $L_2(\Omega)$ are established to aid the engineer's intuition. The infinite dimensionality of function spaces is shown to require that limit concepts be introduced in the function space setting. As a tool to describe a variety of functions, methods of constructing Fourier sine and cosine series and the theory of their convergence are developed.

In Chapter 6, differential equations of model continuum problems that are encountered in fluid dynamics, elasticity, and heat transfer are derived, using conservation laws of mechanics, multiple integral theorems, and the concept of the material derivative. While many engineering applications have fundamentally different physical properties, it is shown that the governing equations for each of these fields fall into one of four different forms of second-order partial differential equation: elliptic (equilibrium), parabolic (heat), hyperbolic (wave), and eigenvalue equations.

Second-order linear partial differential equations are classified in Chapter 7, from a mathematical point of view, using characteristic variables. It is shown that the same three basic forms of equations obtained in Chapter 6, from an applications point of view, are obtained using a mathematical criteria. The Cauchy–Kowalewski theorem that gives a theoretical method for analyzing second-order partial differential equations is used in this chapter. The concepts of stability of solutions and well-posed problems are also introduced and related to the physical behavior of engineering systems.

Methods of solving second-order partial differential equations derived in Chapter 6 and classified in Chapter 7 are developed and illustrated in Chapter 8. Solution methods such as the method of separation of variables and D'Alembert's formula are presented and shown to be broadly applicable.

Partial differential equations that govern the mechanics of a broad range of engineering systems are formulated in Chapter 9, in terms of linear operators. Basic concepts of linear operators are introduced and properties such as symmetry, boundedness, and positive definiteness are defined and related to physical characteristics of applications. Sturm–Liouville problems that arise in separation of variables and associated eigenvalue problems are studied in detail and completeness of their eigenfunctions is established. Separation of variables and eigenfunction expansion methods introduced in Chapter 8 for second-order partial differential equations are extended to general problems of engineering analysis. Green's functions are introduced for boundary-value problems and used to establish completeness of eigenfunctions of broad classes of operators. It is shown that the methods developed for model problems in Chapters 1 through 8 are in fact broadly applicable in engineering analysis, with mathematical properties that support both theoretical and computational methods.

Variational methods for solving boundary-value problems that have become standard tools in the applied mathematics community are introduced and developed in Chapter 10. Concepts of convergence in energy are shown to be physically natural and, in the case of positive-bounded below operators of mechanics, to yield stronger results than conventional convergence in the mean. The equivalence between solving boundary-value problems and minimizing energy functionals is established, as the foundation for modern computational methods in engineering mechanics. Basic ideas of the calculus of variations are introduced and used to establish criteria for identifying natural boundary conditions that may be ignored during variational solution of boundary-value problems. The Ritz method for creating approximate solutions of boundary-value problems is developed and its convergence properties are established for both equilibrium and eigenvalue problems. Finally, the Galerkin method is introduced and shown to be an extension of the Ritz method for broad classes of applications.

In order to provide the reader with experience in formulation and numerical application of variational methods, moderate scale applications are presented in Chapter 11. The governing equations in these applications are shown to satisfy criteria for application of variational methods and approximating functions are selected to be both theoretically correct and practical for computation. Analytical and numerical results for second-order and higher-order ordinary and partial differential equations are presented and analyzed, to illustrate convergence characteristics of the Ritz and Galerkin methods.

Chapter 12 introduces basic ideas of modern finite element methods, as a means to systematically construct approximating functions that are well suited for application of variational methods. Finite element approximations are presented and illustrated with applications to beams and plates. Numerical comparisons with applications developed in Chapter 11 are presented, to provide the reader with confidence that systematic finite element methods can be used to obtain sound approximate solutions of realistic boundary-value problems.

Acknowledgments

Many colleagues and students have been instrumental in guiding the development of this text, only a few of whom can be noted. Professors Royce Beckett, Darrell Penrod, and Kwan Rim encouraged and contributed to the development of a two-course sequence on advanced mathematical methods for continuum mechanics at The University of Iowa in the late 1960s, which was the genesis of this text. The late Professor Marvin Stippes of the University of Illinois suggested a balance between the underlying principles of mechanics, methods of deriving governing equations, and methods of solving equations, which the authors have attempted to maintain. Professor Louis Landweber suggested topical coverage to support fluid mechanics and provided constructive suggestions based on his teaching of courses with an early version of the text. Drs. Norman Coleman and Edward Daggit taught parts of courses using the text, contributed materials on which several sections are based, and significantly influenced the selection of underlying mathematical concepts used in the text. A steady stream of students in many disciplines who took courses using versions of this text over the past twenty years provided feedback to the authors that permitted text refinement to meet the needs of the intended audience.

The final Macintosh Microsoft Word version of the manuscript, from which this text was printed, was masterfully prepared by Alan Kallmeyer, who took two courses using the text. In addition to precise and clear setting of the manuscript, he made numerous suggestions that improved clarity of presentation and found numerous errors that would otherwise have crept into the text.

1

MATRICES

A variety of mathematical methods from the fields of linear algebra, vector analysis, and ordinary and partial differential equations form the foundation for methods of engineering mathematics. While the full theory of engineering sciences rests primarily on the theory of differential equations, matrix methods play a key role in analysis and solution of algebraic equations and differential equations. It is the purpose of this chapter to provide a summary of matrix methods and supporting techniques, which are of value both in their own right and in preparation for the study of differential equations that arise in engineering science. Applications of the basic ideas and methods to relatively simple problems are presented in this chapter to aid the reader in developing intuition and background for use of these techniques. As more complex problems and techniques arise, the coherence of this body of techniques will become apparent. The scope of this chapter, therefore, includes a review of basic matrix theory and provides an introduction to techniques that will serve as the foundation for later study of more mathematically challenging problems.

1.1 LINEAR EQUATIONS

Linear equations occur in virtually all branches of applied science and engineering; e.g., electrical networks, structures, kinematics and dynamics of machines, curve fitting in statistics, and transportation problems, in the general form

$$\begin{aligned} a_{11}x_1 + a_{12}x_2 + \dots + a_{1\alpha}x_\alpha &= b_1 \\ a_{21}x_1 + a_{22}x_2 + \dots + a_{2\alpha}x_\alpha &= b_2 \\ \cdots\cdots\cdots\cdots&\cdots\cdots \\ a_{\alpha 1}x_1 + a_{\alpha 2}x_2 + \dots + a_{\alpha\alpha}x_\alpha &= b_\alpha \end{aligned} \tag{1.1.1}$$

where a_{ij} and b_i are constants and x_i are unknowns.

Examples of Linear Equations

Example 1.1.1

Consider the problem of determining **member loads** in the space truss of Fig. 1.1.1 (a). Summing forces that act on node D [see Fig. 1.1.1 (b)] in the x, y, and z directions yields the **equilibrium equations**

$$\sum F_x = -\frac{4.5}{12.5} F_A + \frac{5}{13} F_B + 1200 = 0$$

$$\sum F_y = -\frac{10}{12.5} F_A - \frac{12}{13} F_B - \frac{8}{10} F_C = 0$$

$$\sum F_z = \frac{6}{12.5} F_A - \frac{6}{10} F_C = 0$$

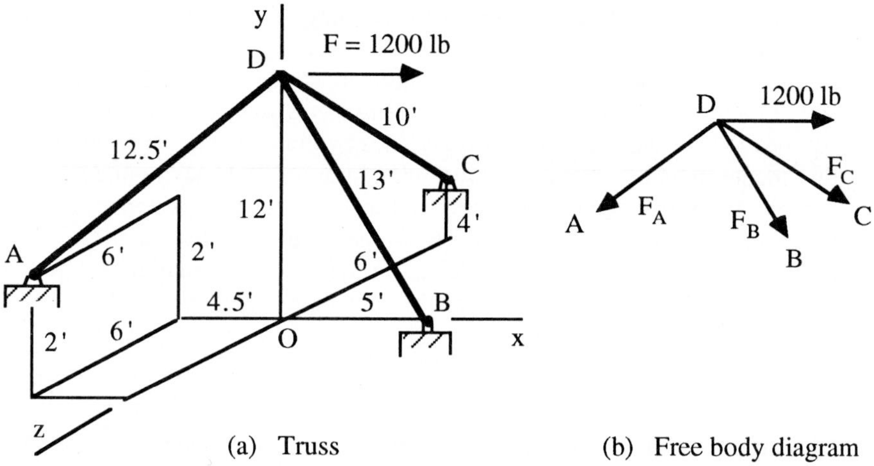

Figure 1.1.1 Determination of Member Loads in Space Frame

These linear equations may be solved for the unknown forces F_A, F_B, and F_C in the members. ■

Example 1.1.2

Consider the one-dimensional **steady-state heat transfer** problem shown in Fig. 1.1.2 (a), where a plane wall separates two fluids that are at different temperatures. Heat transfer occurs by **convection** from the hot fluid with temperature $t_{\infty,1}$ to the left surface of the wall, which is at temperature $t_{s,1}$. Heat flows by **conduction** through the wall to its right surface, which is at temperature $t_{s,2}$, and by convection to the cold fluid with temperature $t_{\infty,2}$.

For **steady-state conditions**, the linear temperature distribution in the wall is

$$t(x) = (t_{s,2} - t_{s,1}) \frac{x}{L} + t_{s,1}$$

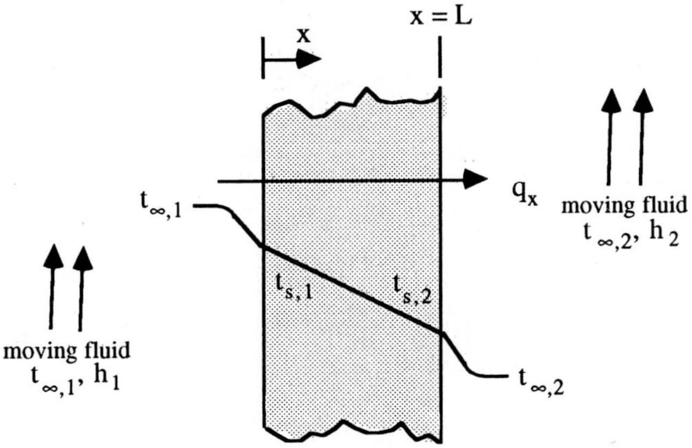

Figure 1.1.2 Heat Transfer From Fluids Through a Plane Wall

From **Fourier's law**, the **heat transfer rate** due to conduction in the wall is

$$q_x = -kA\frac{dt}{dx} = \frac{kA}{L}(t_{s,1} - t_{s,2})$$

where A and k are the area of the wall normal to the direction of heat transfer and the **thermal conductivity** of the wall material, respectively. For convection,

$$\frac{|t_s - t_\infty|}{|q_x|} = \frac{1}{hA}$$

where h is a **convection heat transfer coefficient**. The heat transfer rate may be determined from separate consideration of each mode of heat transfer; i.e.,

$$\frac{t_{\infty,1} - t_{s,1}}{(1/h_1A)} = q_x$$

$$\frac{t_{s,1} - t_{s,2}}{(L/kA)} = q_x$$

$$\frac{t_{s,2} - t_{\infty,2}}{(1/h_2A)} = q_x$$

These three linear equations determine the unknowns $t_{s,1}$, $t_{s,2}$, and q_x, once $t_{\infty,1}$, $t_{\infty,2}$, h_1, h_2, k, and A are known. ■

Example 1.1.3

Consider the multiple standpipe fluid system shown in Fig. 1.1.3. The variable y denotes elevation and y = 0 is the reference level. The ith standpipe has cross-sectional area A_i and the density of the fluid is γ.

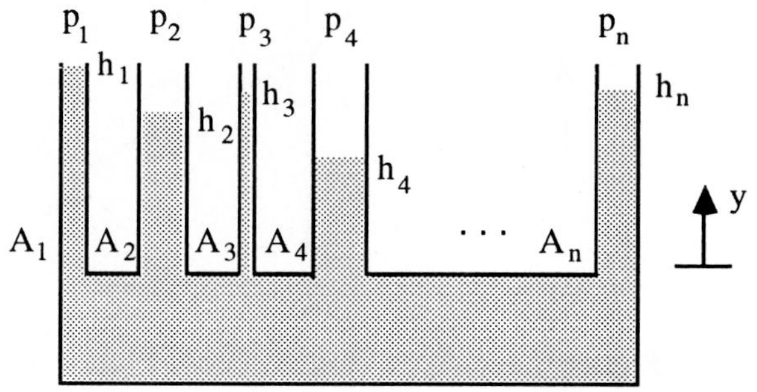

Figure 1.1.3 Static Fluid System

Each standpipe has a known pressure p_i that acts on the fluid free surface. At y = 0, the pressure in all standpipes must be equal; i.e.,

$$\gamma h_i + p_i = \gamma h_1 + p_1, \quad i = 2, 3, \ldots, n$$

which are n − 1 linear equations in n unknowns, h_i, i = 1, 2, . . . , n. Conservation of mass of the incompressible fluid requires that

$$\sum_{i=1}^{n} A_i h_i = V$$

where V is the known volume of fluid in the system. With γ, p_i, A_i, and V known, these relations provide n linear equations in n unknowns, $h_1, \ldots, h_n$. ■

Solution of Linear Equations

Elementary operations for manipulation of linear equations can be used to find their solution. A simple set of rules allows the following operations, which do not alter the solution:

(1) multiply through an equation by a nonzero constant
(2) add one equation to another equation
(3) interchange two equations

While these elementary operations preserve the solution of a set of equations, they do not by themselves create solutions. It is only through their systematic application that solutions of equations are obtained. Prior to studying general methods of solving linear equations, based on these rules, it is helpful to consider an example.

Example 1.1.4

Consider a system of three equations in three unknowns x_i, i = 1, 2, 3,

$$
\begin{aligned}
2x_1 + 2x_2 + 4x_3 &= -2 \\
x_1 + 2x_2 - x_3 &= 0 \\
3x_2 - 6x_3 &= 3
\end{aligned}
$$

The first equation can be multiplied by the inverse of the coefficient of x_1, which happens to be the largest coefficient of this variable in the equations, to obtain

$$
\begin{aligned}
x_1 + x_2 + 2x_3 &= -1 \\
x_1 + 2x_2 - x_3 &= 0 \\
3x_2 - 6x_3 &= 3
\end{aligned}
$$

In order to eliminate x_1 from the second equation, the first equation can be multiplied by −1 and added to the second equation, to obtain

$$
\begin{aligned}
x_1 + x_2 + 2x_3 &= -1 \\
x_2 - 3x_3 &= 1 \\
3x_2 - 6x_3 &= 3
\end{aligned}
$$

Examining the coefficients of x_2 in the second and third equations, the largest is found in the third equation. The second and third equations can be interchanged to bring the largest coefficient of x_2 in or below the second equation into the second equation. Interchanging the second and third equations and multiplying both sides of the new second equation by 1/3 yields

$$
\begin{aligned}
x_1 + x_2 + 2x_3 &= -1 \\
x_2 - 2x_3 &= 1 \\
x_2 - 3x_3 &= 1
\end{aligned}
$$

Multiplying the second equation by −1 and adding the result to the third equation yields

$$
\begin{aligned}
x_1 + x_2 + 2x_3 &= -1 \\
x_2 - 2x_3 &= 1 \\
-x_3 &= 0
\end{aligned}
$$

The only variable that remains in the third equation is x_3, so

$$x_3 = 0$$

With x_3 known, the second equation may be written in the form

$$x_2 = 1 + 2x_3 = 1$$

Finally, with x_2 and x_3 known, the first equation yields

$$x_1 = -1 - (x_2 + 2x_3) = -2$$

■

The basic idea of the manipulation carried out in Example 1.1.4 is to use elementary operations to reduce coefficients of all variables that appear below the diagonal of the left side of the system of equations to zero. Equations are occasionally interchanged so that the largest coefficient of a variable that occurs on or below the diagonal is brought to the diagonal position. This avoids a division by zero or by a very small constant during the process of creating zero coefficients of the same variable in equations below the diagonal. This process is called **forward elimination**. Having completed this process, the last variable only appears in the last equation and may be easily determined. The next to the last equation may then be solved for the next to last variable. The process is continued until the first variable is determined. This procedure is called **back substitution**. The combination of forward elimination and back substitution that is illustrated here with three equations in three unknowns is called **Gaussian elimination**. It is developed in more generality in Sections 1.3 and 2.3.

EXERCISES 1.1

1. Solve the following equations by forward elimination and back substitution:

(a)
$$\begin{aligned} 2x_1 + x_2 + x_3 &= 1 \\ 4x_1 + x_2 \phantom{{}+ x_3} &= -2 \\ -2x_1 + 2x_2 + x_3 &= 7 \end{aligned}$$

(b)
$$\begin{aligned} 2x_1 - 3x_2 \phantom{{}+ x_3} &= 8 \\ 4x_1 - 5x_2 + x_3 &= 15 \\ 2x_1 \phantom{{}- 5x_2} + 4x_3 &= 1 \end{aligned}$$

(c)
$$\begin{aligned} x_1 + x_2 - x_3 &= 2 \\ 3x_1 + 3x_2 + x_3 &= 2 \\ x_1 \phantom{{}+ 3x_2} + x_3 &= 0 \end{aligned}$$

(d)
$$\begin{aligned} x_1 + 2x_2 + 3x_3 &= 6 \\ 2x_1 - 3x_2 + 2x_3 &= 14 \\ 3x_1 + x_2 - x_3 &= -2 \end{aligned}$$

1.2 MATRIX ALGEBRA

One of the most practical and powerful tools utilized in analysis of linear systems is matrix algebra. Matrix notation and rules of manipulation play an important role in development of coherent theoretical techniques, practical solution techniques, and computational methods for modern high-speed digital computers. Thorough familiarity and proficiency in matrix algebra is required for successful study of virtually all fields of engineering and applied science.

Matrix Equations

Many problems that arise in engineering mathematics require the solution of a system of β **linear equations** in α unknowns; i.e.,

$$\begin{aligned} a_{11}x_1 + a_{12}x_2 + \ldots + a_{1\alpha}x_\alpha &= b_1 \\ a_{21}x_1 + a_{22}x_2 + \ldots + a_{2\alpha}x_\alpha &= b_2 \\ \cdots\cdots\cdots\cdots\cdots\cdots\cdots\cdots & \\ a_{\beta 1}x_1 + a_{\beta 2}x_2 + \ldots + a_{\beta\alpha}x_\alpha &= b_\beta \end{aligned} \tag{1.2.1}$$

where the parameters $a_{11}, a_{12}, \ldots, a_{\beta\alpha}$ and $b_1, b_2, \ldots, b_\beta$ are given real constants and $x_1, \ldots, x_\alpha$ are variables that are to be determined.

If $b_1 = b_2 = \ldots = b_\beta = 0$, then Eq. 1.2.1 is called a system of **homogeneous equations** and $x_1 = x_2 = \ldots = x_\alpha = 0$ is called the **trivial solution**. A solution of the homogeneous system in which at least one of $x_1, x_2, \ldots, x_\alpha$ is not zero is called a **nontrivial solution.**

In order to simplify Eq. 1.2.1, **summation notation** may be used; i.e.,

$$\sum_{j=1}^{\alpha} a_{ij}x_j \equiv a_{ij}x_j = b_i, \quad i = 1, 2, \ldots, \beta \tag{1.2.2}$$

in which a **repeated Latin index** in a product (i.e., a Latin index that appears twice in a single term or a product) is summed over its range. In this case, index j on the left of Eq. 1.2.2 is summed from 1 to α. A repeated Greek index in a product is not summed. This is the reason α and β were selected in Eqs. 1.1.1 and 1.2.2, where no summation was to be implied by repeated indices. Note that an index that is not repeated is not summed and normally appears on both sides of the associated equation; e.g., index i in Eq. 1.2.2. The following illustrate common uses of index notation:

$$\sum_{i=1}^{\alpha} a_{ii} \equiv a_{ii}$$

$$\sum_{i=1}^{\alpha} x_i y_i \equiv x_i y_i$$

Another way of writing a system of linear equations is in terms of matrices.

Definition 1.2.1. A **matrix** is an m×n array of real numbers, arranged in rows and columns as

$$\mathbf{A} = \begin{bmatrix} a_{11} & a_{12} & \cdots & a_{1n} \\ a_{21} & a_{22} & \cdots & a_{2n} \\ \cdots & \cdots & \cdots & \cdots \\ a_{m1} & a_{m2} & \cdots & a_{mn} \end{bmatrix}_{m\times n} \equiv [\, a_{ij} \,]_{m\times n} \tag{1.2.3}$$

where i is the **row index**, j is the **column index**, and a_{ij} is called the i-j **element of matrix A**. The elements $a_{i1}, a_{i2}, \ldots, a_{in}$ constitute the $\mathbf{i^{th}}$ **row** of **A** and $a_{1j}, a_{2j}, \ldots, a_{mj}$ constitute the $\mathbf{j^{th}}$ **column** of **A**. Boldface symbols are used throughout the text to represent matrices. The **dimension of matrix A** in Eq. 1.2.3 is said to be m×n. ■

Matrix Operations

Definition 1.2.2. Equality, **addition**, and **multiplication of matrices** are defined as follows :

(1) Two matrices of the same dimension are equal if all their corresponding elements are equal; i.e.,

$$\mathbf{A} = \mathbf{B} \Leftrightarrow a_{ij} = b_{ij},\ i = 1, 2, \ldots, m,\ j = 1, 2, \ldots, n \tag{1.2.4}$$

(2) The sum of two matrices of the same dimension is the matrix formed by adding corresponding elements; i.e.,

$$\mathbf{A} + \mathbf{B} = \mathbf{C} \Leftrightarrow a_{ij} + b_{ij} = c_{ij},\ i = 1, 2, \ldots, m,\ j = 1, 2, \ldots, n \tag{1.2.5}$$

(3) The product of an m×n matrix **A** and an n×p matrix **B** is the m×p matrix **C**, with

$$c_{ij} = a_{ik}b_{kj},\ i = 1, 2, \ldots, m,\ j = 1, 2, \ldots, p \tag{1.2.6}$$

■

Before continuing with a study of matrix algebra, it is instructive to see that the system of linear equations in Eq. 1.2.1 with $\alpha = n$ and $\beta = m$ can be written in terms of matrices. In addition to the m×n matrix **A** of Eq. 1.2.2, whose elements are coefficients a_{ij} on the left of Eq. 1.2.1, define the matrices **x** and **b** of variables and right-side terms of Eq. 1.2.1, respectively,

$$\mathbf{x} = \begin{bmatrix} x_1 \\ \vdots \\ x_n \end{bmatrix}, \quad \mathbf{b} = \begin{bmatrix} b_1 \\ \vdots \\ b_n \end{bmatrix} \tag{1.2.7}$$

Then, the system of linear equations of Eq. 1.2.1 may be written as the **matrix equation**

$$\mathbf{Ax} = \mathbf{b} \tag{1.2.8}$$

Expanding the matrix product on the left of Eq. 1.2.8, using the definition of Eq. 1.2.6 and the definition of equality of Eq. 1.2.4, Eq. 1.2.1 or Eq. 1.2.2 is recovered.

Matrices **x** and **b** of Eq. 1.2.7 contain only one column, so they are called **column matrices**. Since column matrices play a special role in matrix and vector algebra, they are represented by lower-case boldface letters. Boldface capital letters are reserved for matrices with more than one column and row.

Example 1.2.1

The sum of the 2×2 matrices

$$\mathbf{A} = \begin{bmatrix} 3 & 1 \\ 2 & 5 \end{bmatrix}, \quad \mathbf{B} = \begin{bmatrix} 4 & 3 \\ 1 & 2 \end{bmatrix}$$

is the 2×2 matrix

$$\mathbf{C} = \mathbf{A} + \mathbf{B} = \begin{bmatrix} 3+4 & 1+3 \\ 2+1 & 5+2 \end{bmatrix} = \begin{bmatrix} 7 & 4 \\ 3 & 7 \end{bmatrix}$$

■

Example 1.2.2

The product of 2×2 and 2×3 matrices

$$\mathbf{A} = \begin{bmatrix} 3 & 1 \\ 2 & 5 \end{bmatrix}, \quad \mathbf{B} = \begin{bmatrix} 4 & 3 & 1 \\ 1 & 2 & 1 \end{bmatrix}$$

is the 2×3 matrix

$$\mathbf{C} = \mathbf{AB} = \begin{bmatrix} (3\times4 + 1\times1) & (3\times3 + 1\times2) & (3\times1 + 1\times1) \\ (2\times4 + 5\times1) & (2\times3 + 5\times2) & (2\times1 + 5\times1) \end{bmatrix}$$

$$= \begin{bmatrix} 13 & 11 & 4 \\ 13 & 16 & 7 \end{bmatrix}$$

■

Clearly only matrices with the same number of rows and columns can be added. Similarly, the product **AB** is defined only when the number of columns in **A** equals the number of rows in **B**. Using the algebraic properties of real numbers, presuming the operations indicated are defined, matrix operations can be shown to satisfy the following laws:

Associative law of addition:

$$(\mathbf{A} + \mathbf{B}) + \mathbf{C} = \mathbf{A} + (\mathbf{B} + \mathbf{C}) \tag{1.2.9}$$

Commutative law of addition:

$$\mathbf{A} + \mathbf{B} = \mathbf{B} + \mathbf{A} \tag{1.2.10}$$

Associative law of multiplication:

$$(\mathbf{AB})\mathbf{C} = \mathbf{A}(\mathbf{BC}) \tag{1.2.11}$$

Distributive laws:

$$\begin{aligned} \mathbf{A}(\mathbf{B} + \mathbf{C}) &= \mathbf{AB} + \mathbf{AC} \\ (\mathbf{B} + \mathbf{C})\mathbf{A} &= \mathbf{BA} + \mathbf{CA} \end{aligned} \tag{1.2.12}$$

Example 1.2.3

Note that the matrix product **BA** in Example 1.2.2 can not be constructed, since the number of columns of matrix **B** does not match the number of rows of matrix **A**. Multiplication of matrices is not commutative, since in general **AB** ≠ **BA**, even if the matrix product makes sense both ways. For example, with the 2×2 matrices of Example 1.2.1,

$$\mathbf{AB} = \begin{bmatrix} 13 & 11 \\ 13 & 16 \end{bmatrix} \neq \begin{bmatrix} 18 & 19 \\ 7 & 11 \end{bmatrix} = \mathbf{BA}$$

This is the reason it was necessary to state two distributive laws in Eq. 1.2.12; i.e.,

$$\mathbf{A}(\mathbf{B} + \mathbf{C}) \neq (\mathbf{B} + \mathbf{C})\mathbf{A}$$ ■

Definition 1.2.3. The **zero matrix 0** is a matrix, all of whose elements are 0. If **A** and **0** have appropriate numbers of rows and columns,

$$\begin{aligned} \mathbf{A} + \mathbf{0} &= \mathbf{A} \\ \mathbf{0A} = \mathbf{A0} &= \mathbf{0} \end{aligned} \tag{1.2.13}$$

The **identity matrix** $\mathbf{I} = [\, \delta_{ij} \,]$ is the n×n matrix with elements

$$\delta_{ij} = \begin{cases} 1, & i = j \\ 0, & i \neq j \end{cases}$$

It has the properties

$$\begin{aligned} \mathbf{AI} &= \mathbf{A} \\ \mathbf{IA} &= \mathbf{A} \end{aligned} \tag{1.2.14}$$

for any matrix **A** and identity matrix of appropriate dimension. The δ_{ij} defined above is called the **Kronecker delta**.

A **diagonal matrix D** is an n×n matrix with zero elements off the diagonal. That is, the elements of a diagonal matrix can be written as $a_{\alpha\beta} = d_\alpha \delta_{\alpha\beta}$, where d_α, $\alpha = 1, 2, \ldots, n$, are the diagonal elements (recall that repeated Greek indices are not summed). ■

Definition 1.2.4. The **transpose** $\mathbf{A}^T$ of a matrix **A** is the matrix formed by interchanging the rows and columns of **A**. If **A** is m×n, then $\mathbf{A}^T$ is n×m. The elements a_{ij}^T of $\mathbf{A}^T$ are

$$a_{ij}^T = a_{ji}, \quad i = 1, \ldots, n, \quad j = 1, \ldots, m \tag{1.2.15}$$

■

Example 1.2.4

The transpose of the 2×3 matrix

$$\mathbf{A} = \begin{bmatrix} 4 & 3 & 1 \\ 1 & 2 & 1 \end{bmatrix}_{2\times3}$$

is the 3×2 matrix

$$\mathbf{A}^T = \begin{bmatrix} 4 & 1 \\ 3 & 2 \\ 1 & 1 \end{bmatrix}_{3\times2}$$

■

Theorem 1.2.1. The transpose of the product of two matrices is the product of their transposes in the opposite order; i.e.,

$$(\mathbf{AB})^T = \mathbf{B}^T\mathbf{A}^T \tag{1.2.16}$$

■

To prove this, let $\mathbf{C} = \mathbf{AB}$; i.e., $c_{ik} = a_{ij}b_{jk}$. From Eq. 1.2.15, $c_{ik}^T = c_{ki} = a_{kj}b_{ji} = b_{ji}a_{kj} = b_{ij}^T a_{jk}^T$. Therefore, $(\mathbf{AB})^T = \mathbf{C}^T = \mathbf{B}^T\mathbf{A}^T$. ■

Example 1.2.5

From Example 1.2.1,

$$\mathbf{A} = \begin{bmatrix} 3 & 1 \\ 2 & 5 \end{bmatrix}, \quad \mathbf{B} = \begin{bmatrix} 4 & 3 \\ 1 & 2 \end{bmatrix}, \quad \mathbf{C} = \mathbf{AB} = \begin{bmatrix} 13 & 11 \\ 13 & 16 \end{bmatrix}$$

$$\mathbf{C}^T = (\mathbf{AB})^T = \begin{bmatrix} 13 & 13 \\ 11 & 16 \end{bmatrix}$$

$$\mathbf{B}^T\mathbf{A}^T = \begin{bmatrix} 4 & 1 \\ 3 & 2 \end{bmatrix}\begin{bmatrix} 3 & 2 \\ 1 & 5 \end{bmatrix} = \begin{bmatrix} 13 & 13 \\ 11 & 16 \end{bmatrix}$$

■

Definition 1.2.5. Multiplication of a matrix A by a scalar c is defined as the matrix whose elements are the elements of **A** multiplied by c. That is, the elements of $\mathbf{B} = c\mathbf{A}$ are $b_{ij} = ca_{ij}$. The **negative of a matrix A**, denoted $-\mathbf{A}$, is the matrix formed by multiplying each element of **A** by -1; i.e., $-\mathbf{A}$ has elements $-a_{ij}$. **Subtraction of matrices** that have the same number of rows and columns is defined in terms of addition as

$$\mathbf{A} - \mathbf{B} \equiv \mathbf{A} + (-\mathbf{B}) \tag{1.2.17}$$

■

Definition 1.2.6. An n×n matrix **A** is **symmetric** if it is equal to its transpose; i.e., if $\mathbf{A} = \mathbf{A}^T$, equivalently, $a_{ij} = a_{ji}$. An n×n matrix **A** is **skew symmetric** if it is equal to the negative of its transpose; i.e., if $\mathbf{A} = -\mathbf{A}^T$. ■

Example 1.2.7

The 2×2 matrix

$$\mathbf{A} = \begin{bmatrix} 3 & 1 \\ 1 & 2 \end{bmatrix}$$

is symmetric, since

$$\mathbf{A}^T = \begin{bmatrix} 3 & 1 \\ 1 & 2 \end{bmatrix} = \mathbf{A}$$

The 2×2 matrix

$$\mathbf{B} = \begin{bmatrix} 0 & -1 \\ 1 & 0 \end{bmatrix}$$

is skew symmetric, since

$$-\mathbf{B}^T = -\begin{bmatrix} 0 & 1 \\ -1 & 0 \end{bmatrix} = \begin{bmatrix} 0 & -1 \\ 1 & 0 \end{bmatrix} = \mathbf{B}$$

■

Note that if an n×n matrix **A** is skew symmetric, then $a_{\alpha\alpha} = -a_{\alpha\alpha}$, which implies $a_{\alpha\alpha} = 0$, $\alpha = 1, \ldots, n$. Hence the diagonal elements of a skew symmetric matrix are zero.

EXERCISES 1.2

1. Evaluate the following matrix products:

(a) $\begin{bmatrix} 1 & 2 \\ 1 & -1 \end{bmatrix}\begin{bmatrix} 1 & 0 & 1 \\ 1 & -1 & 1 \end{bmatrix}$ (b) $\begin{bmatrix} 1 & 2 \\ 3 & 6 \end{bmatrix}\begin{bmatrix} 6 & -2 \\ -3 & 1 \end{bmatrix}$

(c) $\begin{bmatrix} a_1 & a_2 & . & . & . & a_n \end{bmatrix} \begin{bmatrix} b_1 \\ b_2 \\ . \\ . \\ . \\ b_n \end{bmatrix}$

(d) $\begin{bmatrix} c_1 & 0 \\ 0 & c_2 \end{bmatrix} \begin{bmatrix} a_{11} & a_{12} \\ a_{21} & a_{22} \end{bmatrix}$ (e) $\begin{bmatrix} a_{11} & a_{12} \\ a_{21} & a_{22} \end{bmatrix} \begin{bmatrix} c_1 & 0 \\ 0 & c_2 \end{bmatrix}$

2. If the product $\mathbf{A}(\mathbf{BC})$ is defined, show that it is of the form

$$\mathbf{A}(\mathbf{BC}) = [\, a_{ir} b_{rs} c_{sj} \,]$$

and that $\mathbf{A}(\mathbf{BC}) = (\mathbf{AB})\mathbf{C}$.

3. If $\mathbf{A}$ and $\mathbf{B}$ are n×n matrices, when is it true that

$$(\mathbf{A} + \mathbf{B})(\mathbf{A} - \mathbf{B}) = \mathbf{AA} - \mathbf{BB}?$$

Give an example in which this relation does not hold.

4. If $y_i = b_{ij} x_j$ and $z_i = a_{ij} y_j$, show that $z_i = c_{ij} x_j$, where $c_{ij} = a_{ik} b_{kj}$.

5. If $a_{ij} = a_{ji}$ and $b_{ij} = -b_{ji}$, show that $a_{ij} b_{ij} = 0$.

6. Using the Kronecker delta δ_{ij}, show that $a_{ij}\delta_{jk} = a_{ik} = \delta_{ij} a_{jk}$, where i, j, and k take on the values 1, 2, . . . , n.

7. Construct an example to show that matrix multiplication is not, in general, commutative.

8. If $\mathbf{A}$ and $\mathbf{B}$ are n×n matrices and $\mathbf{AB} = \mathbf{0}$, does this imply that $\mathbf{A} = \mathbf{0}$ or $\mathbf{B} = \mathbf{0}$?

9. Prove that Eqs. 1.2.9 through 1.2.12 are valid.

10. If $\mathbf{A}$ is an n×m matrix and $\mathbf{y}^i$, i = 1, . . . , k, are m×1 column matrices, show that

$$\mathbf{A}[\, \mathbf{y}^1, \mathbf{y}^2, \ldots, \mathbf{y}^k \,] = [\, \mathbf{Ay}^1, \mathbf{Ay}^2, \ldots, \mathbf{Ay}^k \,]$$

1.3 MATRIX INVERSE AND ELEMENTARY MATRIX OPERATIONS

If **A** of Eq. 1.2.8 is a nonzero 1×1 matrix (i.e., if it is a nonzero scalar) then both sides of Eq. 1.2.8 could be multiplied by 1/**A** to solve for the scalar variables **x**. Note, however, that while addition and multiplication of matrices are defined in Section 1.2, no concept of division by a matrix is introduced. The general concept of division by a matrix makes no sense and must be avoided.

Matrix Inverse

Definition 1.3.1. An n×n matrix **A** is called a **nonsingular matrix** if there is an n×n matrix $\mathbf{A}^{-1}$, called the **inverse** of **A**, such that

$$\begin{aligned} \mathbf{A}\mathbf{A}^{-1} &= \mathbf{I} \\ \mathbf{A}^{-1}\mathbf{A} &= \mathbf{I} \end{aligned} \tag{1.3.1}$$

Otherwise, it is called a **singular matrix**. ■

To see that the same matrix $\mathbf{A}^{-1}$ satisfies both of Eq. 1.3.1, let $\mathbf{A}^{-1}$ satisfy the first of Eq. 1.3.1 and let the n×n matrix **B** satisfy

$$\mathbf{B}\mathbf{A} = \mathbf{I}$$

Multiplying both sides on the right by $\mathbf{A}^{-1}$ and using Eq. 1.2.14,

$$\mathbf{B}\mathbf{A}\mathbf{A}^{-1} = \mathbf{I}\mathbf{A}^{-1} = \mathbf{A}^{-1}$$

Multiplying both sides of the first of Eq. 1.3.1 on the left by **B** and using Eq. 1.2.14,

$$\mathbf{B}\mathbf{A}\mathbf{A}^{-1} = \mathbf{B}\mathbf{I} = \mathbf{B}$$

Thus,

$$\mathbf{B} = \mathbf{A}^{-1}$$

and the second of Eq. 1.3.1 follows from the first. ■

Example 1.3.1

For the 2×2 matrix

$$\mathbf{A} = \begin{bmatrix} 1 & 0 \\ 2 & 1 \end{bmatrix}$$

expansion of the matrix multiplication

$$\begin{bmatrix} 1 & 0 \\ -2 & 1 \end{bmatrix} \begin{bmatrix} 1 & 0 \\ 2 & 1 \end{bmatrix} = \begin{bmatrix} 1 & 0 \\ 0 & 1 \end{bmatrix}$$

verifies that

$$\mathbf{A}^{-1} = \begin{bmatrix} 1 & 0 \\ -2 & 1 \end{bmatrix}$$

■

Example 1.3.2

Consider the 2×2 matrix

$$\mathbf{A} = \begin{bmatrix} 1 & 0 \\ 2 & 0 \end{bmatrix}$$

Let **B** be a 2×2 matrix with as yet unknown entries. If **B** were to be the inverse of **A**, then

$$\mathbf{BA} = \begin{bmatrix} b_{11} + 2b_{12} & 0 \\ b_{21} + 2b_{22} & 0 \end{bmatrix} = \begin{bmatrix} 1 & 0 \\ 0 & 1 \end{bmatrix} = \mathbf{I}$$

Clearly, there are no entries possible in **B** to satisfy this equation, since $0 \neq 1$ in the second row and second column. Thus, this 2×2 matrix **A** has no inverse. This elementary example clearly shows that not all square matrices have inverses. ■

Theorem 1.3.1. If **A** is a nonsingular n×n matrix, then for any n×1 matrix **b**, Eq. 1.2.8 has a unique solution,

$$\mathbf{x} = \mathbf{A}^{-1}\mathbf{b} \tag{1.3.2}$$

■

Theorem 1.3.1 may be proved by substituting **x** from Eq. 1.3.2 into Eq. 1.2.8 and using Eqs. 1.3.1 and 1.2.11 to verify that

$$\mathbf{A}(\mathbf{A}^{-1}\mathbf{b}) = (\mathbf{A}\mathbf{A}^{-1})\mathbf{b} = \mathbf{Ib} = \mathbf{b}$$

Thus, **x** of Eq. 1.3.2 is a solution of Eq. 1.2.8. To see that it is the unique solution (i.e., the only solution) assume there is a second solution **y** of Eq. 1.2.8; i.e.,

$$\mathbf{Ay} = \mathbf{b}$$

Subtracting this equation from Eq. 1.2.8,

$$\mathbf{Ax} - \mathbf{Ay} = \mathbf{b} - \mathbf{b} = \mathbf{0}$$

Using Eq. 1.2.12 and multiplying both sides on the left by $\mathbf{A}^{-1}$,

$$\mathbf{A}^{-1}\mathbf{A}(\mathbf{x}-\mathbf{y}) = \mathbf{I}(\mathbf{x}-\mathbf{y}) = \mathbf{x} - \mathbf{y} = \mathbf{0}$$

Thus, $\mathbf{y} = \mathbf{x}$ and $\mathbf{x}$ of Eq. 1.3.2 is indeed the unique solution. ■

The method of proving uniqueness of the solution of Eq. 1.2.8 used above is quite generally applicable. A second solution is assumed and it is then shown that the two solutions are equal; i.e., the solution is in fact unique.

A simple method of finding $\mathbf{A}^{-1}$ for an n×n nonsingular matrix $\mathbf{A}$ is to write the first of Eq. 1.3.1 as

$$\mathbf{A}[\mathbf{y}^1, \ldots, \mathbf{y}^n] = [\mathbf{e}^1, \ldots, \mathbf{e}^n] \tag{1.3.3}$$

where $\mathbf{y}^i$ is the ith column of $\mathbf{A}^{-1}$ and

$$\mathbf{e}^k = [\delta_{ik}]$$

is a column matrix with 1 in the kth row and 0 in the other rows.

Equating the kth columns of both sides of Eq. 1.3.3 yields

$$\mathbf{A}\mathbf{y}^k = \mathbf{e}^k \tag{1.3.4}$$

Solving Eq. 1.3.4 by direct numerical manipulation (see Section 2.3) for $k = 1, 2, \ldots, n$ yields the columns of $\mathbf{A}^{-1}$; i.e.,

$$\mathbf{A}^{-1} = [\mathbf{y}^1, \ldots, \mathbf{y}^n] \tag{1.3.5}$$

Theorem 1.3.2. If a square matrix $\mathbf{A}$ is nonsingular, so is $\mathbf{A}^T$ and

$$(\mathbf{A}^T)^{-1} = (\mathbf{A}^{-1})^T \tag{1.3.6}$$

■

To prove Theorem 1.3.2, recall the second of Eq. 1.3.1; i.e.,

$$\mathbf{A}^{-1}\mathbf{A} = \mathbf{I}$$

Taking the transpose of both sides and using Eq. 1.2.16 and the fact that $\mathbf{I}$ is symmetric,

$$(\mathbf{A}^{-1}\mathbf{A})^T = \mathbf{A}^T(\mathbf{A}^{-1})^T = \mathbf{I}^T = \mathbf{I}$$

By the first of Eq. 1.3.1 in Definition 1.3.1, the matrix $(\mathbf{A}^{-1})^T$ is the inverse of $\mathbf{A}^T$. Thus, $\mathbf{A}^T$ is nonsingular and Eq. 1.3.6 is valid. ■

Elementary Matrices

The elementary operations for manipulation of linear equations in Section 1.1 can be interpreted in terms of matrix operations. The starting point is the system $\mathbf{Ax} = \mathbf{b}$; e.g.,

$$\mathbf{Ax} \equiv \begin{bmatrix} 2 & 1 & 1 \\ 4 & 1 & 0 \\ -2 & 2 & 1 \end{bmatrix} \begin{bmatrix} x_1 \\ x_2 \\ x_3 \end{bmatrix} = \begin{bmatrix} 1 \\ -2 \\ 7 \end{bmatrix} \equiv \mathbf{b} \tag{1.3.7}$$

There are three steps in the **forward elimination algorithm**:

(i) Add −2 times the first equation to the second.
(ii) Add +1 times the first equation to the third.
(iii) Add +3 times the second equation to the third.

to obtain

$$\mathbf{Ux} \equiv \begin{bmatrix} 2 & 1 & 1 \\ 0 & -1 & -2 \\ 0 & 0 & -4 \end{bmatrix} \begin{bmatrix} x_1 \\ x_2 \\ x_3 \end{bmatrix} = \begin{bmatrix} 1 \\ -4 \\ -4 \end{bmatrix} \equiv \mathbf{y} \tag{1.3.8}$$

The coefficient matrix in Eq. 1.3.8 is **upper triangular**; i.e., all the entries below the main diagonal are zero. The right side, which is a new vector **y**, is derived from the original vector **b** by the same steps that transformed **A** to **U**. Thus, in terms of matrix operations, **Gaussian elimination** is the following algorithm:

(1) Start with **A** and **b** of Eq. 1.3.7.
(2) Apply steps (i), (ii), and (iii) of forward elimination, in that order, to **A** and **b**.
(3) End with **U** and **y**.
(4) Solve **Ux** = **y** by back substitution.

A matrix that multiplies **A** on the left to accomplish step (i) of the forward elimination algorithm is denoted by $\mathbf{E}_{21}$, indicating by the subscripts that it changes row 2 by adding a multiple of row 1, to produce a zero in the (2, 1) position in the coefficient matrix. Because the multiple of the first row was −2, the matrix that carries out this operation is obtained by applying the operation to the 3×3 identity; i.e.,

$$\mathbf{E}_{21} = \begin{bmatrix} 1 & 0 & 0 \\ -2 & 1 & 0 \\ 0 & 0 & 1 \end{bmatrix}$$

Similarly, steps (ii) and (iii) of the forward elimination algorithm can be carried out by multiplying on the left by

$$\mathbf{E}_{31} = \begin{bmatrix} 1 & 0 & 0 \\ 0 & 1 & 0 \\ 1 & 0 & 1 \end{bmatrix}$$

and

$$\mathbf{E}_{32} = \begin{bmatrix} 1 & 0 & 0 \\ 0 & 1 & 0 \\ 0 & 3 & 1 \end{bmatrix}$$

Then,

$$\mathbf{E}_{32}\mathbf{E}_{31}\mathbf{E}_{21}\mathbf{A} = \mathbf{U} \tag{1.3.9}$$

Since the same operations apply to terms on the right of Eq. 1.3.7,

$$\mathbf{E}_{32}\mathbf{E}_{31}\mathbf{E}_{21}\mathbf{b} = \mathbf{y} \tag{1.3.10}$$

The matrices $\mathbf{E}_{ij}$ used above to carry out forward elimination are called **elementary matrices**. The elementary matrices may be multiplied to find a single matrix that transforms **A** and **b** into **U** and **y**; i.e.,

$$\mathbf{E} \equiv \mathbf{E}_{32}\mathbf{E}_{31}\mathbf{E}_{21} = \begin{bmatrix} 1 & 0 & 0 \\ -2 & 1 & 0 \\ -5 & 3 & 1 \end{bmatrix} \tag{1.3.11}$$

A single step, say step (i), is not hard to undo to recover **I**. Just add twice the first row of $\mathbf{E}_{21}$ to the second. The matrix that carries out this operation is

$$\mathbf{E}_{21}^{-1} = \begin{bmatrix} 1 & 0 & 0 \\ 2 & 1 & 0 \\ 0 & 0 & 1 \end{bmatrix}$$

The product of $\mathbf{E}_{21}$ and $\mathbf{E}_{21}^{-1}$, taken in either order, is the identity; i.e.,

$$\begin{aligned} \mathbf{E}_{21}^{-1}\mathbf{E}_{21} &= \mathbf{I} \\ \mathbf{E}_{21}\mathbf{E}_{21}^{-1} &= \mathbf{I} \end{aligned} \tag{1.3.12}$$

Similarly, the second and third elementary matrices can be inverted by adding the negative of what was added in steps (ii) and (iii); i.e.,

$$\mathbf{E}_{31}^{-1} = \begin{bmatrix} 1 & 0 & 0 \\ 0 & 1 & 0 \\ -1 & 0 & 1 \end{bmatrix}, \quad \mathbf{E}_{32}^{-1} = \begin{bmatrix} 1 & 0 & 0 \\ 0 & 1 & 0 \\ 0 & -3 & 1 \end{bmatrix}$$

Premultiplication of Eq. 1.3.9 by $\mathbf{E}_{32}^{-1}$, $\mathbf{E}_{31}^{-1}$, and $\mathbf{E}_{21}^{-1}$, in this order, yields

$$\mathbf{A} = \mathbf{E}_{21}^{-1}\mathbf{E}_{31}^{-1}\mathbf{E}_{32}^{-1}\mathbf{U} \equiv \mathbf{LU} \tag{1.3.13}$$

where

$$\mathbf{L} = \mathbf{E}_{21}^{-1}\mathbf{E}_{31}^{-1}\mathbf{E}_{32}^{-1} \tag{1.3.14}$$

Carrying out the multiplication,

$$\mathbf{L} = \begin{bmatrix} 1 & 0 & 0 \\ 2 & 1 & 0 \\ 0 & 0 & 1 \end{bmatrix}\begin{bmatrix} 1 & 0 & 0 \\ 0 & 1 & 0 \\ -1 & 0 & 1 \end{bmatrix}\begin{bmatrix} 1 & 0 & 0 \\ 0 & 1 & 0 \\ 0 & -3 & 1 \end{bmatrix} = \begin{bmatrix} 1 & 0 & 0 \\ 2 & 1 & 0 \\ -1 & -3 & 1 \end{bmatrix} \tag{1.3.15}$$

Note that **L** is **lower triangular**, with unit values on the diagonal. Furthermore, the entries below the diagonal are exactly the multiples 2, −1, and −3 that were used in the three elimination steps. This is not an accident, but will be seen to be a general property of Gaussian elimination. It leads to factorization of a square matrix **A** as a product of lower and upper triangular matrices, called **LU factorization.**

To see the value of LU factorization in solving Eq. 1.3.7, define a column matrix **y** such that **Ux** = **y**. Substituting this into Eq. 1.3.7,

$$\mathbf{Ax} = \mathbf{LUx} = \mathbf{Ly} = \mathbf{b} \tag{1.3.16}$$

Since **L** in Eq. 1.3.15 is lower triangular, Eq. 1.3.16 may be solved first for y_1, then for y_2, and finally for y_3, to obtain

$$\mathbf{y} = \begin{bmatrix} 1 \\ -4 \\ -4 \end{bmatrix}$$

This process is called **forward elimination.** Now, **x** may be recovered from

$$\mathbf{Ux} = \mathbf{y}$$

by **back substitution,** since **U** of Eq. 1.3.8 is upper triangular, to obtain

$$\mathbf{x} = \begin{bmatrix} -1 \\ 2 \\ 1 \end{bmatrix}$$

EXERCISES 1.3

1. Let **A** be a 4×3 matrix. Find the elementary matrix **E**, as a premultiplier of **A**, that performs the following row operations on **A**:

   (a) Multiply the second row of **A** by −3.

   (b) Add four times the third row of **A** to the fourth row of **A**.

   (c) Interchange the first and the third rows of **A**.

2. Show that $\mathbf{A} = \begin{bmatrix} 1 & 2 & 3 \\ 0 & 2 & 1 \\ 1 & 0 & 3 \end{bmatrix}$ is nonsingular, by writing it as a product of elementary matrices.

3. Let $\mathbf{A} = \begin{bmatrix} 1 & 2 & -1 \\ 2 & 0 & 1 \\ 1 & 2 & 3 \\ 3 & 0 & 2 \end{bmatrix}$. Find the elementary matrix, as a postmultiplier of **A**, that performs the following column operations on **A**:

(a) Multiply the third column of **A** by –2.

(b) Interchange the first and third columns of **A**.

(c) Add –3 times the second column of **A** to the first column of **A**.

4. Prove that $\mathbf{A} = \begin{bmatrix} a & b \\ c & d \end{bmatrix}$ is nonsingular if and only if $ad - bc \neq 0$.

1.4 DETERMINANTS

Determinant of a Matrix

Definition 1.4.1. Every n×n matrix has associated with it a determinant. The **determinant of a matrix A**, written as $|\mathbf{A}|$, or

$$|\mathbf{A}| = \begin{vmatrix} a_{11} & a_{12} & \cdots & a_{1n} \\ a_{21} & a_{22} & \cdots & a_{2n} \\ \cdots & \cdots & \cdots & \cdots \\ a_{n1} & a_{n2} & \cdots & a_{nn} \end{vmatrix}$$

is the real number

$$|\mathbf{A}| \equiv e_{i_1 i_2 i_3 \ldots i_n} a_{1i_1} a_{2i_2} \ldots a_{ni_n} \tag{1.4.1}$$

where summation is taken over all repeated indices i_j and the **permutation symbol** is

$$e_{i_1 i_2 \ldots i_n} = \begin{cases} 0 \text{ if any pair of subscripts are equal} \\ 1 \text{ if } i_1, i_2, \ldots, i_n \text{ is an even permutation of } 1, 2, \ldots, n \\ -1 \text{ if } i_1, i_2, \ldots, i_n \text{ is an odd permutation of } 1, 2, \ldots, n \end{cases}$$

■

A **permutation** of the integers 1, 2,..., n is an ordering of the integers. There are, for example, six different permutations of the integers 1, 2, and 3: 1, 2, 3; 2, 3, 1; 3, 1, 2; 1, 3, 2; 2, 1, 3; and 3, 2, 1. An inversion of the order of a pair of integers changes the permutation. If it takes an even number of inversions of order to change a given permutation to the normal order 1, 2, 3, . . . , n, then it is said to be an **even permutation**. If an odd number of inversions is required to restore the integers to normal order, then it is said to be an **odd permutation**. Hence, 1, 2, 3; 2, 3, 1; and 3, 1, 2 are even permutations, while 1, 3, 2; 2, 1, 3; and 3, 2, 1 are odd permutations. It can be shown that evenness and oddness are independent of the specific set of inversions used to change the permutation to normal order [1].

The definition of determinant can be stated somewhat differently, as follows: Form all possible products of n elements from **A**, selecting one element from each row, so that no two elements come from the same column. Then multiply each such product by +1 or −1, according to whether the column subscripts of the elements form an even or an odd permutation of the integers 1 to n. Finally, add all such terms. The resulting sum is the value of the determinant. Since for each product there are n ways of selecting a factor from the first row, n − 1 ways of selecting a factor from the second row, n − 2 ways of selecting a factor from the third row, etc., the number of terms in the sum that gives the value of the determinant is n!.

The above definition can be changed by interchanging the words "row" with "column," throughout. In other words, starting from the definition, it can be shown [2] that

$$|\mathbf{A}| = e_{i_1 i_2 i_3 \ldots i_n} a_{i_1 1} a_{i_2 2} a_{i_3 3} \ldots a_{i_n n} \equiv |\mathbf{A}^T| \tag{1.4.2}$$

Example 1.4.1

Equation 1.4.1, for a 3×3 determinant expansion, is

$$\begin{vmatrix} a_{11} & a_{12} & a_{13} \\ a_{21} & a_{22} & a_{23} \\ a_{31} & a_{32} & a_{33} \end{vmatrix} = e_{ijk} a_{1i} a_{2j} a_{3k} = a_{11}a_{22}a_{33} + a_{12}a_{23}a_{31}$$

$$+ a_{13}a_{21}a_{32} - a_{11}a_{23}a_{32} - a_{12}a_{21}a_{33} - a_{13}a_{22}a_{31} \quad \blacksquare$$

Properties of Determinants

Theorem 1.4.1. The following properties of determinants are valid [2]:

(i) If every element in a given row (or column) of a square matrix is zero, its determinant is zero; e.g.,

$$\begin{vmatrix} a & b \\ 0 & 0 \end{vmatrix} = 0$$

(ii) If every element in a given row (or column) of a square matrix is multiplied by the same number k, the determinant is multiplied by k; e.g.,

$$\begin{vmatrix} ka & kb \\ c & d \end{vmatrix} = k \begin{vmatrix} a & b \\ c & d \end{vmatrix}$$

(iii) If any pair of rows (or columns) of a square matrix is interchanged, the sign of its determinant is changed; e.g.,

$$\begin{vmatrix} a & b \\ c & d \end{vmatrix} = - \begin{vmatrix} c & d \\ a & b \end{vmatrix} = - \begin{vmatrix} b & a \\ d & c \end{vmatrix}$$

(iv) If two rows (or columns) of a square matrix are proportional, its determinant is zero; e.g.,

$$\begin{vmatrix} a & b \\ \alpha a & \alpha b \end{vmatrix} = 0, \quad \begin{vmatrix} a & \alpha a \\ b & \alpha b \end{vmatrix} = 0$$

(v) If each element of a given row (or column) of a square matrix can be written as the sum of two terms, then its determinant can be written as the sum of two determinants, each of which contains one of the terms in the corresponding row (or column); e.g.,

$$\begin{vmatrix} a+e & b+f \\ c & d \end{vmatrix} = \begin{vmatrix} a & b \\ c & d \end{vmatrix} + \begin{vmatrix} e & f \\ c & d \end{vmatrix} = (ad - bc) + (ed - fc)$$

(vi) If to each element of a given row (or column) of a square matrix is added k times the corresponding element of another row (or column), the value of its determinant is unchanged; e.g.,

$$\begin{vmatrix} a & b \\ c & d \end{vmatrix} = \begin{vmatrix} a & (b+ka) \\ c & (d+kc) \end{vmatrix} = ad - bc$$

■

Theorem 1.4.2. The determinant of the product of two square matrices of the same dimension is the product of their determinants; i.e.,

$$|\mathbf{AB}| = |\mathbf{A}||\mathbf{B}| = |\mathbf{B}||\mathbf{A}| \tag{1.4.3}$$

■

To prove Theorem 1.4.2, form the sum

$$e_{j_1 j_2 \cdots j_n} a_{i_1 j_1} a_{i_2 j_2} \cdots a_{i_n j_n}$$

If $i_1, i_2, \ldots, i_n = 1, 2, \ldots, n$ is in normal order, then by definition this expression gives the value of the determinant of **A**. If any pair of the i's have the same value, then the value of the expression is zero, since it will then represent a determinant with two rows equal. If $i_1, i_2, \ldots, i_n$ is a permutation of 1, 2, . . . , n, then the expression gives plus or minus the determinant of **A**, depending on whether it takes an even or an odd number of interchanges of rows to arrive at the expression for | **A** |. Thus,

$$e_{j_1j_2\ldots j_n} a_{i_1j_1} a_{i_2j_2} \cdots a_{i_nj_n} = |\mathbf{A}| e_{i_1i_2\ldots i_n}$$

Similarly, from Eq. 1.4.2,

$$e_{j_1j_2\ldots j_n} a_{j_1i_1} a_{j_2i_2} \cdots a_{j_ni_n} = |\mathbf{A}| e_{i_1i_2\ldots i_n}$$

The desired result is obtained by noting that

$$\begin{aligned}
|\mathbf{A}||\mathbf{B}| &= |\mathbf{A}| e_{i_1i_2\ldots i_n} b_{i_11} b_{i_22} \cdots b_{i_nn} \\
&= e_{j_1j_2\ldots j_n} a_{j_1i_1} a_{j_2i_2} \cdots a_{j_ni_n} b_{i_11} b_{i_22} \cdots b_{i_nn} \\
&= e_{j_1j_2\ldots j_n} (a_{j_1i_1} b_{i_11}) (a_{j_2i_2} b_{i_22}) \cdots (a_{j_ni_n} b_{i_nn}) \\
&= e_{j_1j_2\ldots j_n} (\mathbf{AB})_{j_11} (\mathbf{AB})_{j_22} \cdots (\mathbf{AB})_{j_nn} \\
&= |\mathbf{AB}|
\end{aligned}$$

■

Example 1.4.2

Using determinant expansions,

$$|\mathbf{A}| = \begin{vmatrix} a_{11} & a_{12} \\ a_{21} & a_{22} \end{vmatrix} = a_{11}a_{22} - a_{12}a_{21}$$

$$|\mathbf{B}| = \begin{vmatrix} b_{11} & b_{12} \\ b_{21} & b_{22} \end{vmatrix} = b_{11}b_{22} - b_{12}b_{21}$$

$$\begin{aligned}
|\mathbf{AB}| &= \begin{vmatrix} (a_{11}b_{11} + a_{12}b_{21}) & (a_{11}b_{12} + a_{12}b_{22}) \\ (a_{21}b_{11} + a_{22}b_{21}) & (a_{21}b_{12} + a_{22}b_{22}) \end{vmatrix} \\
&= (a_{11}b_{11} + a_{12}b_{21}) (a_{21}b_{12} + a_{22}b_{22}) \\
&\qquad - (a_{11}b_{12} + a_{12}b_{22}) (a_{21}b_{11} + a_{22}b_{21}) \\
&= a_{11}b_{11}a_{21}b_{12} + a_{11}b_{11}a_{22}b_{22} + a_{12}b_{21}a_{21}b_{12} + a_{12}b_{21}a_{22}b_{22} \\
&\qquad - (a_{11}b_{12}a_{21}b_{11} + a_{11}b_{12}a_{22}b_{21} + a_{12}b_{22}a_{21}b_{11} + a_{12}b_{22}a_{22}b_{21}) \\
&= a_{11}a_{22} (b_{11}b_{22} - b_{12}b_{21}) - a_{12}a_{21} (b_{11}b_{22} - b_{12}b_{21})
\end{aligned}$$

$$= (a_{11}a_{22} - a_{12}a_{21}) (b_{11}b_{22} - b_{12}b_{21})$$

$$= |\mathbf{A}||\mathbf{B}|$$

■

Another way of expanding the determinant of a square matrix is the so-called **expansion by cofactors**. Starting from the definition,

$$|\mathbf{A}| = e_{i_1 i_2 \ldots i_n} a_{1i_1} a_{2i_2} \ldots a_{ni_n}$$

$$= a_{1i_1} e_{i_1 i_2 \ldots i_n} a_{2i_2} \ldots a_{ni_n}$$

$$= a_{1i_1} A_{1i_1}$$

where

$$A_{1i_1} \equiv e_{i_1 i_2 \ldots i_n} a_{2i_2} \ldots a_{ni_n}$$

is the **cofactor** of the element in the first row and i_1^{th} column. In general,

$$A_{ji_j} \equiv e_{i_1 \ldots i_{j-1} i_j i_{j+1} \ldots i_n} a_{1i_1} \ldots a_{j-1\, i_{j-1}} a_{j+1\, i_{j+1}} \ldots a_{n\, i_n} \tag{1.4.4}$$

is the cofactor of the element in the jth row and i_j^{th} column. Note that there is no sum implied by the term A_{ji_j}, since $i_j \neq j$; i.e., $A_{ji_j} \neq A_{jj}$. The expansion by cofactors of elements of the αth row is given by

$$|\mathbf{A}| = a_{\alpha i_j} A_{\alpha i_j} \tag{1.4.5}$$

Similarly, expansion by cofactors of elements of the αth column yields

$$|\mathbf{A}| = a_{i_j \alpha} A_{i_j \alpha} \tag{1.4.6}$$

The cofactors are, except possibly for a sign change, determinants of $(n-1) \times (n-1)$ square matrices that are formed from **A** by deleting one row and one column. Beginning with the definition of the cofactor of Eq. 1.4.4,

$$A_{ji_j} = (-1)^{j-1} e_{i_j i_1 i_2 \ldots i_{j-1} i_{j+1} \ldots i_n} a_{1i_1} a_{2i_2} \ldots a_{j-1\, i_{j-1}} a_{j+1\, i_{j+1}} \ldots a_{n\, i_n}$$

Here the subscript i_j is moved ahead of the others by $j-1$ inversions, hence requiring the factor $(-1)^{j-1}$. If it takes p inversions to put $i_1, i_2, \ldots, i_{j-1}, i_{j+1}, \ldots i_n$ in normal order, except for the missing integer i_j, then it takes $p + i_j - 1$ inversions to put $i_j, i_1, i_2, \ldots i_{j-1}, i_{j+1}, \ldots, i_n$ in normal order. Hence,

$$e_{i_j i_1 i_2 \ldots i_{j-1} i_{j+1} \ldots i_n} = (-1)^{i_j - 1} e_{k_1 k_2 \ldots k_{n-1}}$$

where $k_1, k_2, \ldots, k_{n-1}$ is the set of n − 1 integers that are obtained from $i_1, i_2, \ldots, i_{j-1}, i_{j+1}, \ldots i_n$ by replacing any integer greater than i_j by $i_j - 1$. It now follows from Eq. 1.4.4 and the fact that $(-1)^{j-1+i_j-1} = (-1)^{j+i_j}$, that

$$A_{ji_j} = (-1)^{j+i_j} e_{k_1 k_2 \ldots k_{n-1}} b_{1k_1} b_{2k_2} \ldots b_{n-1\, k_{n-1}}$$

where the b's are the elements of an $(n-1) \times (n-1)$ matrix that is formed from **A** by omitting the jth row and i_j^{th} column. This analysis yields the following result.

Theorem 1.4.3. The cofactor A_{ji_j} of a square matrix **A** is $(-1)^{j+i_j}$ times the determinant of the matrix formed from **A** by omitting the jth row and i_j^{th} column. ■

Example 1.4.3

The cofactor expansion of a 3×3 determinant of Eq. 1.4.5 with $\alpha = 1$, the sign over an element showing the associated sign for its cofactor, is

$$\begin{vmatrix} \overset{+}{a_{11}} & \overset{-}{a_{12}} & \overset{+}{a_{13}} \\ \overset{-}{a_{21}} & \overset{+}{a_{22}} & \overset{-}{a_{23}} \\ \overset{+}{a_{31}} & \overset{-}{a_{32}} & \overset{+}{a_{33}} \end{vmatrix} = a_{11} \begin{vmatrix} a_{22} & a_{23} \\ a_{32} & a_{33} \end{vmatrix} - a_{12} \begin{vmatrix} a_{21} & a_{23} \\ a_{31} & a_{33} \end{vmatrix} + a_{13} \begin{vmatrix} a_{21} & a_{22} \\ a_{31} & a_{32} \end{vmatrix}$$

$$= a_{11}(a_{22}a_{33} - a_{23}a_{32}) - a_{12}(a_{21}a_{33} - a_{23}a_{31})$$

$$+ a_{13}(a_{21}a_{32} - a_{22}a_{31})$$ ■

Matrix Inverse and Cramer's Rule

The next question is, What happens if a cofactor expansion is written using the elements of a given row, but the cofactors of a different row? In other words, what is the value of $a_{ji_k} A_{ki_k}$ where $j \neq k$? Using the definition of the cofactor,

$$\begin{aligned} a_{ji_k} A_{ki_k} &= a_{ji_k} e_{i_1 i_2 \ldots i_n} a_{1i_1} \ldots a_{ji_j} \ldots a_{k-1\, i_{k-1}} a_{k+1\, i_{k+1}} \ldots a_{ni_n} \\ &= e_{i_1 i_2 \ldots i_n} a_{1i_1} \ldots a_{ji_j} \ldots a_{ji_k} \ldots a_{ni_n} \\ &= 0 \end{aligned}$$

since this expression is the expansion of a determinant with two equal rows. This result and the expansion of | **A** | in Eq. 1.4.5 can be combined in the single statement,

$$a_{ik} A_{jk} = |\mathbf{A}|\, \delta_{ij}$$

where δ_{ij} is the Kronecker delta of Definition 1.2.3.

If $|\mathbf{A}| \neq 0$, then

$$a_{ik}\frac{A_{jk}}{|\mathbf{A}|} = a_{ik}\left(\frac{1}{|\mathbf{A}|}A_{kj}^{T}\right) = \delta_{ij}$$

In matrix form, this is

$$\mathbf{A}\left(\frac{1}{|\mathbf{A}|}[\,A_{ij}\,]^{T}\right) = \mathbf{I}$$

This shows that for a square matrix **A**, with $|\mathbf{A}| \neq 0$, there exists an inverse and that

$$\mathbf{A}^{-1} = \frac{1}{|\mathbf{A}|}[\,A_{ij}\,]^{T} \tag{1.4.7}$$

Example 1.4.4

Consider the 2×2 matrix

$$\mathbf{A} = \begin{bmatrix} a & b \\ c & d \end{bmatrix}$$

and assume that $|\mathbf{A}| = ad - bc \neq 0$. The cofactor matrix of **A** is

$$[\,A_{ij}\,] = \begin{bmatrix} d & -c \\ -b & a \end{bmatrix}$$

and

$$\mathbf{A}^{-1} = \frac{1}{|\mathbf{A}|}[\,A_{ij}\,]^{T} = \frac{1}{(ad-bc)}\begin{bmatrix} d & -b \\ -c & a \end{bmatrix}$$

To confirm this, expansion yields

$$\mathbf{A}\mathbf{A}^{-1} = \frac{1}{(ad-bc)}\begin{bmatrix} a & b \\ c & d \end{bmatrix}\begin{bmatrix} d & -b \\ -c & a \end{bmatrix} = \frac{1}{(ad-bc)}\begin{bmatrix} ad-bc & -ab+ab \\ cd-cd & -bc+ad \end{bmatrix}$$

$$= \mathbf{I}$$

■

If a matrix has an inverse, then

$$\mathbf{A}\mathbf{A}^{-1} = \mathbf{I}$$

and

$$|\mathbf{A}||\mathbf{A}^{-1}| = |\mathbf{A}\mathbf{A}^{-1}| = |\mathbf{I}| = 1$$

Therefore, $|\mathbf{A}| \neq 0$. These results prove the following theorem.

Theorem 1.4.4. A square matrix **A** has an inverse if and only if $|\mathbf{A}| \neq 0$. ■

The foregoing results may be used to solve systems of linear algebraic equations. Suppose that

$$\mathbf{Ax} = \mathbf{b} \tag{1.4.8}$$

is a system of n linear algebraic equations in n unknowns and that $|\mathbf{A}| \neq 0$. Since $\mathbf{A}$ is nonsingular, the system of equations has a unique solution, which can be found by multiplication on the left by $\mathbf{A}^{-1}$. Using Eq. 1.4.7, this solution is

$$x_i = \frac{A_{ji}\, b_j}{|\mathbf{A}|} \tag{1.4.9}$$

which is called **Cramer's rule.**

Example 1.4.5

Obtain the solution of the equations of Example 1.1.4; i.e.,

$$\begin{aligned} 2x_1 + 2x_2 + 4x_3 &= -2 \\ x_1 + 2x_2 - \;\; x_3 &= 0 \\ 3x_2 - 6x_3 &= 3 \end{aligned}$$

using Cramer's rule. The determinant of $\mathbf{A}$ is

$$|\mathbf{A}| = \begin{vmatrix} 2 & 2 & 4 \\ 1 & 2 & -1 \\ 0 & 3 & -6 \end{vmatrix} = 6, \quad \mathbf{b} = \begin{bmatrix} -2 \\ 0 \\ 3 \end{bmatrix}$$

The cofactor matrix of $\mathbf{A}$ is

$$[\,A_{ij}\,] = \begin{bmatrix} -9 & 6 & 3 \\ 24 & -12 & -6 \\ -10 & 6 & 2 \end{bmatrix}$$

Thus,

$$x_1 = \frac{A_{j1}b_j}{|\mathbf{A}|} = \frac{1}{6}[\,-9 \;\; 24 \;\; -10\,]\begin{bmatrix} -2 \\ 0 \\ 3 \end{bmatrix} = \frac{-12}{6} = -2$$

$$x_2 = \frac{A_{j2}b_j}{|\mathbf{A}|} = \frac{1}{6}[\,6 \;\; -12 \;\; 6\,]\begin{bmatrix} -2 \\ 0 \\ 3 \end{bmatrix} = \frac{6}{6} = 1$$

$$x_3 = \frac{A_{j3}b_j}{|\mathbf{A}|} = \frac{1}{6}[\,3 \;\; -6 \;\; 2\,]\begin{bmatrix} -2 \\ 0 \\ 3 \end{bmatrix} = \frac{0}{6} = 0$$

which is the same as that obtained using Gaussian elimination in Example 1.1.4. ■

In case Eq. 1.4.8 is homogeneous, a solution is clearly the trivial one $x_1 = x_2 = \ldots = x_n = 0$. The preceding result shows that this is the only possible solution if **A** is nonsingular. This proves the following important result.

Theorem 1.4.5. A set of linear homogeneous equations $\mathbf{Ax} = \mathbf{0}$ with a square coefficient matrix can have a nontrivial solution only if $|\mathbf{A}| = 0$. ■

Example 1.4.6

For the system

$$\begin{aligned} x_1 + 2x_2 &= 0 \\ 2x_1 + 4x_2 &= 0 \end{aligned}$$

$|\mathbf{A}| = 0$. In this case, $x_1 = -2x_2$ is a solution for arbitrary x_2. Thus, there are infinitely many solutions. ■

EXERCISES 1.4

1. Prove the first three properties of determinants in Theorem 1.4.1.

2. Determine those values of λ for which the following equations can have a nontrivial solution :

$$\begin{aligned} 3x_1 + x_2 - \lambda x_3 &= 0 \\ 4x_1 - x_2 - 3x_3 &= 0 \\ 2\lambda x_1 + 3x_2 + \lambda x_3 &= 0 \end{aligned}$$

3. If **A** and **B** are nonsingular n×n matrices, show that

$$(\mathbf{AB})^{-1} = \mathbf{B}^{-1}\mathbf{A}^{-1}$$

4. Construct the inverses of the matrices

$$\mathbf{A} = \begin{bmatrix} 1 & 2 & 1 \\ 2 & 1 & 4 \\ 1 & 3 & 1 \end{bmatrix}, \quad \mathbf{B} = \begin{bmatrix} 3 & 1 & 2 \\ 2 & 1 & 2 \\ 1 & 2 & 2 \end{bmatrix}, \quad \mathbf{C} = \begin{bmatrix} 2 & 1 & 3 \\ 0 & 1 & 2 \\ 1 & 0 & 3 \end{bmatrix}$$

5. If **A**, **B**, and **C** are n×n matrices, with **A** nonsingular and $\mathbf{AB} = \mathbf{AC}$, show that $\mathbf{B} = \mathbf{C}$.

6. The **right inverse** matrix **R** of an m×n matrix **A** ($m \le n$) is defined by

$$\mathbf{A}_{m\times n}\mathbf{R}_{n\times m} = \mathbf{I}_{m\times m}$$

Find the right inverse of

$$\mathbf{A} = \begin{bmatrix} 2 & 0 & 0 \\ 0 & 3 & 0 \end{bmatrix}$$

Similarly, the **left inverse** matrix **L** of an m×n matrix **B** (m ≥ n) is defined by

$$\mathbf{L}_{n\times m}\mathbf{B}_{m\times n} = \mathbf{I}_{n\times n}$$

Find the left inverse of

$$\mathbf{B} = \begin{bmatrix} 2 & 0 \\ 0 & 3 \\ 0 & 0 \end{bmatrix}$$

7. If $\mathbf{B} = \mathbf{P}^{-1}\mathbf{A}\mathbf{P}$, show that $|\,\mathbf{B}\,| = |\,\mathbf{A}\,|$.

8. If every row of **A** adds up to zero and **x** is a column matrix of ones, what is **Ax**? What is $|\,\mathbf{A}\,|$?

2

LINEAR ALGEBRA

The algebraic properties of real numbers and matrices form the foundation for generalizations and abstractions that permit systematic analysis and solution of large-scale engineering problems. Concepts of linear algebra are developed in this chapter, using familiar concepts of vector algebra and matrices. The goal here is to develop a mastery of the concepts and methods of linear algebra in a comfortable setting, prior to developing methods that encounter the technical complexities of differential equations and infinite dimensional spaces.

2.1 VECTOR ALGEBRA

In preparation for developing concepts and tools of linear algebra that form the foundation for more advanced methods presented in this text, it is helpful to review familiar concepts of vector algebra and to relate them to matrices. Geometric definitions of vector algebra are first recalled and related to operations with Cartesian components of vectors. It is then shown that all of these comfortable operations with vectors can be represented as matrix operations. The value of this form of vector algebra is twofold. First, it provides a geometric foundation for concepts of linear algebra that are essential in engineering analysis. Second, it reduces vector operations to a form that can be easily implemented on a digital computer.

Geometric Vectors

The concept of a vector may be introduced in a very general geometric setting, with no requirement for identification of a reference frame. In this setting a **geometric vector**, or simply a **vector**, is defined as the directed line segment from one point to another point in space. Its direction is therefore established and its magnitude is defined to be the distance between the points that are connected by the vector. This geometric setting for vector analysis, together with algebraic operations of addition, multiplication by a scalar, scalar product, and vector product, form the classical foundation of **vector analysis**.

Vector $\vec{a}$ in Fig. 2.1.1, beginning at point A and ending at point B, is denoted by the notation $\rightarrow$, in its geometric sense. The **magnitude of a vector** $\vec{a}$ is its length (the distance between A and B) and is denoted by a, or $|\vec{a}|$. Note that the magnitude of a vector is positive if points A and B do not coincide and is zero only when they coincide. A vector with zero length is denoted $\vec{0}$ and is called the **zero vector**.

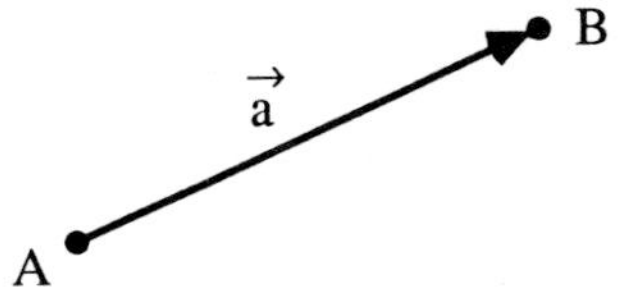

Figure 2.1.1 Vector From Point A to Point B

Definition 2.1.1. Multiplication of a vector $\vec{a}$ by a scalar $\alpha \geq 0$ is defined as a vector in the same direction as $\vec{a}$, but having magnitude αa. A **unit vector,** having a length of a unity, in the direction $\vec{a} \neq \vec{0}$ is $(1/a)\vec{a}$. Multiplication of a vector $\vec{a}$ by a negative scalar $\beta < 0$ is defined as the vector with magnitude $|\beta| a$ and direction opposite to that of $\vec{a}$. The **negative of a vector** is obtained by multiplying the vector by −1. It is the vector with the same magnitude but opposite direction. ■

Example 2.1.1

Let points A and B in Fig. 2.1.1 be located in an **orthogonal reference frame,** as shown in Fig. 2.1.2. The distance between points A and B, with coordinates (A_x, A_y, A_z) and (B_x, B_y, B_z), respectively, is the magnitude of $\vec{a}$; i.e.,

$$|\vec{a}| = [(B_x - A_x)^2 + (B_y - A_y)^2 + (B_z - A_z)^2]^{1/2}$$

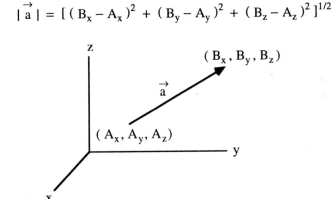

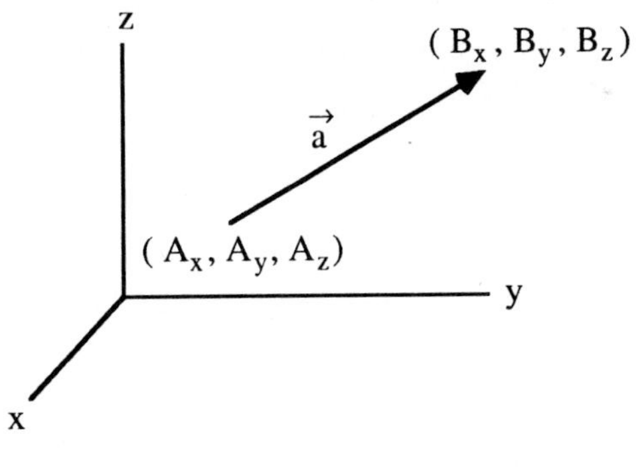

Figure 2.1.2 Vector Located in Orthogonal Reference Frame ■

Definition 2.1.2. Addition of vectors $\vec{a}$ and $\vec{b}$ is according to the **parallelogram rule**, as shown in Fig. 2.1.3. The parallelogram used in this construction is formed in the plane that contains the intersecting vectors $\vec{a}$ and $\vec{b}$. The **vector sum** is written as

$$\vec{a} + \vec{b} = \vec{c} \qquad (2.1.1)$$

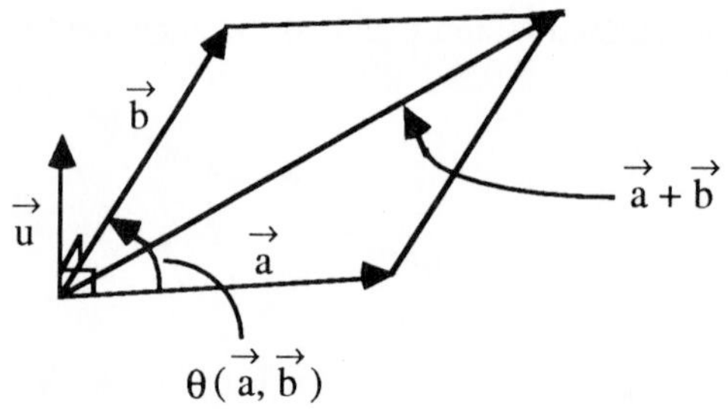

Figure 2.1.3 Addition of Vectors ■

Addition of vectors and multiplication of vectors by scalars obey the following rules [3]:

$$\begin{aligned} \vec{a} + \vec{b} &= \vec{b} + \vec{a} \\ \alpha(\vec{a} + \vec{b}) &= \alpha\vec{a} + \alpha\vec{b} \\ (\alpha + \beta)\vec{a} &= \alpha\vec{a} + \beta\vec{a} \end{aligned} \tag{2.1.2}$$

where α and β are scalars.

Orthogonal reference frames are used extensively in representing vectors. Use in this text is limited to **right-hand** x-y-z orthogonal reference frames; i.e., frames with mutually orthogonal x, y, and z axes that are ordered by the finger structure of the right hand, as shown in Fig. 2.1.4. Such a frame is called a **Cartesian reference frame**. For indexing of variables, the x-y-z frame may be denoted by an x_1-x_2-x_3 frame, as shown in Fig. 2.1.4.

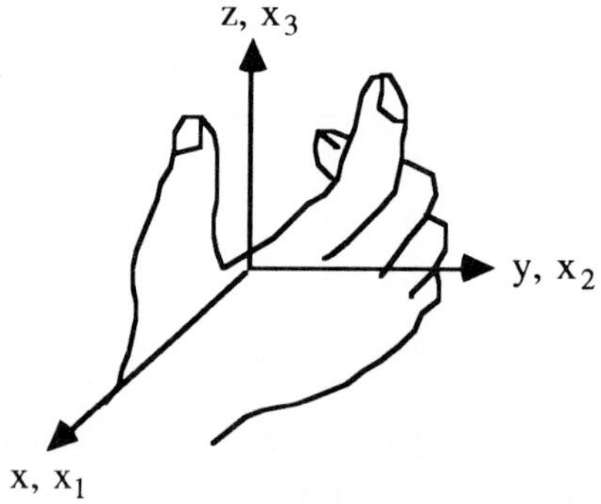

Figure 2.1.4 Right-Hand Orthogonal Reference Frame

A vector $\vec{a}$ can be resolved into components a_1, a_2, and a_3 along the x_1, x_2, and x_3 axes of a Cartesian reference frame, as shown in Fig. 2.1.5. These components are called the **Cartesian components of the vector. Unit coordinate vectors** $\vec{i}$, $\vec{j}$, and $\vec{k}$ are directed along the x_1, x_2, and x_3 axes, respectively, as shown in Fig. 2.1.5. In vector notation,

$$\vec{a} = a_1 \vec{i} + a_2 \vec{j} + a_3 \vec{k} \tag{2.1.3}$$

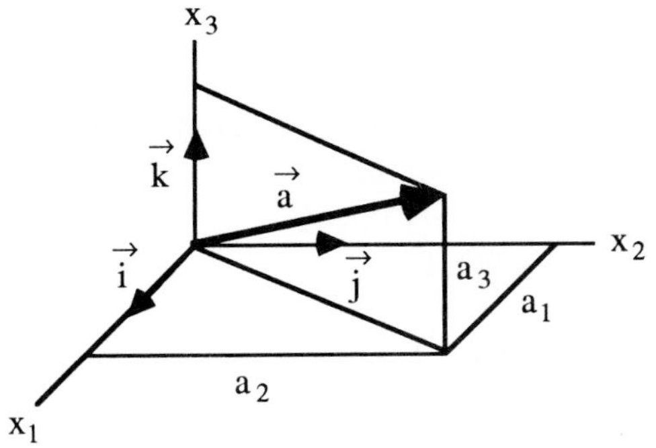

Figure 2.1.5 Components of a Vector

Addition of vectors $\vec{a}$ and $\vec{b}$ may be expressed in terms of their components, using Eq. 2.1.1, as

$$\begin{aligned} \vec{c} = \vec{a} + \vec{b} &= (a_1 + b_1)\vec{i} + (a_2 + b_2)\vec{j} + (a_3 + b_3)\vec{k} \\ &\equiv c_1 \vec{i} + c_2 \vec{j} + c_3 \vec{k} \end{aligned} \tag{2.1.4}$$

where c_1, c_2, and c_3 are the Cartesian components of vector $\vec{c}$. Thus, addition of vectors occurs component by component. Using this idea, three vectors $\vec{a}$, $\vec{b}$, and $\vec{c}$ may be added, to show that

$$(\vec{a} + \vec{b}) + \vec{c} = \vec{a} + (\vec{b} + \vec{c}) \tag{2.1.5}$$

Definition 2.1.3. Denote the angle from vector $\vec{a}$ to vector $\vec{b}$ in the plane that contains them by $\theta(\vec{a}, \vec{b})$, with counterclockwise as positive about the normal $\vec{u}$ to the plane of the vectors that points toward the viewer, as shown in Fig. 2.1.3. The **scalar product** of two vectors $\vec{a}$ and $\vec{b}$ is defined as the product of their magnitude and the cosine of the angle between them; i.e.,

$$\vec{a} \cdot \vec{b} = ab \cos \theta(\vec{a}, \vec{b}) \tag{2.1.6}$$

■

This definition is in purely geometric terms, so it is independent of the reference frame in which the vectors are represented. Note that if two vectors $\vec{a}$ and $\vec{b}$ are nonzero (i.e., $a \neq 0$ and $b \neq 0$) then their scalar product is zero if and only if $\cos \theta(\vec{a}, \vec{b}) = 0$. Two nonzero vectors are said to be **orthogonal** if their scalar product is zero. Since $\theta(\vec{b}, \vec{a}) = 2\pi - \theta(\vec{a}, \vec{b})$ and $\cos(2\pi - \theta) = \cos\theta$, the order of terms appearing on the right side of Eq. 2.1.6 is immaterial. Thus,

$$\vec{a} \cdot \vec{b} = \vec{b} \cdot \vec{a} \tag{2.1.7}$$

Based on the definition of the scalar product, the following identities hold for the unit coordinate vectors $\vec{i}$, $\vec{j}$, and $\vec{k}$:

$$\begin{aligned} \vec{i} \cdot \vec{j} &= \vec{j} \cdot \vec{k} = \vec{k} \cdot \vec{i} = 0 \\ \vec{i} \cdot \vec{i} &= \vec{j} \cdot \vec{j} = \vec{k} \cdot \vec{k} = 1 \end{aligned} \tag{2.1.8}$$

Furthermore, for any vector $\vec{a}$,

$$\vec{a} \cdot \vec{a} = aa \cos 0 = a^2$$

While not obvious on geometric grounds, the scalar product satisfies the relation [2]

$$(\vec{a} + \vec{b}) \cdot \vec{c} = \vec{a} \cdot \vec{c} + \vec{b} \cdot \vec{c} \tag{2.1.9}$$

Using Eq. 2.1.9 and the identities of Eq. 2.1.8, a direct calculation yields

$$\begin{aligned} \vec{a} \cdot \vec{b} &= (a_1 \vec{i} + a_2 \vec{j} + a_3 \vec{k}) \cdot (b_1 \vec{i} + b_2 \vec{j} + b_3 \vec{k}) \\ &= a_1 b_1 + a_2 b_2 + a_3 b_3 \end{aligned} \tag{2.1.10}$$

Definition 2.1.4. The **vector product** of two vectors $\vec{a}$ and $\vec{b}$ is defined as the vector

$$\vec{a} \times \vec{b} = ab \sin \theta(\vec{a}, \vec{b}) \, \vec{u} \tag{2.1.11}$$

where $\vec{u}$ is a unit vector that is orthogonal (perpendicular) to the plane that contains vectors $\vec{a}$ and $\vec{b}$, taken in the positive right-hand coordinate direction, as shown in Fig. 2.1.6. ■

If the viewer were behind the plane of Fig. 2.1.6 that is formed by vectors $\vec{a}$ and $\vec{b}$, the unit normal to the plane would be $-\vec{u}$ and the counterclockwise angle from $\vec{a}$ to $\vec{b}$

would be $2\pi - \theta(\vec{a}, \vec{b})$. Then the vector product would be

$$\vec{a} \times \vec{b} = ab \sin(2\pi - \theta(\vec{a}, \vec{b}))(-\vec{u})$$

$$= ab \sin \theta(\vec{a}, \vec{b})\, \vec{u}$$

since $\sin(2\pi - \theta) = -\sin\theta$. This is the same as Eq. 2.1.11, so the viewpoint does not influence evaluation of the vector product. Since the definition of vector product is purely geometric, the result is independent of the reference frame in which the vectors are represented.

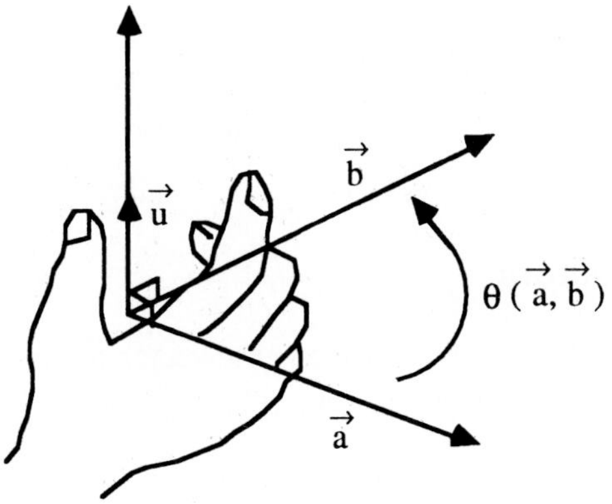

Figure 2.1.6 Vector Product

Since reversal of the order of the vectors $\vec{a}$ and $\vec{b}$ in Eq. 2.1.11 yields the opposite direction for the unit vector $\vec{u}$,

$$\vec{b} \times \vec{a} = -\vec{a} \times \vec{b} \tag{2.1.12}$$

Analogous to Eq. 2.1.9, the vector product satisfies [3]

$$(\vec{a} + \vec{b}) \times \vec{c} = \vec{a} \times \vec{c} + \vec{b} \times \vec{c} \tag{2.1.13}$$

Since $\theta(\vec{a}, \vec{a}) = 0$, for any vector $\vec{a}$,

$$\vec{a} \times \vec{a} = a^2 \sin 0\, \vec{u} = \vec{0}$$

From the definition of unit coordinate vectors and vector product, the following identities are valid:

$$\vec{i} \times \vec{i} = \vec{j} \times \vec{j} = \vec{k} \times \vec{k} = \vec{0}$$

$$\vec{i} \times \vec{j} = -\vec{j} \times \vec{i} = \vec{k}$$

$$\vec{j} \times \vec{k} = -\vec{k} \times \vec{j} = \vec{i} \tag{2.1.14}$$

$$\vec{k} \times \vec{i} = -\vec{i} \times \vec{k} = \vec{j}$$

Using the identities of Eq. 2.1.14 and the property of vector product of Eq. 2.1.13, the vector product of two vectors may be expanded and written in terms of their components as

$$\vec{c} = \vec{a} \times \vec{b} = (a_2b_3 - a_3b_2)\vec{i} + (a_3b_1 - a_1b_3)\vec{j} + (a_1b_2 - a_2b_1)\vec{k}$$

$$\equiv c_1\vec{i} + c_2\vec{j} + c_3\vec{k} \tag{2.1.15}$$

Algebraic Vectors

Recall from Eq. 2.1.3 that a geometric vector $\vec{a}$ can be written in component form in a Cartesian x_1-x_2-x_3 frame as

$$\vec{a} = a_1\vec{i} + a_2\vec{j} + a_3\vec{k}$$

The geometric vector $\vec{a}$ is thus uniquely defined by its Cartesian components, which may be written in column matrix notation as

$$\mathbf{a} \equiv \begin{bmatrix} a_1 \\ a_2 \\ a_3 \end{bmatrix} = [\, a_1, a_2, a_3 \,]^T \tag{2.1.16}$$

This is called the **algebraic representation of a geometric vector.**

Note that the algebraic representation of vectors is dependent on the Cartesian reference frame selected; i.e., the unit vectors $\vec{i}$, $\vec{j}$, $\vec{k}$. Some of the purely geometric properties of vectors are thus lost and properties of the reference frame that is used in defining algebraic components come into play.

Definition 2.1.5. An **algebraic vector** is defined as a column matrix. When an algebraic vector represents a geometric vector in three-dimensional space, it has three components. Algebraic vectors with more than three components will also be employed. In case $\mathbf{a} = [\, a_1, \ldots, a_n \,]^T$, the algebraic vector $\mathbf{a}$ is called an **n-vector** and is said to belong to **n-dimensional real space**, denoted R^n. ■

If two geometric vectors $\vec{a}$ and $\vec{b}$ are represented in algebraic form as

$$\mathbf{a} = [\, a_1, a_2, a_3 \,]^T \tag{2.1.17}$$

$$\mathbf{b} = [\, b_1, b_2, b_3 \,]^T \tag{2.1.18}$$

then their vector sum $\vec{c} = \vec{a} + \vec{b}$, from Eq. 2.1.4, is represented in algebraic form by

$$\mathbf{c} = \mathbf{a} + \mathbf{b} \tag{2.1.19}$$

Example 2.1.2

The algebraic representation of geometric vectors

$$\vec{a} = \vec{i} + 2\vec{j} + 3\vec{k}$$

$$\vec{b} = -\vec{i} + \vec{j} - \vec{k}$$

is

$$\mathbf{a} = [\, 1, 2, 3 \,]^T$$

$$\mathbf{b} = [\, -1, 1, -1 \,]^T$$

The algebraic representation of the sum $\vec{c} = \vec{a} + \vec{b}$ is

$$\mathbf{c} = \mathbf{a} + \mathbf{b} = [\, 0, 3, 2 \,]^T$$ ■

Two geometric vectors are equal; i.e., $\vec{a} = \vec{b}$, if and only if the Cartesian components of the vectors are equal; i.e., $\mathbf{a} = \mathbf{b}$. Multiplication of a vector $\vec{a}$ by a scalar α occurs component by component, so the geometric vector $\alpha\vec{a}$ is represented by the algebraic vector $\alpha\mathbf{a}$.

Since there is a one-to-one correspondence between geometric vectors and 3×1 algebraic vectors that are formed from their Cartesian components in a specified Cartesian reference frame, no distinction other than notation will be made between them in the remainder of this text.

The scalar product of vectors $\vec{a}$ and $\vec{b}$ may be expressed in algebraic form, using the result of Eq. 2.1.10, as

$$\vec{a} \cdot \vec{b} = a_1 b_1 + a_2 b_2 + a_3 b_3 = \mathbf{a}^T \mathbf{b} \tag{2.1.20}$$

Example 2.1.3

The scalar product of vectors $\vec{a}$ and $\vec{b}$ (or **a** and **b**) in Example 2.1.2 is

$$\vec{a} \cdot \vec{b} = \mathbf{a}^T\mathbf{b} = [\,1, 2, 3\,]\begin{bmatrix} -1 \\ 1 \\ -1 \end{bmatrix} = -2$$

From the definition of scalar product,

$$\mathbf{a}^T\mathbf{b} = ab\cos\theta(\mathbf{a}, \mathbf{b})$$

and, with $a = (\mathbf{a}^T\mathbf{a})^{1/2} = \sqrt{14}$ and $b = (\mathbf{b}^T\mathbf{b})^{1/2} = \sqrt{3}$, $\cos\theta(\mathbf{a}, \mathbf{b}) = -2/\sqrt{42}$. Thus, $\theta(\mathbf{a}, \mathbf{b}) = 1.88$ or 4.40 rad. ■

A skew symmetric matrix $\tilde{\mathbf{a}}$, associated with a 3×1 algebraic vector **a**, is defined as

$$\tilde{\mathbf{a}} \equiv \begin{bmatrix} 0 & -a_3 & a_2 \\ a_3 & 0 & -a_1 \\ -a_2 & a_1 & 0 \end{bmatrix} \tag{2.1.21}$$

Note that an overhead ~ (called tilde) indicates that the components of vector **a** are used to generate a skew symmetric 3×3 matrix $\tilde{\mathbf{a}}$.

The **vector product** $\vec{c} = \vec{a} \times \vec{b}$, which is expanded in component form in Eq. 2.1.15, can be written in algebraic vector form as

$$\mathbf{c} = \tilde{\mathbf{a}}\mathbf{b} = \begin{bmatrix} a_2b_3 - a_3b_2 \\ a_3b_1 - a_1b_3 \\ a_1b_2 - a_2b_1 \end{bmatrix} \tag{2.1.22}$$

A direct computation verifies that

$$\tilde{\mathbf{a}}\mathbf{b} = -\tilde{\mathbf{b}}\mathbf{a}$$

which agrees with the vector product property of Eq. 2.1.12.

Example 2.1.4

The algebraic representation of the vector product $\vec{c} = \vec{a} \times \vec{b}$ in Example 2.1.2 is

$$\mathbf{c} = \tilde{\mathbf{a}}\mathbf{b} = \begin{bmatrix} 0 & -3 & 2 \\ 3 & 0 & -1 \\ -2 & 1 & 0 \end{bmatrix}\begin{bmatrix} -1 \\ 1 \\ -1 \end{bmatrix} = \begin{bmatrix} -5 \\ -2 \\ 3 \end{bmatrix}$$ ■

For later use, it is helpful to recall a fundamental property of the ~ operation; i.e.,

$$\tilde{\mathbf{a}}^T = \begin{bmatrix} 0 & a_3 & -a_2 \\ -a_3 & 0 & a_1 \\ a_2 & -a_1 & 0 \end{bmatrix} = -\tilde{\mathbf{a}} \tag{2.1.23}$$

The reduction of geometric concepts of vector algebra to matrix algebra in this section provides insights and computational tools that are needed later in the text and in applications. The reader is encouraged to return to this intuitively clear relationship between three-dimensional geometry and matrix algebra when concepts of higher dimensional analysis become more complex.

EXERCISES 2.1

1. Verify Eq. 2.1.5.

2. Verify Eq. 2.1.9.

3. Verify Eq. 2.1.15.

4. For given vectors $\vec{a} = \vec{i} + 2\vec{j} + 3\vec{k}$ and $\vec{b} = 4\vec{j} + 5\vec{k}$, find $\vec{a} + \vec{b}$, $\vec{a} \cdot \vec{b}$, and $\vec{a} \times \vec{b}$. Also find $\mathbf{a} + \mathbf{b}$, $\mathbf{a}^T\mathbf{b}$, and $\tilde{\mathbf{a}}\mathbf{b}$, using algebraic representations $\mathbf{a}$ and $\mathbf{b}$ of $\vec{a}$ and $\vec{b}$, respectively.

5. Repeat Exercise 4 for the vectors $\vec{a} = 5\vec{i} + 2\vec{j} + \vec{k}$ and $\vec{b} = 2\vec{i} + 6\vec{j} + 4\vec{k}$.

2.2 VECTOR SPACES

Matrix analysis, which is reviewed in Chapter 1, is of importance in its own right and serves as a model for more general methods of engineering analysis. For this reason, it is important to obtain proficiency in viewing linear calculations and matrix analysis in a vector space setting. Vector space concepts and techniques that generalize methods presented in Section 2.1 provide a rational setting that allows a clear view of the forest, without getting bogged down in all the notational trees that arise in complex problems. The contents of this section should be viewed by the reader as both tools for use in computational algebra and a conceptual base for methods of engineering analysis that follow.

n-Dimensional Real Space (R^n)

It is conventional to speak of an n×1 matrix as a **column vector**, or simply as a **vector**. The collection of all such vectors is called **n-dimensional real space** and is denoted R^n. A vector in R^n is thus

$$\mathbf{x} = [\, x_1, x_2, \ldots, x_n \,]^T$$

A row matrix $[\, x_1, x_2, \ldots, x_n \,]$, which is the transpose of the column matrix $\mathbf{x}$, is called a **row vector** and is denoted by $\mathbf{x}^T$.

Vectors in R^n follow the same laws of addition and multiplication by a scalar as three-dimensional vectors in Section 2.1. These laws are typical of an algebraic system called a vector space.

Definition 2.2.1. A **vector space** V is a set of vectors with operations of addition and multiplication by a real scalar that satisfy the following **vector space postulates:**

(1) **Closure under addition**: For every pair of vectors $\mathbf{x}$ and $\mathbf{y}$ in V there is a unique sum denoted by $\mathbf{x} + \mathbf{y}$ in V.

For vectors $\mathbf{x}$, $\mathbf{y}$, and $\mathbf{z}$ in V,

(2) $(\mathbf{x} + \mathbf{y}) + \mathbf{z} = \mathbf{x} + (\mathbf{y} + \mathbf{z})$

(3) $\mathbf{x} + \mathbf{y} = \mathbf{y} + \mathbf{x}$

(4) A zero vector $\mathbf{0}$ exists, such that $\mathbf{x} + \mathbf{0} = \mathbf{x}$, for all $\mathbf{x}$.

(5) A negative vector $-\mathbf{x}$ exists for all $\mathbf{x}$, such that $\mathbf{x} + (-\mathbf{x}) = \mathbf{0}$.

(6) **Closure under multiplication by a scalar**: For every real scalar α and every vector $\mathbf{x}$ in V there is a unique vector $\alpha\mathbf{x}$ in V.

For real scalars α and β and vectors $\mathbf{x}$ and $\mathbf{y}$ in V,

(7) $\alpha(\mathbf{x} + \mathbf{y}) = \alpha\mathbf{x} + \alpha\mathbf{y}$

(8) $(\alpha + \beta)\mathbf{x} = \alpha\mathbf{x} + \beta\mathbf{x}$

(9) $(\alpha\beta)\mathbf{x} = \alpha(\beta\mathbf{x})$

(10) $1\mathbf{x} = \mathbf{x}$ ■

Note that there need not be any concept of magnitude of vectors or products of vectors as part of the definition of a vector space. These and additional operations and properties may be associated with the underlying vector space, to enrich its structure and make it more useful for specific engineering applications. It is important to realize, however, that many of these add-on properties make sense only if the basic vector space algebraic properties are well defined.

In order to specify a vector space, and since many different vector spaces are used in engineering analysis, a set V of vectors with operations of addition and multiplication by a real scalar that satisfy all the properties of Definition 2.2.1 must be defined. For example, the set R^1 of real numbers is a vector space, where addition and multiplication by a scalar are the usual addition and multiplication of real numbers. In this case, a real number plays the role of both a vector and a scalar.

Example 2.2.1

Consider the set* of vectors

$$E = \{ \mathbf{x} \in R^n : x_1 = 0 \}$$

---

* **Set notation** used in this text is limited to defining **subsets** of the form

$$\{ \mathbf{x} \in S : \text{condition satisfied by } \mathbf{x} \}$$

where S is a given set and the condition following the colon must be satisfied by each $\mathbf{x}$ in the subset.

with addition and multiplication by a scalar defined by matrix operations of Section 1.2. Properties (2) – (5) and (7) – (10) of Definition 2.2.1 hold, since they are properties of addition and multiplication by scalars in R^n. Property 2.2.1 holds, since if $\mathbf{x} \in E$ and $\mathbf{y} \in E$, then $x_1 = y_1 = 0$. Thus, for $\mathbf{z} = \mathbf{x} + \mathbf{y}$, $z_1 = x_1 + y_1 = 0$, so $\mathbf{z} \in E$ and E is closed under addition. Likewise, if $\mathbf{x} \in E$, then for $\mathbf{z} = \alpha\mathbf{x}$ where α is any scalar, $z_1 = \alpha x_1 = 0$, so $\mathbf{z} \in E$ and E is closed under multiplication by a scalar.

To make this example more concrete, let n = 3. Then $E = \{ \mathbf{x} \in R^3: x_1 = 0 \}$ is a collection of vectors that lie in the x_2-x_3 plane. ■

Example 2.2.1 illustrates that when a candidate vector space is defined as a subset of a known vector space, then properties (2) – (5) and (7) – (10) of Definition 2.2.1 automatically follow. Therefore, in this case, only properties (1) and (6) need to be verified.

Example 2.2.2

As a second example, let

$$F = \{ \mathbf{x} \in R^n: x_1 = 1 \}$$

For $\mathbf{x} \in F$ and $\mathbf{y} \in F$, $x_1 = y_1 = 1$. Defining $\mathbf{z} = \mathbf{x} + \mathbf{y}$, $z_1 = x_1 + y_1 = 2$. Hence, $\mathbf{z} \notin F$, so F is not closed under the operation of addition. Thus, it is not a vector space. ■

Linear Independence of Vectors

Definition 2.2.2. A set of vectors $\mathbf{x}^1, \mathbf{x}^2, \ldots, \mathbf{x}^n$ is said to be **linearly dependent** if there exist scalars $c_1, c_2, \ldots, c_n$ that are not all zero, such that

$$c_i\mathbf{x}^i = \mathbf{0} \tag{2.2.1}$$

where summation notation is used. If a set of vectors is not linearly dependent, then it is said to be **linearly independent**. Equivalently, a set of vectors $\mathbf{x}^1, \mathbf{x}^2, \ldots, \mathbf{x}^n$ is linearly independent if $c_i\mathbf{x}^i = \mathbf{0}$ implies that $c_i = 0$, for all i. ■

The last observation of Definition 2.2.2 forms a convenient means of testing for linearly independence. Simply write down Eq. 2.2.1 and treat it as a set of equations in the variables c_i, with the known vectors $\mathbf{x}^i$ as coefficients. If the only solution is $c_1 = \ldots = c_n = 0$, then the $\mathbf{x}^1, \mathbf{x}^2, \ldots, \mathbf{x}^n$ are linearly independent.

If any vector $\mathbf{x}^k$ among a set $\mathbf{x}^1, \mathbf{x}^2, \ldots, \mathbf{x}^n$ is zero, then the set is linearly dependent. To see that this is true, let $\mathbf{x}^k = \mathbf{0}$, choose $c_i = \delta_{ik}$, and note that $c_i\mathbf{x}^i = \delta_{ik}\mathbf{x}^i = \mathbf{x}^k = \mathbf{0}$. Since $c_k = 1$, not all the c_i are zero and the $\mathbf{x}^i$ are linearly dependent.

Example 2.2.3

Consider the vector space R^3 and vectors $\mathbf{x}^1 = [\,1, 0, 0\,]^T$, $\mathbf{x}^2 = [\,1, 1, 0\,]^T$, and $\mathbf{x}^3 = [\,1, 1, 1\,]^T$. To determine whether $\mathbf{x}^1$, $\mathbf{x}^2$, and $\mathbf{x}^3$ are linearly independent, form the equation

$$c_1\mathbf{x}^1 + c_2\mathbf{x}^2 + c_3\mathbf{x}^3 = \begin{bmatrix} 1 & 1 & 1 \\ 0 & 1 & 1 \\ 0 & 0 & 1 \end{bmatrix} \begin{bmatrix} c_1 \\ c_2 \\ c_3 \end{bmatrix} = \mathbf{0}$$

Since the determinant of the coefficient matrix is 1, it is nonsingular and $c_1 = c_2 = c_3 = 0$ is the only solution. Thus, the vectors are linearly independent. ■

Example 2.2.4

Consider again the vector space R^3 and vectors

$$\mathbf{x}^1 = \begin{bmatrix} 1 \\ 2 \\ -1 \end{bmatrix}, \quad \mathbf{x}^2 = \begin{bmatrix} 3 \\ 6 \\ 0 \end{bmatrix}, \quad \mathbf{x}^3 = \begin{bmatrix} 0 \\ 1 \\ 0 \end{bmatrix}, \quad \mathbf{x}^4 = \begin{bmatrix} 2 \\ 5 \\ 0 \end{bmatrix}$$

To check for linear independence, form

$$c_i\mathbf{x}^i = \begin{bmatrix} 1 & 3 & 0 & 2 \\ 2 & 6 & 1 & 5 \\ -1 & 0 & 0 & 0 \end{bmatrix} \begin{bmatrix} c_1 \\ c_2 \\ c_3 \\ c_4 \end{bmatrix} = \begin{bmatrix} 0 \\ 0 \\ 0 \end{bmatrix}$$

A solution may be obtained for c_1, c_3, and c_4 as functions of c_2,

$$\begin{aligned} c_1 &= 0 \\ c_4 &= -\,3c_2/2 \\ c_3 &= -\,(\,6c_2 + 5c_4\,) \;=\; 3c_2/2 \end{aligned}$$

which is valid for any nonzero value of c_2; e.g., $c_2 = 1$. Thus, the $\mathbf{x}^i$ are not linearly independent; i.e., they are linearly dependent. ■

Definition 2.2.3. A set of vectors $\mathbf{x}^1, \mathbf{x}^2, \ldots, \mathbf{x}^n$ in a vector space V is said to **span the vector space** if every vector in the space can be written as a **linear combination** of the set; i.e., for every vector $\mathbf{x}$ in V, there exist scalars $c_1, c_2, \ldots, c_n$, such that

$$\mathbf{x} = c_i\mathbf{x}^i \tag{2.2.2}$$

■

Example 2.2.5

Consider the vectors $\mathbf{x}^1 = [\,1, 2, 1\,]^T$, $\mathbf{x}^2 = [\,1, 0, 2\,]^T$, and $\mathbf{x}^3 = [\,1, 0, 0\,]^T$ in R^3. To find out whether $\{\,\mathbf{x}^1, \mathbf{x}^2, \mathbf{x}^3\,\}$ spans R^3, choose any vector $\mathbf{a} = [\,a_1, a_2, a_3\,]^T$ in

R^3 and attempt to find c_1, c_2, and c_3 so that $\mathbf{a} = c_1\mathbf{x}^1 + c_2\mathbf{x}^2 + c_3\mathbf{x}^3$; i.e.,

$$c_i\mathbf{x}^i = \begin{bmatrix} 1 & 1 & 1 \\ 2 & 0 & 0 \\ 1 & 2 & 0 \end{bmatrix} \begin{bmatrix} c_1 \\ c_2 \\ c_3 \end{bmatrix} = \begin{bmatrix} a_1 \\ a_2 \\ a_3 \end{bmatrix} \equiv \mathbf{a}$$

Direct manipulation yields the solution

$$c_1 = \frac{a_2}{2}, \quad c_2 = \frac{2a_3 - a_2}{4}, \quad c_3 = \frac{4a_1 - a_2 - a_3}{4}$$

Thus, $\{ \mathbf{x}^1, \mathbf{x}^2, \mathbf{x}^3 \}$ spans R^3. ■

A set of vectors $\mathbf{x}^1, \mathbf{x}^2, \ldots, \mathbf{x}^n$ may span a vector space and still not be linearly independent. However, if the vectors are linearly dependent, a subset can be selected that is linearly independent and also spans the same space. To see that this is true, suppose $\mathbf{x}^1, \mathbf{x}^2, \ldots, \mathbf{x}^n$ span the space and are linearly dependent. Then there exists a set of scalars $\gamma_1, \gamma_2, \ldots, \gamma_n$ that are not all zero, such that

$$\gamma_i\mathbf{x}^i = \mathbf{0}$$

Without loss of generality, assume that $\gamma_n \neq 0$. Then $\mathbf{x}^n$ can be written in terms of $\mathbf{x}^1, \mathbf{x}^2, \ldots, \mathbf{x}^{n-1}$ as

$$\mathbf{x}^n = -\frac{\gamma_1}{\gamma_n}\mathbf{x}^1 - \frac{\gamma_2}{\gamma_n}\mathbf{x}^2 - \ldots - \frac{\gamma_{n-1}}{\gamma_n}\mathbf{x}^{n-1}$$

For any vector $\mathbf{x}$ in the vector space, there exist c_i, $i = 1, \ldots, n$, such that

$$\begin{aligned} \mathbf{x} &= c_i\mathbf{x}^i \\ &= c_1\mathbf{x}^1 + c_2\mathbf{x}^2 + \ldots + c_{n-1}\mathbf{x}^{n-1} \\ &\quad + c_n\left[-\frac{\gamma_1}{\gamma_n}\mathbf{x}^1 - \frac{\gamma_2}{\gamma_n}\mathbf{x}^2 - \ldots - \frac{\gamma_{n-1}}{\gamma_n}\mathbf{x}^{n-1} \right] \end{aligned}$$

Therefore, the subset $\mathbf{x}^1, \mathbf{x}^2, \ldots, \mathbf{x}^{n-1}$ spans the space. If this subset is linearly dependent, the process can be repeated. Eventually, a subset $\mathbf{x}^1, \mathbf{x}^2, \ldots, \mathbf{x}^m$, with $m < n$, will be obtained that spans the space and is linearly independent.

Basis and Dimension of Vector Spaces

Definition 2.2.4. A **basis of a vector space** is a set of linearly independent vectors that spans the space. ■

Example 2.2.6

The vectors $\mathbf{x}^1 = [\ 1, 0, \ldots, 0\ ]^T$, $\mathbf{x}^2 = [\ 0, 1, \ldots, 0\ ]^T, \ldots, \mathbf{x}^n = [\ 0, 0, \ldots, 1\ ]^T$ span R^n, since any vector $\mathbf{a} = [\ a_1, a_2, \ldots, a_n\ ]^T$ can be written as

$$\mathbf{a} = a_1\mathbf{x}^1 + a_2\mathbf{x}^2 + \ldots + a_n\mathbf{x}^n = a_i\mathbf{x}^i$$

Note that the vectors $\mathbf{x}^i$ are linearly independent, since

$$c_1\mathbf{x}^1 + c_2\mathbf{x}^2 + \ldots + c_n\mathbf{x}^n = [\ c_1, c_2, \ldots, c_n\ ]^T = \mathbf{0}$$

implies that $c_1 = c_2 = \ldots = c_n = 0$. Therefore, the set of vectors $\mathbf{x}^1, \mathbf{x}^2, \ldots, \mathbf{x}^n$ is a basis for the vector space R^n. ■

It has already been shown that from any finite set of vectors that spans a vector space, a linearly independent subset can be selected that spans the vector space. Such a subset forms a basis for the vector space. There may be many bases for the same vector space. For example, one basis for the vector space R^n was given in Example 2.2.6. Another basis for R^n consists of the vectors

$$\begin{aligned}
\mathbf{y}^1 &= [\ 1, 0, 0, \ldots, 0\ ]^T \\
\mathbf{y}^2 &= [\ 1, 1, 0, \ldots, 0\ ]^T \\
\mathbf{y}^3 &= [\ 1, 1, 1, \ldots, 0\ ]^T \\
&\ldots\ldots\ldots\ldots\ldots\ldots \\
\mathbf{y}^n &= [\ 1, 1, 1, \ldots, 1\ ]^T
\end{aligned}$$

Note that both this basis and the basis of Example 2.2.6 have n vectors.

To see if all bases have the same number of vectors, suppose that $\mathbf{x}^1, \mathbf{x}^2, \ldots, \mathbf{x}^n$ and $\mathbf{y}^1, \mathbf{y}^2, \ldots, \mathbf{y}^m$ are bases of a vector space. Since the $\mathbf{x}^i$ span the space, $\mathbf{y}^1$ can be expressed as a linear combination of the $\mathbf{x}^i$. That is,

$$\mathbf{y}^1 = c_1\mathbf{x}^1 + c_2\mathbf{x}^2 + \ldots + c_n\mathbf{x}^n$$

where at least one of the c_i is not zero. Otherwise, $\mathbf{y}^1 = \mathbf{0}$ and the set $\mathbf{y}^1, \mathbf{y}^2, \ldots, \mathbf{y}^m$ would be linearly dependent. Assume $c_n \neq 0$ (otherwise renumber the $\mathbf{x}^i$ so that it is). Then,

$$\mathbf{x}^n = \frac{1}{c_n}(\ \mathbf{y}^1 - c_1\mathbf{x}^1 - c_2\mathbf{x}^2 - \ldots - c_{n-1}\mathbf{x}^{n-1}\ )$$

Therefore, the set of vectors $\mathbf{y}^1, \mathbf{x}^1, \mathbf{x}^2, \ldots, \mathbf{x}^{n-1}$ spans the vector space and $\mathbf{y}^2$ can be expressed as a linear combination of this set. That is,

$$\mathbf{y}^2 = b_1\mathbf{y}^1 + \gamma_1\mathbf{x}^1 + \ldots + \gamma_{n-1}\mathbf{x}^{n-1}$$

where at least one of the γ_i is not zero. Otherwise, $\mathbf{y}^1$ and $\mathbf{y}^2$ would be linearly dependent. Thus, one of the $\mathbf{x}^i$, whose coefficient is not zero, can be written in terms of $\mathbf{y}^1$, $\mathbf{y}^2$, and the $n-2$ other $\mathbf{x}^i$. These vectors thus span the space. After repeating this process m times, a set of vectors $\mathbf{y}^1, \mathbf{y}^2, \ldots, \mathbf{y}^m$, with $n - m \geq 0$, of the $\mathbf{x}^i$ remain that span the space. Otherwise, the $\mathbf{y}^i$ are linearly dependent. This argument can be applied with the roles of the $\mathbf{x}^i$ and $\mathbf{y}^i$ interchanged, leading to the conclusion that $m - n \geq 0$. The result $n = m$ follows. Thus, all bases of the same vector space have the same number of vectors. This result justifies the following definition.

Definition 2.2.5. The **dimension of a vector space** is the minimun number of nonzero vectors that span the space. A vector space is called a **finite-dimensional vector space** if it can be spanned by a finite number of vectors. ■

To find the dimension of a finite-dimensional vector space, it is sufficient to demonstrate that a set of vectors is a basis and to count the number of vectors in the set. Thus, Example 2.2.6 verifies that the space R^n of $n\times1$ column vectors is an n-dimensional vector space.

Theorem 2.2.1. The number of vectors in every basis of a finite-dimensional vector space is the same and is equal to the dimension of the space. Further, every vector in the space can be represented by a unique linear combination of vectors in a basis. ■

To prove the last part of Theorem 2.2.1, suppose that some vector $\mathbf{x}$ has two representations in terms of the basis $\mathbf{x}^1, \mathbf{x}^2, \ldots, \mathbf{x}^n$; i.e.,

$$\mathbf{x} = a_i\mathbf{x}^i = b_i\mathbf{x}^i$$

Then,

$$\mathbf{0} = \mathbf{x} - \mathbf{x} = (a_i - b_i)\,\mathbf{x}^i$$

Since the $\mathbf{x}^i$ are linearly independent, $a_i - b_i = 0$, for all i, and the representation is indeed unique. ■

Suppose that an n-dimensional vector space V has a basis $\mathbf{x}^1, \mathbf{x}^2, \ldots, \mathbf{x}^n$. Relative to this basis, every vector in the space has a unique representation in terms of n scalars. There is, therefore, a one-to-one correspondence between the vector space V and the vector space R^n of $n\times1$ column vectors. Furthermore, if $\mathbf{x} = a_i\mathbf{x}^i$ and $\mathbf{y} = b_i\mathbf{x}^i$, then $\mathbf{x} + \mathbf{y} = (a_i + b_i)\,\mathbf{x}^i$. If α is a scalar, $\alpha\mathbf{x} = (\alpha a_i)\,\mathbf{x}^i$. Therefore, addition of vectors and multiplication of vectors by a scalar are equivalent to carrying out these operations on column matrices (i.e., vectors in R^n) of the coefficients that represent the vectors in terms of a basis. This conclusion permits limiting the study of finite-dimensional vector spaces to a study of the space R^n. Note that this is just a generalization of the conclusion drawn in Section 2.1 that geometric vectors in three-dimensional space can be represented by column vectors in R^3.

Subspaces

Definition 2.2.6. Let V be a vector space and let S be a **subset** of V; i.e., if $\mathbf{x} \in S$, then $\mathbf{x} \in V$. If S is a vector space with respect to the same operations as those in V, then S is called a **subspace** of V. ■

Example 2.2.7

An example of a subspace S of a vector space V is the space of all possible linear combinations of a subset $\{ \mathbf{x}^1, \ldots, \mathbf{x}^n \}$ of vectors from V; i.e.,

$$S = \{ \mathbf{x} \in V: \mathbf{x} = c_i\mathbf{x}^i, c_i \in R \}$$

If the subset $\{ \mathbf{x}^1, \ldots, \mathbf{x}^n \}$ is linearly independent and is not a basis of V, then the subspace S is a **proper subspace** of V; i.e., it is not the whole space. A specific example of a proper subspace is the subspace S of R^3 that is spanned by the vectors $[1, -1, 0]^T$ and $[0, 1, -1]^T$; i.e.,

$$S = \left\{ \mathbf{x} \in R^3: \mathbf{x} = c_i\mathbf{x}^i = c_1\begin{bmatrix} 1 \\ -1 \\ 0 \end{bmatrix} + c_2\begin{bmatrix} 0 \\ 1 \\ -1 \end{bmatrix} = \begin{bmatrix} c_1 \\ -c_1 + c_2 \\ -c_2 \end{bmatrix}, c_i \in R \right\}$$

This two-dimensional subspace consists of the plane that contains the two given vectors and the origin. ■

Theorem 2.2.2. Let V be a vector space and let S be a nonempty subset of V, with addition and multiplication by a scalar defined as in V. Then S is a subspace of V if and only if the following conditions hold:

(1) If $\mathbf{x}$ and $\mathbf{y}$ are any vectors in S, then $\mathbf{x} + \mathbf{y}$ is in S.
(2) If α is any real number and $\mathbf{x}$ is any vector in S, then $\alpha\mathbf{x}$ is in S. ■

The proof of Theorem 2.2.2 follows from Definition 2.2.1 of vector space and the observation already made in Example 2.2.1.

Example 2.2.8

Consider the system of homogeneous linear equations $\mathbf{Ax} = \mathbf{0}$, where $\mathbf{A}$ is an m×n matrix and $\mathbf{x} \in R^n$. A solution consists of a vector $\mathbf{x} = [x_1, x_2, \ldots, x_n]^T$; i.e., a vector in R^n. Thus, the set of all solutions is a subset of R^n. It may be shown that this is a subspace of R^n by verifying conditions (1) and (2) of Theorem 2.2.2.

Let $\mathbf{x}^1$ and $\mathbf{x}^2$ be solutions of $\mathbf{Ax} = \mathbf{0}$. Then $\mathbf{x}^1 + \mathbf{x}^2$ is a solution, because $\mathbf{A}(\mathbf{x}^1 + \mathbf{x}^2) = \mathbf{Ax}^1 + \mathbf{Ax}^2 = \mathbf{0} + \mathbf{0} = \mathbf{0}$. Also, if $\mathbf{x}$ is a solution, then $\alpha\mathbf{x}$ is a solution because $\mathbf{A}(\alpha\mathbf{x}) = \alpha(\mathbf{Ax}) = \alpha\mathbf{0} = \mathbf{0}$.

It should be noted that the set of all solutions to the system of nonhomogeneous linear equations $\mathbf{Ax} = \mathbf{c}$, $\mathbf{c} \neq \mathbf{0}$, is not a subspace of R^n. ■

EXERCISES 2.2

1. Determine the dimension of the vector space spanned by each of the following sets of vectors:

   (a) $[1, 1, 0]^T$, $[1, 0, 0]^T$, $[0, 1, 1]^T$
   (b) $[1, 0, 0]^T$, $[0, 1, 0]^T$, $[0, 0, 1]^T$, $[1, 1, 1]^T$
   (c) $[1, 1, 1]^T$, $[1, 0, 0]^T$, $[1, 2, 1]^T$

2. Determine whether the vector $[6, 1, -6, 2]^T$ is in the vector space spanned by the vectors $[1, 1, -1, 1]^T$, $[-1, 0, 1, 1]^T$, and $[1, -1, -1, 0]^T$.

3. Let S be the subspace of R^3 spanned by $[1, 2, 2]^T$, $[3, 2, 1]^T$, $[11, 10, 7]^T$, and $[7, 6, 4]^T$. Find a basis for S. What is the dimension of S?

4. Which of the following subsets of R^3 are subspaces?

   (a) S consists of all vectors of the form $[a, b, 1]^T$ and a and b are real.
   (b) S consists of all vectors of the form $[a, b, 0]^T$ and a and b are real.
   (c) S consists of all vectors of the form $[a, b, c]^T$ and a, b, and c are real, where $a + b = c$.
   (d) S consists of all vectors of the form $[a, b, c]^T$ and a, b, and c are real, where $a = b$.

5. Show that the collection of all polynomials of degree three or less; i.e.,

$$P = \{ a + bx + cx^2 + dx^3 : a, b, c, \text{ and } d \text{ are real and } -\infty < x < \infty \}$$

   is a four-dimensional vector space.

6. Are the vectors $[0, -1, 0]^T$, $[0, 1, -1]^T$, and $[1, -2, 1]^T$ linearly dependent? Can $[-2, 1, -3]^T$ be expressed as a linear combination of these vectors? Express your result both geometrically and in terms of solutions of linear algebraic equations.

7. Prove that in an n-dimensional vector space, any set of $n + 1$ vectors is linearly dependent.

8. Prove Theorem 2.2.2.

9. What is the dimension of the subspace of R^4 spanned by the vectors

$$\mathbf{x}^1 = \begin{bmatrix} 1 \\ 2 \\ 1 \\ 3 \end{bmatrix}, \quad \mathbf{x}^2 = \begin{bmatrix} 2 \\ 1 \\ 2 \\ 1 \end{bmatrix}, \quad \mathbf{x}^3 = \begin{bmatrix} 3 \\ 0 \\ 3 \\ -1 \end{bmatrix}$$

10. Which of the following subsets of R^n is a vector space?

(a) $\{ \mathbf{x} \in R^n: \mathbf{x} + [\,1, 0, \ldots, 0\,]^T = \mathbf{0} \}$
(b) $\{ \mathbf{x} \in R^n: \mathbf{x}^T[\,1, 0, \ldots, 0\,]^T = 0 \}$
(c) $\{ \mathbf{x} \in R^n: \mathbf{x} = \alpha\mathbf{y} + \beta\mathbf{z},\ \mathbf{y}$ and $\mathbf{z} \in R^n$ and α and β are arbitrary real scalars $\}$

11. Which of the following subsets of R^3 is a subspace?

(a) $\{ \mathbf{x} \in R^3: x_1 = 1 \}$
(b) $\{ \mathbf{x} \in R^3: \mathbf{x} = \alpha\,[\,1, 1, 0\,]^T + \beta\,[\,2, 2, 0\,]^T, \alpha$ and β are real $\}$
(c) $\{ \mathbf{x} \in R^3: x_1 x_2 = 0 \}$
(d) $\{ \mathbf{x} \in R^3: x_1 + 2x_2 - 3x_3 = 0 \}$

If it is a subspace, what is its dimension?

12. Given a basis $\{ [\,1, 0, 0\,]^T, [\,0, 2, 2\,]^T, [\,0, -1, 1\,]^T \}$ of R^3, represent a vector $[\,a, b, c\,]^T \in R^3$ by a linear combination of vectors in the basis. Is the representation unique?

13. Which of the following subsets of R^n are subspaces of R^n?

(a) $\{ \mathbf{x} \in R^n: \mathbf{x}^T\mathbf{y} = 0$, for a given $\mathbf{y} \in R^n \}$
(b) $\{ \mathbf{x} \in R^n: \mathbf{x}^T\mathbf{x} \le 1 \}$

14. Show that the collection F of all real functions defined on the interval $0 \le x \le 1$ is a vector space, where addition and multiplication by a scalar are defined by

$$\begin{aligned} (f + g)(x) &= f(x) + g(x) \\ (\alpha f)(x) &= \alpha f(x) \end{aligned}$$

where f and g are in F and α is a real number.

2.3 RANK OF MATRICES AND SOLUTION OF MATRIX EQUATIONS

The matrix theory outlined in Section 1.2 and the theory of finite-dimensional vector spaces of Section 2.2 may now be exploited to develop both the basic theory and techniques for solution of matrix equations.

A common problem in engineering analysis is to find a set of n real variables $x_1, x_2, \ldots, x_n$ that represent physical quantities of displacement, velocity, stress, pressure, temperature, etc., and satisfy a set of m **linear equations**

$$\left.\begin{aligned} a_{11}x_1 + a_{12}x_2 + \ldots + a_{1n}x_n &= c_1 \\ a_{21}x_1 + a_{22}x_2 + \ldots + a_{2n}x_n &= c_2 \\ \cdots\cdots\cdots\cdots\cdots\cdots\cdots & \\ a_{m1}x_1 + a_{m2}x_2 + \ldots + a_{mn}x_n &= c_m \end{aligned}\right\} \text{ no sum on n} \tag{2.3.1}$$

where the parameters $a_{11}, a_{21}, \ldots, a_{mn}$ and $c_1, c_2, \ldots, c_m$ are given real constants. To **solve a linear equation** requires the following:

(1) Determine whether a solution exists.
(2) Find a solution, if one exists.
(3) Determine whether the solution found is unique.

The most conclusive method of demonstrating that a solution exists is to construct a solution; i.e., success in step (2) implies success in step (1). However, if no solution exists, all attempts to find a solution, no matter how time-consuming or costly, will fail. It is thus prudent to first consider the question of existence of a solution. Finally, if there are multiple solutions, the engineer should be aware of this possibility (step (3)) and may want to know all possible solutions.

In order to simplify Eq. 2.3.1, matrix notation is used; i.e.,

$$\mathbf{Ax} = \mathbf{c} \tag{2.3.2}$$

where $\mathbf{A} = [\, a_{ij} \,]$ is the given m×n coefficient matrix, $\mathbf{c} = [\, c_i \,]$ is a given m×1 column matrix, and $\mathbf{x} = [\, x_i \,]$ is the n×1 column matrix of variables.

Rank of a Matrix

Definition 2.3.1. Let $\mathbf{A} = [\, a_{ij} \,]$ be an m×n matrix. The **rows of a matrix A** are denoted $\mathbf{a}_1^r = [\, a_{11}, a_{12}, \ldots, a_{1n} \,], \ldots, \mathbf{a}_m^r = [\, a_{m1}, a_{m2}, \ldots, a_{mn} \,]$. Considered as **row vectors** in R^n, they span a subspace of R^n that is called the **row space of a matrix A**. Similarly, the **columns of a matrix A**, $\mathbf{a}_1^c = [\, a_{11}, a_{21}, \ldots, a_{m1} \,]^T, \ldots, \mathbf{a}_m^c = [\, a_{1n}, a_{2n}, \ldots, a_{mn} \,]^T$, span the **column space** of A, which is a subspace of R^m. ■

Example 2.3.1

Consider again the vectors of Example 2.2.5,

$$\mathbf{x}^1 = [\, 1, 2, 1 \,]^T, \quad \mathbf{x}^2 = [\, 1, 0, 2 \,]^T, \quad \mathbf{x}^3 = [\, 1, 0, 0 \,]^T$$

If they are columns of matrix **A**, then

$$\mathbf{A} = \begin{bmatrix} 1 & 1 & 1 \\ 2 & 0 & 0 \\ 1 & 2 & 0 \end{bmatrix}$$

Since it was shown in Example 2.2.5 that these vectors span the space R^3, the column space of **A** is R^3. ■

Definition 2.3.2. The dimension of the row space of **A** is called the **row rank** of **A**; i.e., the number of linearly independent rows of **A**. Similarly, the **column rank** of **A** is the dimension of the column space of **A**; i.e., the number of linearly independent columns of **A**. ■

Example 2.3.2

In Example 2.3.1, the dimension of the column space of **A** is 3, since **A** has 3 linearly independent columns. Thus, the column rank of **A** is 3. A simple calculation shows that the rows of **A** are also linearly independent. Thus, the row rank of A is also 3. ■

Elementary Row Operations

Definition 2.3.3. An **elementary row operation** is one of the following operations on a matrix:

(1) the interchange of a pair of rows
(2) the multiplication of a row by a nonzero scalar
(3) the addition of one row to another row

These elementary row operations can be performed on an m×n matrix by multiplying on the left by **elementary matrices** that are obtained by performing the same operations on m×m identity matrices. ■

The reader may recall that the concept of elementary matrices was introduced for 3×3 matrices in Section 1.3.

Example 2.3.3

The following are elementary matrices that implement elementary row operations on a 3×3 matrix:

$$\mathbf{E}_1 = \begin{bmatrix} 1 & 0 & 0 \\ 0 & -2 & 0 \\ 0 & 0 & 1 \end{bmatrix}, \quad \mathbf{E}_2 = \begin{bmatrix} 0 & 0 & 1 \\ 0 & 1 & 0 \\ 1 & 0 & 0 \end{bmatrix}, \quad \mathbf{E}_3 = \begin{bmatrix} 1 & 1 & 0 \\ 0 & 1 & 0 \\ 0 & 0 & 1 \end{bmatrix}$$

Matrix $\mathbf{E}_1$ is of type (2) — multiply the second row of the identity matrix by −2. Matrix $\mathbf{E}_2$ is of type (1) — exchange the first and the third rows of the identity matrix. Matrix $\mathbf{E}_3$ is of type (3) — addition of the second row of the identity matrix to the first row. ■

Note that an elementary matrix is nonsingular because the identity matrix can be obtained by multiplying an elementary matrix on the left by an elementary matrix that reverses the operation. Similarly, products of elementary matrices are nonsingular. In particular, if

$$\mathbf{E} = \mathbf{E}_1\mathbf{E}_2 \ldots \mathbf{E}_k$$

then

$$\mathbf{E}^{-1} = \mathbf{E}_k^{-1}\mathbf{E}_{k-1}^{-1} \ldots \mathbf{E}_1^{-1}$$

Theorem 2.3.1. If $\mathbf{B} = \mathbf{EA}$, where $\mathbf{E}$ is a product of elementary matrices, then the row rank of $\mathbf{A}$ and the row rank of $\mathbf{B}$ are equal. ■

To prove Theorem 2.3.1, suppose $\mathbf{B} = \mathbf{EA}$. Then the rows of $\mathbf{B}$ are obtained from the rows of $\mathbf{A}$ by the three elementary row operations. Thus, each row of $\mathbf{B}$ is a linear combination of the rows of $\mathbf{A}$. Hence, the row space of $\mathbf{B}$ is contained in the row space of $\mathbf{A}$. On the other hand, since $\mathbf{A} = \mathbf{E}^{-1}\mathbf{B}$, the row space of $\mathbf{A}$ is contained in the row space of $\mathbf{B}$. Hence, the row spaces of $\mathbf{A}$ and $\mathbf{B}$ are identical and the row rank of $\mathbf{A}$ and row rank of $\mathbf{B}$ are equal. ■

Example 2.3.4

To find the row rank of the matrix

$$\mathbf{A} = \begin{bmatrix} 1 & 3 & 2 \\ 2 & 1 & 0 \\ 3 & 4 & 2 \end{bmatrix}$$

note that

$$\mathbf{EA} = \mathbf{B} = \begin{bmatrix} 1 & 3 & 2 \\ 2 & 1 & 0 \\ 0 & 0 & 0 \end{bmatrix}$$

where $\mathbf{E}$ is the product of elementary matrices that (1) multiply the third row by -1, (2) add the first row to the third row, and (3) add the second row to the third row; i.e.,

$$\mathbf{E} = \begin{bmatrix} 1 & 0 & 0 \\ 0 & 1 & 0 \\ 0 & 1 & 1 \end{bmatrix} \begin{bmatrix} 1 & 0 & 0 \\ 0 & 1 & 0 \\ 1 & 0 & 1 \end{bmatrix} \begin{bmatrix} 1 & 0 & 0 \\ 0 & 1 & 0 \\ 0 & 0 & -1 \end{bmatrix} = \begin{bmatrix} 1 & 0 & 0 \\ 0 & 1 & 0 \\ 1 & 1 & -1 \end{bmatrix}$$

Hence the row rank of $\mathbf{A}$ is two. ■

Theorem 2.3.2. The row rank and the column rank of an $m \times n$ matrix $\mathbf{A}$ are equal. ■

For a proof of Theorem 2.3.2, the reader is referred to Ref. 1 (p. 74).

Since the row and column rank of a matrix are equal, they will be referred to simply as the **rank of the matrix**.

Example 2.3.5

Consider the matrix

$$\mathbf{A} = \begin{bmatrix} 4 & 3 & 1 \\ 1 & 2 & 1 \end{bmatrix}$$

Combinations of elementary operations yield

$$\mathbf{E}_1\mathbf{A} = \begin{bmatrix} 1 & -1 \\ 0 & 1 \end{bmatrix} \begin{bmatrix} 4 & 3 & 1 \\ 1 & 2 & 1 \end{bmatrix} = \begin{bmatrix} 3 & 1 & 0 \\ 1 & 2 & 1 \end{bmatrix}$$

$$\mathbf{E}_2\mathbf{E}_1\mathbf{A} = \begin{bmatrix} 1 & 0 \\ -2 & 1 \end{bmatrix} \begin{bmatrix} 3 & 1 & 0 \\ 1 & 2 & 1 \end{bmatrix} = \begin{bmatrix} 3 & 1 & 0 \\ -5 & 0 & 1 \end{bmatrix}$$

Hence, both the row and column rank of **A** are two. ■

If both sides of the matrix of Eq. 2.3.2 are multiplied on the left by a product of elementary matrices, then

$$\mathbf{A}^*\mathbf{x} \equiv \mathbf{EAx} = \mathbf{Ec} \equiv \mathbf{c}^* \qquad (2.3.3)$$

This equation has the same solutions as Eq. 2.3.2, even with the new coefficient matrix $\mathbf{A}^*$ and new matrix $\mathbf{c}^*$.

Solution of Matrix Equations

Elementary operations form the foundation for a powerful method of solving linear equations, which is now outlined for Eq. 2.3.1. To begin with, if the first column of **A** is zero, rename variables (interchange columns of **A**) so that some $a_{i1} \neq 0$. Then make a row interchange to obtain $a_{11} \neq 0$. Now multiply the first equation by $1/a_{11}$. Next, add $-a_{21}$ times the first equation to the second equation. This makes the coefficient of x_1 in the second equation zero. Repeat the process until the coefficient of x_1 in all equations but the first are zero.

The next step is to again rename variables, if necessary, so that there is a nonzero element in the second column in row two or below and to perform elementary row operations so that the coefficient of x_2 in the second equation is one and is zero in all other equations. After a finite number of such steps, the system of equations is reduced to

$$\begin{aligned}
x_1^* \qquad\qquad\quad + a_{1\,r+1}^*\, x_{r+1}^* + \ldots + a_{1n}^*\, x_n^* &= c_1^* \\
x_2^* \qquad\quad + a_{2\,r+1}^*\, x_{r+1}^* + \ldots + a_{2n}^*\, x_n^* &= c_2^* \\
\cdots\cdots\cdots\cdots\cdots\cdots \\
x_r^* + a_{r\,r+1}^*\, x_{r+1}^* + \ldots + a_{rn}^*\, x_n^* &= c_r^* \\
0 &= c_{r+1}^* \\
\cdots\; \\
0 &= c_m^*
\end{aligned} \qquad (2.3.4)$$

where the x^*_i are simply a reordering of the original variables.

The process defined here for obtaining the reduced form of Eq. 2.3.4 is called

Gauss–Jordan reduction. As will become clear, this reduced form of linear equations yields a complete knowledge of the solution; i.e., existence, construction of a solution, and uniqueness of the solution.

The rank of the reduced coefficient matrix in Eq. 2.3.4 is r. Renaming of variables (i.e., interchanging columns of the matrix) does not influence row rank. Since only elementary operations are used in the reduction, Theorem 2.3.1 shows that the row rank; hence, the rank, of matrix **A** is r.

From the reduced form of Eq 2.3.4, it is clear that if one or more of $c^*_{r+1}, \ldots, c^*_m$ is not zero, there is no solution. If $c^*_{r+1} = c^*_{r+2} = \ldots = c^*_m = 0$, then

$$\begin{aligned} x_1^* &= c_1^* - a_{1\,r+1}^* x_{r+1}^* - \ldots - a_{1n}^* x_n^* \\ x_2^* &= c_2^* - a_{2\,r+1}^* x_{r+1}^* - \ldots - a_{2n}^* x_n^* \\ &\ldots\ldots\ldots\ldots\ldots\ldots\ldots\ldots \\ x_r^* &= c_r^* - a_{r\,r+1}^* x_{r+1}^* - \ldots - a_{rn}^* x_n^* \end{aligned} \tag{2.3.5}$$

In this case the equations are said to be **consistent**. If $r = n$ and the equations are consistent, there is a unique solution. If $r < n$ and the equations are consistent, $x^*_1, x^*_2, \ldots, x^*_r$ can be written in terms of $x^*_{r+1}, \ldots, x^*_n$, which can be assigned arbitrarily. In this case, an $n - r$ parameter **family of solutions** has been found. Certainly, if $r = m = n$, a solution exists and is unique. Thus, Eq. 2.3.1 has been solved.

In the special case $m = n = r$, only elementary row operations are needed to reduce Eq. 2.3.2 to

$$\mathbf{E}_p \ldots \mathbf{E}_2\mathbf{E}_1\mathbf{A}\mathbf{x} = \mathbf{I}\mathbf{x} = \mathbf{E}_p \ldots \mathbf{E}_2\mathbf{E}_1\mathbf{c}$$

Thus,

$$\mathbf{A}^{-1} = \mathbf{E}_p \ldots \mathbf{E}_2\mathbf{E}_1$$

and

$$\mathbf{x} = \mathbf{A}^{-1}\mathbf{c} = \mathbf{E}_p \ldots \mathbf{E}_2\mathbf{E}_1\mathbf{c}$$

where $\mathbf{E}_1, \mathbf{E}_2, \ldots, \mathbf{E}_p$ are the elementary matrices used in the Gauss–Jordan reduction.

Associated with Eq. 2.3.2, an $m \times (n+1)$ matrix may be formed by appending the right side **c** as an additional column to the coefficient matrix **A**; i.e.,

$$\mathbf{B} \equiv [\,\mathbf{A}; \mathbf{c}\,]$$

This is called the **augmented matrix** associated with Eq. 2.3.2.

The reduced form of the system of equations in Eq. 2.3.4 can be written in matrix form as

$$\mathbf{A}^*\mathbf{x}^* = \mathbf{c}^*$$

where

$$\mathbf{A}^* = \begin{bmatrix} 1 & 0 & \dots & 0 & a^*_{1\,r+1} & a^*_{1\,r+2} & \dots & a^*_{1n} \\ 0 & 1 & \dots & 0 & a^*_{2\,r+1} & a^*_{2\,r+2} & \dots & a^*_{2n} \\ \dots & \dots & \dots & \dots & \dots & \dots & \dots & \dots \\ 0 & 0 & \dots & 1 & a^*_{r\,r+1} & a^*_{r\,r+2} & \dots & a^*_{rn} \\ 0 & 0 & \dots & 0 & 0 & 0 & \dots & 0 \\ \dots & \dots & \dots & \dots & \dots & \dots & \dots & \dots \\ 0 & 0 & \dots & 0 & 0 & 0 & \dots & 0 \end{bmatrix}$$

The augmented matrix $\mathbf{B}^* = [\,\mathbf{A}^*; \mathbf{c}^*\,]$ for the reduced equation is thus

$$\mathbf{B}^* = \begin{bmatrix} 1 & 0 & \dots & 0 & a^*_{1\,r+1} & a^*_{1\,r+2} & \dots & a^*_{1n} & c^*_1 \\ 0 & 1 & \dots & 0 & a^*_{2\,r+1} & a^*_{2\,r+2} & \dots & a^*_{2n} & c^*_2 \\ \dots & \dots & \dots & \dots & \dots & \dots & \dots & \dots & \dots \\ 0 & 0 & \dots & 1 & a^*_{r\,r+1} & a^*_{r\,r+2} & \dots & a^*_{rn} & c^*_r \\ 0 & 0 & \dots & 0 & 0 & 0 & \dots & 0 & c^*_{r+1} \\ \dots & \dots & \dots & \dots & \dots & \dots & \dots & \dots & \dots \\ 0 & 0 & \dots & 0 & 0 & 0 & \dots & 0 & c^*_m \end{bmatrix}$$

Clearly, $\mathbf{A}^*$ is of rank r. If $c^*_{r+1} = c^*_{r+2} = \dots = c^*_m = 0$, then the system of equations has a solution. The rank of $\mathbf{B}^*$ can be greater than r only if some $c^*_{r+1}, \dots, c^*_m$ is different from zero, in which case the system of equations has no solution. These results may be summarized as follows.

Theorem 2.3.3. A system of linear equations $\mathbf{Ax} = \mathbf{c}$ has a solution if and only if the rank of the augmented matrix $\mathbf{B} = [\,\mathbf{A}; \mathbf{c}\,]$ is equal to the rank of the coefficient matrix $\mathbf{A}$. ■

A special case that is of importance in applications is the following.

Theorem 2.3.4. If $\mathbf{A}$ is a square n×n matrix and the rank of $\mathbf{A}$ is n, a unique solution of the equation $\mathbf{Ax} = \mathbf{c}$ exists for all $\mathbf{c} \in R^n$. ■

Example 2.3.6

Consider the system of linear equations

$$\begin{aligned} x_1 - x_2 + 2x_3 &= 1 \\ x_3 &= 1 \\ 3x_1 - 3x_2 + 7x_3 &= 4 \\ 5x_1 - 5x_2 + 12x_3 &= 7 \end{aligned}$$

The augmented matrix is

$$\mathbf{B} = \begin{bmatrix} 1 & -1 & 2 & 1 \\ 0 & 0 & 1 & 1 \\ 3 & -3 & 7 & 4 \\ 5 & -5 & 12 & 7 \end{bmatrix}$$

After Gauss–Jordan reduction,

$$\mathbf{B}^* = \begin{bmatrix} 1 & -1 & 0 & -1 \\ 0 & 0 & 1 & 1 \\ 0 & 0 & 0 & 0 \\ 0 & 0 & 0 & 0 \end{bmatrix}$$

Thus, the rank of **B** is the same as the rank of **A** and a solution exists. A solution can be determined as $x_3 = 1$ and $x_1 = x_2 - 1$, for any value of x_2. Thus, the solution is not unique. ■

Example 2.3.7

Consider a system of linear equations that is the same as in Example 2.3.6, but with the right side of the third equation changed from 4 to 5. The augmented matrix **B** in this case is

$$\mathbf{B} = \begin{bmatrix} 1 & -1 & 2 & 1 \\ 0 & 0 & 1 & 1 \\ 3 & -3 & 7 & 5 \\ 5 & -5 & 12 & 7 \end{bmatrix}$$

After Gauss–Jordan reduction,

$$\mathbf{B}^* = \begin{bmatrix} 1 & -1 & 0 & -1 \\ 0 & 0 & 1 & 1 \\ 0 & 0 & 0 & 1 \\ 0 & 0 & 0 & 0 \end{bmatrix}$$

Thus, the rank of **B** is 3, whereas the rank of **A** is only 2. Theorem 2.3.3 shows that there is no solution to this system of equations. ■

It is interesting that only a slight difference in the equations of Examples 2.3.6 and 2.3.7 leads to drastically different results. In one case there are infinitely many solutions, whereas in the other case there are none.

A case of particular importance is that in which $c_1 = c_2 = \ldots = c_m = 0$ in Eq. 2.3.1. In this case, the system of linear equations is homogeneous. Adding a column of zeros to the coefficient matrix cannot affect its rank, so the augmented matrix has the same rank as the coefficient matrix. By Theorem 2.3.3, such a system always has a solution. This is not surprising, since $x_1 = x_2 = \ldots = x_n = 0$ is a solution, called the **trivial solution**. It is important to know whether a homogeneous system has a nontrivial solution; i.e., a solution in which at least one of the x_i is not zero.

Theorem 2.3.5. A system of m homogeneous linear algebraic equations $\mathbf{Ax} = \mathbf{0}$ in n unknowns has a nontrivial solution; i.e., $\mathbf{x} \neq \mathbf{0}$, if $m < n$. A system of n homogeneous linear algebraic equations in n unknowns has a **nontrivial solution** if and only if the determinant of the coefficient matrix is zero; i.e., if the coefficient matrix is singular. ■

The proof of this theorem is left as an exercise.

Example 2.3.8

Let the coefficient matrix of a system of two homogeneous equations in three unknowns be, from Example 2.3.5,

$$\mathbf{A} = \begin{bmatrix} 4 & 3 & 1 \\ 1 & 2 & 1 \end{bmatrix}$$

Here, $m = 2$ and $n = 3$, so $m < n$. Thus, a nontrivial solution of $\mathbf{Ax} = \mathbf{0}$ exists as $x_2 = -3x_1$, $x_3 = 5x_1$, for any $x_1 \neq 0$.

For a system of three homogeneous linear equations in three unknowns ($m = 3$, $n = 3$), let the coefficient matrix be

$$\mathbf{A} = \begin{bmatrix} 1 & 3 & 2 \\ 2 & 1 & 0 \\ 3 & 4 & 2 \end{bmatrix}$$

From Example 2.3.4, $\mathbf{A}$ has rank 2. Thus, a nontrivial solution of $\mathbf{Ax} = \mathbf{0}$ exists.

Finally, the coefficient matrix $\mathbf{A}$ of Example 2.3.1 is

$$\mathbf{A} = \begin{bmatrix} 1 & 1 & 1 \\ 2 & 0 & 0 \\ 1 & 2 & 0 \end{bmatrix}$$

Since $|\mathbf{A}| = 1 \neq 0$, there is only the trivial solution for $\mathbf{Ax} = \mathbf{0}$. ■

Deleting some, but not all, of the rows or columns of an m×n matrix yields a **submatrix** of **A**. A well-known result, whose proof may be found in Ref. 1 (p. 98), is the following.

Theorem 2.3.6. The rank of a matrix is the dimension of the largest square submatrix with a nonzero determinant. ■

Example 2.3.9

For the matrix

$$\mathbf{A} = \begin{bmatrix} 1 & 3 & 3 & 2 \\ 2 & 6 & 9 & 5 \\ -1 & -3 & 3 & 0 \end{bmatrix}$$

every 3×3 submatrix has zero determinant. However, deleting the third row and second and fourth columns,

$$\begin{vmatrix} 1 & 3 \\ 2 & 9 \end{vmatrix} = 9 - 6 = 3 \neq 0$$

Thus, the rank of **A** is two. ■

Example 2.3.10

Consider small displacement of the rigid bar of Fig. 2.3.1, where $k_1 > 0$ is the stiffness of a linear spring, $k_2 > 0$ is the stiffness of a torsional spring, and the forces f_1, f_2, and f_3 are given. Equating the total vertical force $f_1 + f_2 + f_3$ to the product of the spring constant k_1 and its vertical displacement $(x_1 + x_2)/2$, an equilibrium equation $k_1(x_1 + x_2) = 2(f_1 + f_2 + f_3)$ is obtained. A second equilibrium equation is obtained by equating torque, $5f_1 - 5f_3$, due to the external forces to the product of k_2 and rotation of the torsional spring, which is approximately $(x_2 - x_1)/10$; i.e., $k_2(x_2 - x_1) = 50(f_1 - f_3)$. These equations may be written in matrix form as

$$\mathbf{Ax} \equiv \begin{bmatrix} k_1 & k_1 \\ k_2 & -k_2 \end{bmatrix} \begin{bmatrix} x_1 \\ x_2 \end{bmatrix} = \begin{bmatrix} 2(f_1 + f_2 + f_3) \\ 50(f_3 - f_1) \end{bmatrix} \equiv \mathbf{c} \qquad (2.3.6)$$

Note that $|\mathbf{A}| = -2k_1k_2 \neq 0$, so Eq. 2.3.6 has a unique solution, for any values of f_1, f_2, and f_3.

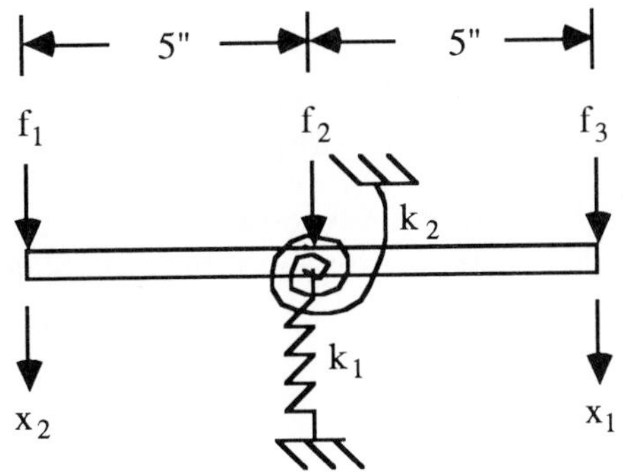

Figure 2.3.1 Elastically Supported Bar ■

EXERCISES 2.3

1. For each of the following matrices, verify Theorem 2.3.2 by determining their row and column ranks:

$$\text{(a)} \begin{bmatrix} 1 & 2 & 3 \\ -1 & 2 & 1 \\ 3 & 1 & 2 \end{bmatrix} \quad \text{(b)} \begin{bmatrix} 1 & -2 & -1 \\ 2 & -1 & 3 \\ 7 & -8 & 3 \end{bmatrix} \quad \text{(c)} \begin{bmatrix} 1 & -2 & -1 \\ 2 & -1 & 3 \\ 7 & -8 & 3 \\ 5 & -7 & 0 \end{bmatrix}$$

2. Find a solution of the following system of equations:

$$\begin{aligned} x_1 + x_2 + 2x_3 &= -1 \\ x_1 - 2x_2 + x_3 &= -5 \\ 3x_1 + x_2 + x_3 &= 3 \end{aligned}$$

if one exists, by using the Gauss–Jordan reduction method.

3. Prove Theorem 2.3.5.

4. Determine if the following system of equations has a solution:

$$\begin{aligned} 2x_1 + x_2 - x_3 + x_4 &= -2 \\ x_1 - x_2 - x_3 + x_4 &= 1 \\ x_1 - 4x_2 - 2x_3 + 2x_4 &= 6 \\ 4x_1 + x_2 - 4x_3 + 3x_4 &= -1 \end{aligned}$$

5. Determine the values of λ for which the following system has a nontrivial solution:

$$\begin{aligned} 9x_1 - 3x_2 \qquad &= \lambda x_1 \\ -3x_1 + 12x_2 - 3x_3 &= \lambda x_2 \\ -3x_2 + 9x_3 &= \lambda x_3 \end{aligned}$$

Find a nontrivial solution for each of these values of λ.

6. Find a nontrivial solution of the following system of equations:

$$\begin{aligned} x_1 + x_2 - x_3 + x_4 &= 0 \\ -x_1 + 2x_2 + x_3 - x_4 &= 0 \\ 2x_1 - x_2 + 3x_3 - 2x_4 &= 0 \end{aligned}$$

7. If $\mathbf{x}^0$ is a solution of $\mathbf{Ax} = \mathbf{0}$ and $\mathbf{x}^1$ is a solution of $\mathbf{Ax} = \mathbf{c}$, show that $k\mathbf{x}^0 + \mathbf{x}^1$ is also a solution of $\mathbf{Ax} = \mathbf{c}$, for any value of k. State a criterion for uniqueness of the solution of $\mathbf{Ax} = \mathbf{c}$.

8. Consider the matrix

$$\mathbf{A} = \begin{bmatrix} 1 & 1 & 1 & 1 \\ 2 & 1 & 2 & 0 \\ 1 & 0 & 1 & -1 \end{bmatrix}$$

(a) What is the rank of **A**?

(b) Solve the matrix equation $\mathbf{Ax} = [\ 1, 1, 1\ ]^T$.

(c) Solve the matrix equation $\mathbf{Ax} = [\ 0, 0, 0\ ]^T$.

9. Consider the system of equations

$$\begin{aligned} x_1 + 2x_2 \qquad &= 1 \\ x_1 - \ x_2 + x_3 &= 4 \\ 3x_2 - x_3 &= c \end{aligned}$$

(a) For what values of c does a solution exist?

(b) If a solution exists, is it unique?

(c) For values of c for which a solution exists, find all solutions.

10. For what values of the constant c does the following system of equations have a solution?

$$\begin{bmatrix} 1 & 0 & 0 \\ 1 & 1 & 1 \\ 0 & 1 & 1 \end{bmatrix} \begin{bmatrix} x_1 \\ x_2 \\ x_3 \end{bmatrix} = \begin{bmatrix} c \\ 0 \\ c \end{bmatrix}$$

11. For what values of c_1, c_2, c_3, and c_4 does the following system of equations have a solution?

$$\begin{bmatrix} 1 & 1 & 0 \\ 0 & 1 & 0 \\ 0 & 1 & 1 \\ 0 & 0 & 0 \end{bmatrix} \begin{bmatrix} x_1 \\ x_2 \\ x_3 \end{bmatrix} = \begin{bmatrix} c_1 \\ c_2 \\ c_3 \\ c_4 \end{bmatrix}$$

12. Let $\mathbf{A} = [\ a_{ij}\ ]_{n\times n}$ be a symmetric matrix. Show that $\mathbf{Ax} = \mathbf{0}$, $\mathbf{x} \in R^n$, has a nontrivial solution if and only if the columns of **A** are linearly dependent.

13. Show that the following equation has a unique solution, without solving the problem numerically:

$$\mathbf{Ax} = \begin{bmatrix} 1 & 2 & -1 \\ -1 & 1 & 1 \\ 1 & 5 & 3 \end{bmatrix} \begin{bmatrix} x_1 \\ x_2 \\ x_3 \end{bmatrix} = \begin{bmatrix} 1 \\ 3 \\ 5 \end{bmatrix}$$

14. For a given matrix equation $\mathbf{Ax} = \mathbf{c}$, let **A** be an n×m matrix, n > m, with columns $\mathbf{a}^c_i$, i = 1, 2, . . . , m, that are linearly independent. Find condition(s) that the vector $\mathbf{c} \in R^n$ must satisfy so that the matrix equation will have a solution. (Hint: Write **Ax** in terms of the $\mathbf{a}^c_i$.)

2.4 SCALAR PRODUCT AND NORM

The scalar product of two geometric vectors in three-dimensional vector analysis was defined geometrically in Section 2.1 and shown to be

$$\vec{x} \cdot \vec{y} = \mathbf{x}^T\mathbf{y} = x_1y_1 + x_2y_2 + x_3y_3 \tag{2.4.1}$$

where x_i and y_i are Cartesian components of $\vec{x}$ and $\vec{y}$. The scalar product can be viewed as a function that assigns to each pair of vectors **x** and **y** of R^3 a real number. The definition of scalar product can be extended to more general vector spaces by defining it to have the basic properties of the scalar product in R^3 from Section 2.1.

Scalar Product on a Vector Space

Definition 2.4.1. Let V be a vector space. A **scalar product** on V is a real valued function (• , •) of pairs of vectors in V with the following properties:

$$\begin{aligned}
&(1)\quad (\mathbf{x}, \mathbf{y}) = (\mathbf{y}, \mathbf{x}) \\
&(2)\quad (\mathbf{x}, \mathbf{y}+\mathbf{z}) = (\mathbf{x}, \mathbf{y}) + (\mathbf{x}, \mathbf{z}) \\
&(3)\quad (\alpha\mathbf{x}, \mathbf{y}) = \alpha(\mathbf{x}, \mathbf{y}) \\
&(4)\quad (\mathbf{x}, \mathbf{x}) \geq 0 \\
&(5)\quad (\mathbf{x}, \mathbf{x}) = 0, \text{ if and only if } \mathbf{x} = \mathbf{0}
\end{aligned} \tag{2.4.2}$$

for any vectors **x**, **y**, and **z** in V and any real scalar α. ■

Note that the scalar product in Eq. 2.4.1 has all the basic properties of the usual scalar product in R^3. A natural generalization of this scalar product to R^n is

$$(\mathbf{x}, \mathbf{y}) \equiv \mathbf{x}^T\mathbf{y} = x_iy_i \tag{2.4.3}$$

Example 2.4.1

The product of Eq. 2.4.3 is not the only scalar product in R^n; e.g, let $\mathbf{x} = [x_1, x_2]^T$ and $\mathbf{y} = [y_1, y_2]^T$ be in R^2. Define

$$[\mathbf{x}, \mathbf{y}] \equiv x_1y_1 - x_1y_2 - x_2y_1 + 2x_2y_2$$

Then $[\mathbf{x}, \mathbf{x}] = x_1^2 - 2x_1x_2 + 2x_2^2 = (x_1 - x_2)^2 + x_2^2 > 0$ if $\mathbf{x} \neq 0$. The remaining properties of Definition 2.4.1 can also be verified, which is left as an exercise. Hence, [• , •] is also a scalar product on R^2. ■

Norm and Distance in a Vector Space

Recall that, in three-dimensional vector analysis, the scalar product of a vector with itself is the square of the length of the vector. Using this concept, the **length of a vector** may be extended to a general vector space.

Definition 2.4.2. The **norm** of a vector in a vector space V that has a scalar product (• , •) is

$$\| \mathbf{x} \| \equiv \sqrt{(\mathbf{x}, \mathbf{x})} \qquad (2.4.4)$$

■

Example 2.4.2

Find the norm and length of the vector $\mathbf{x} = [\,1, 2, 3\,]^T$ in R^3. The norm of **x** is defined in Eq. 2.4.4 as

$$\| \mathbf{x} \| = \sqrt{(1^2 + 2^2 + 3^2)} = \sqrt{14}$$

The length of $\vec{x}$ is

$$| \vec{x} | = \sqrt{x_i x_i} = \sqrt{14}$$

Therefore, $\| \mathbf{x} \|$ is the length of **x**. ■

The norm has most of the properties that are usually associated with length or distance. From the postulates for scalar product in Definition 2.4.1,

$$\begin{aligned} \| \mathbf{x} \| &\geq 0 \\ \| \mathbf{x} \| &= 0, \text{ if and only if } \mathbf{x} = \mathbf{0} \\ \| \alpha \mathbf{x} \| &= | \alpha | \, \| \mathbf{x} \| \end{aligned} \qquad (2.4.5)$$

Note that a different scalar product will give a different norm, and hence one vector space may have many different norms. Thus, the concept of distance in a general vector space is dependent on the norm used. This might be considered as analogous to using different units of length in the physical world. Unfortunately general norms are not proportional to one another, so there is not a simple scale conversion factor that relates norms.

Theorem 2.4.1 (Schwartz Inequality). For any vectors **x** and **y** in a vector space V with scalar product (• , •) and associated norm $\| \bullet \|$ defined by Eq. 2.4.4,

$$| (\mathbf{x}, \mathbf{y}) | \leq \| \mathbf{x} \| \, \| \mathbf{y} \| \qquad (2.4.6)$$

■

To prove Theorem 2.4.1, let α be any scalar and form

$$\begin{aligned} 0 \leq \| \mathbf{x} + \alpha \mathbf{y} \|^2 &= (\mathbf{x} + \alpha \mathbf{y}, \mathbf{x} + \alpha \mathbf{y}) \\ &= (\mathbf{x}, \mathbf{x}) + (\alpha \mathbf{y}, \mathbf{x}) + (\mathbf{x}, \alpha \mathbf{y}) + (\alpha \mathbf{y}, \alpha \mathbf{y}) \\ &= \| \mathbf{x} \|^2 + \alpha (\mathbf{y}, \mathbf{x}) + \alpha (\mathbf{x}, \mathbf{y}) + \alpha^2 \| \mathbf{y} \|^2 \end{aligned}$$

This can be written as

$$0 \le \|\mathbf{x}\|^2 + 2\alpha(\mathbf{x}, \mathbf{y}) + \alpha^2 \|\mathbf{y}\|^2 \equiv g(\alpha)$$

which must hold for all α. This is a quadratic expression in the real variable α, with real coefficients. If the **discriminant** $4(\mathbf{x}, \mathbf{y})^2 - 4\|\mathbf{x}\|^2\|\mathbf{y}\|^2$ of the quadratic expression in α were positive, then the quadratic formula shows that the associated quadratic equation $g(\alpha) = 0$ has two distinct roots. But, as shown by the dashed and solid line graphs of $g(\alpha)$ in Fig. 2.4.1, this implies that there are values of α for which the quadratic expression would be negative. The discriminant must therefore be less than or equal to zero. Thus, $4(\mathbf{x}, \mathbf{y})^2 - 4\|\mathbf{x}\|^2\|\mathbf{y}\|^2 \le 0$, or $|(\mathbf{x}, \mathbf{y})| \le \|\mathbf{x}\|\,\|\mathbf{y}\|$.

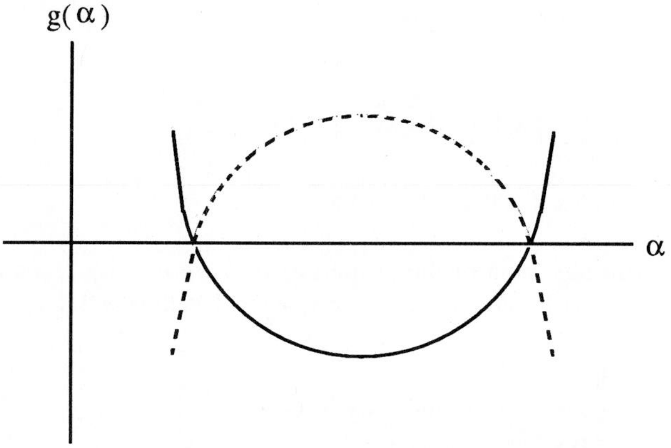

Figure 2.4.1 Graph of Quadratic Expressions With Two Distinct Roots ■

Example 2.4.3

For $\mathbf{x}$ and $\mathbf{y}$ in R^2, the Schwartz inequality is

$$|(\mathbf{x}, \mathbf{y})| = |(x_1y_1 + x_2y_2)| \le \sqrt{x_1^2 + x_2^2}\sqrt{y_1^2 + y_2^2}$$

Squaring both sides, this is

$$x_1^2y_1^2 + 2x_1y_1x_2y_2 + x_2^2y_2^2 \le x_1^2y_1^2 + x_1^2y_2^2 + x_2^2y_1^2 + x_2^2y_2^2$$

which is equivalent to the relation

$$0 \le (x_1y_2 - x_2y_1)^2$$

Note that this relation is an equality for $\mathbf{x} \ne \mathbf{0} \ne \mathbf{y}$ if and only if $\mathbf{x} = \alpha\mathbf{y}$ for some scalar α. ■

The Schwartz inequality for the vector space R^n, with the scalar product defined in Eq. 2.4.3, is

$$|(\mathbf{x},\mathbf{y})| = |x_i y_i| \leq \left(\sum_{i=1}^{n} x_i^2\right)^{1/2} \left(\sum_{i=1}^{n} y_i^2\right)^{1/2} = \|\mathbf{x}\|\,\|\mathbf{y}\| \qquad (2.4.7)$$

The Schwartz inequality leads to another important inequality.

Theorem 2.4.2 (Triangle Inequality). For any vectors **x** and **y** in a vector space V that has a scalar product (• , •) and associated norm || • || defined by Eq. 2.4.4,

$$\|\mathbf{x}+\mathbf{y}\| \leq \|\mathbf{x}\| + \|\mathbf{y}\| \qquad (2.4.8)$$

■

To prove Theorem 2.4.2, note that

$$\begin{aligned}\|\mathbf{x}+\mathbf{y}\|^2 &= (\mathbf{x}+\mathbf{y},\mathbf{x}+\mathbf{y})\\ &= \|\mathbf{x}\|^2 + \|\mathbf{y}\|^2 + (\mathbf{x},\mathbf{y}) + (\mathbf{y},\mathbf{x})\\ &\leq \|\mathbf{x}\|^2 + \|\mathbf{y}\|^2 + 2\,|(\mathbf{x},\mathbf{y})|\\ &\leq \|\mathbf{x}\|^2 + \|\mathbf{y}\|^2 + 2\,\|\mathbf{x}\|\,\|\mathbf{y}\|\\ &= (\|\mathbf{x}\| + \|\mathbf{y}\|)^2\end{aligned}$$

where the Schwartz inequality has been used. Thus, Eq. 2.4.8 follows. ■

In a vector space V, a vector is sometimes referred to as a **point**. The **distance between two points** in V is then defined as the norm of the difference of the vectors; i.e.,

$$d(\mathbf{x},\mathbf{y}) = \|\mathbf{x}-\mathbf{y}\| \qquad (2.4.9)$$

This definition of distance satisfies all the usual properties associated with distance in R^3; i.e., for any **x**, **y**, and **z** in V,

(1) The distance between two points is positive unless the points coincide; i.e.,

$$d(\mathbf{x},\mathbf{y}) = \|\mathbf{x}-\mathbf{y}\| \geq 0$$

$$d(\mathbf{x},\mathbf{y}) = \|\mathbf{x}-\mathbf{y}\| = 0, \text{ if and only if } \mathbf{x}=\mathbf{y}$$

(2) The distance function is symmetric; i.e.,

$$d(\mathbf{x},\mathbf{y}) = \|\mathbf{x}-\mathbf{y}\| = \|\mathbf{y}-\mathbf{x}\| = d(\mathbf{y},\mathbf{x})$$

(3) The triangle inequality is satisfied; i.e.,

$$d(\mathbf{x},\mathbf{y}) = \|\mathbf{x}-\mathbf{y}\| \leq \|\mathbf{x}-\mathbf{z}\| + \|\mathbf{z}-\mathbf{y}\| = d(\mathbf{x},\mathbf{z}) + d(\mathbf{z},\mathbf{y})$$

A vector space in which distance between points is defined with the above three properties is called a **metric space** and the distance function d (• , •) is called a **metric**. These ideas justify the use of $\| \mathbf{x} - \mathbf{y} \|$ as a distance.

Orthonormal Bases

It has already been seen that there are many bases in an n-dimensional vector space, all of which are sets of n linearly independent vectors that span the space. If a scalar product is defined, certain bases with special properties that are useful in analysis and computation can be defined.

Definition 2.4.3. Two vectors are **orthogonal** if their scalar product is zero. A vector is said to be **normalized** if its norm is one. A set of vectors $\mathbf{x}^1, \mathbf{x}^2, \ldots, \mathbf{x}^n$ is **orthonormal** if

$$(\mathbf{x}^i, \mathbf{x}^j) = \delta_{ij} \tag{2.4.10}$$

where δ_{ij} is the Kronecker delta of Definition 1.2.3. ■

Let $\{ \mathbf{x}^i \}$ be a set of n orthonormal vectors that span an n-dimensional vector space V. They must be linearly independent, since if $a_i \mathbf{x}^i = \mathbf{0}$, taking the scalar product of both sides with $\mathbf{x}^j$ yields

$$0 = (\mathbf{x}^j, \mathbf{0}) = (\mathbf{x}^j, a_i \mathbf{x}^i) = a_i \delta_{ij} = a_j$$

Hence, $a_j = 0$ for all j and the set $\{ \mathbf{x}^i \}$ is linearly independent. Therefore, the set of vectors $\{ \mathbf{x}^i \}$ is a basis for V.

Having an orthonormal basis is very convenient for computation because the representation of any vector in the space is easy to find. Suppose $\mathbf{x}$ is a vector in V with the representation

$$\mathbf{x} = a_i \mathbf{x}^i \tag{2.4.11}$$

in terms of an orthonormal basis $\{ \mathbf{x}^i \}$. Taking the scalar product of both sides with $\mathbf{x}^j$,

$$\begin{aligned} (\mathbf{x}^j, \mathbf{x}) &= a_i (\mathbf{x}^j, \mathbf{x}^i) \\ &= a_i \delta_{ji} \\ &= a_j \end{aligned} \tag{2.4.12}$$

Thus, the components a_i of the vector $\mathbf{a} = [a_1, \ldots, a_n]^T$ in R^n that represents $\mathbf{x}$ are easily determined. Consider another vector $\mathbf{y}$ in V, with the representation

$$\mathbf{y} = b_j \mathbf{x}^j$$

where $\mathbf{b} = [b_1, \ldots, b_n]^T$ represents $\mathbf{y}$. The scalar product of $\mathbf{x}$ and $\mathbf{y}$ in V is

$$
\begin{aligned}
(\mathbf{x}, \mathbf{y}) &= (a_i\mathbf{x}^i, b_j\mathbf{x}^j) \\
&= a_ib_j(\mathbf{x}^i, \mathbf{x}^j) \\
&= a_ib_j\delta_{ij} \\
&= a_ib_i
\end{aligned}
$$

These relations establish a one-to-one correspondence between the n-dimensional vector space V and the coefficients of Eq. 2.4.11 in the space R^n, since the scalar product in R^n for a pair of vectors $\mathbf{a} = [a_1, a_2, \ldots, a_n]^T$ and $\mathbf{b} = [b_1, b_2, \ldots, b_n]^T$ is a_ib_i. Furthermore, under this correspondence, scalar product is preserved; i.e.,

$$(\mathbf{x}, \mathbf{y}) = a_ib_i = \mathbf{a}^T\mathbf{b} \tag{2.4.13}$$

Consider now the possibility of changing the representation of a vector space V by a change of basis. Let $\{ \mathbf{x}^i \}$ and $\{ \mathbf{y}^i \}$ be two bases for the same vector space. Then, any vector $\mathbf{w}$ in R^n has a representation in terms of each basis; i.e.,

$$\mathbf{w} = b_i\mathbf{x}^i \tag{2.4.14}$$

$$\mathbf{w} = c_i\mathbf{y}^i \tag{2.4.15}$$

Each vector $\mathbf{x}^j$ in the basis $\{ \mathbf{x}^i \}$ has a representation in terms of the $\{ \mathbf{y}^i \}$, denoted as

$$\mathbf{x}^j = a_{ij}\mathbf{y}^i \tag{2.4.16}$$

Substituting this relation into Eq. 2.4.14 and using Eq. 2.4.15,

$$\mathbf{w} = c_i\mathbf{y}^i = b_j\mathbf{x}^j = b_ja_{ij}\mathbf{y}^i = a_{ij}b_j\mathbf{y}^i$$

Since the representation of a vector in terms of a given basis is unique,

$$c_i = a_{ij}b_j \tag{2.4.17}$$

Let $\mathbf{b}$ be the column matrix with elements b_i, $\mathbf{c}$ be the column matrix with elements c_i, and $\mathbf{A}$ be the square matrix with elements a_{ij}. Then the change of representation of Eq. 2.4.17 can be written in matrix form as

$$\mathbf{c} = \mathbf{A}\mathbf{b} \tag{2.4.18}$$

Thus, the change in coefficients of a vector in V, associated with a change in basis, is defined by a matrix transformation.

Consider the important special case of a change from one orthonormal basis $\{ \mathbf{x}^i \}$ to another orthonormal basis $\{ \mathbf{y}^i \}$. Taking the scalar product of both sides of Eq. 2.4.16 with $\mathbf{x}^i$ and using Eq. 2.4.16,

$$\begin{aligned}\delta_{ij} &= (\mathbf{x}^i, \mathbf{x}^j) = (a_{ki}\mathbf{y}^k, a_{mj}\mathbf{y}^m) \\ &= a_{ki}a_{mj}(\mathbf{y}^k, \mathbf{y}^m) = a_{ki}a_{mj}\delta_{km} \\ &= a_{ki}a_{kj} \\ &= a^T_{ik}a_{kj}\end{aligned}$$

In terms of operations with the matrix $\mathbf{A} = [\, a_{ij} \,]$, this is

$$(\mathbf{A})^T\mathbf{A} = \mathbf{I} \tag{2.4.19}$$

Thus,

$$\mathbf{A}^{-1} = \mathbf{A}^T \tag{2.4.20}$$

Definition 2.4.4. A matrix **A** for which $\mathbf{A}^{-1} = \mathbf{A}^T$ is called an **orthogonal matrix**. ■

Example 2.4.4

Find the transformation matrix **A** between the orthonormal bases

$$\mathbf{y}^1 = [\,1, 0, 0\,]^T, \quad \mathbf{y}^2 = [\,0, 1, 0\,]^T, \quad \mathbf{y}^3 = [\,0, 0, 1\,]^T$$

and

$$\mathbf{x}^1 = \frac{1}{\sqrt{3}}[\,1, 1, 1\,]^T, \ \mathbf{x}^2 = \frac{1}{\sqrt{2}}[\,-1, 1, 0\,]^T, \ \mathbf{x}^3 = \frac{1}{\sqrt{6}}[\,-1, -1, 2\,]^T$$

It is easy to write the $\mathbf{x}^i$ in terms of the $\mathbf{y}^i$ as

$$\mathbf{x}^1 = \frac{1}{\sqrt{3}}(\mathbf{y}^1 + \mathbf{y}^2 + \mathbf{y}^3)$$

$$\mathbf{x}^2 = \frac{1}{\sqrt{2}}(-\mathbf{y}^1 + \mathbf{y}^2)$$

$$\mathbf{x}^3 = \frac{1}{\sqrt{6}}(-\mathbf{y}^1 - \mathbf{y}^2 + 2\mathbf{y}^3)$$

Thus, the matrix $\mathbf{A} = [\, a_{ij} \,]$ defined in Eq. 2.4.16, in this case, is

$$\mathbf{A} = \begin{bmatrix} \frac{1}{\sqrt{3}} & \frac{1}{\sqrt{3}} & \frac{1}{\sqrt{3}} \\ -\frac{1}{\sqrt{2}} & \frac{1}{\sqrt{2}} & 0 \\ -\frac{1}{\sqrt{6}} & -\frac{1}{\sqrt{6}} & \frac{2}{\sqrt{6}} \end{bmatrix}$$

Direct manipulation verifies that $\mathbf{A}^T\mathbf{A} = \mathbf{I}$. ■

Example 2.4.5

A rotation of the x_1-x_2 plane in R^3 about the x_3-axis through an angle θ, as shown in Fig. 2.4.2, can be viewed as the orthogonal transformation; i.e.,

$$\begin{bmatrix} x_1 \\ x_2 \\ x_3 \end{bmatrix} = \begin{bmatrix} \cos\theta & -\sin\theta & 0 \\ \sin\theta & \cos\theta & 0 \\ 0 & 0 & 1 \end{bmatrix} \begin{bmatrix} x_1' \\ x_2' \\ x_3' \end{bmatrix} \equiv \mathbf{Ax}'$$

where x_1', x_2', and x_3' are components of the vector $\mathbf{x}$ in the x_1'-x_2'-x_3' frame. Using trigonometric identities,

$$\mathbf{A}^T\mathbf{A} = \begin{bmatrix} \cos^2\theta + \sin^2\theta & 0 & 0 \\ 0 & \cos^2\theta + \sin^2\theta & 0 \\ 0 & 0 & 1 \end{bmatrix} = \mathbf{I}$$

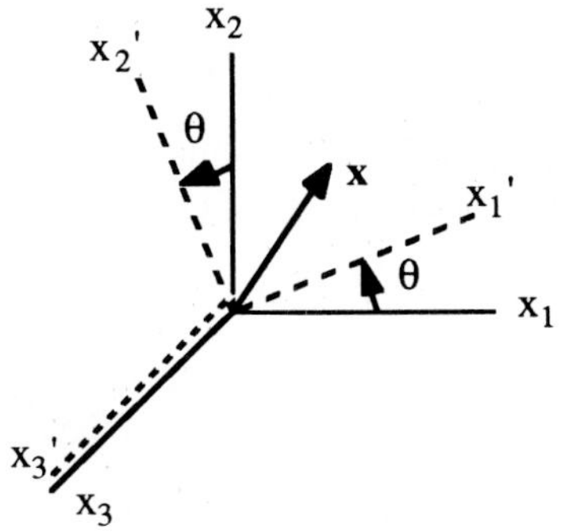

Figure 2.4.2 Rotation of the x_1-x_2 Plane About the x_3-Axis ■

One of the important properties of an orthogonal transformation is that it leaves the scalar product invariant. That is, if $\mathbf{x} = \mathbf{Ax}'$ and $\mathbf{y} = \mathbf{Ay}'$, $\mathbf{x}' = \mathbf{A}^T\mathbf{x}$ and $\mathbf{y}' = \mathbf{A}^T\mathbf{y}$, and

$$(\mathbf{x}', \mathbf{y}') = (\mathbf{A}^T\mathbf{x})^T \mathbf{A}^T\mathbf{y} = \mathbf{x}^T\mathbf{A}\mathbf{A}^T\mathbf{y} = \mathbf{x}^T\mathbf{I}\mathbf{y} = (\mathbf{x}, \mathbf{y}) \qquad (2.4.21)$$

Further, the orthogonal transformation is norm preserving; i.e.,

$$\| \mathbf{x}' \| = (\mathbf{x}', \mathbf{x}')^{1/2} = (\mathbf{x}, \mathbf{x})^{1/2} = \| \mathbf{x} \|$$

Since the angle between two vectors is given by $\cos\theta = (\mathbf{x}, \mathbf{y})/(\|\mathbf{x}\| \, \|\mathbf{y}\|)$, an orthogonal transformation is also angle preserving.

The scalar product can be used to obtain an analytical criterion for linear dependence of a set of m vectors $\{ \mathbf{x}^1, \mathbf{x}^2, \ldots, \mathbf{x}^m \}$ in a vector space V. Set a linear combination of the $\mathbf{x}^i$ to zero, as in Eq. 2.2.1, to check for linear independence; i.e., $c_i\mathbf{x}^i = \mathbf{0}$. By successively forming the scalar products of $\mathbf{x}^1, \ldots, \mathbf{x}^m$ with both sides, the constants c_i must satisfy

$$\begin{aligned}
c_1(\mathbf{x}^1, \mathbf{x}^1) + c_2(\mathbf{x}^1, \mathbf{x}^2) + \ldots + c_m(\mathbf{x}^1, \mathbf{x}^m) &= 0 \\
c_1(\mathbf{x}^2, \mathbf{x}^1) + c_2(\mathbf{x}^2, \mathbf{x}^2) + \ldots + c_m(\mathbf{x}^2, \mathbf{x}^m) &= 0 \\
\cdots\cdots\cdots\cdots\cdots\cdots\cdots\cdots\cdots\cdots \\
c_1(\mathbf{x}^m, \mathbf{x}^1) + c_2(\mathbf{x}^m, \mathbf{x}^2) + \ldots + c_m(\mathbf{x}^m, \mathbf{x}^m) &= 0
\end{aligned} \tag{2.4.22}$$

This system of equations in $\mathbf{c} = [c_1, \ldots, c_m]^T$ can be written in matrix form as

$$\mathbf{Gc} = \mathbf{0} \tag{2.4.23}$$

where the m×m matrix **G** is called the **Gram matrix** of $\{\mathbf{x}^1, \ldots, \mathbf{x}^m\}$, defined as

$$\mathbf{G} \equiv \begin{bmatrix} (\mathbf{x}^1, \mathbf{x}^1) & (\mathbf{x}^1, \mathbf{x}^2) & \ldots & (\mathbf{x}^1, \mathbf{x}^m) \\ (\mathbf{x}^2, \mathbf{x}^1) & (\mathbf{x}^2, \mathbf{x}^2) & \ldots & (\mathbf{x}^2, \mathbf{x}^m) \\ \vdots & \vdots & & \vdots \\ (\mathbf{x}^m, \mathbf{x}^1) & (\mathbf{x}^m, \mathbf{x}^2) & \ldots & (\mathbf{x}^m, \mathbf{x}^m) \end{bmatrix} \tag{2.4.24}$$

The set of m equations in the m variables c_i in Eq. 2.4.23 has a nontrivial solution if and only if the determinant of the matrix of coefficients vanishes; i.e.,

$$|\mathbf{G}| = 0 \tag{2.4.25}$$

The determinant $|\mathbf{G}|$ is called the **Gram determinant**, or **Gramian**, of $\{\mathbf{x}^1, \ldots, \mathbf{x}^m\}$. This result may be stated in terms of linear dependence of the $\mathbf{x}^i$ as follows.

Theorem 2.4.3. A set of vectors in a vector space V that has a scalar product $(\bullet, \bullet)$ is linearly dependent if and only if its Gramian is zero. Conversely, they are linearly independent if and only if their Gramian is not zero. ■

Example 2.4.6

Use Theorem 2.4.3 to determine whether the following vectors in R^4 are linearly dependent:

$$\mathbf{x}^1 = [1, 1, 1, 0]^T, \quad \mathbf{x}^2 = [1, -1, 0, 0]^T, \quad \mathbf{x}^3 = [0, 0, 1, 1]^T$$

The Gramian of these vectors is

$$|\mathbf{G}| = \begin{vmatrix} 3 & 0 & 1 \\ 0 & 2 & 0 \\ 1 & 0 & 2 \end{vmatrix} = 12 - 2 = 10 \neq 0$$

Therefore, $\mathbf{x}^1$, $\mathbf{x}^2$, and $\mathbf{x}^3$ are linearly independent. ■

Consider a basis $\{ \mathbf{x}^1, \ldots, \mathbf{x}^m \}$ of a vector space V that may not be orthogonal. If $\mathbf{x}$ is a vector in V, it can be written as a linear combination of the $\mathbf{x}^i$; i.e.,

$$\mathbf{x} = a_j \mathbf{x}^j \tag{2.4.26}$$

Taking the scalar product of both sides with $\mathbf{x}^i$, i = 1, 2, . . . , m, yields

$$(\mathbf{x}^i, \mathbf{x}^j)\, a_j = (\mathbf{x}^i, \mathbf{x})$$

or, in matrix form,

$$\mathbf{Ga} = [\, (\mathbf{x}^1, \mathbf{x}), \ldots, (\mathbf{x}^m, \mathbf{x}) \,]^T \equiv \mathbf{b} \tag{2.4.27}$$

Thus, finding the expansion of a vector in terms of a basis of V requires the solution of a matrix equation, Eq. 2.4.27, whose coefficient matrix is the Gram matrix. By Theorem 2.4.3, $\mathbf{G}$ is nonsingular, so a unique solution exists.

Note that if the basis $\{ \mathbf{x}^i \}$ is orthonormal, $(\mathbf{x}^i, \mathbf{x}^j) = \delta_{ij}$ and the Gram matrix is the identity; i.e., $\mathbf{G} = \mathbf{I}$. In this very special case, Eq. 2.4.27 reduces to $a_i = (\mathbf{x}^i, \mathbf{x})$, which is the same as Eq. 2.4.12. The value of having an orthonormal basis thus becomes clear.

Gram–Schmidt Orthonormalization

As noted earlier, orthonormal bases have very attractive properties. It is now to be shown that from any n linearly independent vectors that span a vector space V, an orthonormal basis can be constructed. Let $\{ \mathbf{y}^i \}$ be a basis whose vectors are not necessarily orthonormal. First normalize $\mathbf{y}^1$, to obtain

$$\mathbf{x}^1 = \frac{1}{\| \mathbf{y}^1 \|} \mathbf{y}^1$$

Next, form $\mathbf{x}^2 = a_{21}\mathbf{x}^1 + a_{22}\mathbf{y}^2$ and choose a_{21} and a_{22} so that $(\mathbf{x}^1, \mathbf{x}^2) = 0$ and $\| \mathbf{x}^2 \| = 1$. This leads to two equations for a_{21} and a_{22}. The result is

$$a_{21} = -\, a_{22}\, (\mathbf{x}^1, \mathbf{y}^2)$$

$$a_{22} = [\, \| \mathbf{y}^2 \|^2 - (\mathbf{x}^1, \mathbf{y}^2)^2 \,]^{-1/2}$$

Thus,

$$\mathbf{x}^2 = a_{22}\, [\, \mathbf{y}^2 - (\mathbf{x}^1, \mathbf{y}^2)\, \mathbf{x}^1 \,]$$

By mathematical induction, continue constructing an orthonormal set of vectors $\mathbf{x}^1, \mathbf{x}^2, \ldots, \mathbf{x}^\alpha$, $\alpha \le n$, where

$$\mathbf{x}^{\alpha} = a_{\alpha\alpha} [\, \mathbf{y}^{\alpha} - (\mathbf{x}^1, \mathbf{y}^{\alpha})\, \mathbf{x}^1 - (\mathbf{x}^2, \mathbf{y}^{\alpha})\, \mathbf{x}^2 - \ldots - (\mathbf{x}^{\alpha-1}, \mathbf{y}^{\alpha})\, \mathbf{x}^{\alpha-1} \,]$$

$$a_{\alpha\alpha} = [\, \| \mathbf{y}^{\alpha} \|^2 - (\mathbf{x}^1, \mathbf{y}^{\alpha})^2 - (\mathbf{x}^2, \mathbf{y}^{\alpha})^2 - \ldots - (\mathbf{x}^{\alpha-1}, \mathbf{y}^{\alpha})^2 \,]^{-1/2} \tag{2.4.28}$$

This process is continued until all the $\mathbf{y}^i$ are used and n vectors $\mathbf{x}^i$ are computed. The process could terminate in less than n steps only if some $\mathbf{x}^i = \mathbf{0}$. This is impossible, since it would imply that a nontrivial linear combination of the $\mathbf{y}^i$ is the zero vector, contradicting the linear independence of the $\mathbf{y}^i$. This procedure is called the **Gram–Schmidt orthonormalization process** and Eq. 2.4.28 is called the **Gram–Schmidt recurrence relation.**

Example 2.4.7

Let $\{ \mathbf{y}^1, \mathbf{y}^2, \mathbf{y}^3 \} = \{ [\, 1, 1, 1 \,]^T, [\, -1, 0, -1 \,]^T, [\, -1, 2, 3 \,]^T \}$ be a basis for the vector space R^3, with the scalar product defined in Eq. 2.4.1. To construct an orthonormal basis, first let

$$\mathbf{x}^1 = \frac{1}{\| \mathbf{y}^1 \|} \mathbf{y}^1 = \left[\frac{1}{\sqrt{3}}, \frac{1}{\sqrt{3}}, \frac{1}{\sqrt{3}} \right]^T$$

Then, with

$$a_{22} = [\, \| \mathbf{y}^2 \|^2 - (\mathbf{x}^1, \mathbf{y}^2)^2 \,]^{-1/2}$$

$$= \left(2 - \frac{4}{3} \right)^{-1/2} = \sqrt{\frac{3}{2}}$$

$$\mathbf{x}^2 = a_{22} [\, \mathbf{y}^2 - (\mathbf{x}^1, \mathbf{y}^2)\, \mathbf{x}^1 \,]$$

$$= \sqrt{\frac{3}{2}} \left([\, -1, 0, -1 \,]^T + \frac{2}{\sqrt{3}} \left[\frac{1}{\sqrt{3}}, \frac{1}{\sqrt{3}}, \frac{1}{\sqrt{3}} \right]^T \right)$$

$$= \left[-\frac{1}{\sqrt{6}}, \frac{2}{\sqrt{6}}, -\frac{1}{\sqrt{6}} \right]^T$$

Finally,

$$a_{33} = [\, \| \mathbf{y}^3 \|^2 - (\mathbf{x}^1, \mathbf{y}^3)^2 - (\mathbf{x}^2, \mathbf{y}^3)^2 \,]^{-1/2}$$

$$= \left(14 - \frac{16}{3} - \frac{4}{6} \right)^{-1/2} = \frac{1}{2\sqrt{2}}$$

$$\mathbf{x}^3 = a_{33}\,[\,\mathbf{y}^3 - (\mathbf{x}^1, \mathbf{y}^3)\,\mathbf{x}^1 - (\mathbf{x}^2, \mathbf{y}^3)\,\mathbf{x}^2\,]$$

$$= \frac{1}{2\sqrt{2}}\left([-1, 2, 3]^T - \frac{4}{\sqrt{3}}\left[\frac{1}{\sqrt{3}}, \frac{1}{\sqrt{3}}, \frac{1}{\sqrt{3}}\right]^T - \frac{2}{\sqrt{6}}\left[-\frac{1}{\sqrt{6}}, \frac{2}{\sqrt{6}}, -\frac{1}{\sqrt{6}}\right]^T \right)$$

$$= \left[-\frac{1}{\sqrt{2}}, 0, \frac{1}{\sqrt{2}}\right]^T$$

■

A final result associated with the scalar product that is of great theoretical and computational importance is the following.

Theorem 2.4.4. Let $\mathbf{x}^i$, i = 1, . . . , n, be a set of linearly independent vectors in an n-dimensional vector space V with scalar product (• , •). Then the only vector in V that is orthogonal to each of these vectors is the zero vector. ■

To prove Theorem 2.4.4, let **y** be a vector in V such that

$$(\mathbf{x}^i, \mathbf{y}) = 0, \quad i = 1, \ldots, n \tag{2.4.29}$$

Since the $\mathbf{x}^i$ are linearly independent, they form a basis for V. Thus, there is a unique set of constants a_i, i = 1, . . . , n, such that

$$\mathbf{y} = a_j \mathbf{x}^j \tag{2.4.30}$$

Substituting Eq. 2.4.30 into Eq. 2.4.29,

$$(\mathbf{x}^i, a_j\mathbf{x}^j) = (\mathbf{x}^i, \mathbf{x}^j)\, a_j = 0, \quad i = 1, \ldots, n \tag{2.4.31}$$

The coefficient matrix in Eq. 2.4.31 is the Gram matrix of the linearly independent vectors $\mathbf{x}^i$, i = 1, . . . , n, so it has rank n. Therefore, $a_j = 0$, j = 1, . . . , n. But, from Eq. 2.4.30, $\mathbf{y} = \mathbf{0}$, which was to be shown. ■

Example 2.4.8

The vectors $\mathbf{x}^1 = [1, 1, 1]^T$, $\mathbf{x}^2 = [1, -1, 0]^T$, and $\mathbf{x}^3 = [0, 0, 1]^T$ are linearly independent. Let **y** be a vector in R^3. Then, **y** is orthogonal to $\mathbf{x}^1$, $\mathbf{x}^2$, and $\mathbf{x}^3$ if

$$(\mathbf{x}^1, \mathbf{y}) = y_1 + y_2 + y_3 = 0$$

$$(\mathbf{x}^2, \mathbf{y}) = y_1 - y_2 = 0$$

$$(\mathbf{x}^3, \mathbf{y}) = y_3 = 0$$

The only solution, consistent with Theorem 2.4.4, is

$$y_1 = y_2 = y_3 = 0$$ ■

EXERCISES 2.4

1. Verify that the scalar product of Example 2.4.1 satisfies the properties of Definition 2.4.1.

2. Prove the following statements for a vector space with a scalar product:

   (a) The parallelogram rule

   $$\|\mathbf{x}+\mathbf{y}\|^2 + \|\mathbf{x}-\mathbf{y}\|^2 = 2\|\mathbf{x}\|^2 + 2\|\mathbf{y}\|^2$$

   (b) The Pythagorean rule

   $$\|\mathbf{x}+\mathbf{y}\|^2 = \|\mathbf{x}\|^2 + \|\mathbf{y}\|^2, \text{ if } (\mathbf{x}, \mathbf{y}) = 0$$

   (c) $\|\mathbf{x}-\mathbf{y}\| \geq |\,\|\mathbf{x}\| - \|\mathbf{y}\|\,|$

3. Show that the Schwartz inequality is an equality if and only if the vectors **x** and **y** are proportional; i.e., $\mathbf{y} = \alpha\mathbf{x}$ for some scalar α.

4. Show that the triangle inequality is an equality if and only if the two vectors **x** and **y** are proportional and the constant of proportionality is a nonnegative real number.

5. Test the following set of vectors for linear independence and construct from it an orthonormal basis for R^4: $[1, 0, 1, 0]^T$, $[1, -1, 0, 1]^T$, $[0, 1, -1, 1]^T$, $[1, -1, 1, -1]^T$.

6. (a) Let **x**, **y**, and **z** be vectors in R^3, such that **x** and **y** are orthogonal and **x** and **z** are orthogonal. Prove that **x** and any vector in R^3 of the form $a\mathbf{y} + b\mathbf{z}$, where a and b are real numbers (i.e., a linear combination of **y** and **z**) are orthogonal.

   (b) Let **x** be a vector in R^3. Prove that the set of all vectors **y** in R^3 such that **x** and **y** are orthogonal is a subspace of R^3.

7. Prove that if **x** is in R^3 and $(\mathbf{x}, \mathbf{y}) = 0$ for all **y** in R^3, then $\mathbf{x} = \mathbf{0}$.

8. Show that the matrix equation

   $$\mathbf{A}\mathbf{x} \equiv \begin{bmatrix} 1 & 1 & 1 \\ 1 & 1 & -1 \\ 3 & 3 & -5 \end{bmatrix} \begin{bmatrix} x_1 \\ x_2 \\ x_3 \end{bmatrix} = \begin{bmatrix} 3 \\ 1 \\ 1 \end{bmatrix} \equiv \mathbf{c}$$

   possesses a one-parameter family of solutions and verify directly that the vector **c** is orthogonal to all solutions of $\mathbf{A}^T\mathbf{y} = \mathbf{0}$.

9. Show that if **A** and **B** are orthogonal n×n matrices, then $\mathbf{A}^T\mathbf{B}$ is an orthogonal matrix.

10. If **A** is an orthogonal n×n matrix and $\mathbf{c} \in R^n$, solve $\mathbf{AAx} = \mathbf{c}$ for **x**. Find the solution in terms of $\mathbf{A}^T$.

11. In R^3, find all vectors that are orthogonal to both $[\ 1,\ 1,\ 1\ ]^T$ and $[\ 1, -1, 0\ ]^T$. Produce, from these vectors, a mutually orthogonal system of unit vectors; i.e., an orthonormal basis for R^3.

12. If **u** is a unit vector, show that $\mathbf{Q} = \mathbf{I} - 2\mathbf{u}\mathbf{u}^T$ is an orthogonal matrix, known as a **Householder transformation**. Compute **Q** explicitly when $\mathbf{u} = (\ 1/\sqrt{3}\ )[\ 1, 1, 1\ ]^T$.

13. Show that $\mathbf{x} - \mathbf{y}$ is orthogonal to $\mathbf{x} + \mathbf{y}$ if and only if $\|\ \mathbf{x}\ \| = \|\ \mathbf{y}\ \|$.

3

EIGENVALUE PROBLEMS AND QUADRATIC FORMS

The focus of this chapter is on eigenvalue problems and quadratic forms. Eigenvalue problems arise naturally in the study of vibration and stability. They also arise in mathematical techniques such as separation of variables and eigenvector expansion methods that are used in solving ordinary and partial differential equations in engineering analysis.

Analytical mechanics is based on concepts of energy associated with deformation or displacement of discrete masses or a continuum. In applications for which the force-displacement relation is linear, energy is expressed as a sum of terms that involve products of the displacement variables. Such expressions are called quadratic forms. As is shown in this chapter, and to a greater extent in later chapters of the text, the solution of linear equations is often equivalent to minimization of certain quadratic forms. This equivalence serves as the foundation for many of the modern computational methods in mechanics, such as the finite element method.

3.1 EIGENVALUE PROBLEMS

Example 3.1.1

Consider the compound column of Fig. 3.1.1(a). A slightly deflected position is sought [see Fig. 3.1.1(b)], in which both members of the column are in equilibrium, as a criteria for onset of buckling; i.e., if P is increased from a value for which a slightly deflected position is possible, the column will buckle. Summing counterclockwise moments on both elements, using free-body diagrams of Figs. 3.1.1(c) and (d),

$$-k_2(\theta_2-\theta_1)+k_1\theta_1-P\ell\theta_1=0$$

$$k_2(\theta_2-\theta_1)\qquad\qquad-P\ell\theta_2=0$$

In matrix form, with $\mathbf{x}=[\,\theta_1,\theta_2\,]^T$, this is

$$\mathbf{Ax} \equiv \begin{bmatrix} k_1 + k_2 & -k_2 \\ -k_2 & k_2 \end{bmatrix} \begin{bmatrix} \theta_1 \\ \theta_2 \end{bmatrix} = P\ell \begin{bmatrix} \theta_1 \\ \theta_2 \end{bmatrix} \equiv \lambda \mathbf{x}$$

where $\lambda = P\ell$.

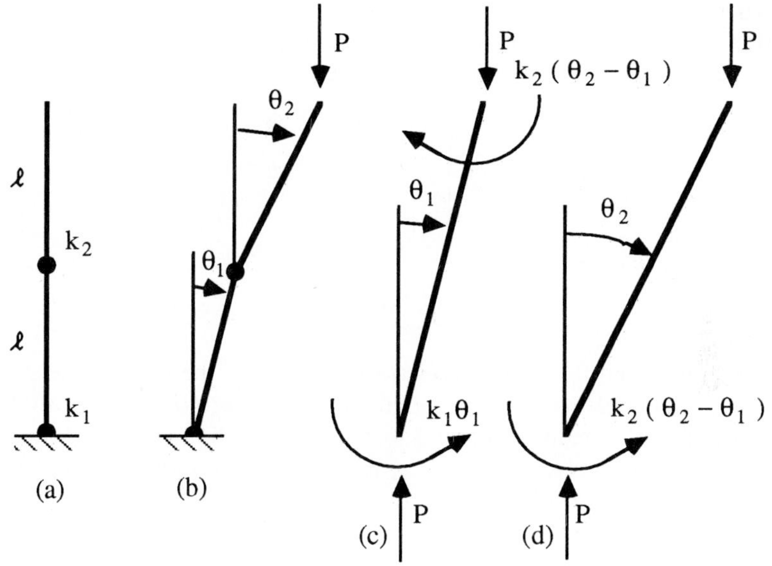

Figure 3.1.1 Buckling of Compound Column ■

Eigenvalues and Eigenvectors

A frequently encountered problem, as illustrated in Example 3.1.1, is determining values of a constant λ for which there is a nontrivial solution of a homogeneous set of equations of the form

$$\mathbf{Ax} = \lambda \mathbf{x} \tag{3.1.1}$$

where $\mathbf{A}$ is an n×n real matrix, $\mathbf{x}$ is an n×1 column vector (real or complex), and λ is a constant (real or complex).

Definition 3.1.1. Equation 3.1.1 for λ and $\mathbf{x}$ is known as an **eigenvalue problem**. Values of λ for which solutions $\mathbf{x} \neq \mathbf{0}$ of Eq. 3.1.1 exist are called **eigenvalues**, or **characteristic values**, of the matrix $\mathbf{A}$. The corresponding nonzero solutions $\mathbf{x}$ are called **eigenvectors** of the matrix $\mathbf{A}$. ■

In many applications in which such problems arise, the matrix $\mathbf{A}$ is symmetric.

Characteristic Equations

Equation 3.1.1 may be written as

$$(\mathbf{A} - \lambda\mathbf{I})\,\mathbf{x} = \mathbf{0} \tag{3.1.2}$$

where **I** is the n×n identity matrix. According to Theorem 2.3.5, this homogeneous equation for **x** has nontrivial solutions if and only if the determinant of the coefficient matrix $\mathbf{A} - \lambda\mathbf{I}$ vanishes. That is, if and only if λ satisfies

$$|\mathbf{A} - \lambda\mathbf{I}| \equiv \begin{vmatrix} (a_{11} - \lambda) & a_{12} & \cdots & a_{1n} \\ a_{21} & (a_{22} - \lambda) & \cdots & a_{2n} \\ \vdots & \vdots & & \vdots \\ a_{n1} & a_{n2} & \cdots & (a_{nn} - \lambda) \end{vmatrix}$$

$$= c_0\lambda^n + c_1\lambda^{n-1} + \ldots + c_n = 0 \tag{3.1.3}$$

which is called the **characteristic equation** of the eigenvalue problem of Eq. 3.1.1. The left side of Eq. 3.1.3 is called the **characteristic polynomial** for the eigenvalue problem. This condition requires that λ is the solution of a polynomial equation of degree n. The n solutions $\lambda_1, \lambda_2, \ldots, \lambda_n$, which need not all be distinct, are the eigenvalues of the matrix **A**.

Example 3.1.2

The characteristic equation for the buckling problem of Example 3.1.1 is

$$\begin{vmatrix} k_1 + k_2 - \lambda & -k_2 \\ -k_2 & k_2 - \lambda \end{vmatrix} = \lambda^2 - (k_1 + 2k_2)\,\lambda + k_1k_2 = 0$$

There are two eigenvalues, λ_1 and λ_2. Recalling that $P = \lambda/\ell$, the buckling loads of the compound column of Example 3.1.1 are

$$P_1 = \left(k_1 + 2k_2 - \sqrt{k_1^2 + 4k_2^2} \right) \Big/ \, 2\ell$$

$$P_2 = \left(k_1 + 2k_2 + \sqrt{k_1^2 + 4k_2^2} \right) \Big/ \, 2\ell$$

where $P_1 < P_2$. Substituting P_1 into the equilibrium equations of Example 3.1.1,

$$\begin{bmatrix} \dfrac{k_1}{2} + \dfrac{\sqrt{k_1^2 + 4k_2^2}}{2} & -k_2 \\ -k_2 & -\dfrac{k_1}{2} + \dfrac{\sqrt{k_1^2 + 4k_2^2}}{2} \end{bmatrix} \begin{bmatrix} \theta_1 \\ \theta_2 \end{bmatrix} = \begin{bmatrix} 0 \\ 0 \end{bmatrix}$$

From the first equation,

$$\theta_2 = \frac{k_1 + \sqrt{k_1^2 + 4k_2^2}}{2k_2}\,\theta_1$$

for any value of θ_1. A simple check shows that the second equation is also satisfied by this solution, which is expected, since the matrix was caused to be singular. Setting $\theta_1 = 1$ yields an eigenvector corresponding to P_1,

$$\mathbf{x}^1 = \begin{bmatrix} \theta_1^1 \\ \theta_2^1 \end{bmatrix} = \begin{bmatrix} 1 \\ \dfrac{k_1 + \sqrt{k_1^2 + 4k_2^2}}{2k_2} \end{bmatrix}$$

Similarly, an eigenvector for P_2 is

$$\mathbf{x}^2 = \begin{bmatrix} \theta_1^2 \\ \theta_2^2 \end{bmatrix} = \begin{bmatrix} 1 \\ \dfrac{k_1 - \sqrt{k_1^2 + 4k_2^2}}{2k_2} \end{bmatrix}$$

These eigenvectors are called **buckling modes** of the structure. Noting that $\sqrt{k_1^2 + 4k_2^2} > k_1$, the forms of buckling modes $\mathbf{x}^1$ and $\mathbf{x}^2$ are shown in Figs. 3.1.2(a) and (b), respectively.

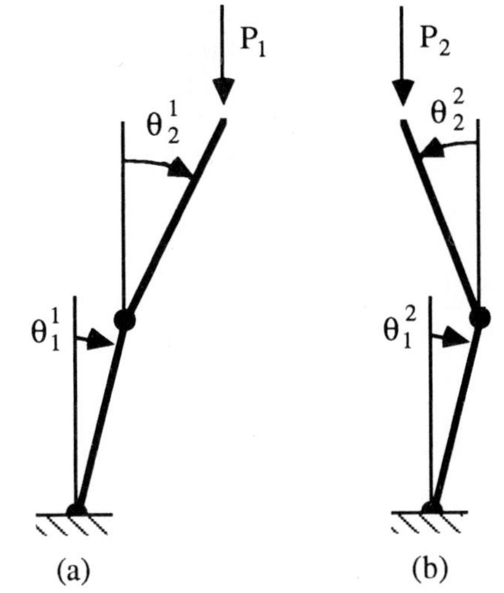

Figure 3.1.2 Buckling Modes of Compound Column

It is physically clear that, since $0 < P_1 < P_2$, if P is increased slowly from zero, buckling in the mode shown in Fig. 3.1.2(a) will occur when P reaches P_1. ■

Example 3.1.3

The characteristic equation for the 3×3 matrix

$$\mathbf{A} = \begin{bmatrix} 1 & 2 & 0 \\ 2 & 2 & 2 \\ 0 & 2 & 3 \end{bmatrix}$$

is

$$|\mathbf{A} - \lambda\mathbf{I}| = \begin{vmatrix} 1-\lambda & 2 & 0 \\ 2 & 2-\lambda & 2 \\ 0 & 2 & 3-\lambda \end{vmatrix}$$

$$= (1-\lambda)(2-\lambda)(3-\lambda) - 4(3-\lambda) - 4(1-\lambda)$$

$$= \lambda^3 - 6\lambda^2 + 3\lambda + 10 = 0$$

This equation has three solutions,

$$\lambda_1 = -1, \quad \lambda_2 = 2, \quad \lambda_3 = 5$$ ■

The sum of the diagonal elements of a matrix **A** is called the **trace** of **A** and is denoted by tr(**A**). It is shown in Ref. 4 that the coefficients of the characteristic equation of Eq. 3.1.3, for an n×n matrix, are given by

$$\begin{aligned} c_0 &= 1 \\ c_1 &= -t_1 \\ c_2 &= -(1/2)(c_1 t_1 + t_2) \\ c_3 &= -(1/3)(c_2 t_1 + c_1 t_2 + t_3) \\ &\cdots\cdots\cdots\cdots \\ c_n &= -(1/n)(c_{n-1} t_1 + c_{n-2} t_2 + \ldots + c_1 t_{n-1} + t_n) \end{aligned} \tag{3.1.4}$$

and that

$$\sum_{i=1}^{n} \lambda_i = \mathrm{tr}(\mathbf{A})$$

where

$$t_k = \mathrm{tr}(\mathbf{A}^k), \quad k = 1, \ldots, n$$

and the kth **power of a matrix A** is

$$\mathbf{A}^k = \mathbf{A}\,\mathbf{A} \ldots \mathbf{A} \quad \text{(k-times)}$$

Equations 3.1.4 are called **Bocher's formulas** for the coefficients of the characteristic polynomial.

Example 3.1.4

Applying Eq. 3.1.4 to the matrix **A** of Example 3.1.3,

$$t_1 = \mathrm{tr}(\mathbf{A}) = 1 + 2 + 3 = 6$$

and, with

$$\mathbf{A}^2 = \begin{bmatrix} 5 & 6 & 4 \\ 6 & 12 & 10 \\ 4 & 10 & 13 \end{bmatrix}, \quad \mathbf{A}^3 = \begin{bmatrix} 17 & 30 & 24 \\ 30 & 56 & 54 \\ 24 & 54 & 59 \end{bmatrix}$$

$$t_2 = \mathrm{tr}(\mathbf{A}^2) = 5 + 12 + 13 = 30$$

$$t_3 = \mathrm{tr}(\mathbf{A}^3) = 17 + 56 + 59 = 132$$

Eq. 3.1.4 yields

$$\begin{aligned} c_0 &= 1 \\ c_1 &= -6 \\ c_2 &= -(1/2)[(-6) \times 6 + 30] = 3 \\ c_3 &= -(1/3)[3 \times 6 + (-6) \times 30 + 132] = 10 \end{aligned}$$

Hence, the characteristic equation for **A** is

$$\lambda^3 - 6\lambda^2 + 3\lambda + 10 = 0$$

which agrees with Example 3.1.3. Furthermore, the trace of **A** equals the sum of all eigenvalues λ_i [1]; i.e., from the result of Example 3.1.3,

$$\lambda_1 + \lambda_2 + \lambda_3 = -1 + 2 + 5 = 6 = \mathrm{tr}(\mathbf{A})$$ ■

From Eq. 1.4.2, the determinant of a matrix is equal to the determinant of its transpose. Thus, $|\mathbf{A} - \lambda\mathbf{I}| = |\mathbf{A}^T - \lambda\mathbf{I}|$ and the eigenvalues of **A** are also the eigenvalues of $\mathbf{A}^T$.

Eigenvector Orthogonality for Symmetric Matrices

Corresponding to each eigenvalue λ_k, there exists at least one eigenvector $\mathbf{x}^k \neq \mathbf{0}$ from Eq. 3.1.1, or Eq. 3.1.2, which is determined only to within an arbitrary multiplicative constant; i.e., $\alpha\mathbf{x}^k$ is also an eigenvector, for any constant $\alpha \neq 0$. Let λ_α and λ_β be two distinct eigenvalues of a symmetric matrix $\mathbf{A}$ and denote corresponding eigenvectors by $\mathbf{x}^\alpha$ and $\mathbf{x}^\beta$, respectively; i.e., $\lambda_\alpha \neq \lambda_\beta$ and

$$\mathbf{A}\mathbf{x}^\alpha = \lambda_\alpha \mathbf{x}^\alpha \tag{3.1.5}$$

$$\mathbf{A}\mathbf{x}^\beta = \lambda_\beta \mathbf{x}^\beta \tag{3.1.6}$$

Taking the transpose of Eq. 3.1.5 and multiplying both sides on the right by $\mathbf{x}^\beta$,

$$\mathbf{x}^{\alpha T}\mathbf{A}^T\mathbf{x}^\beta = \lambda_\alpha \mathbf{x}^{\alpha T}\mathbf{x}^\beta \tag{3.1.7}$$

Multiplying both sides of Eq. 3.1.6 on the left by $\mathbf{x}^{\alpha T}$,

$$\mathbf{x}^{\alpha T}\mathbf{A}\mathbf{x}^\beta = \lambda_\beta \mathbf{x}^{\alpha T}\mathbf{x}^\beta \tag{3.1.8}$$

Subtracting Eq. 3.1.7 from Eq. 3.1.8 and noting that for a symmetric matrix $\mathbf{A}^T = \mathbf{A}$,

$$(\lambda_\beta - \lambda_\alpha)\,\mathbf{x}^{\alpha T}\mathbf{x}^\beta = \mathbf{x}^{\alpha T}(\mathbf{A} - \mathbf{A}^T)\,\mathbf{x}^\beta = 0 \tag{3.1.9}$$

Since $\lambda_\alpha \neq \lambda_\beta$, the following important result holds.

Theorem 3.1.1. Eigenvectors $\mathbf{x}^i$ and $\mathbf{x}^j$ of a real symmetric matrix that correspond to distinct eigenvalues $\lambda_i \neq \lambda_j$ are orthogonal; i.e.,

$$\mathbf{x}^{iT}\mathbf{x}^j = 0 \tag{3.1.10}$$

Furthermore, the eigenvalues and eigenvectors of a symmetric matrix are real. ■

To prove the second part of Theorem 3.1.1, suppose that $\lambda = \alpha + i\beta$ is a root of Eq. 3.1.3, where α and β are real and $i = \sqrt{-1}$. Let $\mathbf{x} \neq \mathbf{0}$ be a corresponding eigenvector; i.e.,

$$\mathbf{A}\mathbf{x} = \lambda\mathbf{x}$$

which may be complex. Taking the **complex conjugate** (denoted by *) of both sides of this equation yields

$$\mathbf{A}\mathbf{x}^* = \lambda^*\mathbf{x}^*$$

since the conjugate of a product is the product of the conjugates and $\mathbf{A}$ is real. Thus,

$\lambda^* = \alpha - i\beta$ is also an eigenvalue of **A**, with $\mathbf{x}^*$ as the associated eigenvector. Multiplying the first relation on the left by $\mathbf{x}^{*T}$ and multiplying the transpose of the second relation on the right by **x** yields

$$\begin{aligned} \mathbf{x}^{*T}\mathbf{A}\mathbf{x} &= \lambda\mathbf{x}^{*T}\mathbf{x} \\ \mathbf{x}^{*T}\mathbf{A}^T\mathbf{x} &= \lambda^*\mathbf{x}^{*T}\mathbf{x} \end{aligned} \tag{3.1.11}$$

Subtracting the second of Eq. 3.1.11 from the first and using symmetry of **A**,

$$(\lambda - \lambda^*)\,\mathbf{x}^{*T}\mathbf{x} = 0 \tag{3.1.12}$$

Since the product $\mathbf{x}^{*T}\mathbf{x} = x_j^* x_j = (a_j - ib_j)(a_j + ib_j) = a_ja_j + b_jb_j \neq 0$ is real, it follows that $\lambda - \lambda^* = 2i\beta = 0$, so λ must be real. Accordingly, all eigenvalues of **A** are real.

Let the eigenvector **x** be written in the form

$$\mathbf{x} = \mathbf{y} + i\mathbf{z}$$

where **y** and **z** are real vectors. Since **A** and λ are real, $\mathbf{Ax} = \lambda\mathbf{x}$ can be written as

$$\mathbf{Ay} - \lambda\mathbf{y} = -i(\mathbf{Az} - \lambda\mathbf{z})$$

where the left side is real and the right side is imaginary. Thus, both sides are zero; i.e.,

$$\mathbf{Ay} - \lambda\mathbf{y} = \mathbf{0} = \mathbf{Az} - \lambda\mathbf{z}$$

so the real vectors **y** and **z** are eigenvectors corresponding to the real eigenvalue λ. Thus, the eigenvectors of a symmetric matrix may be chosen to be real. ■

Example 3.1.5

Consider the real symmetric matrix

$$\mathbf{A} = \begin{bmatrix} 0 & 0 & -2 \\ 0 & -2 & 0 \\ -2 & 0 & 3 \end{bmatrix}$$

The characteristic equation for its eigenvalues is

$$|\mathbf{A} - \lambda\mathbf{I}| = -(\lambda+2)(\lambda-4)(\lambda+1) = 0$$

Once an eigenvalue is found, it may be substituted into

$$(\mathbf{A} - \lambda\mathbf{I})\,\mathbf{x} = \begin{bmatrix} -\lambda & 0 & -2 \\ 0 & -2-\lambda & 0 \\ -2 & 0 & 3-\lambda \end{bmatrix} \begin{bmatrix} x_1 \\ x_2 \\ x_3 \end{bmatrix} = \mathbf{0}$$

to determine associated eigenvectors $\mathbf{x}$. The eigenvalues and corresponding eigenvectors for this 3×3 matrix are

$$\begin{aligned} \lambda_1 &= -2, & \mathbf{x}^1 &= [\,0, a, 0\,]^T \\ \lambda_2 &= 4, & \mathbf{x}^2 &= [\,-b/2, 0, b\,]^T \\ \lambda_3 &= -1, & \mathbf{x}^3 &= [\,2c, 0, c\,]^T \end{aligned}$$

where any nonzero scalars a, b, and c may be selected. Note that $\mathbf{x}^1$, $\mathbf{x}^2$, and $\mathbf{x}^3$ are mutually orthogonal. ■

If an eigenvalue λ_1 of a symmetric matrix is a **multiple root** of **multiplicity** s; i.e., if the left side of Eq. 3.1.3 contains the factor $(\lambda - \lambda_1)^s$, then there are s linearly independent eigenvectors associated with λ_1. The proof of this result is postponed to Section 3.2 (see Theorem 3.2.2).

Example 3.1.6

For the symmetric matrix

$$\mathbf{A} = \begin{bmatrix} 0 & 1 & 0 \\ 1 & 0 & 0 \\ 0 & 0 & 1 \end{bmatrix}$$

the characteristic equation is

$$\begin{aligned} |\mathbf{A} - \lambda\mathbf{I}| &= \begin{vmatrix} -\lambda & 1 & 0 \\ 1 & -\lambda & 0 \\ 0 & 0 & 1-\lambda \end{vmatrix} \\ &= \lambda^2(1-\lambda) - (1-\lambda) \\ &= -(\lambda+1)(\lambda-1)^2 = 0 \end{aligned}$$

Thus, there is a multiple root

$$\lambda_1 = \lambda_2 = 1$$

Substituting this result into Eq. 3.1.2,

$$\begin{bmatrix} -1 & 1 & 0 \\ 1 & -1 & 0 \\ 0 & 0 & 0 \end{bmatrix} \begin{bmatrix} x_1 \\ x_2 \\ x_3 \end{bmatrix} = \mathbf{0}$$

The rank of the coefficient matrix is one, so two linearly independent eigenvectors are obtained for the multiple root $\lambda_1 = \lambda_2$,

$$\mathbf{x}^1 = [\,1, 1, 0\,]^T$$

$$\mathbf{x}^2 = [\,1, 1, 1\,]^T$$

For $\lambda_3 = -1$, Eq. 3.1.2 is

$$\begin{bmatrix} 1 & 1 & 0 \\ 1 & 1 & 0 \\ 0 & 0 & 2 \end{bmatrix} \begin{bmatrix} x_1 \\ x_2 \\ x_3 \end{bmatrix} = \mathbf{0}$$

and the associated eigenvector is

$$\mathbf{x}^3 = [\,1, -1, 0\,]^T$$

Note that $\mathbf{x}^3$ is orthogonal to both $\mathbf{x}^1$ and $\mathbf{x}^2$. However, $\mathbf{x}^1$ and $\mathbf{x}^2$ are not orthogonal. ■

Eigenvector Expansion

By the Gram–Schmidt orthogonalization process of Section 2.4, it is possible to choose s orthonormal eigenvectors of a symmetric matrix $\mathbf{A}$ that correspond to an eigenvalue of multiplicity s. By Theorem 3.1.1, all these eigenvectors are orthogonal to all other eigenvectors. Thus, if multiple roots of Eq. 3.1.3 are counted separately, exactly n mutually orthogonal eigenvectors may be found. By virtue of the results of Section 2.2, this set of vectors forms a basis for R^n. Thus, any vector in R^n can be expressed as a linear combination of n eigenvectors of a symmetric n×n matrix.

In particular, if each of the n orthogonal eigenvectors has been multiplied by the inverse of its norm, so that it is a unit vector, the resulting set of vectors is orthonormal. These vectors may be denoted by $\mathbf{x}^1, \mathbf{x}^2, \ldots, \mathbf{x}^n$, with

$$\mathbf{x}^{iT}\mathbf{x}^k = \delta_{ik} \tag{3.1.13}$$

Thus, the ith coefficient in the representation, called an **eigenvector expansion**,

$$\mathbf{x} = a_k\mathbf{x}^k \tag{3.1.14}$$

for a vector $\mathbf{x}$ in R^n is obtained by forming the scalar product of $\mathbf{x}^i$ with both sides of Eq. 3.1.14,

$$a_i = \mathbf{x}^{iT}\mathbf{x} \tag{3.1.15}$$

Consider now the nonhomogeneous equation

$$\mathbf{A}\mathbf{x} - \beta\mathbf{x} = \mathbf{c} \tag{3.1.16}$$

where **A** is a real symmetric matrix. If Eq. 3.1.16 has a solution, then it can be expressed as a linear combination of the eigenvectors of **A**. Suppose that n orthonormal eigenvectors $\mathbf{x}^1, \mathbf{x}^2, \ldots, \mathbf{x}^n$ are known, satisfying the equations

$$\mathbf{A}\mathbf{x}^\alpha = \lambda_\alpha \mathbf{x}^\alpha, \quad \alpha = 1, \ldots, n$$

Since the vectors $\{ \mathbf{x}^i \}$ form a basis of R^n, the solution of Eq. 3.1.16 can be expressed in the form

$$\mathbf{x} = a_k \mathbf{x}^k \tag{3.1.17}$$

where the constants a_k are to be determined. Substituting Eq. 3.1.17 into Eq. 3.1.16,

$$\sum_{k=1}^{n} (\lambda_k - \beta) a_k \mathbf{x}^k = \mathbf{c} \tag{3.1.18}$$

By forming the scalar product of any $\mathbf{x}^\alpha$ with both sides of Eq. 3.1.18,

$$(\lambda_\alpha - \beta) a_\alpha = \mathbf{x}^{\alpha T}\mathbf{c}, \quad \alpha = 1, 2, \ldots, n \tag{3.1.19}$$

Hence, if β is not an eigenvalue, the solution of Eq. 3.1.16 is obtained in the form

$$\mathbf{x} = \sum_{k=1}^{n} \frac{\mathbf{x}^{kT}\mathbf{c}}{(\lambda_k - \beta)} \mathbf{x}^k \tag{3.1.20}$$

Thus, a unique solution of the nonhomogeneous equation of Eq. 3.1.16 is obtained if β is not an eigenvalue. If $\lambda_p = \beta$, no solution exists unless the vector **c** is orthogonal to all eigenvectors that correspond to λ_p. If this condition is satisfied, Eq. 3.1.19 shows that the corresponding coefficient (or coefficients) a_p may be chosen arbitrarily, so infinitely many solutions exist. In particular, if $\beta = 0$, Eq. 3.1.16 reduces to

$$\mathbf{A}\mathbf{x} = \mathbf{c}$$

which was studied previously. This equation has a unique solution, unless $\beta = 0$ is an eigenvalue of **A**; i.e., unless the equation $\mathbf{A}\mathbf{x} = \mathbf{0}$ has nontrivial solutions. In this exceptional situation, no solution exists unless **c** is orthogonal to all vectors **x** that satisfy $\mathbf{A}\mathbf{x} = \mathbf{0}$, in which case infinitely many solutions exist. This result is a special case of the **Theorem of the Alternative** [1] for symmetric matrices. This existence criterion, in the more general case $\beta = \lambda_p$, can also be obtained by replacing **A** by $\mathbf{A} - \lambda_p\mathbf{I}$ and noticing that the latter matrix is symmetric.

Example 3.1.7

Consider the matrix equation of Example 2.3.10; i.e.,

$$\mathbf{Ax} \equiv \begin{bmatrix} k_1 & k_1 \\ k_2 & -k_2 \end{bmatrix} \begin{bmatrix} x_1 \\ x_2 \end{bmatrix} = \begin{bmatrix} 2\,(f_1 + f_2 + f_3) \\ 50\,(f_3 - f_1) \end{bmatrix} \equiv \mathbf{c} \tag{3.1.21}$$

The characteristic equation for the eigenvalues of matrix **A** is

$$\begin{vmatrix} k_1 - \lambda & k_1 \\ k_2 & -k_2 - \lambda \end{vmatrix} = -k_1k_2 - k_1\lambda + k_2\lambda + \lambda^2 - k_1k_2 = 0 \tag{3.1.22}$$

Thus,

$$\lambda = \frac{(k_1 - k_2) \pm \sqrt{(k_1 - k_2)^2 + 8k_1k_2}}{2}$$

Consider the case in which $k_1 = k_2 = 1$. Then the matrix **A** is symmetric,

$$\lambda_1 = -\sqrt{2}, \quad \lambda_2 = \sqrt{2} \tag{3.1.23}$$

and the associated orthogonal eigenvectors are

$$\mathbf{x}^1 = \begin{bmatrix} 1 \\ -1-\sqrt{2} \end{bmatrix}, \quad \mathbf{x}^2 = \begin{bmatrix} 1 \\ \sqrt{2}-1 \end{bmatrix} \tag{3.1.24}$$

An eigenvector expansion for the solution of Eq. 3.1.21 may be written in the form

$$\mathbf{x} = a_i\mathbf{x}^i$$

and Eq. 3.1.21 becomes

$$\mathbf{Ax} = a_i\mathbf{Ax}^i = \sum_{i=1}^{2} a_i\lambda_i\mathbf{x}^i = \mathbf{c}$$

Taking the scalar product of both sides of this equation with $\mathbf{x}^\alpha$,

$$a_\alpha\lambda_\alpha\mathbf{x}^{\alpha T}\mathbf{x}^\alpha = \mathbf{x}^{\alpha T}\mathbf{c} \tag{3.1.25}$$

Thus, the solution is

$$\mathbf{x} = \sum_{j=1}^{2} \left[\frac{\mathbf{x}^{jT}\mathbf{c}}{\lambda_j\mathbf{x}^{jT}\mathbf{x}^j} \right] \mathbf{x}^j \tag{3.1.26}$$

Consider the case $f_1 = f_2 = 10$ and $f_3 = 0$. Then, $\mathbf{c} = [\,40, -500\,]^T$ and

$$\mathbf{x} = \left(\frac{-125\sqrt{2} - 135}{\sqrt{2} + 1} \right) \begin{bmatrix} 1 \\ -1 - \sqrt{2} \end{bmatrix} + \left(\frac{135 - 125\sqrt{2}}{\sqrt{2} - 1} \right) \begin{bmatrix} 1 \\ \sqrt{2} - 1 \end{bmatrix}$$

$$= \begin{bmatrix} -230 \\ 270 \end{bmatrix} \tag{3.1.27}$$

■

Another method of solving linear and nonlinear equations is **Galerkin's method.** It is illustrated here with matrix equations. In order to solve the equation

$$\mathbf{Ax} = \mathbf{c} \tag{3.1.28}$$

where $\mathbf{x}$ is in R^n and $\mathbf{A}$ is an $n \times n$ symmetric matrix, first define the **residual** as

$$\mathbf{r} = \mathbf{Ax} - \mathbf{c}$$

By Theorem 2.4.4, Eq. 3.1.28 is satisfied if and only if $\mathbf{r}^T\mathbf{x}^j = 0$, for all eigenvectors $\mathbf{x}^j$ of the matrix $\mathbf{A}$, since they form a basis of R^n. That is,

$$\mathbf{x}^{jT}\mathbf{Ax} - \mathbf{x}^{jT}\mathbf{c} = 0, \quad j = 1, \ldots, n$$

As before, let $\mathbf{x} = a_i\mathbf{x}^i$ and substitute it into Eq. 3.1.28, to obtain

$$a_i\mathbf{x}^{jT}\mathbf{Ax}^i - \mathbf{x}^{jT}\mathbf{c} = \sum_{i=1}^{n} a_i\lambda_i\mathbf{x}^{jT}\mathbf{x}^i - \mathbf{x}^{jT}\mathbf{c} = 0, \quad j = 1, \ldots, n \tag{3.1.29}$$

Since the $\mathbf{x}^i$ are orthogonal, Eq. 3.1.29 reduces to

$$a_\alpha\lambda_\alpha\mathbf{x}^{\alpha T}\mathbf{x}^\alpha - \mathbf{x}^{\alpha T}\mathbf{c} = 0, \quad \alpha = 1, \ldots, n$$

which is the result obtained in Eq. 3.1.25. Thus, in this case, the Galerkin method yields the same result as the eigenvector expansion method.

EXERCISES 3.1

1. Show that the eigenvalue problem

$$\begin{aligned} x_1 - 2x_2 &= \lambda x_1 \\ x_1 - x_2 &= \lambda x_2 \end{aligned}$$

does not have a real nontrivial solution for x_1 and x_2, for any value of λ.

2. (a) Determine the eigenvalues λ_1 and λ_2 and the corresponding normalized eigenvectors $\mathbf{x}^1$ and $\mathbf{x}^2$ of the matrix

$$\mathbf{A} = \begin{bmatrix} 5 & 2 \\ 2 & 2 \end{bmatrix}$$

(b) Verify that $\mathbf{x}^1$ and $\mathbf{x}^2$ are orthogonal.

(c) If $\mathbf{y} = [\,1, 1\,]^T$, determine α_1 and α_2 so that

$$\mathbf{y} = \alpha_1 \mathbf{x}^1 + \alpha_2 \mathbf{x}^2$$

(d) Use the results of part (a), together with Eq. 3.1.20, to obtain the solution of the following set of equations:

$$\begin{aligned} 5x_1 + 2x_2 &= \lambda x_1 + 2 \\ 2x_1 + 2x_2 &= \lambda x_2 + 1 \end{aligned}$$

3. For the linear elastic system shown;

(a) Write the static equilibrium equations in matrix form.

(b) For $k_1 = k_2 = k_3 = 1$ lb/in., construct an eigenvector expansion for the solution.

(c) For what forces f_1 and f_2 does the problem have a solution?

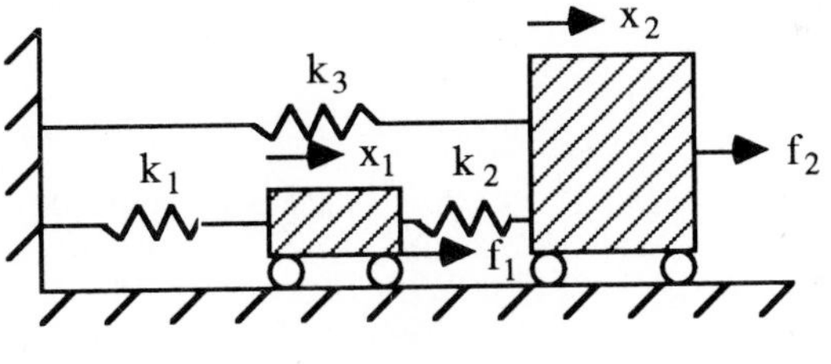

4. If λ is an eigenvalue of a matrix $\mathbf{A}$, with an associated eigenvector $\mathbf{x}$, prove that λ^k (k-th power of λ) is an eigenvalue of the matrix $\mathbf{A}^k = \mathbf{A}\,\mathbf{A} \ldots \mathbf{A}$ (k products of $\mathbf{A}$) with the associated eigenvector $\mathbf{x}$, where k is any positive integer.

5. Find the eigenvalues and eigenvectors of $\mathbf{A}^2$ and verify the result of Exercise 4, where

$$\mathbf{A} = \begin{bmatrix} 2 & 2 & 3 \\ 1 & 2 & 1 \\ 2 & -2 & 1 \end{bmatrix}$$

6. Show that a square matrix is singular if and only if it has zero as an eigenvalue.

7. Let **A** be a symmetric n×n matrix and $\mathbf{A}\mathbf{x}^{\alpha} = \lambda_{\alpha}\mathbf{x}^{\alpha}$, $\alpha = 1, 2, \ldots, n$, where all eigenvalues are distinct.

   (a) Write the eigenvector expansion of a vector $\mathbf{c} \in R^n$.

   (b) Is the eigenvector expansion unique or not? Why?

3.2 QUADRATIC FORMS

Matrix Quadratic Forms

Definition 3.2.1. A linear combination of products of variables x_i of the form

$$\begin{aligned} J \equiv\ & a_{11}x_1^{\,2} + a_{22}x_2^{\,2} + \ldots + a_{nn}x_n^{\,2} + a_{12}x_1x_2 + \ldots \\ & + a_{13}x_1x_3 + \ldots + a_{n,n-2}x_nx_{n-2} + a_{n,n-1}x_nx_{n-1} \end{aligned} \tag{3.2.1}$$

is called a **quadratic form** in $x_1, x_2, \ldots, x_n$, where the a_{ij} are real. The quadratic form of Eq. 3.2.1 can be written in matrix form as

$$\begin{aligned} J &= \mathbf{x}^T\mathbf{A}\mathbf{x} \\ &= [\, x_1, x_2, \ldots, x_n \,] \begin{bmatrix} a_{11} & a_{12} & \ldots & a_{1n} \\ a_{21} & a_{22} & \ldots & a_{2n} \\ \cdot\cdot & \cdot\cdot & \cdot\cdot\cdot & \cdot\cdot \\ a_{n1} & a_{n2} & \ldots & a_{nn} \end{bmatrix} \begin{bmatrix} x_1 \\ x_2 \\ \cdot\cdot \\ x_n \end{bmatrix} \\ &= a_{ij}x_ix_j \end{aligned} \tag{3.2.2}$$

where $\mathbf{x} = [\, x_1, x_2, \ldots, x_n \,]^T$ and $\mathbf{A} = [\, a_{ij} \,]$ is the **matrix of the quadratic form.** ■

Example 3.2.1

Construct the quadratic form $J = \mathbf{x}^T\mathbf{A}\mathbf{x}$ for the matrix

$$\mathbf{A} = \begin{bmatrix} 1 & 3 \\ 2 & 1 \end{bmatrix}$$

By Eq. 3.2.2,

$$J = [\, x_1, x_2 \,] \begin{bmatrix} 1 & 3 \\ 2 & 1 \end{bmatrix} \begin{bmatrix} x_1 \\ x_2 \end{bmatrix} = x_1^2 + 5x_1x_2 + x_2^2$$

■

For a general n×n matrix **B** that is not necessarily symmetric, consider the quadratic form

$$J = \mathbf{x}^T\mathbf{B}\mathbf{x}$$

Defining a symmetric matrix, called the **symmetrized matrix** associated with **B**,

$$\mathbf{B}^* = (1/2)(\mathbf{B} + \mathbf{B}^T) \tag{3.2.3}$$

the related quadratic form

$$J^* \equiv \mathbf{x}^T\mathbf{B}^*\mathbf{x} = (1/2)(\mathbf{x}^T\mathbf{B}\mathbf{x} + \mathbf{x}^T\mathbf{B}^T\mathbf{x})$$

can be formed. Since $\mathbf{x}^T\mathbf{B}^T\mathbf{x}$ is a scalar, it is equal to its transpose, so

$$\mathbf{x}^T\mathbf{B}^T\mathbf{x} = (\mathbf{x}^T\mathbf{B}^T\mathbf{x})^T = (\mathbf{B}^T\mathbf{x})^T\mathbf{x} = \mathbf{x}^T\mathbf{B}\mathbf{x} \tag{3.2.4}$$

Substituting Eq. 3.2.4 into the definition of J*,

$$J^* = \mathbf{x}^T\mathbf{B}^*\mathbf{x} = \mathbf{x}^T\mathbf{B}\mathbf{x} = J \tag{3.2.5}$$

This result may be summarized as follows.

Theorem 3.2.1. A nonsymmetric matrix **B** in a quadratic form can always be replaced by the symmetrized matrix of Eq. 3.2.3, without affecting the value of the quadratic form. ■

The fact that a quadratic form can always be written with a symmetric matrix is of fundamental importance. It provides the basis for restricting attention to symmetric matrices in large classes of applications.

Example 3.2.2

Theorem 3.2.1 may be illustrated using Example 3.2.1, with

$$\mathbf{A}^* = \frac{1}{2}(\mathbf{A} + \mathbf{A}^T) = \begin{bmatrix} 1 & \frac{5}{2} \\ \frac{5}{2} & 1 \end{bmatrix}$$

Then,

$$J^* = \mathbf{x}^T\mathbf{A}^*\mathbf{x} = x_1^2 + 5x_1x_2 + x_2^2 = J$$ ■

Strain Energy

The **energy** stored in an elastic mechanical element, such as the **elastic spring** shown in Fig. 3.2.1, is of interest in many applications. The force required to deform the spring from its free length is f = kx, where k is called the **spring constant** and x is the deformation of the spring. Integrating for the work done in stretching the spring from its free length by an amount x,

$$U \equiv wk = \int_0^x f\, d\xi = \int_0^x k\xi\, d\xi = \frac{1}{2} kx^2 \tag{3.2.6}$$

This work is equal to the energy stored in the spring, called its **strain energy**.

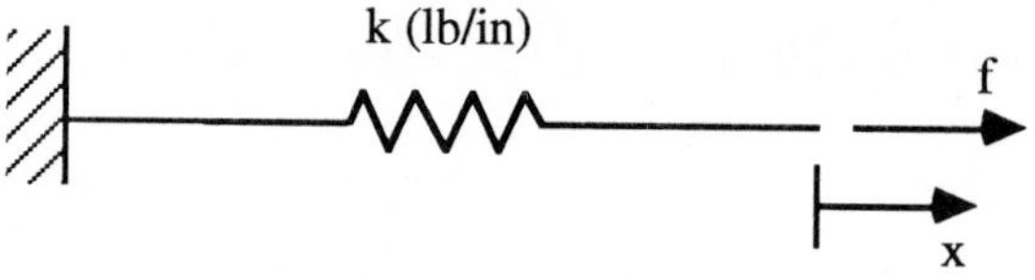

Figure 3.2.1 Linear Spring

Example 3.2.3

Consider the elastic system in Exercise 3 of Section 3.1. The sum of the strain energies stored in the three springs is

$$\begin{aligned} U &= \frac{1}{2} k_1 x_1^2 + \frac{1}{2} k_2 (x_2 - x_1)^2 + \frac{1}{2} k_3 x_2^2 \\ &= \frac{1}{2} (k_1 + k_2) x_1^2 - k_2 x_1 x_2 + \frac{1}{2} (k_2 + k_3) x_2^2 \\ &= \frac{1}{2} [x_1, x_2] \begin{bmatrix} k_1 + k_2 & -k_2 \\ -k_2 & k_2 + k_3 \end{bmatrix} \begin{bmatrix} x_1 \\ x_2 \end{bmatrix} \equiv \frac{1}{2} \mathbf{x}^T \mathbf{K} \mathbf{x} \end{aligned} \tag{3.2.7}$$

which is a quadratic form. The matrix **K** is called the system **stiffness matrix.** ■

Trusses in structural engineering [5] consist of tension or compression members called **bars**, as shown in Fig. 3.2.2, that deform as stiff elastic springs. The force-deflection relation is [6]

$$\delta\ell = f\ell / AE$$

where ℓ is the length of the bar, $\delta\ell$ is the change in length, f is the applied load, A is the cross-sectional area of the bar, and E is **Young's modulus** of the material. This relation may be rewritten as

$$f = (AE/\ell)\, \delta\ell$$

so $k = AE/\ell$ is the spring constant for the bar. Thus, the strain energy of the bar is

$$U = (1/2) (AE/\ell) (\delta\ell)^2 \tag{3.2.8}$$

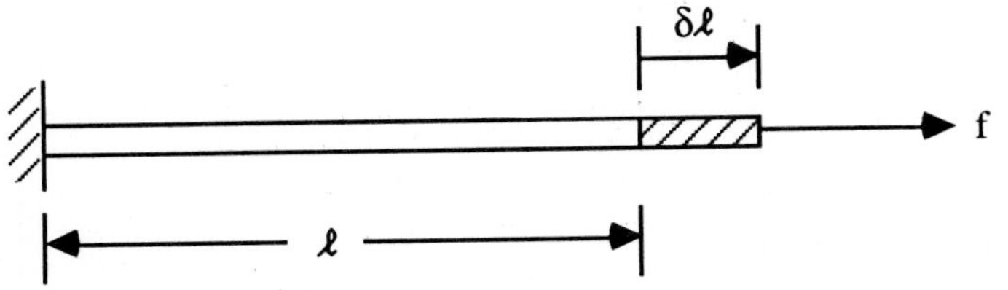

Figure 3.2.2 Deflection of a Bar

Example 3.2.4

Consider the simple three bar truss structure of Fig. 3.2.3. From trigonometric relations and using small displacement approximations,

$$\delta \ell_1 = \frac{\sqrt{2}}{2}(x_1 + x_2)$$

$$\delta \ell_2 = x_1$$

$$\delta \ell_3 = \frac{\sqrt{2}}{2}(x_1 - x_2)$$

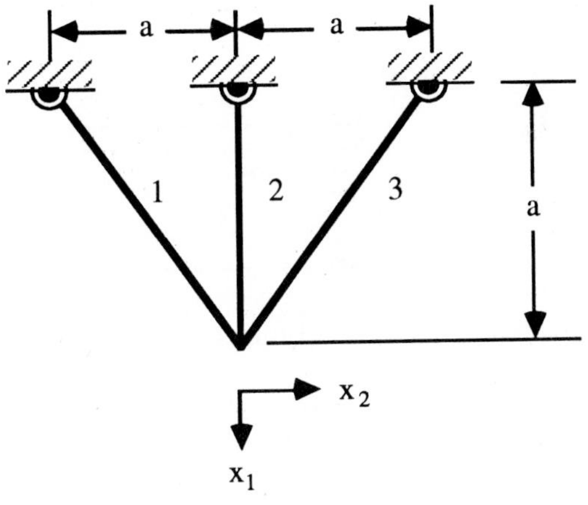

Figure 3.2.3 Three Bar Truss

Since $\ell_1 = \ell_3 = \sqrt{2}a$ and $\ell_2 = a$, using Eq. 3.2.8, the total strain energy is

$$U = \frac{AE}{2a}\left[\frac{(x_1 + x_2)^2}{2\sqrt{2}} + x_1^2 + \frac{(x_1 - x_2)^2}{2\sqrt{2}}\right]$$

$$= \frac{AE}{2a}\left[\left(\frac{\sqrt{2}}{2} + 1 \right) x_1^2 + \frac{\sqrt{2}}{2} x_2^2 \right]$$

$$= \frac{1}{2} [\, x_1, x_2 \,] \begin{bmatrix} \frac{AE}{a}\left(\frac{\sqrt{2}}{2} + 1 \right) & 0 \\ 0 & \frac{AE}{a}\frac{\sqrt{2}}{2} \end{bmatrix} \begin{bmatrix} x_1 \\ x_2 \end{bmatrix}$$

$$\equiv \frac{1}{2} \mathbf{x}^T \mathbf{K} \mathbf{x} \tag{3.2.9}$$

■

These elementary examples of strain energy as quadratic forms typify considerations that arise throughout the study of mechanics of deformable bodies.

Example 3.2.5

Analysis of the **bending of beams** becomes important, because many structures are constructed of, or can be approximated by assemblages of **beams**. Therefore, consider the **beam element** of Fig. 3.2.4, with its four displacement coordinates $\mathbf{x} = [\, x_1, x_2, x_3, x_4 \,]^T$.

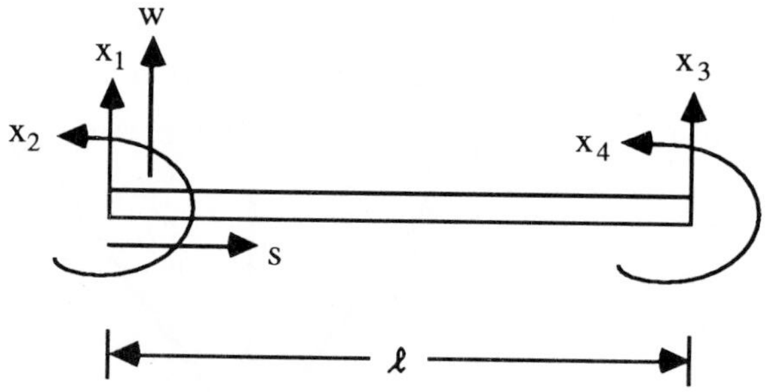

Figure 3.2.4 Beam Element

Since no lateral loads are applied to the beam and its Young's modulus E and its centroidal **cross-sectional moment of inertia** I are constant, the **beam bending equation** [6] is $EIw^{(4)} = 0$, where w is beam deflection. Thus, the bent shape of the beam, which is defined by w(s), is a cubic polynomial. To have $w(0) = x_1$, $w'(0) = x_2$, $w(\ell) = x_3$, and $w'(\ell) = x_4$, the displacement must be

$$w(s) = \frac{x_1}{\ell^3}(\, 2s^3 - 3\ell s^2 + \ell^3 \,) + \frac{x_2}{\ell^2}(\, s^3 - 2\ell s^2 + \ell^2 s \,)$$

$$- \frac{x_3}{\ell^3}(\, 2s^3 - 3\ell s^2 \,) + \frac{x_4}{\ell^2}(\, s^3 - \ell s^2 \,) \tag{3.2.10}$$

The strain energy of a bent beam is [6]

$$U = \frac{1}{2}\int_0^{\ell} EI\,(w'')^2\,ds \tag{3.2.11}$$

Substituting from Eq. 3.2.10 into Eq. 3.2.11 and integrating to obtain a quadratic expression in $\mathbf{x} = [x_1, x_2, x_3, x_4]^T$, it is left to the reader to verify that

$$U = \frac{1}{2}\mathbf{x}^T \frac{EI}{\ell^3}\begin{bmatrix} 12 & 6\ell & -12 & 6\ell \\ 6\ell & 4\ell^2 & -6\ell & 2\ell^2 \\ -12 & -6\ell & 12 & -6\ell \\ 6\ell & 2\ell^2 & -6\ell & 4\ell^2 \end{bmatrix}\mathbf{x} \equiv \frac{1}{2}\mathbf{x}^T\mathbf{K}^b\mathbf{x} \tag{3.2.12}$$

where $\mathbf{K}^b$ is the **beam element stiffness matrix.** ■

Canonical (Diagonal Matrix) Forms

In many applications, it is desirable to make a change of variables, writing $x_1, x_2, \ldots, x_n$ as linear combinations of variables $x_1', x_2', \ldots, x_n'$. Let $\mathbf{x} = [x_1, \ldots, x_n]^T$ be expressed in terms of $\mathbf{x}' = [x_1', \ldots, x_n']^T$ by the transformation

$$\mathbf{x} = \mathbf{Q}\mathbf{x}' \tag{3.2.13}$$

where $\mathbf{Q}$ is an n×n matrix. Substituting Eq. 3.2.13 into Eq. 3.2.2, where $\mathbf{A}$ is symmetric,

$$\begin{aligned} J &= \mathbf{x}^T\mathbf{A}\mathbf{x} = (\mathbf{Q}\mathbf{x}')^T\mathbf{A}\mathbf{Q}\mathbf{x}' = \mathbf{x}'^T\mathbf{Q}^T\mathbf{A}\mathbf{Q}\mathbf{x}' \\ &\equiv \mathbf{x}'^T\mathbf{A}'\mathbf{x}' \end{aligned} \tag{3.2.14}$$

The matrix $\mathbf{A}'$ of the quadratic form in $\mathbf{x}'$, which is also symmetric, is thus

$$\mathbf{A}' = \mathbf{Q}^T\mathbf{A}\mathbf{Q} \tag{3.2.15}$$

It is often desirable to simplify the quadratic form J in $\mathbf{x}'$ so that it involves only squares of the variables x_i'. To do this, the transformation matrix $\mathbf{Q}$ must be chosen so that $\mathbf{Q}^T\mathbf{A}\mathbf{Q}$ is a diagonal matrix; i.e., $a_{ij}' = 0$ if $i \neq j$. If the eigenvalues and corresponding eigenvectors of the symmetric matrix $\mathbf{A}$ are known, a matrix $\mathbf{Q}$ that has this property can be easily constructed. Suppose that the eigenvalues of $\mathbf{A}$ are $\lambda_1, \lambda_2, \ldots, \lambda_n$, repeated roots of the characteristic equation being numbered separately, and that the corresponding orthonormal eigenvectors are $\mathbf{x}^1, \mathbf{x}^2, \ldots, \mathbf{x}^n$; i.e.,

$$\mathbf{A}\mathbf{x}^{\alpha} = \lambda_{\alpha}\mathbf{x}^{\alpha}, \quad \alpha = 1, \ldots, n \tag{3.2.16}$$

Let a matrix **Q** be constructed so that $\mathbf{x}^1, \mathbf{x}^2, \ldots, \mathbf{x}^n$ are the columns of **Q**; i.e.,

$$\mathbf{Q} = [\, \mathbf{x}^1, \mathbf{x}^2, \ldots, \mathbf{x}^n \,] \tag{3.2.17}$$

Then, from Eq. 3.2.16,

$$\begin{aligned} \mathbf{AQ} &= [\, \lambda_1 \mathbf{x}^1, \lambda_2 \mathbf{x}^2, \ldots, \lambda_n \mathbf{x}^n \,] \\ &= \mathbf{Q} \begin{bmatrix} \lambda_1 & 0 & \ldots & 0 \\ 0 & \lambda_2 & \ldots & 0 \\ \vdots & \vdots & & \vdots \\ 0 & 0 & \ldots & \lambda_n \end{bmatrix} \end{aligned} \tag{3.2.18}$$

Multiplying both sides of Eq. 3.2.18 on the left by $\mathbf{Q}^T$,

$$\mathbf{Q}^T\mathbf{AQ} = \mathbf{Q}^T\mathbf{Q} \begin{bmatrix} \lambda_1 & 0 & \ldots & 0 \\ 0 & \lambda_2 & \ldots & 0 \\ \vdots & \vdots & & \vdots \\ 0 & 0 & \ldots & \lambda_n \end{bmatrix} \tag{3.2.19}$$

Since **Q** is made up of orthonormal column vectors $\mathbf{x}^j$,

$$\mathbf{Q}^T\mathbf{Q} = \left[\sum_{\alpha=1}^{n} x_\alpha^i x_\alpha^j \right] = [\, \mathbf{x}^{iT}\mathbf{x}^j \,] = [\, \delta_{ij} \,] = \mathbf{I} \tag{3.2.20}$$

so $\mathbf{Q}^T = \mathbf{Q}^{-1}$ and **Q** is an orthogonal matrix. Therefore, the transformed matrix **A**' of Eq. 3.2.15 is a diagonal matrix and the quadratic form of Eq. 3.2.14, in terms of the variable **x**', is

$$J = \mathbf{x}'^T \begin{bmatrix} \lambda_1 & 0 & \ldots & 0 \\ 0 & \lambda_2 & \ldots & 0 \\ \vdots & \vdots & & \vdots \\ 0 & 0 & \ldots & \lambda_n \end{bmatrix} \mathbf{x}' = \lambda_i \, (x_i')^2 \tag{3.2.21}$$

which is the desired result. This is called the **canonical form of the quadratic form** J.

Modal Matrices

A matrix whose columns are n linearly independent eigenvectors of a given n×n matrix **A** is called a **modal matrix** of **A**. When these n eigenvectors are orthonormal, the modal matrix is orthogonal. Thus, the matrix **Q** of Eq. 3.2.17 is an **orthogonal modal matrix** of **A**. Since **Q** is orthogonal, the variable **x**' is related to the variable **x**, in accordance with Eq. 3.2.13, as

$$\mathbf{x}' = \mathbf{Q}^{-1}\mathbf{x} = \mathbf{Q}^{T}\mathbf{x} \tag{3.2.22}$$

or, in component form,

$$x_i' = (\mathbf{x}^i, \mathbf{x}), \quad i = 1, 2, \ldots, n$$

Example 3.2.6

To illustrate the preceding reduction in a specific numerical case, consider the quadratic form

$$J = 25x_1^2 + 34x_2^2 + 41x_3^2 - 24x_2x_3$$

The symmetric matrix **A** of the form is

$$\mathbf{A} = \begin{bmatrix} 25 & 0 & 0 \\ 0 & 34 & -12 \\ 0 & -12 & 41 \end{bmatrix}$$

The characteristic equation $| \mathbf{A} - \lambda\mathbf{I} | = 0$ is

$$(25 - \lambda) (\lambda^2 - 75\lambda + 1250) = 0$$

from which the eigenvalues are

$$\lambda_1 = \lambda_2 = 25, \quad \lambda_3 = 50$$

With $\lambda = \lambda_1 = \lambda_2 = 25$, the equation $\mathbf{A}\mathbf{x} - \lambda\mathbf{x} = \mathbf{0}$ becomes

$$\begin{aligned} 0x_1 &= 0 \\ 9x_2 - 12x_3 &= 0 \\ -12x_2 + 16x_3 &= 0 \end{aligned}$$

The general solution is $x_1 = c_1$, $x_2 = c_2$, $x_3 = 3c_2/4$. In vector form, $\mathbf{x} = c_1\mathbf{x}^1 + c_2\mathbf{x}^2$, where $\mathbf{x}^1 = [1, 0, 0]^T$ and $\mathbf{x}^2 = [0, 4/5, 3/5]^T$, which are orthonormal eigenvectors. In a similar way, a normalized eigenvector corresponding to $\lambda_3 = 50$ is found to be

$$\mathbf{x}^3 = [0, 3/5, -4/5]^T$$

The orthonormal modal matrix **Q** of Eq. 3.2.17 is

$$\mathbf{Q} = \begin{bmatrix} 1 & 0 & 0 \\ 0 & \frac{4}{5} & \frac{3}{5} \\ 0 & \frac{3}{5} & -\frac{4}{5} \end{bmatrix}$$

and the coordinates defined by Eq. 3.2.22 are

$$\begin{aligned} x_1' &= x_1 \\ x_2' &= (4/5)\,x_2 + (3/5)\,x_3 \\ x_3' &= (3/5)\,x_2 - (4/5)\,x_3 \end{aligned}$$

With this choice of the coordinates, Eq. 3.2.21 yields

$$J \equiv 25x_1'^2 + 25x_2'^2 + 50x_3'^2$$ ■

The tools needed to establish an assertion that was made in Section 3.1 are now available.

Theorem 3.2.2. A real symmetric matrix **A** with an eigenvalue λ_1 of multiplicity s has s linearly independent eigenvectors corresponding to λ_1. ■

To prove Theorem 3.2.2, suppose first that **A** is a symmetric n×n matrix, such that the characteristic equation

$$|\mathbf{A} - \lambda\mathbf{I}| = 0$$

has $(\lambda - \lambda_1)^2$ as a factor and let $\mathbf{x}^1$ be one normalized eigenvector of **A** that corresponds to λ_1. Then, if **Q** is any n×n orthogonal matrix having $\mathbf{x}^1$ as its first column, it follows that

$$\mathbf{Q}^T\mathbf{A}\mathbf{Q} = \begin{bmatrix} -\mathbf{x}^{1T}\rightarrow \\ \cdots \\ \cdots \end{bmatrix} \mathbf{A} \begin{bmatrix} \cdot & \cdot & \cdot \\ | & \cdot & \cdot \\ \mathbf{x}^1 & \cdot & \cdot \\ \downarrow & \cdot & \cdot \\ \cdot & \cdot & \cdot \\ \cdot & \cdot & \cdot \end{bmatrix} = \begin{bmatrix} -\mathbf{x}^{1T}\rightarrow \\ \cdots \\ \cdots \end{bmatrix} \begin{bmatrix} \cdot & \cdot & \cdot \\ | & \cdot & \cdot \\ \lambda_1\mathbf{x}^1 & \cdot & \cdot \\ \downarrow & \cdot & \cdot \\ \cdot & \cdot & \cdot \\ \cdot & \cdot & \cdot \end{bmatrix}$$

Since each vector whose elements comprise a row of $\mathbf{Q}^T$, except the first, is orthogonal to the column $\lambda_1\mathbf{x}^1$, each element of the first column of the product, except the leading element, will vanish. Thus, the result will be of the form

$$\mathbf{Q}^T\mathbf{A}\mathbf{Q} = \begin{bmatrix} \lambda_1 & \alpha_{12} & \dots & \alpha_{1n} \\ 0 & \alpha_{22} & \dots & \alpha_{2n} \\ \vdots & \vdots & & \vdots \\ 0 & \alpha_{n2} & \dots & \alpha_{nn} \end{bmatrix}$$

Since symmetry of $\mathbf{A}$ implies symmetry of $\mathbf{Q}^T\mathbf{A}\mathbf{Q}$, the elements $\alpha_{12}, \ldots, \alpha_{1n}$ must also vanish. Hence, if $\mathbf{Q}$ is any orthogonal matrix having $\mathbf{x}^1$ as its first column, the product $\mathbf{Q}^T\mathbf{A}\mathbf{Q} = \mathbf{Q}^{-1}\mathbf{A}\mathbf{Q}$ is of the form

$$\mathbf{Q}^{-1}\mathbf{A}\mathbf{Q} = \begin{bmatrix} \lambda_1 & 0 & \dots & 0 \\ 0 & \alpha_{22} & \dots & \alpha_{2n} \\ \vdots & \vdots & & \vdots \\ 0 & \alpha_{n2} & \dots & \alpha_{nn} \end{bmatrix} \tag{3.2.23}$$

Thus,

$$\mathbf{Q}^{-1}\mathbf{A}\mathbf{Q} - \lambda\mathbf{I} = \begin{bmatrix} \lambda_1 - \lambda & 0 & \dots & 0 \\ 0 & \alpha_{22} - \lambda & \dots & \alpha_{2n} \\ \vdots & \vdots & & \vdots \\ 0 & \alpha_{n2} & \dots & \alpha_{nn} - \lambda \end{bmatrix} \tag{3.2.24}$$

Next, note that

$$\mathbf{Q}^{-1}\mathbf{A}\mathbf{Q} - \lambda\mathbf{I} = \mathbf{Q}^{-1}(\mathbf{A} - \lambda\mathbf{I})\mathbf{Q}$$

Thus,

$$|\mathbf{Q}^{-1}\mathbf{A}\mathbf{Q} - \lambda\mathbf{I}| = |\mathbf{Q}^{-1}||\mathbf{A} - \lambda\mathbf{I}||\mathbf{Q}| = |\mathbf{A} - \lambda\mathbf{I}|$$

since $|\mathbf{Q}^{-1}||\mathbf{Q}| = |\mathbf{Q}^{-1}\mathbf{Q}| = |\mathbf{I}| = 1$. From Eq. 3.2.24,

$$|\mathbf{A} - \lambda\mathbf{I}| = \begin{vmatrix} \lambda_1 - \lambda & 0 & \dots & 0 \\ 0 & \alpha_{22} - \lambda & \dots & \alpha_{2n} \\ \vdots & \vdots & & \vdots \\ 0 & \alpha_{2n} & \dots & \alpha_{nn} - \lambda \end{vmatrix}$$

$$= (\lambda_1 - \lambda)\begin{vmatrix} \alpha_{22} - \lambda & \dots & \alpha_{2n} \\ \vdots & & \vdots \\ \alpha_{2n} & \dots & \alpha_{nn} - \lambda \end{vmatrix} \tag{3.2.25}$$

By hypothesis, the left side of Eq. 3.2.25 has $(\lambda_1 - \lambda)^2$ as a factor, so it follows that the determinant on the right of Eq. 3.2.25 has $(\lambda_1 - \lambda)$ as a factor; hence, it vanishes when $\lambda = \lambda_1$. Thus, the largest submatrix with a nonzero determinant on the right of Eq. 3.2.25 is of dimension $(n-2) \times (n-2)$, for $\lambda = \lambda_1$. But then the largest possible square submatrix in $\mathbf{A} - \lambda_1\mathbf{I}$ with nonzero determinant has dimension $(n-2) \times (n-2)$ and the matrix $\mathbf{Q}^{-1}\mathbf{A}\mathbf{Q} - \lambda\mathbf{I}$ in Eq. 3.2.24 is of rank $n-2$ or less. Thus, a second eigenvector $\mathbf{x}^2$ can be chosen that is orthogonal to $\mathbf{x}^1$ and normalized. There are thus two orthonormal eigenvectors corresponding to λ_1.

If the multiplicity of λ_1 is greater than two, then by taking $\mathbf{Q}$ to be any orthogonal matrix having two eigenvectors $\mathbf{x}^1$ and $\mathbf{x}^2$ corresponding to λ_1 as its first two columns, it is deduced in an analogous way that

$$\mathbf{Q}^{-1}\mathbf{A}\mathbf{Q} - \lambda\mathbf{I} = \begin{bmatrix} \lambda_1 - \lambda & 0 & 0 & \cdots & 0 \\ 0 & \lambda_1 - \lambda & 0 & \cdots & 0 \\ 0 & 0 & \alpha_{33} - \lambda & \cdots & \alpha_{3n} \\ \vdots & \vdots & \vdots & & \vdots \\ 0 & 0 & \alpha_{3n} & \cdots & \alpha_{nn} - \lambda \end{bmatrix}$$

The same argument leads to the conclusion that the matrix $\mathbf{A} - \lambda\mathbf{I}$ is of rank not greater than $n-3$ when $\lambda = \lambda_1$, so at least a three-parameter family of corresponding eigenvectors exists when the multiplicity of λ is at least three.

By induction, if λ_1 is an eigenvalue of a symmetric matrix $\mathbf{A}$ of multiplicity s, then the rank of the matrix $\mathbf{A} - \lambda\mathbf{I}$ is not greater than $n - s$ when $\lambda = \lambda_1$. Thus, at least s linearly independent eigenvectors corresponding to λ_1 exist. However, the rank cannot be less than $n - s$. Otherwise, more than s linearly independent eigenvectors would correspond to λ_1, in which case the total number of linearly independent eigenvectors corresponding to all eigenvalues would exceed the dimension of n-space, which is impossible. Thus, if λ_1 is an eigenvalue of an $n \times n$ symmetric matrix $\mathbf{A}$ of multiplicity s, then the rank of the matrix $\mathbf{A} - \lambda\mathbf{I}$ is exactly $n - s$ when $\lambda = \lambda_1$. That is, there exist exactly s linearly independent eigenvectors corresponding to λ_1. ■

Example 3.2.7

The matrix

$$\mathbf{A} = \begin{bmatrix} 1 & 0 & 0 \\ 0 & 0 & 1 \\ 0 & 1 & 0 \end{bmatrix}$$

has the eigenvalues $\lambda_1 = \lambda_2 = 1$ and $\lambda_3 = -1$. Two linearly independent eigenvectors of $\mathbf{A}$ corresponding to λ_1 and λ_2 are

$$\mathbf{x}^1 = \begin{bmatrix} 1 \\ 0 \\ 0 \end{bmatrix}, \quad \mathbf{x}^2 = \begin{bmatrix} 0 \\ 1 \\ 1 \end{bmatrix}$$

and an eigenvector for $\lambda_3 = -1$ is $\mathbf{x}^3 = [\ 0, 1, -1\ ]^T$. ■

EXERCISES 3.2

1. Write each of the following quadratic forms as $J = \mathbf{x}^T\mathbf{A}\mathbf{x}$, where $\mathbf{A}$ is a symmetric matrix:

   (a) $J = -3x_1^2 + 5x_1x_2 - 2x_2^2$

   (b) $J = 2x_1^2 + 3x_1x_2 - 5x_1x_3 + 7x_2x_3$

   (c) $J = 3x_1^2 + x_1x_2 - 2x_1x_3 + x_2^2 - 4x_2x_3 - 2x_3^2$

2. Construct an orthonormal modal matrix corresponding to the matrix

$$\mathbf{A} = \begin{bmatrix} 1 & 0 & 0 \\ 0 & 3 & -1 \\ 0 & -1 & 3 \end{bmatrix}$$

3. Reduce the quadratic form $J = x_1^2 + 3x_2^2 + 3x_3^2 - 2x_2x_3$ to canonical form.

4. Prove that if $\mathbf{A}$ and $\mathbf{B}$ are n×n orthogonal matrices, then $\mathbf{AB}$ is orthogonal. If $\mathbf{A}$ is also nonsingular, show that $\mathbf{A}^{-1}$ is orthogonal.

5. Calculate the total strain energy in the vertical bars of the structure shown, due to displacement x_1 and rotation x_2, where the horizontal beam is presumed to be rigid.

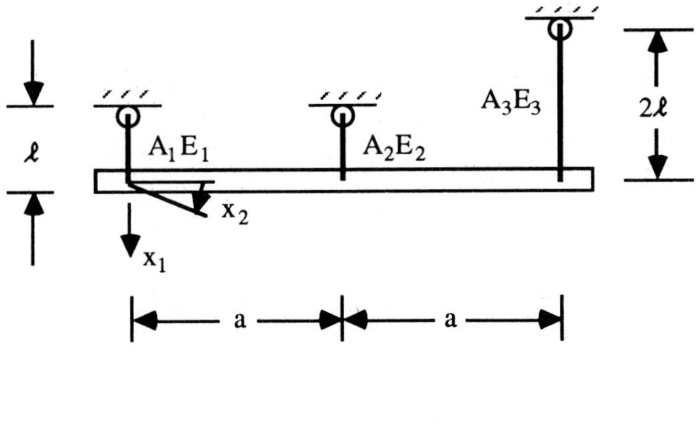

6. Calculate the total strain energy in the bars of the structure shown, due to the symmetric displacement $\mathbf{x} = [\, x_1, x_2, x_3 \,]^T$.

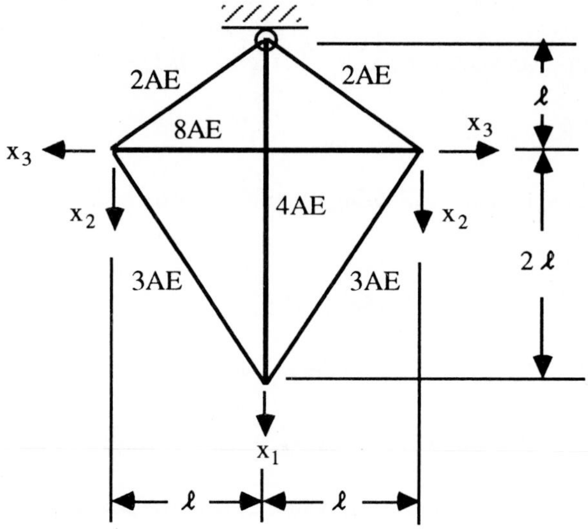

7. If $\mathbf{A}^T = -\mathbf{A}$; i.e., if $\mathbf{A}$ is skew symmetric, show that the quadratic form $\mathbf{x}^T\mathbf{A}\mathbf{x}$ is zero for all $\mathbf{x}$.

8. Find the eigenvalues and eigenvectors of the matrix

$$\mathbf{A} = \begin{bmatrix} 2 & 1 \\ 1 & 2 \end{bmatrix}$$

and find a transformation that reduces $\mathbf{x}^T\mathbf{A}\mathbf{x}$ to a sum of squares.

9. If $\mathbf{A}$ is a symmetric n×n matrix, show that the eigenvalue equation $\mathbf{A}\mathbf{x} = \lambda\mathbf{x}$ can be transformed to $\mathbf{D}\mathbf{y} = \lambda\mathbf{y}$, by the linear transformation $\mathbf{x} = \mathbf{Q}\mathbf{y}$, where $\mathbf{D} = \text{diag}\,(\lambda_1, \ldots, \lambda_n)$ is a diagonal matrix and $\mathbf{Q}$ is the modal matrix of $\mathbf{A}$.

3.3 POSITIVE DEFINITE QUADRATIC FORMS

Positive Definite and Semidefinite Matrices

Definition 3.3.1. If $\mathbf{x}^T\mathbf{A}\mathbf{x} \geq 0$ for all vectors $\mathbf{x}$ in R^n and if $\mathbf{x}^T\mathbf{A}\mathbf{x} = 0$ only if $\mathbf{x} = \mathbf{0}$, then the quadratic form, equivalently the symmetric matrix $\mathbf{A}$, is said to be **positive definite**; i.e.,

$$\begin{aligned} \mathbf{x}^T\mathbf{A}\mathbf{x} &\geq 0, \quad \text{for all } \mathbf{x} \text{ in } R^n \\ \mathbf{x}^T\mathbf{A}\mathbf{x} &= 0, \quad \text{only if } \mathbf{x} = \mathbf{0} \end{aligned} \tag{3.3.1}$$

■

Example 3.3.1

Let

$$J = \mathbf{x}^T\mathbf{A}\mathbf{x} = x_1^2 - 4x_1x_2 + 7x_2^2$$

where

$$\mathbf{A} = \begin{bmatrix} 1 & -2 \\ -2 & 7 \end{bmatrix}$$

This quadratic form can be factored as

$$J = x_1^2 - 4x_1x_2 + 7x_2^2 = (x_1 - 2x_2)^2 + 3x_2^2 \geq 0$$

for all $\mathbf{x}$ in R^2. If $J = 0$, then each of the squared terms must be zero; i.e.,

$$x_1 - 2x_2 = 0, \quad x_2 = 0$$

Thus, $\mathbf{x} = \mathbf{0}$ and the quadratic form is positive definite. ■

Theorem 3.3.1. A symmetric matrix $\mathbf{A}$ is positive definite if and only if all its eigenvalues are positive. ■

To prove Theorem 3.3.1, recall the orthogonal transformation matrix $\mathbf{Q}$ of Eq. 3.2.17, whose columns are orthonormal eigenvectors of $\mathbf{A}$. This transformation defines variables $\mathbf{x}'$ by Eqs. 3.2.13 and 3.2.22,

$$\mathbf{x} = \mathbf{Q}\mathbf{x}'$$

$$\mathbf{x}' = \mathbf{Q}^T\mathbf{x}$$

From Eq. 3.2.21,

$$J \equiv \mathbf{x}^T\mathbf{A}\mathbf{x} = \lambda_i (x_i')^2$$

If $\lambda_i > 0$ for all i, then $J \geq 0$ and $J = 0$ implies all terms in the sum are zero. Hence, $\mathbf{x}' = \mathbf{0}$. But then $\mathbf{x} = \mathbf{Q}\mathbf{x}' = \mathbf{Q}\mathbf{0} = \mathbf{0}$. Thus, the quadratic form is positive definite.

Conversely, if $J > 0$ for all $\mathbf{x} \neq \mathbf{0}$, then $J = \lambda_i (x_i')^2 > 0$ for all $\mathbf{x}' = \mathbf{Q}^T\mathbf{x}$. For proof by contradiction, assume $\lambda_k \leq 0$ for some k. Then let $\mathbf{x}' = [\delta_{ik}]$ and observe that $J = \lambda_i (x_i')^2 = \lambda_i (\delta_{ik})^2 = \lambda_k \leq 0$. This contradicts the hypothesis that J is positive definite; hence, $\lambda_i > 0$, $i = 1, \ldots, n$. ■

A test for positive definiteness can be stated [1] in terms of signs of determinants.

Theorem 3.3.2. Let $\mathbf{S}_i$ be an i×i submatrix of an n×n symmetric matrix $\mathbf{A}$ that is formed from the first i rows and first i columns of $\mathbf{A}$. Submatrix $\mathbf{S}_i$ is called a **principal**

minor of matrix **A**. The matrix **A** is positive definite if and only if the determinants of $\mathbf{S}_1$, $\mathbf{S}_2, \ldots, \mathbf{S}_n$ are all positive. ■

Example 3.3.2

Determine whether the matrix **A** in Example 3.2.5 is positive definite, using both Theorems 3.3.1 and 3.3.2.

First, recall that the matrix

$$\mathbf{A} = \begin{bmatrix} 25 & 0 & 0 \\ 0 & 34 & -12 \\ 0 & -12 & 41 \end{bmatrix}$$

is symmetric and, by the result of Example 3.2.5, $\lambda_1 = \lambda_2 = 25$ and $\lambda_3 = 50$ are all positive. Thus, Theorem 3.3.1 implies **A** is positive definite.

To apply the test of Theorem 3.3.2, determinants of the principal minors of **A** are

$$|\mathbf{S}_1| = a_{11} = 25 > 0$$

$$|\mathbf{S}_2| = \begin{vmatrix} a_{11} & a_{12} \\ a_{21} & a_{22} \end{vmatrix} = \begin{vmatrix} 25 & 0 \\ 0 & 34 \end{vmatrix} = 850 > 0$$

$$|\mathbf{S}_3| = \begin{vmatrix} a_{11} & a_{12} & a_{13} \\ a_{21} & a_{22} & a_{23} \\ a_{31} & a_{32} & a_{33} \end{vmatrix} = \begin{vmatrix} 25 & 0 & 0 \\ 0 & 34 & -12 \\ 0 & -12 & 41 \end{vmatrix} = 31250 > 0$$

Therefore, **A** is positive definite. ■

Definition 3.3.2. A quadratic form with the symmetric matrix **A** is said to be **positive semidefinite** when it takes on only nonnegative values for all values of the variables, but vanishes for some nonzero value of the variable; i.e.,

$$\begin{aligned} \mathbf{x}^T\mathbf{A}\mathbf{x} &\geq 0, \text{ for all } \mathbf{x} \text{ in } R^n \\ \mathbf{x}^T\mathbf{A}\mathbf{x} &= 0, \text{ for some } \mathbf{x} \neq \mathbf{0} \end{aligned} \qquad (3.3.2)$$

■

The preceding arguments lead to the conclusion that a positive semidefinite matrix is singular and possesses only nonnegative eigenvalues, at least one of which is zero.

Example 3.3.3

The strain energy in the three spring system of Exercise 3 in Section 3.1 was shown in Example 3.2.3 to be $U = (1/2)\,\mathbf{x}^T\mathbf{K}\mathbf{x}$, where

$$\mathbf{K} = \begin{bmatrix} k_1 + k_2 & -k_2 \\ -k_2 & k_2 + k_3 \end{bmatrix}$$

Presume for this example that $k_i > 0$, $i = 1, 2, 3$. Using the test of Theorem 3.3.2,

$$|\mathbf{S}_1| = k_1 + k_2 > 0$$

$$|\mathbf{S}_2| = (k_1 + k_2)(k_2 + k_3) - k_2^2$$

$$= k_1k_2 + k_1k_3 + k_2k_3 > 0$$

Thus, **K** is positive definite. Physically, this says that in order to displace the system from its undeformed state; i.e., deforming it requires that a positive amount of work (work done equals strain energy) must be done.

Note that the stiffness matrix of the strain energy quadratic form of Eq. 3.2.9, for the three-bar truss of Example 3.2.4, is positive definite. Since it is a diagonal matrix with positive elements, its eigenvalues are those elements and are positive. ■

Example 3.3.4

It might be expected that the beam stiffness matrix $\mathbf{K}^b$ of Eq. 3.2.12 would be positive definite. In fact, even without examining the matrix, note from Eq. 3.2.11 that since $(w'')^2 \geq 0$, $U \geq 0$. Thus, $\mathbf{K}^b$ is either positive semidefinite or positive definite.

Applying the test of Theorem 3.3.2,

$$|\mathbf{S}_1| = 12EI/\ell^3 > 0$$

$$|\mathbf{S}_2| = 12EI/\ell > 0$$

$$|\mathbf{S}_3| = 0$$

$$|\mathbf{S}_4| = 0$$

Thus, $\mathbf{K}^b$ is not positive definite, but it is positive semidefinite, so there is a vector $\mathbf{x} \neq \mathbf{0}$ such that the strain energy is zero.

Does this say that it is possible to deform the beam without doing a positive amount of work? To see that the answer is no, consider the vectors

$$\mathbf{x}^1 = [\,1, 0, 1, 0\,]^T$$

$$\mathbf{x}^2 = [\,-\ell/2, 1, \ell/2, 1\,]^T$$

Referring to Fig. 3.2.4, note that $\mathbf{x}^1$ defines an undeformed (rigid) translation of the beam and $\mathbf{x}^2$ defines an undeformed (rigid) rotation (within the limits of small displacement linear theory). In fact, direct evaluations show that

$$\mathbf{x}^{1T}\mathbf{K}^b\mathbf{x}^1 = \mathbf{x}^{2T}\mathbf{K}^b\mathbf{x}^2 = 0$$

This result agrees with the physical wisdom that the beam remains undeformed in rigid body displacement, which requires no work.

As a final check with reality, let the beam be cantilevered at its left end; i.e., $x_1 = x_2 = 0$, as shown in Fig. 3.3.1. Denoting $y_1 = x_3$ and $y_2 = x_4$ and carrying out the expansion of Eq. 3.2.12 with $\mathbf{x} = [\,0, 0, y_1, y_2\,]^T$ and $\mathbf{y} = [\,y_1, y_2\,]^T$, for the cantilever beam,

$$U = \frac{1}{2}\mathbf{y}^T \frac{EI}{\ell^3}\begin{bmatrix} 12 & -6\ell \\ -6\ell & 4\ell^2 \end{bmatrix}\mathbf{y} \tag{3.3.3}$$

Direct application of the test of Theorem 3.3.2 shows that this quadratic form is positive definite.

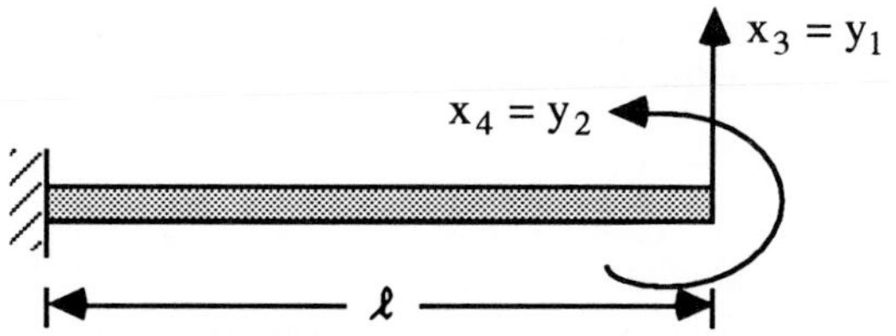

Figure 3.3.1 Cantilever Beam

The reduction from displacements $\mathbf{x}$ in R^4 to $\mathbf{y}$ in R^2 for the cantilever beam is typical of the general concept of defining **boundary conditions** in specific applications. A vector $\mathbf{y} \neq \mathbf{0}$ in R^2 for the cantilever beam defines a deformed beam, consistent with the boundary conditions at the left end. Thus, the strain energy, which is the work required to deform the cantilever beam, must be positive. ■

Energy Scalar Product and Norm

In Section 2.4, the scalar product

$$(\,\mathbf{x}, \mathbf{y}\,) = \mathbf{x}^T\mathbf{y} \tag{3.3.4}$$

was defined on R^n and satisfies all the properties of Definition 2.4.1; i.e., Eq. 2.4.2. Example 2.4.1, however, demonstrated that a different scalar product could be defined on R^2. In fact, many different scalar products are possible.

Theorem 3.3.3. Let $\mathbf{A}$ be an n×n symmetric, positive definite matrix. Then,

$$(\,\mathbf{x}, \mathbf{y}\,)_{\mathbf{A}} \equiv \mathbf{x}^T\mathbf{A}\,\mathbf{y} \tag{3.3.5}$$

is a scalar product on R^n, called the **A-scalar product.** ■

To prove Theorem 3.3.3, it must be shown that $(\bullet, \bullet)_{\mathbf{A}}$ satisfies Eq. 2.4.2. Proceeding step by step, since the transpose of a scalar is the scalar and **A** is symmetric,

$$(\mathbf{x}, \mathbf{y})_{\mathbf{A}} \equiv \mathbf{x}^T\mathbf{A}\mathbf{y} = (\mathbf{x}^T\mathbf{A}\mathbf{y})^T = \mathbf{y}^T\mathbf{A}\mathbf{x} = (\mathbf{y}, \mathbf{x})_{\mathbf{A}}$$

Using the distributive properties of matrix multiplication,

$$(\mathbf{x}, \mathbf{y} + \mathbf{z})_{\mathbf{A}} \equiv \mathbf{x}^T\mathbf{A} (\mathbf{y} + \mathbf{z}) = \mathbf{x}^T\mathbf{A}\mathbf{y} + \mathbf{x}^T\mathbf{A}\mathbf{z} = (\mathbf{x}, \mathbf{y})_{\mathbf{A}} + (\mathbf{x}, \mathbf{z})_{\mathbf{A}}$$

$$(\alpha\mathbf{x}, \mathbf{y})_{\mathbf{A}} \equiv (\alpha\mathbf{x})^T \mathbf{A}\mathbf{y} = \alpha\mathbf{x}^T\mathbf{A}\mathbf{y} = \alpha (\mathbf{x}, \mathbf{y})_{\mathbf{A}}$$

Since **A** is positive definite,

$$(\mathbf{x}, \mathbf{x})_{\mathbf{A}} \equiv \mathbf{x}^T\mathbf{A}\mathbf{x} \geq 0$$

$$(\mathbf{x}, \mathbf{x})_{\mathbf{A}} \equiv \mathbf{x}^T\mathbf{A}\mathbf{x} = 0, \text{ if and only if } \mathbf{x} = \mathbf{0}$$

Thus, $(\bullet, \bullet)_{\mathbf{A}}$ is a scalar product. ■

Example 3.3.5

The scalar product introduced for R^2 in Example 2.4.1 is a quadratic form

$$\begin{aligned} [\mathbf{x}, \mathbf{y}] &\equiv x_1y_1 - x_1y_2 - x_2y_1 + 2x_2y_2 \\ &= \mathbf{x}^T \begin{bmatrix} 1 & -1 \\ -1 & 2 \end{bmatrix} \mathbf{y} \\ &\equiv (\mathbf{x}, \mathbf{y})_{\mathbf{A}} \end{aligned}$$

where

$$\mathbf{A} = \begin{bmatrix} 1 & -1 \\ -1 & 2 \end{bmatrix}$$

is a positive definite matrix. ■

Recall that strain energies of the spring system, three-bar truss, and cantilever beam in Examples 3.3.3 and 3.3.4 were positive definite quadratic forms,

$$U = (1/2) (\mathbf{x}^T\mathbf{K}\mathbf{x}) \tag{3.3.6}$$

The associated scalar product

$$(\mathbf{x}, \mathbf{y})_{\mathbf{K}} \equiv (1/2) (\mathbf{x}^T\mathbf{K}\mathbf{y}) \tag{3.3.7}$$

is called the **energy scalar product** and the norm it defines, by Eq. 2.4.4, is

$$\| \mathbf{x} \|_K \equiv \sqrt{(\mathbf{x}, \mathbf{x})_K} = \sqrt{\frac{1}{2}\mathbf{x}^T\mathbf{K}\mathbf{x}} \tag{3.3.8}$$

which is called the **energy norm.** These concepts of energy scalar product and energy norm can be used with conceptual and computational benefit in applications. They permit what might otherwise be seen as abstract mathematical methods to be viewed as intuitively clear and natural analytical methods in engineering. These concepts are developed in detail in Chapters 9 through 12 of this text.

EXERCISES 3.3

1. Determine whether the quadratic form

$$J = x_1^2 + 2x_2^2 + x_3^2 - 2x_1x_3 + 2x_2x_3$$

is positive definite.

2. Determine which of the following matrices is positive definite:

$$\mathbf{A} = \begin{bmatrix} 2 & 0 & 0 \\ 0 & 3 & -1 \\ 0 & -1 & 1 \end{bmatrix}$$

$$\mathbf{B} = \begin{bmatrix} 0 & 0 & 1 \\ 0 & 1 & 0 \\ 1 & 0 & 0 \end{bmatrix}$$

3. If **A** is a positive definite matrix, show that the matrix $\mathbf{B}^{-1}\mathbf{A}\mathbf{B}$ is positive definite, for any orthogonal matrix **B**. [Hint: Show that the eigenvalues of **A** and $\mathbf{B}^{-1}\mathbf{A}\mathbf{B}$ are the same.]

4. Determine whether the following quadratic form is positive definite:

$$J = \mathbf{x}^T \begin{bmatrix} 2 & 1 & 1 \\ 1 & 4 & 2 \\ 1 & 2 & 4 \end{bmatrix} \mathbf{x}$$

5. Find the eigenvalues and corresponding eigenvectors of the matrix

$$\mathbf{A} = \begin{bmatrix} 0 & 1 & 0 \\ 1 & 1 & 1 \\ 0 & 1 & 0 \end{bmatrix}$$

Is the matrix positive definite? Why?

6. Show that if the columns $\mathbf{b}^1, \ldots, \mathbf{b}^n$ of the matrix $\mathbf{B} = [\,\mathbf{b}^1, \ldots, \mathbf{b}^n\,]$ are linearly independent, then the matrix $\mathbf{B}^T\mathbf{B}$ is positive definite. [Hint: Note that $\mathbf{Bx} = \mathbf{b}^i x_i$ and that $\mathbf{x}^T(\mathbf{B}^T\mathbf{B})\mathbf{x} = (\mathbf{Bx})^T(\mathbf{Bx})$.]

7. Prove that if the n×n symmetric matrix $\mathbf{A}$ is positive definite, then the matrix $\mathbf{Q}^T\mathbf{AQ}$ is positive definite if and only if the n×m matrix $\mathbf{Q}$ has full column rank. [Hint: It may be helpful to define an n×m matrix $\mathbf{Q} = [\,\mathbf{x}^1, \mathbf{x}^2, \ldots, \mathbf{x}^m\,]$ and note that $\mathbf{Qx} = \mathbf{x}^i x_i$ for $\mathbf{x}$ in R^m. Since $\mathbf{Qx} = \mathbf{x}^i x_i$, $\mathbf{Q}$ has full column rank if and only if $\mathbf{Qx} = \mathbf{0}$ implies $\mathbf{x} = \mathbf{0}$.]

8. Show that if a symmetric matrix $\mathbf{A}$ is positive definite, so are $\mathbf{A}^2$ and $\mathbf{A}^{-1}$.

3.4 GENERALIZED EIGENVALUE PROBLEMS

Example 3.4.1

Consider vibration of the rigid bar in Fig. 3.4.1, with mass m and moment of inertia $I = m\ell^2/12$, where the effect of gravity is ignored.

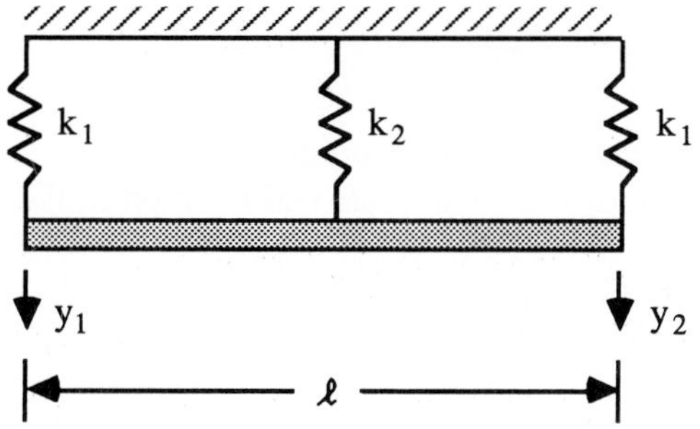

Figure 3.4.1 Vibrating Bar

The equations of motion are obtained, by summing forces in the vertical direction and moments about the center of the bar, as

$$\frac{m(\ddot{y}_1 + \ddot{y}_2)}{2} = -k_1(y_1 + y_2) - \frac{k_2(y_1 + y_2)}{2}$$

$$\frac{m\ell^2}{12}\,\frac{(\ddot{y}_1 - \ddot{y}_2)}{\ell} = \frac{k_1(y_2 - y_1)\,\ell}{2}$$

Equivalent equations can be obtained by multiplying the first equation by 6 and the second equation by 12, adding the second equation to the first, and then subtracting the second from the first. In matrix form, the following equations are obtained:

$$\begin{bmatrix} 4k_1 + k_2 & k_2 \\ k_2 & 4k_1 + k_2 \end{bmatrix} \begin{bmatrix} y_1 \\ y_2 \end{bmatrix} = -\frac{2m}{3} \begin{bmatrix} 2 & 1 \\ 1 & 2 \end{bmatrix} \begin{bmatrix} \ddot{y}_1 \\ \ddot{y}_2 \end{bmatrix}$$

If harmonic motion is to occur, which is the definition of natural vibration, then $y_i(t) = x_i \sin \omega t$, $i = 1, 2$, where the x_i are constants. Substituting this form of solution in the equations of motion, with $\mathbf{x} = [\, x_1, x_2 \,]^T$, and dividing both sides by $\sin \omega t$,

$$\mathbf{Ax} \equiv \begin{bmatrix} 4k_1 + k_2 & k_2 \\ k_2 & 4k_1 + k_2 \end{bmatrix} \mathbf{x} = \lambda \begin{bmatrix} 2 & 1 \\ 1 & 2 \end{bmatrix} \mathbf{x} \equiv \lambda \mathbf{Bx}$$

where $\lambda = 2m\omega^2/3$. Note that this is not quite the standard eigenvalue problem, since $\mathbf{B} \neq \mathbf{I}$. ■

Generalized Eigenvalue Equations

In many applications, a **generalized eigenvalue problem** of the form

$$\mathbf{Ax} = \lambda \mathbf{Bx} \tag{3.4.1}$$

is encountered, where $\mathbf{A}$ and $\mathbf{B}$ are n×n matrices. If Eq. 3.4.1, or equivalently $(\mathbf{A} - \lambda \mathbf{B})\,\mathbf{x} = \mathbf{0}$, is to have a nontrivial solution $\mathbf{x}$, then λ must satisfy the **characteristic equation**

$$|\mathbf{A} - \lambda \mathbf{B}| = 0 \tag{3.4.2}$$

Example 3.4.2

The characteristic equation for the generalized eigenvalue problem of Example 3.4.1 is

$$|\mathbf{A} - \lambda \mathbf{B}| = \begin{vmatrix} 4k_1 + k_2 - 2\lambda & k_2 - \lambda \\ k_2 - \lambda & 4k_1 + k_2 - 2\lambda \end{vmatrix}$$

$$= 3\lambda^2 - (\, 16k_1 + 2k_2 \,)\, \lambda + 16k_1^2 + 8k_1k_2 = 0$$

The solutions are

$$\lambda_1 = 4k_1, \quad \lambda_2 = (\, 4k_1 + 2k_2 \,)/3$$

Recalling from Example 3.4.1 that the natural frequency ω is related to λ by $\lambda = 2m\omega^2/3$,

$$\omega_1 = \sqrt{\frac{6k_1}{m}}, \quad \omega_2 = \sqrt{\frac{2k_1 + k_2}{m}}$$

To find the associated eigenvectors, first substitute λ_1 in Eq. 3.4.1 and solve for $\mathbf{x}^1$; i.e.,

$$\begin{bmatrix} k_2 - 4k_1 & k_2 - 4k_1 \\ k_2 - 4k_1 & k_2 - 4k_1 \end{bmatrix} \mathbf{x}^1 = \mathbf{0}$$

so

$$\mathbf{x}^1 = \begin{bmatrix} 1 \\ -1 \end{bmatrix}$$

is an eigenvector corresponding to λ_1, or ω_1. Similarly, an eigenvector associated with λ_2, or ω_2, is

$$\mathbf{x}^2 = \begin{bmatrix} 1 \\ 1 \end{bmatrix}$$

■

B-Orthogonality of Eigenvectors

Let $\lambda_\alpha \neq \lambda_\beta$ be distinct eigenvalues of Eq. 3.4.1 with symmetric matrices **A** and **B**, corresponding to eigenvectors $\mathbf{x}^\alpha$ and $\mathbf{x}^\beta$; i.e.,

$$\begin{aligned} \mathbf{A}\mathbf{x}^\alpha &= \lambda_\alpha \mathbf{B}\mathbf{x}^\alpha \\ \mathbf{A}\mathbf{x}^\beta &= \lambda_\beta \mathbf{B}\mathbf{x}^\beta \end{aligned} \tag{3.4.3}$$

Forming the matrix products

$$\begin{aligned} (\mathbf{A}\mathbf{x}^\alpha)^T \mathbf{x}^\beta &= \lambda_\alpha (\mathbf{B}\mathbf{x}^\alpha)^T \mathbf{x}^\beta \\ \mathbf{x}^{\alpha T}\mathbf{A}\mathbf{x}^\beta &= \lambda_\beta \mathbf{x}^{\alpha T}\mathbf{B}\mathbf{x}^\beta \end{aligned}$$

and making use of symmetry of **A** and **B**,

$$\begin{aligned} \mathbf{x}^{\alpha T}\mathbf{A}\mathbf{x}^\beta &= \lambda_\alpha \mathbf{x}^{\alpha T}\mathbf{B}\mathbf{x}^\beta \\ \mathbf{x}^{\alpha T}\mathbf{A}\mathbf{x}^\beta &= \lambda_\beta \mathbf{x}^{\alpha T}\mathbf{B}\mathbf{x}^\beta \end{aligned} \tag{3.4.4}$$

Subtracting the first of Eq. 3.4.4 from the second,

$$(\lambda_\beta - \lambda_\alpha)\, \mathbf{x}^{\alpha T}\mathbf{B}\mathbf{x}^\beta = 0 \tag{3.4.5}$$

Since $\lambda_\alpha \neq \lambda_\beta$, $\mathbf{x}^{\alpha T}\mathbf{B}\mathbf{x}^\beta = 0$. This establishes the following theorem.

Theorem 3.4.1. Eigenvectors $\mathbf{x}^i$ and $\mathbf{x}^j$ that correspond to two distinct eigenvalues $\lambda_i \neq \lambda_j$ of the generalized eigenvalue problem $\mathbf{Ax} = \lambda\mathbf{Bx}$, where $\mathbf{A}$ and $\mathbf{B}$ are symmetric, are **orthogonal with respect to B**, or are **B-orthogonal**; i.e.,

$$\mathbf{x}^{iT}\mathbf{B}\mathbf{x}^j = 0 \tag{3.4.6}$$

■

A direct calculation verifies that the eigenvectors of Example 3.4.2 are **B**-orthogonal. If **B** is positive definite, the left side of Eq. 3.4.6 is the **B**-scalar product of $\mathbf{x}^i$ and $\mathbf{x}^j$, as defined in Eq. 3.3.5; i.e.,

$$(\mathbf{x}, \mathbf{x})_{\mathbf{B}} \equiv \mathbf{x}^T\mathbf{B}\mathbf{x} \tag{3.4.7}$$

Note that when Eq. 3.4.6 is satisfied, the eigenvectors $\mathbf{x}^i$ and $\mathbf{x}^j$ are also orthogonal relative to matrix **A**.

Example 3.4.3

For the generalized eigenvalue problem

$$\begin{bmatrix} 1 & 0 \\ 0 & -1 \end{bmatrix} \mathbf{x} = \lambda \begin{bmatrix} 0 & 1 \\ 1 & 0 \end{bmatrix} \mathbf{x}$$

the characteristic equation is

$$\begin{vmatrix} 1 & -\lambda \\ -\lambda & -1 \end{vmatrix} = -1 - \lambda^2 = 0$$

and the eigenvalues are complex; i.e.,

$$\lambda_1 = i, \quad \lambda_2 = -i$$

■

Example 3.4.3 shows that the second result of Theorem 3.1.1 does not extend to the generalized eigenvalue problem. Even though **A** and **B** are symmetric, the eigenvalues may be complex. The following theorem, however, provides a result that holds in many engineering applications.

Theorem 3.4.2. If matrices **A** and **B** are symmetric and if **B** is positive definite, the eigenvalues of Eq. 3.4.1 are real. ■

To prove Theorem 3.4.2, repeat the manipulations that lead to Eq. 3.1.12, but with the generalized eigenvalue problem of Eq. 3.4.1, to obtain

$$(\lambda - \lambda^*)\,\mathbf{x}^{*T}\mathbf{B}\mathbf{x} = 0$$

Writing the possibly complex eigenvector $\mathbf{x} \neq \mathbf{0}$ as $\mathbf{x} = \mathbf{a} + i\mathbf{b}$, direct expansion yields

$$\mathbf{x}^{*T}\mathbf{B}\mathbf{x} = (\,\mathbf{a} - i\mathbf{b}\,)^T\,\mathbf{B}\,(\,\mathbf{a} + i\mathbf{b}\,)$$

$$= \mathbf{a}^T\mathbf{B}\mathbf{a} - i\mathbf{b}^T\mathbf{B}\mathbf{a} + i\mathbf{a}^T\mathbf{B}\mathbf{b} - (i)^2\,\mathbf{b}^T\mathbf{B}\mathbf{b}$$

$$= \mathbf{a}^T\mathbf{B}\mathbf{a} + \mathbf{b}^T\mathbf{B}\mathbf{b} > 0$$

Thus, $\lambda = \lambda^*$ and λ is real. ■

By methods completely analogous to those of Section 2.4, the set of eigenvectors can be orthogonalized and normalized, relative to **B**. If **B** is positive definite, the characteristic equation of Eq. 3.4.2 is of degree n [1]. Hence, if **B** is positive definite, there are always n mutually **B-orthonormal eigenvectors** $\mathbf{x}^1, \mathbf{x}^2, \ldots, \mathbf{x}^n$.

A **B-orthogonal modal matrix Q** associated with Eq. 3.4.1 may be defined as

$$\mathbf{Q} = [\,\mathbf{x}^1, \mathbf{x}^2, \ldots, \mathbf{x}^n\,]$$

where

$$\mathbf{Q}^T\mathbf{B}\mathbf{Q} = \mathbf{I}$$

It follows that

$$\mathbf{A}\mathbf{Q} = \mathbf{B}\mathbf{Q}\begin{bmatrix} \lambda_1 & 0 & \ldots & 0 \\ 0 & \lambda_2 & \ldots & 0 \\ \vdots & \vdots & & \vdots \\ 0 & 0 & \ldots & \lambda_3 \end{bmatrix} \equiv \mathbf{B}\mathbf{Q}\mathbf{D}$$

and

$$\mathbf{Q}^T\mathbf{A}\mathbf{Q} = \mathbf{Q}^T\mathbf{B}\mathbf{Q}\mathbf{D} = \mathbf{I}\mathbf{D} = \mathbf{D}$$

Diagonalization of a Pair of Quadratic Forms

With the change of variables

$$\mathbf{x} = \mathbf{Q}\mathbf{y} \tag{3.4.8}$$

where **Q** is a **B**-orthogonal modal matrix associated with Eq. 3.4.1, **A** is symmetric, and **B** is positive definite, the quadratic forms

$$\begin{aligned} J &= \mathbf{x}^T\mathbf{A}\mathbf{x} \\ K &= \mathbf{x}^T\mathbf{B}\mathbf{x} \end{aligned} \tag{3.4.9}$$

are reduced simultaneously to the canonical forms

$$
\begin{aligned}
J &= \mathbf{y}^T\mathbf{D}\mathbf{y} = \lambda_1 y_1^2 + \lambda_2 y_2^2 + \ldots + \lambda_n y_n^2 \\
K &= \mathbf{y}^T\mathbf{y} = y_1^2 + y_2^2 + y_3^2 + \ldots + y_n^2
\end{aligned}
\tag{3.4.10}
$$

To verify that this is true, note that

$$K = \mathbf{y}^T\mathbf{Q}^T\mathbf{B}\mathbf{Q}\mathbf{y} = \mathbf{y}^T\mathbf{y}$$

and

$$J = \mathbf{y}^T\mathbf{Q}^T\mathbf{A}\mathbf{Q}\mathbf{y} = \mathbf{y}^T\mathbf{Q}^T\mathbf{B}\mathbf{Q}\mathbf{D}\mathbf{y} = \mathbf{y}^T\mathbf{D}\mathbf{y}$$

From the identity $\mathbf{Q}^T\mathbf{B}\mathbf{Q} = \mathbf{I}$, it follows that

$$|\mathbf{Q}| = \pm\frac{1}{\sqrt{|\mathbf{B}|}}$$

Further, this identity shows that $\mathbf{Q}^{-1} = \mathbf{Q}^T\mathbf{B}$. The solution of Eq. 3.4.8 may, therefore, be conveniently written as

$$\mathbf{y} = \mathbf{Q}^T\mathbf{B}\mathbf{x} \tag{3.4.11}$$

Example 3.4.4

Reduce the following quadratic forms simultaneously to canonical form, using Eqs. 3.4.8 to 3.4.11:

$$
\begin{aligned}
J &= \mathbf{x}^T\mathbf{A}\mathbf{x} = (x_1 - 2x_2 + x_3)^2 \\
K &= \mathbf{x}^T\mathbf{B}\mathbf{x} = x_1^2 + 2x_2^2 + x_3^2
\end{aligned}
$$

where

$$\mathbf{A} = \begin{bmatrix} 1 & -2 & 1 \\ -2 & 4 & -2 \\ 1 & -2 & 1 \end{bmatrix}, \quad \mathbf{B} = \begin{bmatrix} 1 & 0 & 0 \\ 0 & 2 & 0 \\ 0 & 0 & 1 \end{bmatrix}$$

The characteristic equation of Eq. 3.4.2 is

$$|\mathbf{A} - \lambda\mathbf{B}| = \begin{vmatrix} 1-\lambda & -2 & 1 \\ -2 & 4-2\lambda & -2 \\ 1 & -2 & 1-\lambda \end{vmatrix} = \lambda^2(\lambda - 4) = 0$$

Thus, the eigenvalues are

$$\lambda_1 = \lambda_2 = 0, \quad \lambda_3 = 4$$

and associated **B**-orthonormal eigenvectors are

$$\mathbf{x}^1 = \frac{\sqrt{2}}{2}[\,1, 0, -1\,]^T$$

$$\mathbf{x}^2 = \frac{1}{2}[\,1, 1, 1\,]^T$$

$$\mathbf{x}^3 = \frac{1}{2}[\,1, -1, 1\,]^T$$

The **B**-orthogonal modal matrix is thus

$$\mathbf{Q} = \begin{bmatrix} \frac{\sqrt{2}}{2} & \frac{1}{2} & \frac{1}{2} \\ 0 & \frac{1}{2} & -\frac{1}{2} \\ -\frac{\sqrt{2}}{2} & \frac{1}{2} & \frac{1}{2} \end{bmatrix}$$

The transformation from **x** to **y** of Eq. 3.4.11 is then

$$\mathbf{y} = \mathbf{Q}^T\mathbf{B}\mathbf{x} = \begin{bmatrix} \frac{\sqrt{2}}{2} & 0 & -\frac{\sqrt{2}}{2} \\ \frac{1}{2} & \frac{1}{2} & \frac{1}{2} \\ \frac{1}{2} & -\frac{1}{2} & \frac{1}{2} \end{bmatrix} \begin{bmatrix} 1 & 0 & 0 \\ 0 & 2 & 0 \\ 0 & 0 & 1 \end{bmatrix} \begin{bmatrix} x_1 \\ x_2 \\ x_3 \end{bmatrix}$$

$$= \left[\, \frac{\sqrt{2}}{2}(x_1 - x_3), \frac{1}{2}(x_1 + 2x_2 + x_3), \frac{1}{2}(x_1 - 2x_2 + x_3) \,\right]^T$$

and the canonical forms of J and K are

$$J = 4y_3^2 = (x_1 - 2x_2 + x_3)^2$$

$$K = y_1^2 + y_2^2 + y_3^2 = x_1^2 + 2x_2^2 + x_3^2$$ ■

Theorem 3.4.3. If the matrices **A** and **B** are symmetric and positive definite, all the eigenvalues of Eq. 3.4.1 are positive. ■

Theorem 3.4.3 may be proved by noting that for $\mathbf{x}^\alpha \neq \mathbf{0}$,

$$\mathbf{A}\mathbf{x}^\alpha = \lambda_\alpha \mathbf{B}\mathbf{x}^\alpha$$

Multiplying both sides on the left by $\mathbf{x}^{\alpha T}$,

$$\mathbf{x}^{\alpha T}\mathbf{A}\mathbf{x}^{\alpha} = \lambda_{\alpha}\mathbf{x}^{\alpha T}\mathbf{B}\mathbf{x}^{\alpha}$$

Since both $\mathbf{x}^{\alpha T}\mathbf{A}\mathbf{x}^{\alpha}$ and $\mathbf{x}^{\alpha T}\mathbf{B}\mathbf{x}^{\alpha}$ are positive if **A** and **B** are positive definite, the same is true of λ_{α}. ■

If **A** and **B** are symmetric and **B** is positive definite, then Theorem 3.2.2 can be extended to the generalized eigenvalue problem of Eq. 3.4.1. In this case, there are n **B**-orthonormal eigenvectors $\{ \mathbf{x}^i \}$, which are linearly independent. Since the eigenvectors $\mathbf{x}^1, \mathbf{x}^2, \ldots, \mathbf{x}^n$ are linearly independent they form a basis for R^n. Thus, any vector **y** in R^n can be expressed as a linear combination of these vectors; i.e.,

$$\mathbf{y} = a_1\mathbf{x}^1 + a_2\mathbf{x}^2 + \ldots + a_n\mathbf{x}^n = a_i\mathbf{x}^i \tag{3.4.12}$$

In order to evaluate a coefficient a_r in Eq. 3.4.12, form the **B**-scalar product of $\mathbf{x}^r$ with both sides of Eq. 3.4.12 and use Eq. 3.4.6 and the fact that the $\mathbf{x}^i$ are **B**-normalized to obtain

$$a_r = (\mathbf{x}^r, \mathbf{y})_{\mathbf{B}} \equiv \mathbf{x}^{rT}\mathbf{B}\mathbf{y}, \quad r = 1, 2, \ldots, n \tag{3.4.13}$$

A case that is frequently encountered in practice (e.g., see Example 3.4.4) is that **B** is a diagonal matrix,

$$\mathbf{B} = \begin{bmatrix} b_1 & 0 & \ldots & 0 \\ 0 & b_2 & \ldots & 0 \\ \vdots & \vdots & & \vdots \\ 0 & 0 & \ldots & b_n \end{bmatrix} \tag{3.4.14}$$

where $b_i > 0$, $i = 1, \ldots, n$. Thus, Eq. 3.4.1 takes the special form

$$\begin{aligned} a_{11}x_1 + a_{12}x_2 + \ldots + a_{1n}x_n &= \lambda b_1 x_1 \\ \ldots\ldots\ldots\ldots\ldots\ldots\ldots\ldots & \\ a_{n1}x_1 + a_{n2}x_2 + \ldots + a_{nn}x_n &= \lambda b_n x_n \end{aligned} \tag{3.4.15}$$

where $a_{ji} = a_{ij}$. The **B**-scalar product $(\mathbf{x}, \mathbf{y})_{\mathbf{B}}$ then takes the form

$$(\mathbf{x}, \mathbf{y})_{\mathbf{B}} = b_1x_1y_1 + b_2x_2y_2 + \ldots + b_nx_ny_n$$

Second-order Linear Differential Equations of Vibration

Matrix theory provides a valuable conceptual and computational tool for solution of systems of **second-order linear differential equations** that arise in a variety of vibrating systems. Consider the linear equations of motion

$$\mathbf{M}\ddot{\mathbf{y}} + \mathbf{K}\mathbf{y} = \mathbf{F} \tag{3.4.16}$$

with a positive definite n×n stiffness matrix **K** and a positive definite n×n mass matrix **M**. There are n **M**-orthonormal eigenvectors $\mathbf{x}^i$ of the generalized eigenvalue problem*

$$\mathbf{K}\mathbf{x} = \lambda\mathbf{M}\mathbf{x} \tag{3.4.17}$$

with the associated eigenvalues $\lambda_i > 0$. The **M**-orthogonal modal matrix **Q** is formed with columns $\mathbf{x}^i$. Then, by Eq. 3.4.8, the transformation

$$\mathbf{y} = \mathbf{Q}\mathbf{z} \tag{3.4.18}$$

reduces Eq. 3.4.16 to

$$\mathbf{M}\mathbf{Q}\ddot{\mathbf{z}} + \mathbf{K}\mathbf{Q}\mathbf{z} = \mathbf{F}$$

Multiplying both sides on the left by $\mathbf{Q}^T$,

$$\mathbf{Q}^T\mathbf{M}\mathbf{Q}\ddot{\mathbf{z}} + \mathbf{Q}^T\mathbf{K}\mathbf{Q}\mathbf{z} = \mathbf{Q}^T\mathbf{F}$$

which can be rewritten as

$$\ddot{\mathbf{z}} + \mathbf{D}\mathbf{z} = \mathbf{Q}^T\mathbf{F} \equiv \bar{\mathbf{F}}$$

where $\mathbf{D} = \mathbf{diag}\,(\lambda_1, \lambda_2, \ldots, \lambda_n)$, or in scalar form,

$$\ddot{z}_\alpha + \lambda_\alpha z_\alpha = \bar{F}_\alpha, \quad \alpha = 1, 2, \ldots, n \tag{3.4.19}$$

which are uncoupled equations of motion in the variables z_α.

The uncoupled dynamic equations of Eq. 3.4.19 can be solved in closed form as

$$z_\alpha(t) = A_\alpha \sin\sqrt{\lambda_\alpha}\,t + B_\alpha \cos\sqrt{\lambda_\alpha}\,t$$

$$+ \frac{1}{\sqrt{\lambda_\alpha}}\int_0^t \bar{F}_\alpha(\tau)\sin\sqrt{\lambda_\alpha}\,(t-\tau)\,d\tau, \quad \alpha = 1, \ldots, n \tag{3.4.20}$$

---

* If $\mathbf{F} = \mathbf{0}$ in Eq. 3.4.16 and if the dynamic response of the system is harmonic (i.e., it is in **natural vibration**) $\mathbf{y}(t) = \mathbf{x}\sin\omega t$ and Eq. 3.4.16 becomes $(-\omega^2\mathbf{M}\mathbf{x} + \mathbf{K}\mathbf{x})\sin\omega t = \mathbf{0}$, for all t. Since $\sin\omega t$ is not always zero if $\omega \neq 0$, with $\lambda = \omega^2$, this yields Eq. 3.4.17. Thus, Eq. 3.4.17 governs natural vibration of the unforced ($\mathbf{F} = \mathbf{0}$) dynamic system.

The solution $\mathbf{y}(t)$ of Eq. 3.4.16 is then computed from Eq. 3.4.18. While this technique requires considerable computation, it provides a direct method for solving rather complex problems. In the case of realistic engineering problems, the dimension n can be of the order of hundreds or even thousands. A high-speed computer is thus required to construct the solution.

Example 3.4.5

Consider vibration of the bar in Example 3.4.1, with external excitation forces $F_1(t)$ and $F_2(t)$, as shown in Fig. 3.4.2.

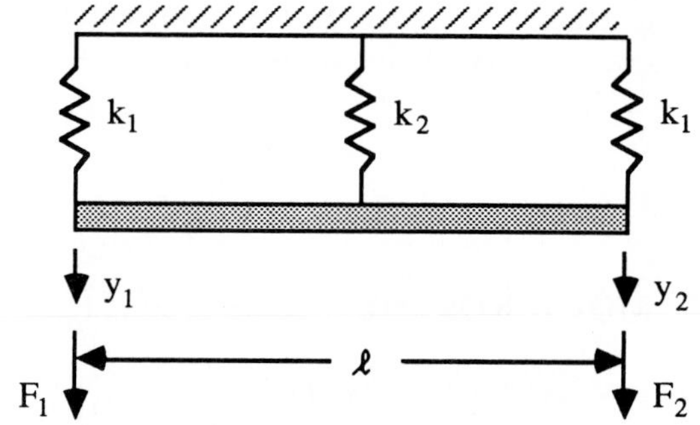

Figure 3.4.2 Vibrating Bar With External Forces

The differential equations of motion, in matrix form, are

$$\frac{2m}{3}\begin{bmatrix} 2 & 1 \\ 1 & 2 \end{bmatrix}\begin{bmatrix} \ddot{y}_1 \\ \ddot{y}_2 \end{bmatrix} + \begin{bmatrix} 4k_1 + k_2 & k_2 \\ k_2 & 4k_1 + k_2 \end{bmatrix}\begin{bmatrix} y_1 \\ y_2 \end{bmatrix} = \begin{bmatrix} F_1 \\ F_2 \end{bmatrix} \qquad (3.4.21)$$

where $\mathbf{M} = \frac{2m}{3}\begin{bmatrix} 2 & 1 \\ 1 & 2 \end{bmatrix}$ and $\mathbf{K} = \begin{bmatrix} 4k_1 + k_2 & k_2 \\ k_2 & 4k_1 + k_2 \end{bmatrix}$. The **M**-orthonormal eigenvectors of Eq. 3.4.17 can be obtained from Example 3.4.2 as

$$\mathbf{x}^1 = \left[\sqrt{\frac{3}{4m}},\ -\sqrt{\frac{3}{4m}}\right]^T, \quad \mathbf{x}^2 = \left[\frac{1}{2\sqrt{m}},\ \frac{1}{2\sqrt{m}}\right]^T$$

with eigenvalues $\lambda_1 = \frac{6k_1}{m}$ and $\lambda_2 = \frac{2k_1 + k_2}{m}$. Thus, the **M**-orthogonal modal matrix is

$$\mathbf{Q} = \begin{bmatrix} \sqrt{\frac{3}{4m}} & \frac{1}{2\sqrt{m}} \\ -\sqrt{\frac{3}{4m}} & \frac{1}{2\sqrt{m}} \end{bmatrix}$$

Then, Eq. 3.4.19 can be written as

$$\ddot{z}_1 + \frac{6k_1}{m} z_1 = \sqrt{\frac{3}{4m}} \left(F_1 - F_2 \right)$$
$$\ddot{z}_2 + \frac{2k_1 + k_2}{m} z_2 = \frac{1}{2\sqrt{m}} \left(F_1 + F_2 \right) \tag{3.4.22}$$

The solutions of Eq. 3.4.22 are obtained from Eq. 3.4.20 as

$$z_1(t) = A_1 \sin \sqrt{\frac{6k_1}{m}}\, t + B_1 \cos \sqrt{\frac{6k_1}{m}}\, t + \sqrt{\frac{m}{6k_1}} \int_0^t \sqrt{\frac{3}{4m}} \left(F_1 - F_2 \right) \sin \sqrt{\frac{6k_1}{m}} \left(t - \tau \right) d\tau \tag{3.4.23}$$

and

$$z_2(t) = A_2 \sin \sqrt{\frac{2k_1 + k_2}{m}}\, t + B_2 \cos \sqrt{\frac{2k_1 + k_2}{m}}\, t + \sqrt{\frac{m}{2k_1 + k_2}} \int_0^t \frac{1}{2\sqrt{m}} \left(F_1 + F_2 \right) \sin \sqrt{\frac{2k_1 + k_2}{m}} \left(t - \tau \right) d\tau \tag{3.4.24}$$

where A_i and B_i, $i = 1, 2$, are arbitrary constants that can be determined from initial conditions. The solution $\mathbf{y}(t)$ is then obtained from

$$\mathbf{y}(t) = \begin{bmatrix} \sqrt{\frac{3}{4m}} & \frac{1}{2\sqrt{m}} \\ -\sqrt{\frac{3}{4m}} & \frac{1}{2\sqrt{m}} \end{bmatrix} \begin{bmatrix} z_1(t) \\ z_2(t) \end{bmatrix} \tag{3.4.25}$$

■

Next, consider a linear dynamic system with **harmonic excitation**, so that the dynamic system equation is

$$\mathbf{M}\ddot{\mathbf{y}} + \mathbf{K}\mathbf{y} = \mathbf{F} \sin \omega t \tag{3.4.26}$$

where ω is the excitation frequency and $\mathbf{F}$ is a constant vector. Using the transformation of Eq. 3.4.18, the system of Eq. 3.4.26 can be decoupled to obtain

$$\ddot{z}_\alpha + \lambda_\alpha z_\alpha = \bar{F}_\alpha \sin \omega t, \quad \alpha = 1, 2, \ldots, n \tag{3.4.27}$$

where $\bar{\mathbf{F}} = \mathbf{Q}^T\mathbf{F}$ and λ_α are eigenvalues of the generalized eigenvalue problem of Eq. 3.4.17. A solution of Eq. 3.4.27 can be obtained by assuming that the steady-state solution (or particular solution) is harmonic and in phase with the loading; i.e.,

$$z_{\alpha_p}(t) = c_\alpha \sin \omega t \tag{3.4.28}$$

in which the amplitude c_α is to be evaluated. Substituting Eq. 3.4.28 into Eq. 3.4.27 leads to

$$-\omega^2 c_\alpha \sin \omega t + \lambda_\alpha c_\alpha \sin \omega t = \bar{F}_\alpha \sin \omega t \tag{3.4.29}$$

Dividing by sin ωt and rearranging,

$$c_\alpha = \frac{\bar{F}_\alpha}{-\omega^2 + \lambda_\alpha}, \quad \alpha = 1, 2, \ldots, n \tag{3.4.30}$$

Thus, the solutions of Eq. 3.4.27 are

$$z_\alpha = \frac{\bar{F}_\alpha}{-\omega^2 + \lambda_\alpha} \sin \omega t, \quad \alpha = 1, 2, \ldots, n \tag{3.4.31}$$

The solution of Eq. 3.4.26 can now be obtained from Eq. 3.4.18.

Equation 3.4.31 reveals the important fact that if the excitation frequency ω is equal to any of the system natural frequencies; i.e., $\omega_\alpha = \sqrt{\lambda_\alpha}$, then resonance occurs and the solution z_α in Eq. 3.4.31 diverges. The associated physical behavior of the system is that the amplitude of motion grows until nonlinear effects dominate or failure of the system occurs.

EXERCISES 3.4

1. Show that if a set of nonzero vectors $\mathbf{x}^1, \mathbf{x}^2, \ldots, \mathbf{x}^n$ in R^n are mutually orthogonal with respect to a positive definite nxn matrix $\mathbf{B}$, then they are linearly independent.

2. Find the eigenvalues and eigenvectors of the generalized eigenvalue problem

$$\mathbf{Ax} \equiv \begin{bmatrix} 1 & 1 \\ 1 & 1 \end{bmatrix} \begin{bmatrix} x_1 \\ x_2 \end{bmatrix} = \lambda \begin{bmatrix} 1 & 0 \\ 0 & 2 \end{bmatrix} \begin{bmatrix} x_1 \\ x_2 \end{bmatrix} \equiv \lambda \mathbf{Bx}$$

Construct a coordinate transformation that reduces the quadratic forms $\mathbf{x}^T\mathbf{Ax}$ and $\mathbf{x}^T\mathbf{Bx}$ to canonical form.

3. Obtain a transformation that simultaneously diagonalizes the quadratic form $J = \mathbf{x}^T\mathbf{A}\mathbf{x}$ and $K = \mathbf{x}^T\mathbf{B}\mathbf{x}$, where the matrices **A** and **B** are given in Exercise 2 of Section 3.3.

4. Find a transformation that reduces the quadratic forms $J = 2x_1^2 + 4x_1x_2 + 2x_2^2$ and $K = 3x_1^2 + 2x_1x_2 + x_2^2$ simultaneously to canonical forms. Find the canonical forms.

5. Write and decouple the equations of motion for a spring-mass system that is similar to that of Exercise 3 of Section 3.1, but with the upper spring removed and $k_2 = k_1$ and $m_2 = m_1$.

6. Write and decouple the equations of motion for a spring-mass system that is similar to that of Exercise 3 of Section 3.1, but with both the upper and lower left springs removed. What is the physical significance of the eigenvalue $\lambda = 0$ in this case?

3.5 MINIMUM PRINCIPLES FOR MATRIX EQUATIONS

The theory of quadratic forms introduced in this chapter for matrix equations plays a central role in modern computational methods in mechanics. More specifically, consider the equation

$$\mathbf{A}\mathbf{x} = \mathbf{c} \tag{3.5.1}$$

where the matrix **A** is symmetric and positive definite. If the vector **x** is a displacement and the vector **c** is the corresponding applied force, then the **virtual work** done by force **c**, acting through a **virtual displacement** $\delta\mathbf{x}$, is

$$\delta W = \mathbf{c}^T\delta\mathbf{x} = \mathbf{x}^T\mathbf{A}\delta\mathbf{x} \tag{3.5.2}$$

Integrating the differential of Eq. 3.5.2,

$$W = (1/2)\,\mathbf{x}^T\mathbf{A}\mathbf{x}$$

Since the work done by the applied forces in causing displacement of a stable mechanical system should be positive if the displacement is different from zero,

$$W = (1/2)\,\mathbf{x}^T\mathbf{A}\mathbf{x} > 0 \tag{3.5.3}$$

if $\mathbf{x} \neq \mathbf{0}$, which is just the statement that the matrix **A** is positive definite. This observation forms the basis for special methods of solving equations whose coefficient matrices are positive definite.

Minimum Functional Theorem

Theorem 3.5.1 (Minimum Functional Theorem). Let **A** be a symmetric positive definite matrix. Then $\mathbf{x}^*$ is the solution of Eq. 3.5.1 if and only if $\mathbf{x}^*$ minimizes

$$F(\mathbf{x}) = \mathbf{x}^T\mathbf{A}\mathbf{x} - 2\mathbf{x}^T\mathbf{c} \tag{3.5.4}$$

■

To prove Theorem 3.5.1, first let $\mathbf{x}^*$ be the solution of Eq. 3.5.1. Substituting $\mathbf{c} = \mathbf{A}\mathbf{x}^*$ into Eq. 3.5.4 and performing an algebraic manipulation,

$$F(\mathbf{x}) = \mathbf{x}^T\mathbf{A}\mathbf{x} - 2\mathbf{x}^T\mathbf{A}\mathbf{x}^* = (\mathbf{x} - \mathbf{x}^*)^T \mathbf{A} (\mathbf{x} - \mathbf{x}^*) - \mathbf{x}^{*T}\mathbf{A}\mathbf{x}^* \tag{3.5.5}$$

Noting that the last term in Eq. 3.5.5 does not depend on $\mathbf{x}$, selection of $\mathbf{x}$ to minimize $F(\mathbf{x})$ requires only that the first term on the right of Eq. 3.5.5 be minimized. Since the matrix $\mathbf{A}$ is positive definite, this first quadratic term is greater than or equal to zero and achieves its minimum value 0 only when $\mathbf{x} - \mathbf{x}^* = \mathbf{0}$. Thus, $\mathbf{x}^*$ minimizes $F(\mathbf{x})$.

Conversely, let $\mathbf{x}^*$ minimize $F(\mathbf{x})$. Note that

$$\begin{aligned} F(\mathbf{x}^* + \tau\mathbf{y}) &= (\mathbf{x}^* + \tau\mathbf{y})^T \mathbf{A} (\mathbf{x}^* + \tau\mathbf{y}) - 2(\mathbf{x}^* + \tau\mathbf{y})^T \mathbf{c} \\ &= \mathbf{x}^{*T}\mathbf{A}\mathbf{x}^* + 2\tau\mathbf{y}^T\mathbf{A}\mathbf{x}^* + \tau^2\mathbf{y}^T\mathbf{A}\mathbf{y} - 2\tau\mathbf{y}^T\mathbf{c} - 2\mathbf{x}^{*T}\mathbf{c} \end{aligned}$$

for any $\mathbf{y} \in R^n$ and $\tau \in R^1$. For any $\mathbf{y}$, $\tau = 0$ minimizes this function. Thus, it is necessary that

$$\left.\frac{dF}{d\tau}\right|_{\tau=0} = 2\mathbf{y}^T\mathbf{A}\mathbf{x}^* - 2\mathbf{y}^T\mathbf{c} = 2(\mathbf{y}, \mathbf{A}\mathbf{x}^* - \mathbf{c}) = 0$$

Since this must hold for any $\mathbf{y} \in R^n$, in particular for each vector $\mathbf{y}^i$ in a basis for R^n, Theorem 2.4.4 implies that $\mathbf{A}\mathbf{x}^* - \mathbf{c} = \mathbf{0}$ and $\mathbf{x}^*$ is the solution of Eq. 3.5.1. ■

Example 3.5.1

Let a symmetric positive definite matrix $\mathbf{A}$ and a vector $\mathbf{c}$ be given as

$$\mathbf{A} = \begin{bmatrix} 2 & 1 & 1 \\ 1 & 3 & 2 \\ 1 & 2 & 3 \end{bmatrix}, \quad \mathbf{c} = \begin{bmatrix} 1 \\ 2 \\ 1 \end{bmatrix}$$

The function of Eq. 3.5.5 is

$$\begin{aligned} F(\mathbf{x}) &= (\mathbf{x} - \mathbf{x}^*)^T \mathbf{A} (\mathbf{x} - \mathbf{x}^*) - \mathbf{x}^{*T}\mathbf{A}\mathbf{x}^* \\ &= [(x_1 - 1/4) + (x_2 - 3/4)]^2 + [(x_1 - 1/4) + (x_3 + 1/4)]^2 \\ &\quad + 2[(x_2 - 3/4) + (x_3 + 1/4)]^2 - 3/2 \end{aligned}$$

The vector $\mathbf{x}$ that minimizes $F(\mathbf{x})$, hence the solution of $\mathbf{A}\mathbf{x} = \mathbf{c}$, is obtained as

$$x_1 = 1/4, \quad x_2 = 3/4, \quad x_3 = -1/4$$

■

Note from Eq. 3.5.3 that if $\mathbf{A}$ is the stiffness matrix of an elastic system, the quadratic term in Eq. 3.5.4 is twice the work done in deforming the system and the function $F(\mathbf{x})$ of Eq. 3.5.4 is twice the **total potential energy** of the system. Theorem 3.5.1, therefore, is a statement of the **principle of minimum total potential energy** for a linear mechanical system.

Ritz Method

A major application of the principle of minimum total potential energy is in construction of an approximate solution of a system of equations with a large number of variables, through a lower dimensional approximation. For example, let the dimension of the vector $\mathbf{x}$ in Eq. 3.5.1 be n and let $\mathbf{x}^1, \mathbf{x}^2, \ldots, \mathbf{x}^m$ be linearly independent vectors in R^n, where $m < n$. Consider an approximation to the solution $\mathbf{x}$ of the form

$$\mathbf{x} \approx a_1\mathbf{x}^1 + a_2\mathbf{x}^2 + \ldots + a_m\mathbf{x}^m = a_i\mathbf{x}^i \tag{3.5.6}$$

The m unknown coefficients a_i are to be selected to best approximate the solution $\mathbf{x}$ of Eq. 3.5.1. As a definition of best approximation, Theorem 3.5.1 can be used. The idea is to select the coefficients a_i to minimize the function of Eq. 3.5.4. This idea is supported by the fact that if $m = n$, the precise solution of Eq. 3.5.1 will be obtained. This technique is called the **Ritz method.**

Substituting Eq. 3.5.6 into Eq. 3.5.4, the function $F(a_i\mathbf{x}^i)$ becomes a function of $\mathbf{a} = [\, a_1, \ldots, a_m \,]^T$; i.e.,

$$\begin{aligned} f(\mathbf{a}) \equiv F(a_i\mathbf{x}^i) &= (\, a_i\mathbf{x}^i \,)^T \mathbf{A} \,(\, a_j\mathbf{x}^j \,) - 2\,(\, a_i\mathbf{x}^i \,)^T \mathbf{c} \\ &= a_i a_j \mathbf{x}^{iT}\mathbf{A}\mathbf{x}^j - 2a_i\mathbf{x}^{iT}\mathbf{c} \end{aligned} \tag{3.5.7}$$

In order for $f(\mathbf{a})$ to be minimum with respect to the coefficients a_i, it is necessary that the derivative of this function with respect to each component a_i be equal to 0, which results in the following equations:

$$\frac{\partial f(\mathbf{a})}{\partial a_k} = 0 = a_j\mathbf{x}^{kT}\mathbf{A}\mathbf{x}^j + a_i\mathbf{x}^{iT}\mathbf{A}\mathbf{x}^k - 2\mathbf{x}^{kT}\mathbf{c}, \quad k = 1, \ldots, m$$

or, noting that $\mathbf{x}^{kT}\mathbf{A}\mathbf{x}^j = \mathbf{x}^{jT}\mathbf{A}\mathbf{x}^k$, in $\mathbf{A}$-scalar product notation as

$$(\, \mathbf{x}^k, \mathbf{x}^j \,)_{\mathbf{A}}\, a_j = \mathbf{x}^{kT}\mathbf{A}\mathbf{x}^j a_j = \mathbf{x}^{kT}\mathbf{c}, \quad k = 1, \ldots, m \tag{3.5.8}$$

Note that these equations are precisely the same as Eq. 3.1.29, which resulted from application of the Galerkin method to the same equation. Therefore, for linear equations with a positive definite coefficient matrix, the Galerkin and Ritz techniques are equivalent.

Equation 3.5.8 can be written in matrix form as

$$\hat{\mathbf{A}}\mathbf{a} = \mathbf{c}^* \tag{3.5.9}$$

where

$$\hat{\mathbf{A}} = [\,(\mathbf{x}^i, \mathbf{x}^j)_{\mathbf{A}}\,]$$
$$\mathbf{c}^* = [\,(\mathbf{x}^i, \mathbf{c})\,] \tag{3.5.10}$$

It is important to note that the matrix $\hat{\mathbf{A}}$ in Eq. 3.5.10 is the Gram matrix of Eq. 2.4.24, constructed with the **A**-scalar product of Eq. 3.3.5. Theorem 2.4.3 guarantees that it is nonsingular, providing the matrix **A** is positive definite and the vectors $\mathbf{x}^i$ are linearly independent.

Minimization of the Rayleigh Quotient

Definition 3.5.1. For a positive definite matrix **A**, the **Rayleigh quotient** is defined for vectors $\mathbf{x} \neq \mathbf{0}$ as

$$R(\mathbf{x}) \equiv \frac{\mathbf{x}^T\mathbf{A}\mathbf{x}}{\mathbf{x}^T\mathbf{x}} \tag{3.5.11}$$

■

Let the Rayleigh quotient take on its minimum value λ at some $\mathbf{x} = \mathbf{x}^* \neq \mathbf{0}$; i.e.,

$$0 < \lambda \equiv \frac{\mathbf{x}^{*T}\mathbf{A}\mathbf{x}^*}{\mathbf{x}^{*T}\mathbf{x}^*} \leq \frac{\mathbf{x}^T\mathbf{A}\mathbf{x}}{\mathbf{x}^T\mathbf{x}} \tag{3.5.12}$$

for all $\mathbf{x} \neq \mathbf{0}$ in R^n. Let $\mathbf{y}$ be any vector in R^n and form the function

$$R(\mathbf{x}^* + \tau\mathbf{y}) = \frac{(\mathbf{x}^* + \tau\mathbf{y})^T\mathbf{A}(\mathbf{x}^* + \tau\mathbf{y})}{(\mathbf{x}^* + \tau\mathbf{y})^T(\mathbf{x}^* + \tau\mathbf{y})} \tag{3.5.13}$$

where τ is a real variable. This function takes on its minimum at $\tau = 0$, so it is necessary that

$$\left.\frac{dR}{d\tau}\right|_{\tau=0} = 0 = \frac{2(\mathbf{x}^{*T}\mathbf{x}^*)(\mathbf{x}^{*T}\mathbf{A}\mathbf{y}) - 2(\mathbf{x}^{*T}\mathbf{A}\mathbf{x}^*)(\mathbf{x}^{*T}\mathbf{y})}{(\mathbf{x}^{*T}\mathbf{x}^*)^2} \tag{3.5.14}$$

Using Eq. 3.5.12 in Eq. 3.5.14,

$$\mathbf{y}^T(\mathbf{A}\mathbf{x}^* - \lambda\mathbf{x}^*) = 0 \tag{3.5.15}$$

Since Eq. 3.5.15 must hold for all $\mathbf{y}$ in R^n, in particular for a basis $\{\mathbf{y}^i\}$ of R^n, Theorem

2.4.4 implies that

$$\mathbf{A}\mathbf{x}^{*} = \lambda \mathbf{x}^{*} \tag{3.5.16}$$

This proves the following result.

Theorem 3.5.2. The vector $\mathbf{x}^{*} \neq \mathbf{0}$ that minimizes the Rayleigh quotient of Eq. 3.5.11, for a symmetric positive definite matrix **A**, is an eigenvector corresponding to the smallest eigenvalue λ of **A**. ■

Let

$$\mathbf{x} \approx a_i \mathbf{x}^i \tag{3.5.17}$$

be an approximation of the eigenvector of the positive definite n×n matrix **A** corresponding to its smallest eigenvalue, where $\mathbf{x}^1, \mathbf{x}^2, \ldots, \mathbf{x}^m$ with $m < n$ are given orthonormal vectors in R^n. The coefficients a_i are to be chosen to minimize the Rayleigh quotient of Eq. 3.5.11; i.e.,

$$r(\mathbf{a}) \equiv R(a_i \mathbf{x}^i) = \frac{a_i a_j \mathbf{x}^{iT} \mathbf{A} \mathbf{x}^j}{a_k a_l \mathbf{x}^{kT} \mathbf{x}^l} = \frac{\mathbf{a}^T \hat{\mathbf{A}} \mathbf{a}}{\mathbf{a}^T \mathbf{a}} \tag{3.5.18}$$

where

$$\hat{\mathbf{A}} \equiv [\, (\mathbf{x}^i, \mathbf{x}^j)_{\mathbf{A}} \,] \tag{3.5.19}$$

Setting the partial derivatives of $r(\mathbf{a})$ with respect to a_α to zero, as a condition for minimization, yields

$$\frac{2\delta_{i\alpha} a_j \hat{a}_{ij} (\mathbf{a}^T \mathbf{a}) - 2 (\mathbf{a}^T \hat{\mathbf{A}} \mathbf{a}) a_\alpha}{(\mathbf{a}^T \mathbf{a})^2} = 0, \quad \alpha = 1, \ldots, m \tag{3.5.20}$$

Since $\hat{\mathbf{A}}$ is the Gram matrix of the linearly independent vectors $\mathbf{x}^i$ with respect to the **A**-scalar product, it is positive definite. Thus, Eq. 3.5.20 may be written in the form

$$\hat{a}_{\alpha j} a_j = \frac{\mathbf{a}^T \hat{\mathbf{A}} \mathbf{a}}{\mathbf{a}^T \mathbf{a}} a_\alpha, \quad \alpha = 1, \ldots, m \tag{3.5.21}$$

Defining

$$\hat{\lambda} = \frac{\mathbf{a}^T \hat{\mathbf{A}} \mathbf{a}}{\mathbf{a}^T \mathbf{a}} \tag{3.5.22}$$

Eq. 3.5.21 may be written in matrix form as

$$\hat{\mathbf{A}}\mathbf{a} = \hat{\lambda}\mathbf{a} \tag{3.5.23}$$

Thus, **a** is an eigenvector of $\hat{\mathbf{A}}$ and $\hat{\lambda}$ is its smallest eigenvalue. This is called the **Ritz method for the eigenvalue problem $\mathbf{Ax} = \lambda\mathbf{x}$.**

The Ritz method is often used for high-dimensional applications; i.e., for n very large, to reduce the dimension of the eigenvalue problem. The dimension of the matrix $\hat{\mathbf{A}}$ in Eq. 3.5.23 is m×m, whereas that in Eq. 3.5.16 is n×n. This method is used in practice with m much less than n.

EXERCISES 3.5

1. Verify the identity of Eq. 3.5.5.

2. Show that setting the derivative of $F(\mathbf{x})$ in Eq. 3.5.4 with respect to each component of the vector **x** equal to zero yields Eq. 3.5.1.

3. Show that if the vector $\mathbf{x}^*$ minimizes the Rayleigh quotient $\dfrac{\mathbf{x}^T\mathbf{A}\mathbf{x}}{\mathbf{x}^T\mathbf{B}\mathbf{x}}$ with **A** and **B** symmetric and positive definite, then it is an eigenvector corresponding to the smallest eigenvalue λ of the generalized eigenvalue problem $\mathbf{Ax} = \lambda\mathbf{Bx}$.

4. Let $\mathbf{x}^1, \mathbf{x}^2, \ldots, \mathbf{x}^m$, $m < n$, be linearly independent vectors in R^n and let **A** and **B** be symmetric positive definite matrices. Find equations for the coefficients of $\mathbf{x} \approx a_i\mathbf{x}^i$ to minimize the Rayleigh quotient of Exercise 3. This is called the **Ritz method for the generalized eigenvalue problem $\mathbf{Ax} = \lambda\mathbf{Bx}$.**

4

INFINITE SERIES

Thus far in this text, only finite-dimensional equations and vector spaces have been encountered. This chapter begins the transition to classes of applications that involve differential equations and their solution spaces, which are infinite dimensional. Before delving into infinite series solution methods for differential equations, a review is made of the theory of infinite series, upon which solution methods presented here and later in the text are based. Certain types of second-order ordinary differential equations describe physical systems and arise as a result of separation of variables in partial differential equations, so they deserve to be studied in their own right. Such equations, with constant and polynomial coefficients, can often be solved by power series techniques, a subject of this chapter. Finally, some special classes of functions that arise as solutions of second-order ordinary differential equations are studied.

4.1 INFINITE SERIES WHOSE TERMS ARE CONSTANTS

Infinite series play a key role in both theoretical and approximate treatment of differential equations that arise in engineering applications. Consider first the problem of attaching meaning to the sum of an infinite number of terms.

Convergence of Series of Nonnegative Constants

Definition 4.1.1. For a sequence of finite constants $\{ u_i \} = \{ u_1, u_2, \ldots \}$, define the nth **partial sum** as

$$s_n = \sum_{i=1}^{n} u_i \tag{4.1.1}$$

which establishes a **sequence of partial sums** $\{ s_n \}$. A **sequence** $\{ s_n \}$, whether partial sums or otherwise, is said to **converge** to s if, for any number $\varepsilon > 0$, there is an integer M, such that if $n \geq M$, then $| s - s_n | < \varepsilon$. If the partial sum s_n converges to a finite limit s as $n \to \infty$, denoted

$$\lim_{n\to\infty} s_n = s \equiv \sum_{i=1}^{\infty} u_i \tag{4.1.2}$$

the **infinite series** $\sum_{i=1}^{\infty} u_i$ is said to be a **convergent series** and to have the value s, called the **sum of the series.** ■

Defining the sum of a convergent infinite series in Eq. 4.1.1 as s, a **necessary condition for convergence** is that

$$\lim_{i \to \infty} u_i = \lim_{i \to \infty} (s_i - s_{i-1}) = \lim_{i \to \infty} s_i - \lim_{i \to \infty} s_{i-1} = s - s = 0$$

This condition, however, is not sufficient to guarantee convergence.

The partial sums s_n may not converge to a single limit, but may oscillate or diverge to $\pm \infty$. In the case $u_i = (-1)^i$,

$$s_n = \sum_{i=1}^{n} u_i = \begin{cases} 1, & n \text{ even} \\ 0, & n \text{ odd} \end{cases}$$

This is an **oscillating series** that does not converge to a limit. For the case $u_i = i$,

$$s_n = \sum_{i=1}^{n} u_i = \frac{n (n + 1)}{2} \tag{4.1.3}$$

so, as $n \to \infty$, $\lim_{n \to \infty} s_n = \infty$. Whenever the sequence of partial sums approaches $\pm \infty$, the infinite series is said to be a **divergent series.**

Example 4.1.1

For the **geometric sequence** with $r \geq 0$; i.e.,

$$u_i = ar^i$$

the partial sums are $s_n = \sum_{i=0}^{n-1} ar^i$. Note that for $n = 1$, $r^0 \equiv 1$, so $s_1 = a$. For $n = 2$, $s_2 = a + ar = a (1 + r) = a (1 - r^2)/(1 - r)$. Assume, for proof by induction, that $s_k = a (1 - r^k)/(1 - r)$. Then, by definition of the partial sum,

$$\begin{aligned} s_{k+1} &= s_k + ar^k \\ &= a (1 - r^k)/(1 - r) + ar^k = a (1 - r^{k+1})/(1 - r) \end{aligned}$$

Thus, the partial sum of the **geometric series** $\sum_{i=0}^{\infty} ar^i$ is

$$\sum_{i=0}^{n-1} u_i \equiv s_n = a\frac{1-r^n}{1-r} \tag{4.1.4}$$

which can also be verified by division of the fraction. Taking the limit as $n \to \infty$, for $r < 1$,

$$\lim_{n\to\infty} s_n = \frac{a}{1-r} \tag{4.1.5}$$

By definition, the geometric series converges for $r < 1$ and its sum is given by

$$\sum_{i=0}^{\infty} ar^i = \frac{a}{1-r} \tag{4.1.6}$$

On the other hand, if $r \geq 1$, the necessary condition $u_i \to 0$ is not satisfied and the geometric series diverges. ■

In practice, it is a matter of extreme importance to be able to tell whether a given series is convergent. Several convergence tests are given, starting with the simple comparison test and working up to more complicated but quite sensitive tests. Proofs of validity of these tests may be found in Refs. 7–9. For the present, consider a series of positive terms, $a_i > 0$.

Theorem 4.1.1 (Comparison Test). If $|u_i| \leq a_i$ for all but a finite number of i and if the a_i form a convergent series, then the series $\sum_{i=1}^{\infty} u_i$ is also convergent. If $v_i \geq b_i \geq 0$ for all but a finite number of i and if the b_i form a divergent series, then the series $\sum_{i=1}^{\infty} v_i$ is also divergent. ■

As a convergent series a_i for application of the comparison test, the geometric series of Example 4.1.1 is already available. Once other series are identified as being either convergent or divergent, they may be used as the known series in this comparison test.

Example 4.1.2

Consider $u^i = 1/(i)^i$, hence the series

$$\sum_{i=1}^{\infty} u_i = 1 + \frac{1}{4} + \frac{1}{27} + \frac{1}{256} + \ldots$$

Since $| u_i | = (1/i)^i \leq (1/2)^i$ for $i > 1$ and the geometric series of Example 4.1.1 with $r = 1/2$ converges, this series converges. ■

Theorem 4.1.2 (Cauchy Root Test). If $a_i \geq 0$ and $(a_i)^{1/i} \leq r < 1$ for all sufficiently large i [i.e., if $\lim_{i\to\infty} (a_i)^{1/i} < 1$] then $\sum_{i=1}^{\infty} a_i$ is convergent. If $(a_i)^{1/i} > 1$ for all sufficiently large i [i.e., if $\lim_{i\to\infty} (a_i)^{1/i} > 1$] then $\sum_{i=1}^{\infty} a_i$ is divergent. If $\lim_{i\to\infty} (a_i)^{1/i} = 1$, the test fails to yield any conclusion. ■

The first part of this Cauchy root test is verified easily by raising $(a_i)^{1/i} \leq r$ to the ith power, noting that

$$a_i \leq r^i < 1 \tag{4.1.7}$$

Since r^i is just the ith term in a convergent geometric series, $\sum_{i=1}^{\infty} a_i$ is convergent by the comparison test. Conversely, if $(a_i)^{1/i} > 1$, the series must diverge. ■

The root test is particularly useful in establishing properties of power series.

Theorem 4.1.3 (Ratio Test). If $a_i \geq 0$ and $a_{i+1}/a_i \leq r < 1$ for all sufficiently large i (i.e., if $\lim_{i\to\infty} \frac{a_{i+1}}{a_i} < 1$) then $\sum_{i=1}^{\infty} a_i$ is convergent. If $a_{i+1}/a_i > 1$ for all sufficiently large i (i.e., if $\lim_{i\to\infty} \frac{a_{i+1}}{a_i} > 1$) then $\sum_{i=1}^{\infty} a_i$ is divergent. If $\lim_{i\to\infty} \frac{a_{i+1}}{a_i} = 1$, the test fails to yield any conclusion. ■

For proof of the ratio test, see Refs. 7–9.

Although not quite so sensitive as the root test, the ratio test is one of the easiest to apply and is widely used.

Example 4.1.3

Consider the series $\sum_{i=1}^{\infty} \frac{i}{2^i}$. Using the ratio test,

$$\frac{a_{i+1}}{a_i} = \frac{(i+1)/2^{i+1}}{i/2^i} = \frac{1}{2}\left(\frac{i+1}{i}\right)$$

Since

$$\frac{a_{i+1}}{a_i} \le \frac{3}{4} < 1$$

for $i \ge 2$, convergence is verified; i.e., $\lim_{i\to\infty} \frac{1}{2}\left(\frac{i+1}{i}\right) = \frac{1}{2} < 1.$ ■

Example 4.1.4

Consider the series $\sum_{i=1}^{\infty} \frac{1}{i}$. Note that

$$\frac{a_{i+1}}{a_i} = \frac{1/(i+1)}{1/i} = \frac{i}{i+1} < 1$$

for $i \ge 1$, but there is no r such that

$$\frac{i}{i+1} \le r < 1$$

for all sufficiently large i. Equivalently, $\lim_{i\to\infty} \frac{a_{i+1}}{a_i} = \lim_{i\to\infty} \frac{i}{i+1} = 1$. The ratio test thus fails and it cannot be determined whether the series converges or not. ■

Theorem 4.1.4 (Integral Test). Let $f(x) \ge 0$ be a continuous, **monotonic decreasing function**; i.e., $f(x) \ge f(y)$ if $y \ge x$. Let $a_i = f(i)$. Then $\sum_{i=1}^{\infty} a_i$ converges if the integral $\int_1^{\infty} f(x)\,dx$ is finite and it diverges if the integral is infinite. ■

Example 4.1.5

For the series of Example 4.1.4, $f(i) = 1/i$. By the integral test,

$$\int_1^{\infty} f(x)\,dx = \int_1^{\infty} (1/x)\,dx$$

$$= \ln x \Big|_1^{\infty} = \infty$$

Thus, the series diverges. Recall that the ratio test failed to yield a conclusion in Example 4.1.4. This suggests that the integral test is more powerful than the ratio test. ■

Example 4.1.6

The **Riemann Zeta function** is defined as

$$\zeta(p) = \sum_{i=1}^{\infty} \frac{1}{i^p}$$

where $p \geq 0$. Define $f(x) = x^{-p}$ and calculate

$$\int_1^{\infty} x^{-p}\, dx = \begin{cases} \left. \dfrac{x^{1-p}}{1-p} \right|_1^{\infty}, & \text{if } p \neq 1 \\ \left. \ln x \right|_1^{\infty}, & \text{if } p = 1 \end{cases}$$

The integral and the series are divergent for $p \leq 1$ and convergent for $p > 1$. ■

Alternating Series

In the foregoing tests, attention has been limited to series with positive terms. Consider next infinite series in which the signs of terms alternate, called **alternating series.** The partial cancellation that occurs due to alternating signs makes convergence more rapid and much easier to identify.

Theorem 4.1.5 (Leibniz Test). With $a_i \geq 0$, form $\sum_{i=1}^{\infty} (-1)^{i+1} a_i$. If, beyond some integer N, the a_i are monotonically decreasing (i.e., $a_k \geq a_i$ for all $i > k > N$) and if $\lim_{i \to \infty} a_i = 0$, then the alternating series converges. ■

Example 4.1.7

The divergent series in Examples 4.1.4 and 4.1.5 is altered so that the signs of terms alternate; i.e.,

$$\sum_{i=1}^{\infty} (-1)^i \frac{1}{i}$$

The term $a_i = \frac{1}{i}$ is monotonically decreasing for $i \geq 1$ and

$$\lim_{i \to \infty} a_i = 0$$

Thus, the alternating series converges. ■

Example 4.1.8

To show that the series $\sum_{i=1}^{\infty} (-1)^i \frac{\ln i}{i}$ converges, it may be shown that all of the conditions of the Leibniz test are satisfied, except possibly for the inequality $\frac{\ln(i+1)}{i+1} < \frac{\ln i}{i}$. The simplest way to establish this is to show that $f(x) = \frac{\ln x}{x}$ is strictly decreasing. This follows, since its derivative is $f'(x) = \frac{1 - \ln x}{x^2} < 0$, for all $x > e$. ■

Absolute Convergence

Definition 4.1.2. Given a sequence of terms u_i, in which u_i may vary in sign, if $\sum_{i=1}^{\infty} | u_i |$ converges, then $\sum_{i=1}^{\infty} u_i$ is said to be **absolutely convergent**. A series is said to be **conditionally convergent** if it converges, but does not converge absolutely. ■

The series in Examples 4.1.7 and 4.1.8 converge conditionally, but they do not converge absolutely. The following property of an absolutely convergent series follows from the comparison test.

Theorem 4.1.6. An absolutely convergent series is convergent and

$$\left| \sum_{i=1}^{\infty} a_i \right| \le \sum_{i=1}^{\infty} | a_i |$$ ■

The tests established for convergence of series with positive terms are immediately applicable as tests for absolute convergence of arbitrary series. The following is a restatement of the ratio test for absolute convergence.

Theorem 4.1.7 (Ratio Test for Absolute Convergence). Let $\sum_{i=1}^{\infty} a_i$ be a series with nonzero terms. If $\lim_{i \to \infty} \left| \frac{a_{i+1}}{a_i} \right| < 1$, then $\sum_{i=1}^{\infty} a_i$ converges absolutely. If

$\lim_{i \to \infty} \left| \frac{a_{i+1}}{a_i} \right| > 1$, then $\sum_{i=1}^{\infty} a_i$ is divergent. If $\lim_{i \to \infty} \left| \frac{a_{i+1}}{a_i} \right| = 1$, the test fails to yield any conclusion. ■

Note that for the case $\lim_{i \to \infty} \left| \frac{a_{i+1}}{a_i} \right| > 1$, the conclusion is not that the series fails to converge absolutely, but that it actually diverges. Another point to note is that the ratio test for absolute convergence never establishes convergence of conditionally convergent series.

EXERCISES 4.1

1. Determine whether the following series converge or not:

   (a) $\sum_{i=2}^{\infty} (\ln i)^{-1}$

   (b) $\sum_{i=1}^{\infty} \frac{1}{2i (2i - 1)}$

   (c) $\sum_{i=2}^{\infty} \frac{1}{i \ln i}$

2. Determine whether

$$\sum_{i=2}^{\infty} \frac{1}{(\ln i)^i}$$

   is convergent or divergent by the Cauchy root test.

3. Test for absolute convergence, conditional convergence, or divergence of the following series:

   (a) $\sum_{i=1}^{\infty} \frac{(-1)^{i-1}}{2i + 3}$

   (b) $\sum_{i=1}^{\infty} (-1)^{i-1} \frac{i^4}{(i + 1)!}$

(c) $\sum_{i=1}^{\infty} (-1)^i \frac{i \ln i}{e^i}$

(d) $\sum_{i=1}^{\infty} (-1)^i \frac{\cos i\alpha}{i^2}$

4. Show by three counterexamples that each of the three conditions of the Leibniz test is needed in the statement of that test; i.e., the alternating of signs, the decreasing nature of a_i, and the limit of a_i being zero.

4.2 INFINITE SERIES WHOSE TERMS ARE FUNCTIONS

The concept of infinite series is next extended to include the possibility that each term u_i may be a function of a real variable x; i.e., $u_i = u_i(x)$. The partial sums thus become functions of the variable x; i.e.,

$$s_n(x) = \sum_{i=1}^{n} u_i(x)$$

The sum of the series, defined as the limit of the partial sums, is also a function of x; i.e.,

$$\sum_{i=1}^{\infty} u_i(x) = s(x) = \lim_{n\to\infty} s_n(x) \tag{4.2.1}$$

Thus far, attention has been focused on the behavior of partial sums as functions of n. Consider now how convergence depends on the value of x.

Definition 4.2.1. A sequence $s_n(x)$ is said to **converge pointwise** to a function s(x) that is defined on an interval [a, b] if

$$\lim_{n\to\infty} s_n(x) = s(x)$$

for each $x \in [a, b]$. Equivalently, for each $\varepsilon > 0$ and $x \in [a, b]$, there exists an integer N, which may be dependent on ε and x, such that

$$| s_n(x) - s(x) | < \varepsilon$$

for all $n > N$. ■

Example 4.2.1

The sequence $s_n(x) = x^n$, $x \in [0, 1]$, converges pointwise to s(x) on [0, 1], where $s(x) = 0$ for $x \in [0, 1)$ and $s(1) = 1$. Note that even though each $s_n(x)$ is continuous, the limit s(x) is not continuous, since it has a discontinuity at x = 1. This elementary example shows that a sequence of continuous functions of a variable x can converge for each value of x to a discontinuous function. ■

Uniform Convergence

Definition 4.2.2. If for each $\varepsilon > 0$ there is an integer N that is independent of x in the interval [a, b], such that

$$| s(x) - s_n(x) | < \varepsilon \tag{4.2.2}$$

for all $n \geq N$ and for all $x \in [a, b]$, then the sequence $\{ s_n(x) \}$ is said to be **uniformly convergent** to s(x) in the interval [a, b]. ■

This condition is illustrated in Fig. 4.2.1. No matter how small ε is taken to be, N can always be chosen large enough so that the magnitude of the difference between s(x) and $s_n(x)$ is less than ε, for all $n > N$ and all $x \in [a, b]$.

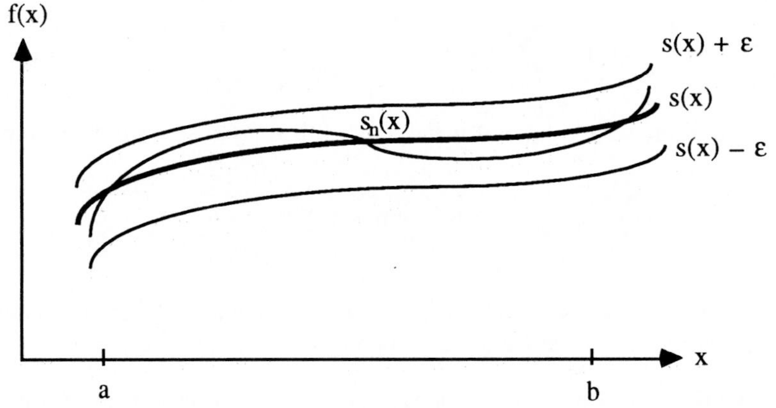

Figure 4.2.1 Uniform Convergence of a Sequence

Example 4.2.2

To see that the sequence $s_n(x) = x^n$ in Example 4.2.1 is not uniformly convergent in [0, 1], choose $\varepsilon = 1/2$. Let $\delta > 0$ be a small constant, so

$$| s(1-\delta) - s_n(1-\delta) | = | s_n(1-\delta) | = s_n(1-\delta) = (1-\delta)^n$$

For any integer N, set

$$(1-\delta)^N = 3/4$$

so

$$\delta = 1 - (3/4)^{1/N} > 0$$

Thus, for any N, it is possible to find an $x = 1 - \delta \in [0, 1]$ such that

$$| s(x) - s_n(x) | = 3/4 > 1/2$$

The sequence thus fails to be uniformly convergent in the interval [0, 1]. ■

Example 4.2.3

Consider the series

$$\sum_{i=1}^{\infty} u_i(x) = \sum_{i=1}^{\infty} \frac{x}{[(i-1)x + 1][ix + 1]} \tag{4.2.3}$$

To show that $s_n(x) = nx(nx + 1)^{-1}$, first note that this expression holds for $s_n(x)$ with n = 1 and 2. For proof by induction, assume that the relation holds for n. It is then to be shown that it holds for n + 1. Since

$$\begin{aligned} s_{n+1}(x) &= s_n(x) + \frac{x}{[nx+1][(n+1)x + 1]} \\ &= \frac{nx}{[nx+1]} + \frac{x}{[nx+1][(n+1)x + 1]} \\ &= \frac{(n+1)x}{(n+1)x + 1} \end{aligned}$$

the proof by induction is complete. Letting n approach infinity, the pointwise convergent sum of the series is

$$s(x) = \lim_{n\to\infty} s_n(x) = \begin{cases} 0, & \text{if } x = 0 \\ 1, & \text{if } x \neq 0 \end{cases} \tag{4.2.4}$$

Thus, the sum has a discontinuity at x = 0. However, $s_n(x)$ is a continuous function of x, $0 \le x \le 1$, for all finite n. As shown in Example 4.2.2, with any fixed ε, Eq. 4.2.2 must be violated for some finite n, so this series does not converge uniformly. ■

The most commonly encountered test for uniform convergence is the Weierstrass M-test.

Theorem 4.2.1 (Weierstrass M-test). If a sequence of nonnegative numbers M_i can be found for which $M_i \ge | u_i(x) |$, for all x in the closed interval [a, b], and for which $\sum_{i=1}^{\infty} M_i$ is convergent, then the series $\sum_{i=1}^{\infty} u_i(x)$ is absolutely and uniformly convergent for $x \in [a, b]$. ■

The proof of the Weierstrass M-test is direct, simple, and instructive. Since $\sum_{i=1}^{\infty} M_i$ converges, for each $\varepsilon > 0$, there is an N such that for $n + 1 \geq N$,

$$\sum_{i=n+1}^{\infty} M_i < \varepsilon$$

Then, with $| u_i(x) | \leq M_i$, for all x in the interval $a \leq x \leq b$,

$$\sum_{i=n+1}^{\infty} | u_i(x) | < \varepsilon$$

By the comparison test, the series $\sum_{i=1}^{\infty} u_i(x)$ converges to a limit s(x), for each $x \in [a, b]$. Furthermore,

$$| s(x) - s_n(x) | = \left| \sum_{i=n+1}^{\infty} u_i(x) \right| \leq \sum_{i=n+1}^{\infty} | u_i(x) | < \varepsilon$$

for $n + 1 > N$ and all $x \in [a, b]$. Thus, $\sum_{i=1}^{\infty} u_i(x)$ is uniformly convergent in [a, b].

Since absolute values of $u_i(x)$ are used in the statement of the Weierstrass M-test, the series $\sum_{i=1}^{\infty} u_i(x)$ is also absolutely convergent. ■

Example 4.2.4

Consider the series

$$\sum_{i=0}^{\infty} e^{-ix} \cos ix = 1 + e^{-x} \cos x + e^{-2x} \cos 2x + \ldots$$

for $a \le x < \infty$, where $a > 0$. Note that

$$| e^{-ix} \cos ix | \le e^{-ix} \le e^{-ia} \equiv M_i$$

and the series of constants

$$\sum_{i=0}^{\infty} M_i = \sum_{i=0}^{\infty} e^{-ia} = \sum_{i=0}^{\infty} (e^{-a})^i = \sum_{i=0}^{\infty} r^i$$

with $r = e^{-a} < 1$ is a convergent geometric series (see Example 4.1.1). Thus, the series of functions converges absolutely and uniformly in $[a, \infty)$, for any $a > 0$. ■

Since the Weierstrass M-test establishes both uniform and absolute convergence, it will necessarily fail for series that are uniformly but not absolutely convergent. A somewhat more delicate test for uniform convergence has been given by Abel [8, 9].

Theorem 4.2.2 (Abel's Test). Let $\sum_{i=1}^{\infty} a_i$ be convergent and

$$u_i(x) = a_i f_i(x) \tag{4.2.5}$$

where for fixed x, $f_i(x)$ is monotonic in i, either increasing or decreasing, and $f_i(x) \ge 0$ is bounded; i.e., $0 \le f_i(x) \le M$, for all $x \in [a, b]$. Then $\sum_{i=1}^{\infty} u_i(x)$ converges uniformly in $[a, b]$. ■

To see why monotonicity in i is important, let $a_i = (-1)^i \frac{1}{i}$, so that $\sum_{i=1}^{\infty} a_i$ converges. Let $f_i(x) = 1$ if i is even and $f_i(x) = 0$ if i is odd on $[a, b]$, which is not monotonic in i. Then, $a_i f_i(x) = \frac{1}{i}$ if i is even and $a_i f_i(x) = 0$ if i is odd. Thus, $\sum_{i=1}^{\infty} a_i f_i(x) = \sum_{i=1, i \text{ even}}^{\infty} \frac{1}{i}$ diverges.

The Abel test is especially useful in analyzing power series with coefficients a_i.

Example 4.2.5

If $\sum_{n=0}^{\infty} a_n$ is a convergent series of constants, since $| x^i | \le M \equiv 1$ and x^i is monotonic in i; i.e., $x^{i+1} \le x^i$ for $x \in [0, 1]$, the power series $\sum_{i=0}^{\infty} a_i x^i$ converges uniformly for $0 \le x \le 1$. ■

Differentiation and Integration of Series

As shown in Examples 4.2.2 and 4.2.3, series $\sum_{i=1}^{\infty} u_i(x)$ exist for which each $u_i(x)$ is continuous (in fact all their derivatives are continuous) and the series converges at each x to a finite sum s(x), but s(x) is not even continuous. This raises important questions about properties of functions that are defined as infinite series of functions. Many methods of solving differential equations are based on assuming a solution y(x) in the form of an infinite series

$$y(x) = \sum_{i=0}^{\infty} a_i \phi_i(x) \tag{4.2.6}$$

where the functions $\phi_i(x)$ are known. The series of Eq. 4.2.6 is **formally differentiated**; i.e., the order of differentiation and infinite summation are reversed, without theoretical justification, to form

$$\frac{dy}{dx} \sim \sum_{i=0}^{\infty} a_i \frac{d\phi_i}{dx} \tag{4.2.7}$$

up to the highest order of derivatives that appear in the differential equation.

Equations 4.2.6, 4.2.7, and any higher-order derivatives that are needed are then substituted into the differential equation. The coefficients a_i are then determined to satisfy the resulting equation. The a_i obtained are then substituted into Eq. 4.2.6 to yield the hoped-for solution. But is the function y(x) of Eq. 4.2.6 that is constructed by this formal (nonrigorous) procedure really a solution of the differential equation?

First, the infinite series on the right of Eq. 4.2.6 might diverge. Or it might converge pointwise, as in Example 4.2.2, to a discontinuous function. Since a discontinuous function does not have a derivative, the function y(x) defined by the infinite series that is constructed cannot possibly be the solution of a differential equation.

Since many methods of solving differential equations use infinite series in a formal approach, similar to that outlined above, great care must be taken to assure that operations with infinite series of functions preserve continuity and differentiability properties that are required of solutions. The following theorem provides valuable tools that will be used in establishing the validity and limits of applicability of infinite series methods for solving differential equations.

Theorem 4.2.3. Let the series $\sum_{i=1}^{\infty} u_i(x)$ be uniformly convergent on an interval [a, b]. Then,

(a) if the individual terms $u_i(x)$ are continuous, the limit

$$f(x) = \sum_{i=1}^{\infty} u_i(x) \tag{4.2.8}$$

is also continuous.

(b) if the individual terms $u_i(x)$ are continuous, the series may be integrated term by term and the sum of the integrals is equal to the integral of the sum; i.e.,

$$\int_a^b f(x)\,dx = \int_a^b \sum_{i=1}^{\infty} u_i(x)\,dx = \sum_{i=1}^{\infty} \int_a^b u_i(x)\,dx \tag{4.2.9}$$

(c) the derivative of the sum f(x) equals the sum of the individual derivatives; i.e.,

$$\frac{d}{dx} f(x) = \frac{d}{dx}\left(\sum_{i=1}^{\infty} u_i(x) \right) = \sum_{i=1}^{\infty} \frac{d}{dx} u_i(x) \tag{4.2.10}$$

provided the following conditions are satisfied:

(i) $u_i(x)$ and $\dfrac{du_i(x)}{dx}$ are continuous in [a, b]

(ii) $\sum_{i=1}^{\infty} \dfrac{du_i(x)}{dx}$ is uniformly convergent in [a, b] ■

For proof, the reader is referred to a calculus text such as Ref. 8 or 9.

The significance of this theorem should not be underestimated. Presume that the functions $u_i(x)$ in the infinite series of Eq. 4.2.8 have been found by some method of approximation or formal solution. Equation 4.2.8 thus defines a continuous function, under the stated conditions. Since no closed-form formula can be found for the sums of most infinite series, this theorem may be the only tool available to assure that the infinite series constructed really defines the solution being sought. Equations 4.2.9 and 4.2.10, under the conditions stated, permit the interchange of order of infinite summation and integration and differentiation, respectively. Note that the conditions (b) that permit interchange of infinite summation and integral are mild. In contrast, the conditions (c) that permit interchange of infinite summation and derivative are much more demanding. This is a disappointing fact of mathematical life, which must be understood and accepted, since it is this interchange that must be justified in most engineering applications of infinite series.

Example 4.2.6

Consider again the series of Example 4.2.4; i.e.,

$$f(x) = \sum_{i=0}^{\infty} e^{-ix} \cos ix \tag{4.2.11}$$

for $0 < a \leq x \leq b < \infty$. It was shown in Example 4.2.4 that the series of Eq. 4.2.11 converges uniformly in [a, b]. Since each term in the series is continuous, Eq. 4.2.9 of Theorem 4.2.3 holds. In this case,

$$\int_a^b f(x)\, dx = \sum_{i=0}^{\infty} \int_a^b e^{-ix} \cos ix\, dx \tag{4.2.12}$$

To verify that f(x) has a derivative, which is given by

$$f'(x) = -\sum_{i=0}^{\infty} ie^{-ix} (\cos ix + \sin ix) \tag{4.2.13}$$

note first that $u_i(x)$ and $\dfrac{du_i(x)}{dx}$ are continuous in $a \leq x \leq b$. To show that the series on the right of Eq. 4.2.13 is uniformly convergent, note that

$$\left| ie^{-ix} (\cos ix + \sin ix) \right| \leq 2ie^{-ia} \equiv M_i$$

By the ratio test,

$$\begin{aligned} \lim_{i\to\infty} \frac{M_{i+1}}{M_i} &= \lim_{i\to\infty} \frac{2 (i + 1) e^{-(i+1)a}}{2ie^{-ia}} \\ &= \lim_{i\to\infty} \left(1 + \frac{1}{i} \right) e^{-a} \\ &= e^{-a} < 1 \end{aligned}$$

Thus, the series $\sum_{i=1}^{\infty} M_i$ converges and, by the Weierstrass M-test, the series on the right of Eq. 4.2.13 is uniformly convergent in [a, b]. Thus, Theorem 4.2.3 assures that f(x) has a derivative and that it is given by Eq. 4.2.13. ■

Taylor Series

Let a function f(x) have a continuous n^{th} derivative in the interval $a \le x \le b$. Integrating this derivative from a to $x \le b$,

$$\int_a^x f^{(n)}(x)\,dx = f^{(n-1)}(x)\Big|_a^x = f^{(n-1)}(x) - f^{(n-1)}(a)$$

Integrating again,

$$\int_a^x \int_a^x f^{(n)}(x)\,dx\,dx = \int_a^x [\, f^{(n-1)}(x) - f^{(n-1)}(a)\,]\,dx$$

$$= f^{(n-2)}(x) - f^{(n-2)}(a) - (x-a) f^{(n-1)}(a)$$

Finally, integrating for the nth time,

$$\int_a^x \cdots \int_a^x f^{(n)}(x)\,(dx)^n = f(x) - f(a) - (x-a) f'(a)$$

$$- \frac{(x-a)^2}{2!} f''(a) - \ldots - \frac{(x-a)^{n-1}}{(n-1)!} f^{(n-1)}(a) \qquad (4.2.14)$$

Note that this expression is exact. No terms have been dropped and no approximations are made. Solving Eq. 4.2.14 for f(x) yields the following fundamental result.

Theorem 4.2.4 (Taylor's Theorem). If f(x) has n continuous derivatives in [a, b], then it is equal to its **Taylor expansion**

$$f(x) = f(a) + (x-a) f'(a) + \frac{(x-a)^2}{2!} f''(a) + \ldots$$

$$+ \frac{(x-a)^{n-1}}{(n-1)!} f^{(n-1)}(a) + R_n \qquad (4.2.15)$$

where the **integral form of the remainder** R_n is given by

$$R_n = \int_a^x \cdots \int_a^x f^{(n)}(x)\,(dx)^n \qquad (4.2.16)$$

■

The remainder term of Eq. 4.2.16 may be put into a more useful form, by using the mean value theorem of calculus (see Theorem 6.1.1 or Refs. 8 and 9); i.e., if a function g(x) is continuous in [a, x], there is a ξ with $a \le \xi \le x$ such that

$$\int_a^x g(x)\,dx = (x - a)\, g(\xi) \tag{4.2.17}$$

Applying this result to Eq. 4.2.16 and integrating n times yields the **Lagrangian form of the remainder**

$$R_n = \frac{(x - a)^n}{n!} f^{(n)}(\xi_n) \tag{4.2.18}$$

where $a \le \xi_n \le x$.

In Taylor's expansion, there are no questions of infinite series convergence, since the series of Eq. 4.2.15 is finite. Thus, the only questions that arise concern the magnitude of the remainder. When the function f(x) is such that

$$\lim_{n\to\infty} R_n = 0 \tag{4.2.19}$$

then Eq. 4.2.15 yields the **Taylor series**

$$f(x) = \sum_{i=0}^{\infty} \frac{(x - a)^i}{i!} f^{(i)}(a) \tag{4.2.20}$$

If the expansion of Eq. 4.2.20 is about the origin (a = 0), Eq. 4.2.20 is known as a **Maclaurin series,**

$$f(x) = \sum_{i=0}^{\infty} \frac{x^i}{i!} f^{(i)}(0) \tag{4.2.21}$$

or **Maclaurin expansion** of f(x).

Example 4.2.7

Let $f(x) = e^x$. Differentiating and evaluating at x = 0, $f^{(n)}(0) = e^0 = 1$, for all n. The remainder of Eq. 4.2.18 is thus

$$R_n = \frac{x^n}{n!}$$

For x = 0, $R_n = 0$. For any given $x \ne 0$, let N be an integer greater than | x |. Then,

$$|R_n| = \left|\frac{x^n}{n!}\right|$$

$$= \left|\frac{1}{\left(\frac{n}{x}\frac{n-1}{x}\cdots\frac{N+1}{x}\right)\left(\frac{N}{x}\cdots\frac{1}{x}\right)}\right|$$

$$\leq \frac{1}{n}\left(\frac{|x|}{\left|\frac{N}{x}\cdots\frac{1}{x}\right|}\right)$$

and $\lim_{n\to\infty} R_n = 0$. Thus, the Maclaurin series for e^x converges and is the **power series**

$$e^x = \sum_{i=0}^{\infty} \frac{x^i}{i!} \tag{4.2.22}$$

This series converges for all x, but it is not uniformly convergent on $-\infty < x < \infty$. Uniform convergence may be verified for $-a \leq x \leq a$, with any $a < +\infty$, by the Weierstrass M-test. ■

Example 4.2.8

Let $f(x) = \ln(1 + x)$. By differentiating,

$$f'(x) = \frac{1}{1+x}$$

$$f^{(i)}(x) = (-1)^{i-1}(i-1)!(1+x)^{-i}$$

The Maclaurin expansion of Eq. 4.2.21 yields

$$\ln(1+x) = \sum_{i=1}^{n}(-1)^{i-1}\frac{x^i}{i} + R_n$$

In this case, the remainder of Eq. 4.2.18 is

$$R_n = \frac{x^n}{n!}f^{(n)}(\xi_n) = \frac{x^n}{n}(-1)^{n-1}(1+\xi_n)^{-n}$$

where $0 \leq \xi_n \leq x$. Thus,

$$|R_n| \leq \frac{x^n}{n}$$

The remainder approaches zero as n approaches ∞, provided $0 \le x \le 1$, so the infinite series

$$\ln(1+x) = \sum_{i=1}^{\infty} (-1)^{i-1} \frac{x^i}{i} \tag{4.2.23}$$

converges for $0 \le x \le 1$. Absolute convergence in $-1 < x < 1$ is easily established by the ratio test; i.e.,

$$\lim_{i\to\infty} \frac{|a_{i+1}|}{|a_i|} = \lim_{i\to\infty} \frac{|x^{i+1}|/(i+1)}{|x^i|/i} = \lim_{i\to\infty} \frac{i\,|x|}{i+1} \le |x| < 1$$

The series is uniformly convergent on $-r \le x \le r$, for any $0 \le r < 1$, since $M_i = r^i$ forms a convergent geometric series and may be used in the Weierstrass M-test. ■

For a function f(x, y) of two variables, the **Taylor expansion** about the point (a, b) is

$$\begin{aligned} f(x, y) &= f(a, b) + (x-a)\frac{\partial f}{\partial x} + (y-b)\frac{\partial f}{\partial y} \\ &+ \frac{1}{2!}\left[(x-a)^2 \frac{\partial^2 f}{\partial x^2} + 2(x-a)(y-b)\frac{\partial^2 f}{\partial x \partial y} + (y-b)^2 \frac{\partial^2 f}{\partial y^2} \right] \\ &+ \frac{1}{3!}\left[(x-a)^3 \frac{\partial^3 f}{\partial x^3} + 3(x-a)^2(y-b)\frac{\partial^3 f}{\partial x^2 \partial y} \right. \\ &\left. + 3(x-a)(y-b)^2 \frac{\partial^3 f}{\partial x \partial y^2} + (y-b)^3 \frac{\partial^3 f}{\partial y^3} \right] + \dots \end{aligned} \tag{4.2.24}$$

with all derivatives evaluated at x = a and y = b.

Using $\mathbf{x} = [x_1, \dots, x_m]^T$, the **Taylor expansion** for m independent variables about the point $\mathbf{x}^0 = [x_1^0, \dots, x_m^0]^T$ takes the form

$$f(\mathbf{x}) = \sum_{i=0}^{\infty} \frac{1}{(i)!} \left[\sum_{j=1}^{m} (x_j - x_j^0) \frac{\partial}{\partial x_j} \right]^i f(\mathbf{x}) \Bigg|_{\mathbf{x}=\mathbf{x}^0} \tag{4.2.25}$$

Power Series

Power series is a special and extremely useful type of infinite series, of the form

$$f(x) = \sum_{i=0}^{\infty} a_i x^i \tag{4.2.26}$$

where the coefficients a_i are constants that are independent of x. This series may readily be tested for convergence by either the root test or the ratio test. If

$$\lim_{i\to\infty} \left| \frac{a_{i+1}}{a_i} \right| = R^{-1} \tag{4.2.27}$$

then

$$\lim_{i\to\infty} \left| \frac{u_{i+1}}{u_i} \right| = \lim_{i\to\infty} \frac{| a_{i+1} x^{i+1} |}{| a_i x^i |} = \frac{|x|}{R} < 1 \tag{4.2.28}$$

and the series converges for $-R < x < R$, which is the **interval of convergence** and R is called the **radius of convergence.** Since the root and ratio tests fail when the limit is unity, the endpoints of the interval require special attention.

Example 4.2.9

For the power series $\sum_{i=0}^{\infty} ix^i$, $a_i = i$ and the radius of convergence is

$$R^{-1} = \lim_{i\to\infty} \left| \frac{a_{i+1}}{a_i} \right| = \lim_{i\to\infty} \left| \frac{i+1}{i} \right| = 1$$

Thus, the series converges for $-1 < x < 1$. Note that at both $x = -1$ and $x = 1$, the series diverges. ■

Theorem 4.2.5. For a power series of the form of Eq. 4.2.26 that satisfies Eq. 4.2.28, the following results hold:

(a) **Uniform and absolute convergence**: The power series is convergent for $R < x < R$ and it is uniformly and absolutely convergent in any interior interval $-r \leq x \leq r$ with $0 < r < R$.

(b) **Continuity**: The function f(x) defined by Eq. 4.2.26 is continuous on any interval $-r \leq x \leq r$ with $0 \leq r < R$.

(c) **Differentiation and integration**: Derivatives and integrals of f(x) in $-r \le x \le r$ may be obtained by interchanging the order of derivative and integral and infinite summation in Eq. 4.2.26, for any r with $0 \le r < R$.

(d) **Uniqueness**: If

$$f(x) = \sum_{i=0}^{\infty} a_i x^i, \quad -R_a < x < R_a$$

$$= \sum_{i=0}^{\infty} b_i x^i, \quad -R_b < x < R_b \tag{4.2.29}$$

with overlapping intervals of convergence that include $x = 0$, then

$$a_i = b_i, \quad i = 1, 2, \ldots \tag{4.2.30}$$

(e) **Addition of power series**: Two power series may be added term by term for every value of x for which both series are convergent.

(f) **Multiplication of power series**: The product of two power series

$$A = \sum_{k=0}^{\infty} a_k x^k = a_0 + a_1 x + \ldots$$

$$B = \sum_{k=0}^{\infty} b_k x^k = b_0 + b_1 x + \ldots \tag{4.2.31}$$

is the power series

$$AB = a_0 b_0 + (a_0 b_1 + a_1 b_0)\, x + (a_0 b_2 + a_1 b_1 + a_2 b_0)\, x^2 + \ldots$$

$$= \sum_{n=0}^{\infty} \left(\sum_{j=0}^{n} a_j b_{n-j} \right) x^n \tag{4.2.32}$$

This series is called the **Cauchy product** of the series A and B. ■

Each part of Theorem 4.2.5 is proved separately in the following.

Part (a) may be proved directly by the Weierstrass M-test, using $| a_i x^i | \le M_i \equiv | a_i | r^i$ and

$$\lim_{i \to \infty} \frac{M_{i+1}}{M_i} = \lim_{i \to \infty} \frac{| a_{i+1} | r^{i+1}}{| a_i | r^i} = \frac{r}{R} < 1$$

Since $u_i(x) = a_i x^i$ is a continuous function of x and $f(x) = \sum_{i=1}^{\infty} a_i x^i$ converges uniformly for $-r \le x \le r$, f(x) is a continuous function in the interval of uniform convergence, which proves part (b).

Since $u_i(x) = a_i x^i$ is continuous and $\sum_{i=1}^{\infty} a_i x^i$ is uniformly convergent, the differentiated series is a power series of continuous functions and

$$\lim_{i\to\infty} \left| \frac{u'_{i+1}}{u'_i} \right| = \lim_{i\to\infty} \left| \frac{(i+1)\, a_{i+1} x^i}{i a_i x^{i-1}} \right| = \frac{|x|}{R} < 1$$

which has the same radius of convergence as the original series. The new factors that are introduced by differentiation (or integration) do not affect either the root or the ratio test. Therefore, the power series may be differentiated or integrated as often as desired, within the interval of uniform convergence. This proves part (c).

To prove part (d), from Eq. 4.2.29,

$$\sum_{i=0}^{\infty} a_i x^i = \sum_{i=0}^{\infty} b_i x^i \tag{4.2.33}$$

on $-R < x < R$, where R is the smaller of R_a and R_b. Setting $x = 0$, to eliminate all but the constant terms, Eq. 4.2.33 reduces to $a_0 = b_0$. Using differentiability of the power series, both sides of Eq. 4.2.33 may be differentiated to obtain

$$\sum_{i=1}^{\infty} i a_i x^{i-1} = \sum_{i=1}^{\infty} i b_i x^{i-1}$$

Evaluating this result at $x = 0$ yields $a_1 = b_1$. By repeating this process i times,

$$a_i = b_i$$

which shows that the two series coincide. Therefore, the power series representation is unique.

Part (e) follows from a rearrangement property of convergent series [8].

Part (f) is verified by multiplying each term of series A by each term of series B in Eq. 4.2.31 and collecting terms of products of the same powers of x, to obtain Eq. 4.2.32. ■

EXERCISES 4.2

1. Find the range of x for uniform convergence of

(a) $\sum_{i=1}^{\infty} \frac{(-1)^{i-1}}{i^x}$

(b) $\sum_{i=1}^{\infty} \frac{1}{i^x}$

2. For what range of x is the geometric series $\sum_{i=0}^{\infty} x^i$ uniformly convergent?

3. For what range of positive values of x is $\sum_{i=0}^{\infty} \frac{1}{(1+x^i)}$

(a) convergent?

(b) uniformly convergent?

4. Show, by a Taylor expansion about $\theta = \pi/2$, that $\sin(\pi/2 + \theta) = \cos\theta$.

5. Show that

(a) $e^x = 1 + \frac{x}{1!} + \frac{x^2}{2!} + \ldots + \frac{x^n}{n!} + \frac{e^{\theta_1 x}}{(n+1)!} x^{n+1}$

(b) $\cos x = 1 - \frac{x^2}{2!} + \frac{x^4}{4!} - \ldots + (-1)^n \frac{x^{2n}}{(2n)!} + (-1)^{n+1} \frac{\cos(\theta_2 x)}{(2n+2)!} x^{2n+2}$

where $0 < \theta_1 < 1$ and $0 < \theta_2 < 1$.

6. Are the following series (i) absolutely convergent, (ii) convergent, or (iii) divergent?

(a) $\sum_{i=1}^{\infty} e^{-i}$

(b) $\sum_{i=1}^{\infty} \frac{1+(1/i)^2}{i}(-1)^i$

7. Find the range of uniform convergence of the series $\sum_{n=1}^{\infty} \frac{n^2}{2^{n-1}} x^{n-1}$, indicating carefully the bounds on the range and whether the series converges at the bounds.

8. Consider the series $\sum_{i=1}^{\infty} (-1)^{i-1} \frac{(x-1)^i}{i}$.

(a) For what values of x does the series converge?

(b) For what values of x is the convergence uniform?

9. Find the Maclaurin series of $f(x) = 1/(1 + x^2)$. [Hint: Substitute $-x^2$ for x.]

10. Develop $1/(a - bx)$ in powers of $(x - c)$, where $(a - cb) \neq 0$ and $b \neq 0$.

11. Find the Maclaurin series of $f(x) = \tan x$. [Hint: $f'(x) = \sec^2 x$.]

4.3 POWER SERIES SOLUTION OF ORDINARY DIFFERENTIAL EQUATIONS

Consider the general second-order linear **homogeneous ordinary differential equation**, written in the form

$$y'' + f(x)\,y' + g(x)\,y = 0 \tag{4.3.1}$$

Definition 4.3.1. A value x_0 of x is called an **ordinary point** of the differential equation of Eq. 4.3.1 if there exists an interval $|x - x_0| \leq R$ in which f(x) and g(x) have convergent power series representations

$$\begin{aligned} f(x) &= \sum_{i=0}^{\infty} a_i (x - x_0)^i \\ g(x) &= \sum_{i=0}^{\infty} b_i (x - x_0)^i \end{aligned} \tag{4.3.2}$$

■

Expansions About Ordinary Points

Theorem 4.3.1. If x_0 is an ordinary point of Eq. 4.3.1, then the differential equation has a power series solution

$$y(x) = \sum_{k=0}^{\infty} c_k (x - x_0)^k \tag{4.3.3}$$

that converges in the interval $| x - x_0 | < R$. ■

To prove Theorem 4.3.1, assume, without loss of generality, that $x_0 = 0$. If not, a new independent variable $\xi = x - x_0$ can be introduced and the equation will then have an ordinary point at $\xi_0 = 0$. The first part of the proof is to carry out manipulations that are actually used in constructing a power series solution. The second part is to prove convergence.

Assume there is a solution that is valid for $| x | \le r < R$, of the form

$$y(x) = \sum_{i=0}^{\infty} c_i x^i \tag{4.3.4}$$

Formally (i.e., without considering the validity of operations performed or convergence of the series)

$$y'(x) = \sum_{i=0}^{\infty} i c_i x^{i-1} = \sum_{j=0}^{\infty} (j + 1) c_{j+1} x^j \tag{4.3.5}$$

where $j = i - 1$,

$$y''(x) = \sum_{i=0}^{\infty} i (i - 1) c_i x^{i-2} = \sum_{k=0}^{\infty} (k + 2) (k + 1) c_{k+2} x^k \tag{4.3.6}$$

and $k = i - 2$. Substituting these series and the series for $f(x)$ and $g(x)$ from Eq. 4.3.2 into Eq. 4.3.1,

$$\sum_{n=0}^{\infty} (n + 2) (n + 1) c_{n+2} x^n + \left(\sum_{i=0}^{\infty} a_i x^i \right) \left(\sum_{i=0}^{\infty} (i + 1) c_{i+1} x^i \right) + \left(\sum_{i=0}^{\infty} b_i x^i \right) \left(\sum_{i=0}^{\infty} c_i x^i \right) = 0$$

Using Eq. 4.2.32 to expand the products of series and combining like powers of x, this becomes

$$\sum_{n=0}^{\infty}\left\{ (n+2)(n+1)c_{n+2} + \sum_{j=0}^{n}(n-j+1)a_j c_{n-j+1} + \sum_{j=0}^{n} b_j c_{n-j} \right\} x^n = 0 \qquad (4.3.7)$$

Since the coefficients of the power series on the right of Eq. 4.3.7 are all zero, part (d) of Theorem 4.2.5 implies that the coefficients of all powers of x on the left of Eq. 4.3.7 must be zero. This yields the **recurrence formula** for c_n,

$$(n+2)(n+1)c_{n+2} = -\sum_{j=0}^{n}[(n-j+1)a_j c_{n-j+1} + b_j c_{n-j}] \qquad (4.3.8)$$

for n = 0, 1, 2, ... Expanding these equations for n = 0, 1, and 2 yields

$$c_2 = -\frac{1}{2}[a_0c_1 + b_0c_0] \qquad (4.3.9)$$

$$c_3 = -\frac{1}{6}[2a_0c_2 + a_1c_1 + b_0c_1 + b_1c_0] \qquad (4.3.10)$$

$$c_4 = -\frac{1}{12}[3a_0c_3 + 2a_1c_2 + a_2c_1 + b_0c_2 + b_1c_1 + b_2c_0] \qquad (4.3.11)$$

The coefficients c_0 and c_1 can be assigned arbitrarily, yielding two arbitrary constants in the **general solution** of the second-order differential equation. Equation 4.3.9 determines c_2 in terms of c_0 and c_1. Equation 4.3.10 then determines c_3 in terms of c_0, c_1, and c_2, with c_2 known from Eq. 4.3.9. Equation 4.3.11 next determines c_4 in terms of c_0 to c_3, with c_2 and c_3 known from Eqs. 4.3.9 and 4.3.10. This process may be continued using Eq. 4.3.8 for n = 3, 4, 5, . . . , to determine c_5, c_6, c_7, . . . , respectively, all in terms of arbitrary c_0 and c_1. This process constructs a general solution of the differential equation, provided the **formal operations** performed on the series (i.e., differentiation, multiplication, addition, etc.) can be justified. This part of the proof provides the procedure that is actually carried out to construct power series solutions of differential equations.

The remainder of the proof of Theorem 4.3.1 is the more technical argument that the procedure developed above yields a convergent series in Eq. 4.3.4 and that its sum is the solution of Eq. 4.3.1. Since the power series for f(x) and g(x) in Eq. 4.3.2 are absolutely convergent, $\sum_{i=0}^{\infty}|a_i||x|^i$ and $\sum_{i=0}^{\infty}|b_i||x|^i$ converge for $|x| \le r < R$. Therefore,

$\lim_{i\to\infty} | a_i | r^i = \lim_{i\to\infty} | b_i | r^i = 0$. Hence, there exist constants M and N such that $| a_i | \le Mr^{-i}$ and $| b_i | \le Nr^{-i}$, for all i. Let K be the larger of M and Nr. Then,

$$\begin{aligned} | a_i | &\le Kr^{-i} \\ | b_i | &\le Kr^{-i-1} \end{aligned} \tag{4.3.12}$$

and from Eq. 4.3.9,

$$\begin{aligned} 2 | c_2 | &\le | c_1 | | a_0 | + | c_0 | | b_0 | \\ &\le 2K | c_1 | + K | c_0 | r^{-1} \end{aligned}$$

so that $| c_2 | \le \gamma_2$, where $2\gamma_2 = K (2 | c_1 | + | c_0 | r^{-1})$. Also, from Eqs. 4.3.10, 4.3.11, and 4.3.12,

$$\begin{aligned} 6 | c_3 | &\le 2\gamma_2 | a_0 | + | c_1 | | a_1 | + | c_1 | | b_0 | + | c_0 | | b_1 | \\ &\le 3\gamma_2 K + 2 | c_1 | Kr^{-1} + | c_0 | Kr^{-2} \\ 12 | c_4 | &\le 3\gamma_3 | a_0 | + 2\gamma_2 | a_1 | + | c_1 | | a_2 | + \gamma_2 | b_0 | \\ &\qquad + | c_1 | | b_1 | + | c_0 | | b_2 | \\ &\le 4\gamma_3 K + 3\gamma_2 Kr^{-1} + 2 | c_1 | Kr^{-2} + | c_0 | Kr^{-3} \end{aligned}$$

Therefore, $| c_3 | \le \gamma_3$ and $| c_4 | \le \gamma_4$, where

$$\begin{aligned} 6\gamma_3 &= K (3\gamma_2 + 2 | c_1 | r^{-1} + | c_0 | r^{-2}) \\ 12\gamma_4 &= K (4\gamma_3 + 3\gamma_2 r^{-1} + 2 | c_1 | r^{-2} + | c_0 | r^{-3}) \end{aligned}$$

Continuing in this way with Eq. 4.3.8, for $i \ge 5$, $| c_i | \le \gamma_i$, where

$$\begin{aligned} (i - 1) i\gamma_i = K [i\gamma_{i-1} &+ (i - 1) \gamma_{i-2} r^{-1} + \ldots \\ &+ 2 | c_1 | r^{-i+2} + | c_0 | r^{-i+1}] \end{aligned} \tag{4.3.13}$$

Next, write Eq. 4.3.13 with i replaced by i – 1 and multiply by r^{-1}, to obtain

$$(i-2)(i-1)\gamma_{i-1}r^{-1} = K[(i-1)\gamma_{i-2}r^{-1} + \ldots$$

$$+ 2|c_1|r^{-i+2} + |c_0|r^{-i+1}] \quad (4.3.14)$$

Subtracting Eq. 4.3.14 from Eq. 4.3.13,

$$(i-1)i\gamma_i - (i-2)(i-1)\gamma_{i-1}r^{-1} = Ki\gamma_{i-1}$$

from which it follows that

$$\frac{\gamma_i}{\gamma_{i-1}} = \frac{i-2}{ir} + \frac{K}{i-1}$$

Now consider the series $\sum_{i=0}^{\infty} \gamma_i x^i$. The ratio test implies that this series converges for $|x| < r$, since

$$\lim_{i\to\infty} \frac{\gamma_i |x|}{\gamma_{i-1}} = \frac{|x|}{r} < 1$$

By the comparison test, $\sum_{i=0}^{\infty} c_i x^i$ converges in the same interval because $|c_i x^i| \le \gamma_i |x|^i$. Since r is any positive number less than R, the series

$$y(x) = \sum_{i=0}^{\infty} c_i x^i$$

converges for $|x| < R$. Thus, all the operations used to define the c_i are valid and Eq. 4.3.4 yields the desired solution. ■

Example 4.3.1

For the differential equation

$$\frac{d^2y}{dx^2} - y = 0 \quad (4.3.15)$$

$f(x) = 0$ and $g(x) = -1$, so $x = 0$ is an ordinary point. Writing the assumed solution in the form

$$y(x) = \sum_{i=0}^{\infty} c_i x^i$$

and obtaining, by differentiation,

$$\frac{d^2y}{dx^2} = \sum_{i=0}^{\infty} i\,(\,i-1\,)\,c_i x^{i-2}$$

The differential equation of Eq. 4.3.15 is then

$$\sum_{i=0}^{\infty} i\,(\,i-1\,)\,c_i x^{i-2} - \sum_{i=0}^{\infty} c_i x^i = 0 \qquad (4.3.16)$$

In order to collect the coefficients of like powers of x, the indices of summation are changed in such a way that exponents of x in the two summations are the same. For this purpose, replace i by $i-2$ in the second summation, so it becomes

$$\sum_{i-2=0}^{\infty} c_{i-2} x^{i-2} = \sum_{i=2}^{\infty} c_{i-2} x^{i-2}$$

and Eq. 4.3.16 becomes

$$\sum_{i=0}^{\infty} i\,(\,i-1\,)\,c_i x^{i-2} - \sum_{i=2}^{\infty} c_{i-2} x^{i-2} = 0$$

Since the first two terms ($i = 0, 1$) of the first summation are zero, the lower limit of summation may be taken as $i = 2$ and the summations combined to obtain

$$i\,(\,i-1\,)\,c_i = c_{i-2}, \quad i = 2, 3, \ldots$$

This is the recurrence formula for the c_i. It expresses each coefficient c_i for which $i \geq 2$ as a multiple of the second preceding coefficient c_{i-2}. The general solution may thus be written in terms of the undetermined constants c_0 and c_1. ■

Returning to the differential equation of Eq. 4.3.1, if f(x) or g(x) are not continuous at x_0, then x_0 is called a **singular point** of the differential equation.

Definition 4.3.2. If x_0 is not a regular point of Eq. 4.3.1, it is called a **singular point**. If $(\,x - x_0\,)\,f(x)$ and $(\,x - x_0\,)^2\,g(x)$ have power series representations in some interval $|\,x - x_0\,| < R$, then x_0 is called a **regular singular point** of the differential equation. ■

Expansions About Singular Points (Frobenius Method)

In a neighborhood of a regular singular point of Eq. 4.3.1, a series solution of the differential equation can still be obtained by the **Frobenius method**, which is presented next. Assume, without loss of generality, that $x_0 = 0$. First multiply the differential equation of Eq. 4.3.1 by x^2, to obtain

$$x^2 y'' + x F(x) y' + G(x) y = 0 \tag{4.3.17}$$

where

$$\begin{aligned} F(x) &= x f(x) = \sum_{i=0}^{\infty} a_i x^i \\ G(x) &= x^2 g(x) = \sum_{i=0}^{\infty} b_i x^i \end{aligned} \tag{4.3.18}$$

and these series converge for $|x| < R$. Assume a solution of the form

$$y(x) = \sum_{i=0}^{\infty} c_i x^{i+\alpha} \tag{4.3.19}$$

where α is a constant that is to be determined. Then, by formally differentiating and manipulating,

$$xy' = \sum_{i=0}^{\infty} (i+\alpha) c_i x^{i+\alpha}$$

$$x^2 y'' = \sum_{i=0}^{\infty} (i+\alpha)(i+\alpha-1) c_i x^{i+\alpha}$$

Substituting these results and Eq. 4.3.18 into Eq. 4.3.17,

$$\sum_{i=0}^{\infty} (i+\alpha)(i+\alpha-1) c_i x^{i+\alpha} + \left(\sum_{i=0}^{\infty} a_i x^i \right)\left(\sum_{i=0}^{\infty} (i+\alpha) c_i x^{i+\alpha} \right) + \left(\sum_{i=0}^{\infty} b_i x^i \right)\left(\sum_{i=0}^{\infty} c_i x^{i+\alpha} \right) = 0$$

Expanding the products of series and combining terms, as in deriving Eq. 4.3.7, and equating coefficients of the various powers of x to zero yields the **recurrence formula**

$$c_0 [\alpha (\alpha - 1) + \alpha a_0 + b_0] = 0 \tag{4.3.20}$$

$$c_i [(\alpha + i) (\alpha + i - 1) + a_0 (\alpha + i) + b_0] + \sum_{j=0}^{i-1} c_j [(\alpha + j) a_{i-j} + b_{i-j}] = 0, \quad i = 1, 2, \ldots \tag{4.3.21}$$

Equation 4.3.20 indicates that c_0 is arbitrary, only if α is a solution of the **indicial equation**

$$\alpha (\alpha - 1) + \alpha a_0 + b_0 = 0 \tag{4.3.22}$$

This equation is quadratic in α, so it has two solutions that are not necessarily distinct. Let α_1 and α_2 be the roots of the indicial equation. Three special cases arise, with properties as follows.

Theorem 4.3.2. Let $x_0 = 0$ be a regular singular point for Eq. 4.3.1, with the series in Eq. 4.3.18 converging if $| x | < R$. Then,

(1) if $\alpha_1 \neq \alpha_2$ and $\alpha_2 - \alpha_1$ is not an integer, Eq. 4.3.21 yields coefficients for two series solutions, one associated with α_1 and the other with α_2; i.e.,

$$\begin{aligned} y_1 &= \sum_{n=0}^{\infty} c_n x^{n+\alpha_1} \\ y_2 &= \sum_{n=0}^{\infty} d_n x^{n+\alpha_2} \end{aligned} \tag{4.3.23}$$

both with radius of convergence R.

(2) if $\alpha_1 = \alpha_2$, the method yields only one convergent series solution, with radius of convergence R. However, if one solution is known, another independent solution can be obtained by a substitution that reduces the problem to a linear first-order equation, to be carried out later.

(3) if $\alpha_2 = \alpha_1 + n$, where n is an integer, Eq. 4.3.20 yields

$$(\alpha_1 + n) (\alpha_1 + n - 1) + a_0 (\alpha_1 + n) + b_0 = 0 \tag{4.3.24}$$

This means that the coefficient of c_n in Eq. 4.3.21 is zero for α_1 and the solution corresponding to the root α_1 fails, unless

$$\sum_{j=0}^{n-1} c_j [(\alpha_1 + j) a_{n-j} + b_{n-j}] = 0 \tag{4.3.25}$$

In the latter case, c_n is arbitrary. Otherwise, the method fails to yield two solutions. One series with a radius of convergence R is obtained, however, using the root α_2. ■

To prove Theorem 4.3.2, write Eq. 4.3.21 as

$$c_i I(\alpha + i) = - \sum_{j=0}^{i-1} c_j [(\alpha + j) a_{i-j} + b_{i-j}] \tag{4.3.26}$$

where

$$I(\alpha) \equiv \alpha (\alpha - 1) + \alpha a_0 + b_0 = (\alpha - \alpha_1) (\alpha - \alpha_2) \tag{4.3.27}$$

Then,

$$\begin{aligned} I(\alpha_1 + i) &= i (i + \alpha_1 - \alpha_2) \\ I(\alpha_2 + i) &= i (i + \alpha_2 - \alpha_1) \end{aligned} \tag{4.3.28}$$

From the second of Eq. 4.3.28 and Eq. 4.3.26,

$$\begin{aligned} i\Big(i - | \alpha_1 - \alpha_2 | \Big) | c_i | &\le | I(\alpha_2 + i) | \, | c_i | \\ &\le \sum_{j=0}^{i-1} | c_j | \Big[\Big(| \alpha_2 | + j \Big) | a_{i-j} | + | b_{i-j} | \Big] \end{aligned}$$

for all $i \ge | \alpha_1 - \alpha_2 |$.

Let $| c_j | = \gamma_j$, for $j < m$, where m is some integer greater than $| \alpha_1 - \alpha_2 |$. Then, for $\alpha_2 \ge \alpha_1$,

$$m\Big(m - | \alpha_1 - \alpha_2 | \Big) | c_m | \le \sum_{j=0}^{m-1} \gamma_j \Big[\Big(| \alpha_2 | + j \Big) | a_{m-j} | + | b_{m-j} | \Big]$$

As in the case of an ordinary point, since the series in Eq. 4.3.18 converge, there exists a constant K such that

$$| a_i | \leq Kr^{-i}$$

$$| b_i | \leq Kr^{-i}$$

Then,

$$m\left(m - | \alpha_1 - \alpha_2 | \right)| c_m | \leq K \sum_{j=0}^{m-1} \gamma_j \left(| \alpha_2 | + j + 1 \right) r^{-m+j}$$

and $| c_m | \leq \gamma_m$, where γ_m is the solution of

$$m\left(m - | \alpha_1 - \alpha_2 | \right)\gamma_m = K \sum_{j=0}^{m-1} \gamma_j \left(| \alpha_2 | + j + 1 \right) r^{-m+j}$$

Further, for $n \geq m$,

$$n\left(n - | \alpha_1 - \alpha_2 | \right)| c_n | \leq K \sum_{j=0}^{n-1} \gamma_j \left(| \alpha_2 | + j + 1 \right) r^{-n+j}$$

and $| c_n | \leq \gamma_n$, where γ_n is defined by

$$n\left(n - | \alpha_1 - \alpha_2 | \right)\gamma_n = K \sum_{j=0}^{n-1} \gamma_j \left(| \alpha_2 | + j + 1 \right) r^{-n+j}$$

Replacing n by $n - 1$ and dividing by r,

$$(n-1)\left(n - 1 - | \alpha_1 - \alpha_2 | \right)\gamma_{n-1} r^{-1} = K \sum_{j=0}^{n-2} \gamma_j \left(| \alpha_2 | + j + 1 \right) r^{-n+j}$$

Subtracting the preceding pair of equations yields

$$n\left(n - | \alpha_1 - \alpha_2 | \right)\gamma_n - (n-1)\left(n - 1 - | \alpha_1 - \alpha_2 | \right)\gamma_{n-1} r^{-1}$$

$$= K\gamma_{n-1}\left(n + | \alpha_2 | \right) r^{-1}$$

or

$$\frac{\gamma_n}{\gamma_{n-1}} = \frac{(n-1)\left(n - 1 - | \alpha_1 - \alpha_2 | \right)}{n\left(n - | \alpha_1 - \alpha_2 | \right) r} + \frac{K\left(n + | \alpha_2 | \right)}{n\left(n - | \alpha_1 - \alpha_2 | \right) r}$$

Consider the series $\sum_{i=0}^{\infty} \gamma_i x^i$. This series converges absolutely for $|x| < r$, by the ratio test for absolute convergence (Theorem 4.1.7), since

$$\lim_{n\to\infty} \frac{\gamma_n |x|}{\gamma_{n-1}} = \frac{|x|}{r} < 1$$

when $|x| < r$. By the comparison test, $\sum_{i=0}^{\infty} c_i x^i$ converges absolutely in the same interval. This implies that it converges absolutely for $|x| < R$, since r is any positive number less than R. This justifies all of the formal operations that were used in deriving the series solution of the differential equation and shows that

$$y(x) = x^{\alpha_2} \sum_{i=0}^{\infty} c_i x^i \tag{4.3.29}$$

where $\alpha_2 \geq \alpha_1$, is a solution of the differential equation that is valid for $|x| < R$ and the coefficients c_i are determined by the recurrence formula of Eq. 4.3.21.

If $\alpha_2 - \alpha_1$ is not zero or a positive integer, then a second independent solution is obtained by the Frobenius method, using the root α_1, and the proof of convergence of the series is essentially the same as that for α_2.

If $\alpha_2 - \alpha_1$ is zero or a positive integer and the method fails to give two independent solutions, let u(x) be a known solution of the differential equation and seek a second solution of the form $y(x) = u(x)\, v(x)$. Upon substituting this into Eq. 4.3.1, the differential equation that v(x) must satisfy is

$$v'' + \left(\frac{2u'}{u} + f \right) v' = 0$$

This is a first-order linear equation in v', with the solution

$$v' = Au^{-2} e^{-\int f(x)\,dx}$$

$$v = A \int u^{-2} e^{-\int f(x)\,dx}\, dx + B$$

Then,

$$y = uv = Au \int u^{-2} e^{-\int f(x)dx}\, dx + Bu \tag{4.3.30}$$

is the general solution of Eq. 4.3.1, which completes the proof of Theorem 4.3.2. ■

Example 4.3.2

To illustrate the Frobenius method, consider the equation

$$x^2 \frac{d^2y}{dx^2} + (x^2 + x)\frac{dy}{dx} - y = 0$$

or

$$y'' + \left(1 + \frac{1}{x}\right)y' - \frac{1}{x^2}y = 0$$

Thus, $x_0 = 0$ is a regular singular point. By direct substitution of

$$y = x^\alpha \sum_{i=0}^{\infty} c_i x^i = \sum_{i=0}^{\infty} c_i x^{i+\alpha}$$

into the above differential equation,

$$(\alpha^2 - 1)c_0 x^\alpha + \sum_{i=1}^{\infty}\{[(\alpha+i)^2 - 1]c_i + (\alpha+i-1)c_{i-1}\}x^{i+\alpha} = 0$$

Hence, the indicial equation is

$$\alpha^2 - 1 = 0$$

with roots $\alpha_1 = -1$ and $\alpha_2 = 1$, and the recurrence formula for c_i is

$$[(\alpha+i)^2 - 1]c_i + (\alpha+i-1)c_{i-1} = 0$$

or

$$(\alpha+i-1)[(\alpha+i+1)c_i + c_{i-1}] = 0, \quad i \geq 1$$

Since α_1 and α_2 differ by an integer, a solution of the required type is assured only when α has the larger value (Case 3 of Theorem 4.3.2).

With $\alpha_2 = 1$, the recurrence formula is

$$i[(i+2)c_i + c_{i-1}] = 0$$

or, since $i \neq 0$, for $i \geq 1$,

$$c_i = -\frac{c_{i-1}}{i+2}$$

Thus,

$$c_1 = -\frac{c_0}{3}, \quad c_2 = \frac{c_0}{3 \cdot 4}, \quad c_3 = -\frac{c_0}{3 \cdot 4 \cdot 5}, \quad \ldots$$

and the solution corresponding to $\alpha_2 = 1$ is

$$y_1(x) = c_0\left(x - \frac{x^2}{3} + \frac{x^3}{3 \cdot 4} - \frac{x^4}{3 \cdot 4 \cdot 5} + \ldots \right)$$

$$= 2c_0 \sum_{i=1}^{\infty} (-1)^{i+1} \frac{x^i}{(i+1)!}$$

$$= 2c_0 \frac{e^{-x} - 1 + x}{x}$$

Using $u(x) = \frac{1}{x}[e^{-x} - 1 + x]$ in Eq. 4.3.30, with $f(x) = 1 + \frac{1}{x}$, yields the general solution

$$y(x) = A\frac{1}{x}[e^{-x} - 1 + x]\int \frac{xe^{-x}\,dx}{[e^{-x} - 1 + x]^2} + B\frac{1}{x}[e^{-x} - 1 + x]$$

■

EXERCISES 4.3

1. For each of the following differential equations, obtain a general solution that is representable by a Maclaurin series:

(a) $\frac{d^2y}{dx^2} + 2y = 0$

(b) $x^2\frac{d^2y}{dx^2} - x\frac{dy}{dx} + y = 0$

2. Use the Frobenius method to obtain the general solution of each of the following differential equations, near $x = 0$:

(a) $x^2\frac{d^2y}{dx^2} + x\frac{dy}{dx} + \left(x^2 - \frac{1}{4}\right)y = 0$

(b) $x(1-x)\dfrac{d^2y}{dx^2} - 2\dfrac{dy}{dx} + 2y = 0$

(c) $x\dfrac{d^2y}{dx^2} - \dfrac{dy}{dx} + 4x^3y = 0$

3. For each of the equations in Exercise 2, give the largest interval inside which convergence of the Frobenius solution is guaranteed.

4. Determine two values of the constant α for which all solutions of the equation

$$x\frac{d^2y}{dx^2} + (x-1)\frac{dy}{dx} - \alpha y = 0$$

are regular at $x = 0$. Obtain the general solution in each of these cases.

5. Solve the following differential equation, using the power series method:

$$y'' + x^2y = 0$$

Find only the first eight terms. What is the radius of convergence?

6. Derive and solve the indicial equation of the Frobenius method for solving the differential equation

$$x^2y'' + (e^x \sin 3x)\, y' - 3y \cos 3x = 0$$

4.4 INTRODUCTION TO SPECIAL FUNCTIONS

As will be seen later in treating problems that are governed by partial differential equations, a few second-order ordinary differential equations arise again and again. It is efficient, therefore, to construct their solutions once and for all. These solutions are called **special functions** of mathematical physics and are studied in a rather extensive literature in their own right [10]. As a simple illustration, consider the second-order differential equation with constant coefficients,

$$\frac{d^2y}{dx^2} + \omega^2 y = 0$$

The general solution is

$$y(x) = A \sin \omega x + B \cos \omega x$$

which could be obtained by power series methods that define power series expansions of the sine and cosine functions. Thus, trigonometric functions could be viewed as the most common of the special functions.

Legendre Polynomials

Another important second-order equation is the **Legendre equation**

$$[(1 - x^2) y']' + \lambda y = 0 \tag{4.4.1}$$

which can be written as

$$y'' - \frac{2x}{1 - x^2} y' + \frac{\lambda}{1 - x^2} y = 0 \tag{4.4.2}$$

Note that $x = 0$ is an ordinary point, while $x = \pm 1$ are regular singular points. The indicial equation at either of the singular points is $\alpha^2 = 0$, so that $\alpha_1 = \alpha_2 = 0$. Hence, a power series solution can be obtained that is valid in a neighborhood of either of the singularities. However, these solutions are not finite at the other singularity, unless λ has certain values. These values of λ can be determined by seeking solutions that are valid near the origin as series in powers of x. Hence, assume

$$y = \sum_{i=0}^{\infty} c_i x^i$$

Differentiating and substituting in the differential equation,

$$(1 - x^2) \sum_{i=2}^{\infty} i(i-1) c_i x^{i-2} - 2x \sum_{i=1}^{\infty} i c_i x^{i-1} + \lambda \sum_{i=0}^{\infty} c_i x^i = 0$$

or

$$\sum_{m=0}^{\infty} [\lambda c_m + (m+2)(m+1) c_{m+2}] x^m - \sum_{m=0}^{\infty} 2m c_m x^m - \sum_{m=0}^{\infty} m(m-1) c_m x^m = 0$$

Setting the coefficients of various powers of x equal to zero, the recurrence formula is

$$c_{m+2} = \frac{m(m+1) - \lambda}{(m+2)(m+1)} c_m, \quad m = 0, 1, 2, \ldots$$

Setting $c_1 = 0$ with $c_0 \neq 0$ yields a series in even powers of x. Setting $c_0 = 0$ with $c_1 \neq 0$ yields a series in odd powers of x. These solutions are linearly independent. Therefore, a linear combination of them will be the general solution of the differential equation. If $\lambda \neq n(n+1)$ for any nonnegative integer n, these solutions will be infinite series that converge for $|x| < 1$. However, it can be shown that they diverge at $x = \pm 1$ [10]. If finite solutions are sought at $x = \pm 1$, λ must equal $n(n+1)$, for some nonnegative integer n. In this case, either the series in even powers or the one in odd powers terminates, depending on whether n is even or odd, giving a polynomial solution. If c_0 or c_1 is adjusted so that the solution takes on the value 1 at x = 1, the **Legendre polynomials** $P_n(x)$ are obtained, the first four of which are $P_0(x) = 1$, $P_1(x) = x$, $P_2(x) = (3x^2 - 1)/2$, and $P_3(x) = (5x^3 - 3x)/2$. For general n,

$$P_n(x) = \sum_{i=0}^{K} \frac{(-1)^i (2n-2i)!}{2^n i! (n-i)! (n-2i)!} x^{n-2i}$$

where K = n/2 or K = (n − 1)/2, depending on whether n is even or odd. The definition of $P_n(x)$ may be restated as the **Rodrigues' formula**

$$P_n(x) = \sum_{i=0}^{K} \frac{(-1)^i}{2^n i! (n-i)!} \frac{d^n}{dx^n} x^{2n-2i}$$

$$= \frac{1}{2^n n!} \frac{d^n}{dx^n} \sum_{i=0}^{n} \frac{(-1)^i n!}{i! (n-i)!} x^{2n-2i}$$

$$= \frac{1}{2^n n!} \frac{d^n}{dx^n} (x^2 - 1)^n$$

The Legendre polynomials are solutions of the Legendre equation,

$$(1 - x^2) y'' - 2xy' + n(n+1) y = 0 \qquad (4.4.3)$$

Since they are polynomials, they are finite in any finite interval. They are, therefore, solutions of the equation

$$[(1 - x^2) y']' + \lambda y = 0 \qquad (4.4.4)$$

with y(1) and y(−1) finite for $\lambda_n = n(n+1)$.

Bessel Functions

A third important second-order differential equation is **Bessel's equation,**

$$x^2y'' + xy' + (x^2 - \nu^2)y = 0 \tag{4.4.5}$$

It has a regular singular point at $x = 0$. If a series solution is sought that is valid in a neighborhood of $x = 0$, the indicial equation is $\alpha^2 - \nu^2 = 0$. The roots are $\alpha_1 = -\nu$ and $\alpha_2 = \nu$. Three distinct cases arise, as follows:

Case 1: If ν is not an integer in Eq. 4.4.5, two independent solutions are

$$\begin{aligned} y_1 &= x^{\nu} \sum_{i=0}^{\infty} c_i x^i \\ y_2 &= x^{-\nu} \sum_{i=0}^{\infty} c_i x^i \end{aligned} \tag{4.4.6}$$

Substituting y_1 into the differential equation,

$$\sum_{i=0}^{\infty} i(i + 2\nu) c_i x^{i+\nu} + \sum_{i=0}^{\infty} c_i x^{i+\nu+2} = 0$$

The coefficient of $x^{\nu+1}$ is $(1 + 2\nu) c_1$. Since this must be zero, $c_1 = 0$ if $\nu \neq -1/2$. The other coefficients are determined by the recurrence formula

$$c_{i+2} = -\frac{c_i}{(i+2)(i+2+2\nu)}$$

Since $c_1 = 0$, all coefficients with odd subscripts are also zero. Even if c_0 is arbitrary, the other coefficients are determined in terms of c_0 as

$$c_{2n} = \frac{(-1)^n c_0}{2^{2n} n! (\nu + 1)(\nu + 2) \ldots (\nu + n)}$$

Thus, a solution of Bessel's equation is

$$J_\nu(x) = x^{\nu} \sum_{n=0}^{\infty} c_n x^n = x^{\nu} \sum_{n=0}^{\infty} \frac{(-1)^n c_0 x^{2n}}{2^{2n} n! (\nu + 1)(\nu + 2) \ldots (\nu + n)} \tag{4.4.7}$$

where c_0 is defined in the literature as $c_0 = [2^{\nu} \Gamma(\nu + 1)]^{-1}$ and Γ is the **Gamma function** [10]. Even with arbitrary c_0, Eq. 4.4.7 yields a solution of Bessel's equation. This is called the **Bessel function of the first kind of order** ν.

If ν is not an integer and $\nu \neq 1/2$, another solution that corresponds to the root $\alpha_1 = -\nu$ is

$$J_{-\nu}(x) = x^{-\nu} \sum_{n=0}^{\infty} \frac{(-1)^n c_0 x^{2n}}{2^{2n}\, n!\,(-\nu+1)(-\nu+2)\ldots(-\nu+n)} \qquad (4.4.8)$$

This is the **Bessel function of the first kind of order** $-\nu$. These two solutions are linearly independent, so the general solution is

$$y = A\, J_\nu(x) + B\, J_{-\nu}(x)$$

where A and B are arbitrary constants.

Case 2: If $\nu = 0$, then the solutions of Eqs. 4.4.7 and 4.4.8 are the same.

Case 3: If $\nu = m$, $m = 1, 2, 3, \ldots$, then $\alpha_2 - \alpha_1 = 2m$ is an integer and the situation arises in which the Frobenius method with $\alpha = \alpha_1$ may not yield a solution. The recurrence formula is

$$b_k + (k+2)(k+2-2m)\, b_{k+2} = 0$$

When $k = 2m - 2$, this equation cannot be used to determine b_{2m}. However, suppose $b_0 = 0$. Then $b_0 = b_2 = \ldots = b_{2m-2} = 0$ and the recurrence formula for $k = 2m - 2$ is satisfied for arbitrary b_{2m}. Hence,

$$b_{k+2} = -\frac{b_k}{(k+2)(k+2-2m)}, \qquad k = 2m, 2m+2, \ldots$$

From this,

$$b_{2m+2n} = \frac{(-1)^n b_{2m}}{2^{2n}\, n!\,(m+1)(m+2)\ldots(m+n)}$$

and a solution of Eq. 4.4.5 is

$$y = x^{-m} b_{2m} \sum_{n=0}^{\infty} \frac{(-1)^n x^{2m+2n}}{2^{2n}\, n!\,(m+1)(m+2)\ldots(m+n)}$$

$$= 2^m m!\, b_{2m} x^m \sum_{n=0}^{\infty} \frac{(-1)^n (x/2)^{2n}}{2^{2n}\, n!\,\Gamma(m+n+1)}$$

$$= 2^m m!\, b_{2m} J_m(x) \qquad (4.4.9)$$

But this solution is proportional to $J_m(x)$. Therefore, if $\nu = 1, 2, 3, \ldots$, no new independent solution is obtained.

EXERCISES 4.4

1. Find the bounded solution $J_0(x)$ of the zeroth-order Bessel equation

$$xy'' + y' + xy = 0$$

that is valid at the origin. Show that the other solution $Y_0(x)$ becomes unbounded of the form $(2/\pi) \ln x$ as x approaches zero.

2. Show that the Frobenius method yields two independent solutions of Bessel's equation of order 1/2 that are valid near the origin, even though the roots of the indicial equation differ by an integer. Show that the general solution can be written as $y = (A \sin x + B \cos x)/\sqrt{x}$.

3. Show that the Frobenius method fails to give a nontrivial solution of the equation

$$x^3 y'' + x^2 y' + y = 0$$

that is valid near the origin. Note that the singularity at $x = 0$ is not a regular singularity.

4. Find the general solution of the Legendre differential equation

$$(1 - x^2) y'' - 2xy' + n(n+1) y = 0$$

with n a nonnegative integer, that is valid near $x = 1$. Show that it becomes unbounded with the form of a constant times $\ln |x - 1|$ as x approaches 1.

5

FUNCTION SPACES AND FOURIER SERIES

While a knowledge of matrices and finite-dimensional linear algebra, as illustrated in Chapters 1 and 2, is necessary in engineering analysis, it is not sufficient. Many important applications deal with a **continuum**, which is a continuous distribution of material over a one-, two-, or three-dimensional physical region in R^1, R^2, or R^3, respectively. In such applications, a function $u(\mathbf{x}, t)$ of the spatial variable $\mathbf{x}$ and possibly time t must be found to represent temperature in a solid or gas, displacement of an elastic structure, velocity of flow of a fluid, or concentration of a gas in a porous material at the point $\mathbf{x}$ and time t. Static problems will, of course, involve only the space variable $\mathbf{x}$. The idea of function spaces of candidate solutions is defined in this chapter. The concept of such a collection of functions is a natural extension of the finite-dimensional vector spaces of variables that arise in discrete models of engineering systems studied in Chapters 1, 2, and 3.

One of the most common methods of describing functions is the use of Fourier sine and cosine series. Fourier series are important in their own right for representation and approximation of functions in a variety of fields of endeavor. Further, they provide a practical basis for constructing approximate solutions of differential equations. Methods of constructing the Fourier series and verifying their convergence are developed in this chapter.

5.1 THE ALGEBRA OF FUNCTION SPACES

Representation of Functions

A **real-valued function** $f(x)$ of the real variable x is defined by a rule by which a value of $f(x)$ corresponds to each value of x in an assigned domain. Three familiar ways of specifying a function are to give

(a) a formula
(b) a graph
(c) a tabulation of values

Suppose that a problem is to be solved in which an unknown function $f(x)$ must be found that satisfies a given differential equation, with sufficient data concerning boundary-values of the function and its derivatives to make the solution unique. This is a situation that is faced again and again in applications.

Suppose that a problem is easy and that by manipulation with familiar functions the equation can be solved to give a formula for the solution f(x). From this formula, a graph or a tabulation of values can be constructed. Any such graph will have properties such as a slope, smooth segments, derivatives, etc. These may or may not be the important properties of f(x). In fact, a graph requires a great deal of information and a tabulation of values of an unknown function would be out of question.

Thus, on the one hand, a symbol f(x) says too little and a graph or tabulation says too much. A function space representation provides a good compromise. Let f(x) and g(x) be two real-valued functions of the same real variable x over the **domain** $a < x < b$. Suppose these functions exist, perhaps as solutions of differential equations, but they are not known; i.e., no formulae, no graphs, and no tabulations of their values are given.

Algebraic Operations With Functions

It is often necessary to add values of functions that have physical significance. **Addition of functions** and **multiplication of a function by a scalar** are defined as

$$\begin{aligned} (f+g)(x) &\equiv f(x) + g(x) \\ (\alpha f)(x) &\equiv \alpha f(x) \end{aligned} \tag{5.1.1}$$

for all real α and all x in $a < x < b$. That is, the value of the newly defined function $(f + g)(x)$ at the point x is just the sum of the values of f(x) and g(x), evaluated at the same point x. Likewise, the function $(\alpha f)(x)$, evaluated at point x, is just the product of the scalar α and the value of f(x) at the point x. Defining the **zero function** as $0(x) = 0$, for all x, it may be verified that the collection of all functions on an interval $a < x < b$ is a vector space, called a **function space**. To show this, the vector space postulates of Section 2.2 must be verified.

To make the idea of a function space more useful in applications, the space of functions is restricted to those having properties required of the solution. That is, the collection of all functions on $a < x < b$ is far too large a class of functions to be very useful. It contains discontinuous functions, wildly oscillating functions, and functions that do not have derivatives. Since something is generally known about the smoothness of the functions encountered in engineering, a more restrictive space of functions that have the desired properties of the solution may be defined.

Example 5.1.1

To illustrate the idea of function spaces, consider the following physical examples:

A vibrating string. Consider a stretched string that vibrates laterally with fixed end points $x = 0$ and $x = \ell$ and denote the displacement as u(x, t), as shown in Fig. 5.1.1. Since the ends of the string are fixed, only functions that satisfy the boundary conditions $u(0, t) = u(\ell, t) = 0$, for all t, can be admitted as candidate solutions. If the string is not to break, the displacement function u(x, t) must at least be continuous. As will be seen later, the physics of the situation requires that the displacement function be even smoother.

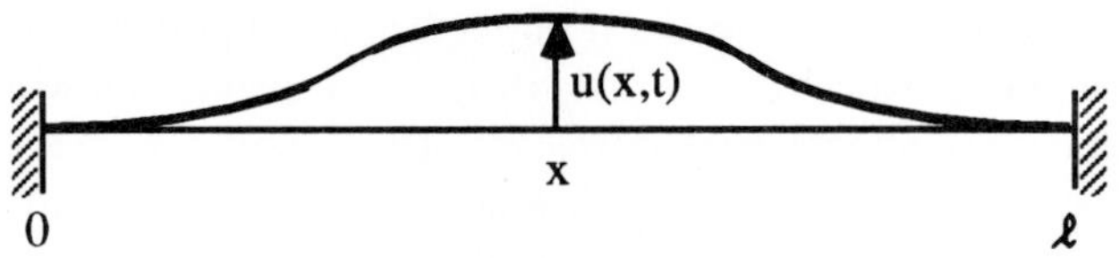

Figure 5.1.1 Vibrating String

A gas. Consider a chamber that occupies a volume $\Omega \subset R^3$ and contains a gas. Let p, ρ, and T be pressure, density, and temperature, respectively. These quantities can vary from point to point in the chamber and also with time. They are, in fact, three functions of $\mathbf{x} = [\, x_1, x_2, x_3 \,]^T \in \Omega$ and t. Thus, a **vector function** must be determined that defines the instantaneous state of the gas; i.e., the set of three functions

$$\mathbf{u}(\mathbf{x}, t) = [\, p(\mathbf{x}, t), \rho(\mathbf{x}, t), T(\mathbf{x}, t) \,]^T \tag{5.1.2}$$

of four variables. As t changes, these functions change, so the trajectory of **u** might be thought of as describing a curve in the function space. This curve represents the history of the state of the gas. It depends on the physical laws of motion of the gas, on its initial state, and on the equation of state of the gas, which is a thermodynamic relationship among p, ρ, and T. Once the equation of state is given, the possibility of **u** ranging through the whole function space is eliminated; i.e., the equation of state confines **u** to a subset of the function space. ■

In applications, confusion between the physical space in which the continuum resides and the function space that is created for purposes of representing candidate solutions must be avoided. For vectors in the physical space, the symbol **x** will generally be used. For vectors in the function space, the notation **u** or **u(x)** is used. The **physical space** is a **domain** Ω (an **open set**) in **3-space** R^3; **2-space** R^2; or the **real line** R^1. The domain may be bounded externally by a surface, a curve, or a pair of points, according to the dimensionality.

Definition 5.1.1. Let the physical space be a domain (an open set) Ω in R^n with coordinates $x_1, x_2, \ldots, x_n$ and let $u_1(\mathbf{x}), \ldots, u_m(\mathbf{x})$ be a set of m functions of these coordinates. A **vector function u(•)** of a **vector variable x** is defined as

$$\mathbf{u}(\mathbf{x}) = [\, u_1(\mathbf{x}), \ldots, u_m(\mathbf{x}) \,]^T \tag{5.1.3}$$

and the **zero vector function** is

$$\mathbf{0}(\mathbf{x}) = [\, 0, 0, \ldots, 0 \,]^T$$

Equality of vector functions

$$\mathbf{u}^1(\mathbf{x}) = [\, u_1{}^1(\mathbf{x}), u_2{}^1(\mathbf{x}), \ldots, u_m{}^1(\mathbf{x}) \,]^T$$

$$\mathbf{u}^2(\mathbf{x}) = [\, u_1{}^2(\mathbf{x}), u_2{}^2(\mathbf{x}), \ldots, u_m{}^2(\mathbf{x}) \,]^T$$

on Ω is defined by

$$u_1{}^1(\mathbf{x}) = u_1{}^2(\mathbf{x}), \ldots, u_m{}^1(\mathbf{x}) = u_m{}^2(\mathbf{x})$$

for all $\mathbf{x}$ in Ω. The **sum of two vector functions** is defined as

$$(\, \mathbf{u}^1 + \mathbf{u}^2 \,)(\mathbf{x}) = [\, u_1{}^1(\mathbf{x}) + u_1{}^2(\mathbf{x}), \ldots, u_m{}^1(\mathbf{x}) + u_m{}^2(\mathbf{x}) \,]^T \tag{5.1.4}$$

for all $\mathbf{x}$ in Ω. **Multiplication of a vector function by a scalar** α is defined as

$$(\, \alpha\mathbf{u} \,)(\mathbf{x}) = [\, \alpha u_1(\mathbf{x}), \ldots, \alpha u_m(\mathbf{x}) \,]^T \tag{5.1.5}$$

for all $\mathbf{x}$ in Ω. Finally, the **negative of a vector function u** is defined as $-\mathbf{u}(\mathbf{x}) = (-1)\mathbf{u}(\mathbf{x})$, so the rule for **subtraction of vector functions** follows as

$$\begin{aligned} (\, \mathbf{u}^1 - \mathbf{u}^2 \,)(\mathbf{x}) &= \mathbf{u}^1(\mathbf{x}) + (-1)\, \mathbf{u}^2(\mathbf{x}) \\ &= [\, u_1{}^1(\mathbf{x}) - u_1{}^2(\mathbf{x}), \ldots, u_m{}^1(\mathbf{x}) - u_m{}^2(\mathbf{x}) \,]^T \end{aligned}$$

for all $\mathbf{x}$ in Ω. The collection of all vector functions defined on Ω, with regularity properties that are preserved by addition and multiplication by a scalar, is called a **function space**, denoted by $F(\Omega)$. ■

By virtue of these definitions, it follows that

$$\begin{aligned} \mathbf{u} + \mathbf{0} &= \mathbf{u} \\ \alpha\, \mathbf{0} &= \mathbf{0} \end{aligned} \tag{5.1.6}$$

Note also that the following algebraic laws are satisfied:

Commutative: $\mathbf{u}^1 + \mathbf{u}^2 = \mathbf{u}^2 + \mathbf{u}^1$

Distributive:

$$\begin{aligned} \alpha\, (\, \mathbf{u}^1 + \mathbf{u}^2 \,) &= \alpha\mathbf{u}^1 + \alpha\mathbf{u}^2 \\ (\, \alpha + \beta \,)\, \mathbf{u} &= \alpha\mathbf{u} + \beta\mathbf{u} \end{aligned} \tag{5.1.7}$$

Associative:

$$\begin{aligned} (\mathbf{u}^1 + \mathbf{u}^2) + \mathbf{u}^3 &= \mathbf{u}^1 + (\, \mathbf{u}^2 + \mathbf{u}^3 \,) \\ (\, \alpha\beta \,)\, \mathbf{u} &= \alpha\, (\, \beta\mathbf{u} \,) = \beta\, (\, \alpha\mathbf{u} \,) \end{aligned}$$

The general conclusion is that, for these operations, the ordinary rules of algebra may be followed. In fact, one of the greatest values of function spaces is that they obey the same rules of algebraic manipulation as R^n. Thus, they permit engineers to take advantage of their algebraic, and even geometric, intuition in the natural physical space R^3.

Vector Spaces of Functions

Theorem 5.1.1. The collection $F(\Omega)$ of all functions defined on a set Ω, with addition and multiplication by a scalar specified in Definition 5.1.1, is a vector space. Any subset of $F(\Omega)$ that is closed under the operations of addition and multiplication by a scalar is itself a vector space. That is, let $D(\Omega)$ be a subset of $F(\Omega)$ and let $\mathbf{f}(\mathbf{x})$ and $\mathbf{g}(\mathbf{x})$ be any functions in $D(\Omega)$. If $(\mathbf{f}+\mathbf{g})(\mathbf{x})$ and $(\alpha\mathbf{f})(\mathbf{x})$ are in $D(\Omega)$, for any scalar α, then $D(\Omega)$ is a vector space. ■

Closure under the operations of multiplication by a scalar and addition that are defined in Definition 5.1.1, the definition of a zero function, and the properties of Eqs. 5.1.6 and 5.1.7 show that $F(\Omega)$ is a vector space. Similarly, if a subset $D(\Omega)$ of $F(\Omega)$ is closed under the operations of multiplication by a scalar and addition, it also inherits the zero function from $F(\Omega)$ (e.g., take $\alpha = 0$ in Eq. 5.1.5) and the properties of Eqs. 5.1.6 and 5.1.7. Thus, $D(\Omega)$ is itself a vector space. ■

Theorem 5.1.2. The following sets are function spaces, hence vector spaces:

$$C^0(\Omega) = \{ f(\mathbf{x}) \in F(\Omega) \text{: } f(\mathbf{x}) \text{ is continuous on } \Omega \} \qquad (5.1.8)$$

$$C^n(\Omega) = \{ f(\mathbf{x}) \in F(\Omega) \text{: } f(\mathbf{x}) \text{ and its first n derivatives are continuous on } \Omega \} \qquad (5.1.9)$$

$$C^\infty(\Omega) = \{ f(\mathbf{x}) \in F(\Omega) \text{: } f(\mathbf{x}) \text{ and all its derivatives are continuous on } \Omega \} \qquad (5.1.10)$$

■

To prove Theorem 5.1.2 for $x \in R^1$, let f(x) and g(x) be in $C^k(\Omega)$. Then, since

$$\left[\frac{d^i}{dx^i} (f+g) \right](x) \equiv \frac{d^i}{dx^i} (f(x) + g(x)) = \frac{d^i f(x)}{dx^i} + \frac{d^i g(x)}{dx^i}$$

for all x in Ω and for all $i \leq k$, $(f+g)(x)$ and its first k derivatives are continuous at each point in Ω, so $(f+g)(x)$ is in $C^k(\Omega)$. Similarly,

$$\left[\frac{d^i}{dx^i} (\alpha f) \right](x) \equiv \frac{d^i}{dx^i} (\alpha f(x)) = \alpha \frac{d^i f(x)}{dx^i}$$

so $(\alpha f)(x)$ is also in $C^k(\Omega)$. Thus, $C^k(\Omega)$ is closed under the operations of addition and multiplication by a scalar. By Theorem 5.1.1, $C^k(\Omega)$ is a vector space.

By replacing ordinary derivatives on R^1 with partial derivatives on R^n, the theorem is proved for $\mathbf{x} \in R^n$. ■

Example 5.1.2

Since the foregoing shows that $C^2(\Omega)$ is a vector space, the set

$$A = \{ f(\mathbf{x}) \in C^2(\Omega) \colon f(\mathbf{x}) = 0 \text{ for } \mathbf{x} \text{ on } \Gamma \} \tag{5.1.11}$$

where Γ is the boundary of Ω, is a vector space if it is closed under addition and multiplication by a scalar. If $f(\mathbf{x})$ and $g(\mathbf{x})$ are in A, then $f(\mathbf{x}) = g(\mathbf{x}) = 0$ for $\mathbf{x}$ on Γ. Thus $(\alpha f)(\mathbf{x}) = \alpha f(\mathbf{x}) = \alpha\, 0 = 0$ and $(f + g)(\mathbf{x}) = f(\mathbf{x}) + g(\mathbf{x}) = 0 + 0 = 0$ for $\mathbf{x}$ on Γ, so $(\alpha f)(\mathbf{x})$ and $(f + g)(\mathbf{x})$ are in A and A is a vector space. ■

The function space A of Example 5.1.2 is especially valuable in solving second-order differential equations with homogeneous boundary conditions; i.e., a **boundary-value problem**. It is the space of **candidate solutions**, in which any of the familiar manipulations in vector spaces such as R^2 are valid.

Example 5.1.3

To illustrate a common pitfall and source of error when vector spaces are employed in the treatment of nonhomogeneous boundary-value problems, consider the set of functions

$$B = \{ f(\mathbf{x}) \in C^2(\Omega) \colon f(\mathbf{x}) = 1 \text{ on } \Gamma \}$$

where Γ is the boundary of Ω. B is a subset of $C^2(\Omega)$, but if $f(\mathbf{x}) \in B$, then $(\alpha f)(\mathbf{x}) = \alpha f(\mathbf{x}) = \alpha(1) = \alpha$ on Γ and, for $\alpha \neq 1$, $(\alpha f)(\mathbf{x})$ is not in B. Similarly, if $f(\mathbf{x})$ and $g(\mathbf{x})$ are in B, $(f + g)(\mathbf{x}) = f(\mathbf{x}) + g(\mathbf{x}) = 1 + 1 = 2$ on Γ, and $(f + g)(\mathbf{x})$ is not in B. This set of functions is, therefore, not a vector space. ■

For an extended discussion of physical and mathematical properties of function spaces in mechanics, the reader is referred to Refs. 11 to 13.

EXERCISES 5.1

1. Show that for $\Omega = (a, b)$, the set of functions $D^n(\Omega) = \{ f(x) \in F(\Omega) \colon f^{(k)}(x)$ is continuous except possibly at a finite number of points x_i, $i = 1, \ldots, m$, for $k = 0, 1, \ldots, n \}$ is a vector space.
2. Determine which of the following sets of functions are vector spaces:

(a) $D_A = \{ f(\mathbf{x}) \in C^1(\Omega) \colon |f(\mathbf{x})| \leq 1, \mathbf{x} \in \Omega \}$

(b) $D_B = \{ f(\mathbf{x}) \in C^0(\Omega) \colon \iint_\Omega f(\mathbf{x})\, d\Omega = 0 \}$

(c) $D_C = \{ f(\mathbf{x}) \in C^0(\Omega) \colon \iint_\Omega f(\mathbf{x})\, d\Omega = 1 \}$

(d) $D_D = \{ f(\mathbf{x}) \in C^0(\Omega): f(\mathbf{x}^1) + f(\mathbf{x}^2) = 0, \mathbf{x}^1, \mathbf{x}^2 \in \Omega \}$

(e) $D_E = \{ f(\mathbf{x}) \in C^0(\Omega): \iint_\Omega f(\mathbf{x})\, d\Omega < 0 \}$

3. Let Ω be an area in R^2 with boundary Γ, over which a membrane (drum head) is stretched and fastened to the curve Γ. The concern is with lateral displacement of the membrane under lateral pressure loading. Define the set D_M of admissible displacements. Is D_M a vector space?

5.2 SCALAR PRODUCT AND NORM IN FUNCTION SPACES

Much as in the study of finite-dimensional vector spaces in Section 2.4, the concepts of scalar product and norm on a function space can be defined. With these concepts, closeness of approximation, orthogonality of functions, and useful analytical identities may be developed. With these tools, the study of function spaces and their use in solving boundary-value problems that arise in engineering applications follows many of the well-understood rules and methods of vector algebra and matrices.

In Eqs. 5.1.4 and 5.1.5, addition and multiplication by a scalar were defined for functions. These definitions are so simple and natural that it is difficult to conceive of replacing them by others. The situation is not so simple in the case of products of functions. The appropriate definition depends on the physical problem involved.

Scalar Product and Norm for Integrable Functions

Definition 5.2.1. In the function space $C^0(a, b)$ of continuous real valued functions of a single variable x, in the bounded domain $a < x < b$, the **scalar product** (u, v) is defined as

$$(u, v) = \int_a^b u(x)\, v(x)\, dx \qquad (5.2.1)$$

More generally, if the physical space is a bounded domain Ω in R^n, if the vector function $\mathbf{u}(\mathbf{x})$ is $\mathbf{u}(\mathbf{x}) = [\, u^1(\mathbf{x}), \ldots, u^m(\mathbf{x})\,]^T$, $\mathbf{x} \in \Omega \subset R^n$, and if each $u_i(\mathbf{x})$ is continuous over Ω, the **scalar product** $(\mathbf{u}, \mathbf{v})$ is defined as

$$(\mathbf{u}, \mathbf{v}) = \iint_\Omega \mathbf{u}(\mathbf{x})^T \mathbf{v}(\mathbf{x})\, dx_1 \ldots dx_n$$

$$= \iint_\Omega (u_i(\mathbf{x})\, v_i(\mathbf{x}))\, dx_1 \ldots dx_n \qquad (5.2.2)$$

■

Using properties of integration, it can be verified that

$$\begin{aligned}
(\mathbf{u}, \mathbf{v}) &= (\mathbf{v}, \mathbf{u}) \\
(\mathbf{u}, \mathbf{v}+\mathbf{w}) &= (\mathbf{u}, \mathbf{v}) + (\mathbf{u}, \mathbf{w}) \\
(\alpha\mathbf{u}, \mathbf{v}) &= \alpha(\mathbf{u}, \mathbf{v}) \\
(\mathbf{u}, \mathbf{u}) &\geq 0 \\
(\mathbf{u}, \mathbf{u}) &= 0, \text{ only if } \mathbf{u} = \mathbf{0}
\end{aligned} \tag{5.2.3}$$

The proofs of the first three relations follow directly from the fact that $\mathbf{u}(\mathbf{x})$, $\mathbf{v}(\mathbf{x})$, and α are real and from linear properties of integration. The fourth property follows from the fact that $u_i^2(\mathbf{x}) \geq 0$, $i = 1, \ldots, m$, for all $\mathbf{x}$ in Ω, so

$$(\mathbf{u}, \mathbf{u}) = \sum_{i=1}^{m} \iint_{\Omega} u_i^2(\mathbf{x})\, dx_1 \ldots dx_n \geq 0$$

Finally, if $(\mathbf{u}, \mathbf{u}) = 0$, then

$$\iint_{\Omega} u_i^2(\mathbf{x})\, dx_1 \ldots dx_n = 0, \quad i = 1, \ldots, m$$

Assume, for purposes of proof by contradiction, that for some i and some point $\mathbf{x}_0 \in \Omega$ $u_i^2(\mathbf{x}_0) = a > 0$. Since the function $u_i(\mathbf{x})$ is continuous, there is an open neighborhood $U \subset \Omega$ of $\mathbf{x}_0$ in which $u_i^2(\mathbf{x}) \geq a/2$. Since U is open and not empty,

$$\iint_{U} dx_1 \ldots dx_n = c > 0$$

Then

$$\begin{aligned}
\iint_{\Omega} u_i^2(\mathbf{x})\, dx_1 \ldots dx_n &\geq \iint_{U} u_i^2(\mathbf{x})\, dx_1 \ldots dx_n \\
&\geq \iint_{U} (a/2)\, dx_1 \ldots dx_n \\
&\geq ac/2 > 0
\end{aligned}$$

But this is a contradiction, so $u_i^2(\mathbf{x}) = 0$ for all $\mathbf{x} \in \Omega$; i.e., $\mathbf{u}(\mathbf{x}) = \mathbf{0}$ in Ω.

Definition 5.2.1 requires that the functions are continuous. In fact, if any functions are finite and integrable, their product is integrable [8]. Thus, Definition 5.2.1 can be extended to spaces of functions that are only integrable, which includes functions with discontinuities. The first four conditions of Eq. 5.2.3 hold for such functions, but the fifth condition may not hold. Note also that Eq. 5.2.1 may be considered a special case of Eq. 5.2.2, with m = n = 1.

Example 5.2.1

Let $\mathbf{x} \in R^2$ and

$$\mathbf{u}(\mathbf{x}) = [\, x_1 - x_2,\; x_1 \,]^T$$

$$\mathbf{v}(\mathbf{x}) = [\, x_1 + x_2,\; x_2 - 1 \,]^T$$

where Ω is the open rectangle

$$\Omega = \{\, \mathbf{x} \in R^2 : 0 < x_1 < 1,\; 1 < x_2 < 2 \,\}$$

Then, by Eq. 5.2.2,

$$\begin{aligned}(\,\mathbf{u}, \mathbf{v}\,) &= \iint_\Omega \mathbf{u}(\mathbf{x})^T \mathbf{v}(\mathbf{x})\, dx_1\, dx_2 \\ &= \int_1^2 \int_0^1 [\,(x_1^2 - x_2^2) + x_1 x_2 - x_1\,]\, dx_1\, dx_2 \\ &= -\frac{7}{4}\end{aligned}$$

■

Definition 5.2.2. On a function space V with scalar product (•, •), the **norm of a vector function u** is defined as

$$\|\,\mathbf{u}\,\| = (\,\mathbf{u}, \mathbf{u}\,)^{1/2} \tag{5.2.4}$$

■

Example 5.2.2

For the vector function **u** of Example 5.2.1,

$$\begin{aligned}\|\,\mathbf{u}\,\| &= (\,\mathbf{u}, \mathbf{u}\,)^{1/2} \\ &= \left[\int_1^2 \int_0^1 [\, u_1^2(\mathbf{x}) + u_2^2(\mathbf{x}) \,]\, dx_1\, dx_2 \right]^{1/2} \\ &= \left[\int_1^2 \int_0^1 [\,(x_1 - x_2)^2 + x_1^2\,]\, dx_1\, dx_2 \right]^{1/2} \\ &= \left[\frac{3}{2} \right]^{1/2}\end{aligned}$$

■

Note that the norm of Definition 5.2.2 is formally identical to the norm of Definition 2.4.2. Thus, for function spaces of continuous functions, it satisfies properties that are given in Eq. 2.4.5; i.e.,

$$\| \mathbf{u} \| \geq 0, \text{ for all } \mathbf{u} \tag{5.2.5}$$

$$\| \mathbf{u} \| = 0, \text{ if and only if } \mathbf{u} = \mathbf{0} \tag{5.2.6}$$

$$\| \alpha\mathbf{u} \| = | \alpha | \, \| \mathbf{u} \|, \text{ where } \alpha \text{ is a scalar} \tag{5.2.7}$$

For the scalar products of Eqs. 5.2.1 and 5.2.2, the resulting norms are

$$\| u \| = \left(\int_a^b u^2(x)\, dx \right)^{1/2} \tag{5.2.8}$$

and

$$\| \mathbf{u} \| = \left(\iint_\Omega \sum_{i=1}^{m} (u_i(\mathbf{x}))^2 \, dx_1 \ldots dx_n \right)^{1/2} \tag{5.2.9}$$

respectively.

The proofs of the Schwartz and triangle inequalities of Section 2.4, in R^n, followed directly from the fact that the norm is defined by a scalar product on R^n. Thus, the same results hold for function spaces of integrable functions.

Theorem 5.2.1 (Schwartz and Triangle Inequalities). For any real valued integrable functions $u(\mathbf{x})$ and $v(\mathbf{x})$ over a domain Ω in R^n, the **Schwartz inequality** is

$$\begin{aligned} | (u, v) | &= \left| \iint_\Omega u(\mathbf{x})\, v(\mathbf{x})\, dx_1 \ldots dx_n \right| \\ &\leq \left[\iint_\Omega u^2(\mathbf{x})\, dx_1 \ldots dx_n \right]^{1/2} \left[\iint_\Omega v^2(\mathbf{x})\, dx_1 \ldots dx_n \right]^{1/2} \\ &= \| u \| \, \| v \| \end{aligned} \tag{5.2.10}$$

or, for vector functions, it is

$$\begin{aligned} | (\mathbf{u}, \mathbf{v}) | &= \left| \iint_\Omega (u_i(\mathbf{x})\, v_i(\mathbf{x}))\, dx_1 \ldots dx_n \right| \\ &\leq \left[\iint_\Omega \sum_{i=1}^{m} (u_i(\mathbf{x}))^2 \, dx_1 \ldots dx_n \right]^{1/2} \left[\iint_\Omega \sum_{i=1}^{m} (v_i(\mathbf{x}))^2 \, dx_1 \ldots dx_n \right]^{1/2} \\ &= \| \mathbf{u} \| \, \| \mathbf{v} \| \end{aligned} \tag{5.2.11}$$

The **triangle inequality** for scalar valued functions is

$$\| u + v \| = \left[\iint_\Omega (u(\mathbf{x}) + v(\mathbf{x}))^2 \, dx_1 \ldots dx_n \right]^{1/2}$$

$$\leq \left[\iint_\Omega u^2(\mathbf{x}) \, dx_1 \ldots dx_n \right]^{1/2} + \left[\iint_\Omega v^2(\mathbf{x}) \, dx_1 \ldots dx_n \right]^{1/2}$$

$$= \| u \| + \| v \| \tag{5.2.12}$$

and, for vector functions, it is

$$\| \mathbf{u} + \mathbf{v} \| = \left[\iint_\Omega \sum_{i=1}^{m} (u_i(\mathbf{x}) + v_i(\mathbf{x}))^2 \, dx_1 \ldots dx_n \right]^{1/2}$$

$$\leq \left[\iint_\Omega \sum_{i=1}^{m} (u_i(\mathbf{x}))^2 \, dx_1 \ldots dx_n \right]^{1/2} + \left[\iint_\Omega \sum_{i=1}^{m} (v_i(\mathbf{x}))^2 \, dx_1 \ldots dx_n \right]^{1/2}$$

$$= \| \mathbf{u} \| + \| \mathbf{v} \| \tag{5.2.13}$$

■

Example 5.2.3

Let $u(x) = x + 1$ and $v(x) = x$, $0 < x < 1$. Their scalar product is

$$(u, v) \equiv \int_0^1 u(x) \, v(x) \, dx = \int_0^1 (x + 1) \, x \, dx$$

$$= \left(\frac{1}{3} x^3 + \frac{1}{2} x^2 \right) \Big|_0^1 = \frac{1}{3} + \frac{1}{2} = 0.8333$$

and their norms are

$$\| u \| = \left[\int_0^1 (x^2 + 2x + 1) \, dx \right]^{1/2} = 1.5275$$

$$\| v \| = \left[\int_0^1 x^2 \, dx \right]^{1/2} = 0.5774$$

To check the Schwartz inequality, note that

$$0.8333 < (1.5275) (0.57735) = 0.8819$$

Similarly, for the triangle inequality,

$$\| u+v \| \equiv \left[\int_0^1 (u(x) + v(x))^2 \, dx \right]^{1/2}$$

$$= \left[\int_0^1 (2x + 1)^2 \, dx \right]^{1/2} = 1.2019$$

and indeed

$$1.2019 < 1.5275 + 0.5774 = 2.1049$$

■

The Space $L_2(\Omega)$ of Square Integrable Functions

With the foregoing inequalities, an important function space can be defined that is used extensively in applications.

Definition 5.2.3. The space of all **square integrable functions** on a domain Ω in R^n is defined as

$$L_2(\Omega) = \left\{ \mathbf{u}(\mathbf{x}) \in F(\Omega): \quad \| \mathbf{u} \| = \left[\iint_\Omega \left(\sum_{i=1}^{m} u_i^2(\mathbf{x}) \right) dx_1 \ldots dx_n \right]^{1/2} < \infty \right\} \tag{5.2.14}$$

The scalar product and norm on $L_2(\Omega)$ are defined by Eqs. 5.2.2 and 5.2.4, denoted $(\bullet, \bullet)_{L_2}$ and $\| \bullet \|_{L_2}$. The zero function in $L_2(\Omega)$ is defined to be any function $\mathbf{v}(\mathbf{x})$ such that $\| \mathbf{v} \| = 0$. ■

The last convention of Definition 5.2.3 is introduced so that the fifth condition of Eq. 5.2.3 holds. The convention is also physically meaningful, since the most pathological functions that arise in most engineering applications may be discontinuous at a finite number of points. Thus, if $\| \mathbf{u} \| = 0$ for such a function, the function is zero except possibly at a finite number of isolated points where it has finite values. It is practical for most engineering applications in which scalar product and norm are used to treat such functions as zero functions.

The set $L_2(\Omega)$ satisfies all postulates for a vector space in Definition 2.2.1, except perhaps closure under addition. Equation 5.2.13, however, resolves this dilemma by showing that if $\| \mathbf{u} \| < \infty$ and $\| \mathbf{v} \| < \infty$, then $\| \mathbf{u} + \mathbf{v} \| \leq \| \mathbf{u} \| + \| \mathbf{v} \| < \infty$. Thus, $L_2(\Omega)$ is closed under addition, so it is a vector space.

Linear Independence and Orthogonality of Functions

The concept of linear dependence of vector functions can also be defined.

Definition 5.2.4. A set of n vector functions $\mathbf{u}^1, \mathbf{u}^2, \ldots, \mathbf{u}^n$ is said to be **linearly dependent** if there exist constants $a_1, a_2, \ldots, a_n$ that are not all zero such that

$$a_1\mathbf{u}^1 + a_2\mathbf{u}^2 + \ldots + a_n\mathbf{u}^n = \mathbf{0} \tag{5.2.15}$$

Otherwise, they are **linearly independent.** A sequence of vector functions $\mathbf{u}^1, \mathbf{u}^2, \ldots$ (infinite in number) is said to be linearly dependent if some finite subset is linearly dependent. Otherwise, it is linearly independent. ■

If the vector functions

$$\mathbf{u}^i(\mathbf{x}) = [\, u_1^i(\mathbf{x}), u_2^i(\mathbf{x}), \ldots, u_m^i(\mathbf{x})\,]^T, \quad i = 1, \ldots, n$$

are defined on a physical domain Ω, then the vector equation of Eq. 5.2.15 is equivalent to

$$\sum_{i=1}^{n} a_i u_j^i(\mathbf{x}) = 0 \tag{5.2.16}$$

for $j = 1, 2, \ldots, m$ and for all $\mathbf{x} \in \Omega$. It is clear that linear dependence is a very special property for vector functions. It means that at each point $\mathbf{x} \in \Omega$, the vectors $\mathbf{u}^i(\mathbf{x})$, $i = 1, \ldots, n$, in R^m are linearly dependent, with the same constants a_i, $i = 1, \ldots, n$, throughout the physical space. This is a very restrictive requirement.

Example 5.2.4

Let

$$\mathbf{u}^1(x) = [\, u_1^1(x), u_2^1(x)\,]^T = [\, x, x^2 + x - 1\,]^T$$

$$\mathbf{u}^2(x) = [\, u_1^2(x), u_2^2(x)\,]^T = [\, 0, 3x^2 + 3x - 3\,]^T$$

$$\mathbf{u}^3(x) = [\, u_1^3(x), u_2^3(x)\,]^T = [\, x, 0\,]^T$$

be vector functions of a single variable on $\Omega = R$. If $a_1 = 3$, $a_2 = -1$, and $a_3 = -3$, then

$$a_i\mathbf{u}^i(x) = \begin{bmatrix} 3x - 3x \\ 3x^2 + 3x - 3 - 3x^2 - 3x + 3 \end{bmatrix} = \mathbf{0}$$

for all x. Thus, $\mathbf{u}^1$, $\mathbf{u}^2$, and $\mathbf{u}^3$ are linearly dependent vector functions. ■

Definition 5.2.5. A set of n vector functions $\mathbf{u}^i(\mathbf{x})$, i = 1, . . . , n, in $L_2(\Omega)$ are said to be **orthonormal functions** if each vector function is orthogonal to all the others and each vector function has unit norm; i.e.,

$$(\mathbf{u}^i, \mathbf{u}^j) = \delta_{ij}, \quad i, j = 1, 2, \ldots, n \tag{5.2.17}$$

■

Example 5.2.5

Consider the trigonometric functions

$$\frac{1}{\sqrt{2\pi}}, \quad \frac{\cos mx}{\sqrt{\pi}}, \quad m = 1, 2, \ldots, \quad \frac{\sin nx}{\sqrt{\pi}}, \quad n = 1, 2, \ldots$$

in $L_2(-\pi, \pi)$. Evaluation of scalar products yields

$$\int_{-\pi}^{\pi} \frac{\sin nx}{\sqrt{\pi}} \frac{\sin mx}{\sqrt{\pi}} dx = \int_{-\pi}^{\pi} \frac{\cos nx}{\sqrt{\pi}} \frac{\cos mx}{\sqrt{\pi}} dx = \delta_{mn}$$

$$\int_{-\pi}^{\pi} \frac{\sin mx}{\sqrt{\pi}} \frac{\cos nx}{\sqrt{\pi}} dx = 0$$

for all m and n. Similarly,

$$\int_{-\pi}^{\pi} \left(\frac{1}{\sqrt{2\pi}} \right)^2 dx = 1$$

$$\int_{-\pi}^{\pi} \frac{1}{\sqrt{2\pi}} \frac{\cos mx}{\sqrt{\pi}} dx = 0$$

$$\int_{-\pi}^{\pi} \frac{1}{\sqrt{2\pi}} \frac{\sin nx}{\sqrt{\pi}} dx = 0$$

for all m and n. Thus, these trigonometric functions are orthonormal in $L_2(-\pi, \pi)$.■

Theorem 5.2.2. If n vector functions are orthonormal in $L_2(\Omega)$, they are linearly independent. ■

To prove Theorem 5.2.2, let

$$a_1\mathbf{u}^1 + a_2\mathbf{u}^2 + \ldots + a_n\mathbf{u}^n = \mathbf{0} \tag{5.2.18}$$

where $\{ \mathbf{u}^i \}$ are orthonormal in $L_2(\Omega)$. Taking the scalar product of both sides of Eq. 5.2.18 with $\mathbf{u}^j$ and employing Eq. 5.2.17, the result is $a_j = 0$. Since the index j is arbitrary, $a_j = 0$, j = 1, . . . , m, and the $\mathbf{u}^i$ are linearly independent. ■

Gram–Schmidt Orthonormalization in $L_2(\Omega)$

For many purposes, it is convenient to transform a given set of n linearly independent vector functions into an orthonormal set. This is the process of orthonormalization, as in Section 2.4. The **Gram–Schmidt orthonormalization process** operates on n linearly independent vector functions $\mathbf{u}^i$, $i = 1, \ldots, n$. Starting with $\mathbf{u}^1$, it is normalized by defining the vector function

$$\mathbf{v}^1 = \frac{1}{\| \mathbf{u}^1 \|} \mathbf{u}^1$$

This is the first vector of the desired orthonormal set. The vector function $\mathbf{u}^1$ is now discarded, since it has been replaced by $\mathbf{v}^1$. For $\mathbf{v}^2$, take a linear combination of $\mathbf{v}^1$ and $\mathbf{u}^2$; i.e.,

$$\mathbf{v}^2 = a_{21}\mathbf{v}^1 + a_{22}\mathbf{u}^2$$

and impose the orthonormality conditions

$$(\mathbf{v}^1, \mathbf{v}^2) = 0, \quad \| \mathbf{v}^2 \| = 1$$

The first of these conditions gives

$$a_{21} + a_{22} (\mathbf{v}^1, \mathbf{u}^2) = 0$$

so

$$a_{21} = - a_{22} (\mathbf{v}^1, \mathbf{u}^2)$$

Substitution into the definition of $\mathbf{v}^2$ yields

$$\mathbf{v}^2 = a_{22} [\mathbf{u}^2 - \mathbf{v}^1 (\mathbf{v}^1, \mathbf{u}^2)]$$

The second orthonormality condition now gives

$$a_{22} = [\| \mathbf{u}^2 \|^2 - (\mathbf{v}^1, \mathbf{u}^2)^2]^{-1/2}$$

yielding the second member of the orthonormal pair $\mathbf{v}^1$ and $\mathbf{v}^2$.

The process continues step by step, demanding nothing more complicated than the calculation of scalar products. The general **Gram–Schmidt recurrence relation** is

$$\mathbf{v}^\alpha = a_{\alpha\alpha} [\mathbf{u}^\alpha - \mathbf{v}^1 (\mathbf{v}^1, \mathbf{u}^\alpha) - \ldots - \mathbf{v}^{\alpha-1} (\mathbf{v}^{\alpha-1}, \mathbf{u}^\alpha)] \tag{5.2.19}$$

$$a_{\alpha\alpha} = [\| \mathbf{u}^\alpha \|^2 - (\mathbf{v}^1, \mathbf{u}^\alpha)^2 - \ldots - (\mathbf{v}^{\alpha-1}, \mathbf{u}^\alpha)^2]^{-1/2} \tag{5.2.20}$$

These formulas serve to determine $\mathbf{v}^\alpha$, after $\mathbf{v}^1, \mathbf{v}^2, \ldots$, and $\mathbf{v}^{\alpha-1}$ have been found. Note that this result is identical to that for R^n in Eq. 2.4.28. Thus, in concept, the Gram–Schmidt orthonormalization process is identical for function spaces and R^n.

Example 5.2.6

Let the physical space be $0 < x < 1$ and the scalar product be defined in Eq. 5.2.1. To orthonormalize the three linearly independent functions

$$u^1(x) = 1, \; u^2(x) = x, \; u^3(x) = x^2$$

the following calculations are carried out:

$$\| u^1 \|^2 = \int_0^1 1^2 \, dx = 1, \text{ so } v^1(x) = 1$$

$$(v^1, u^2) = \int_0^1 x \, dx = \frac{1}{2}, \text{ and } \| u^2 \|^2 = \int_0^1 x^2 \, dx = \frac{1}{3}, \text{ so}$$

$$a_{22} = 2\sqrt{3}, \text{ and } v^2(x) = \sqrt{3}\,(2x - 1)$$

$$(v^1, u^3) = \int_0^1 x^2 \, dx = \frac{1}{3}, \; (v^2, u^3) = \int_0^1 \sqrt{3}\,(2x - 1)\, x^2 \, dx = \frac{\sqrt{3}}{6}$$

$$\text{and } \| u^3 \|^2 = \int_0^1 x^4 \, dx = \frac{1}{5}, \text{ so } a_{33} = 6\sqrt{5} \text{ and}$$

$$v^3(x) = \sqrt{5}\,(6x^2 - 6x + 1)$$

The accuracy of calculations may be checked by applying the test $(v^j, v^i) = \delta_{ji}$ to the final result. ■

The reader should note that most of the basic properties and concepts of vector analysis in the function space $L_2(\Omega)$ are identical to those in R^n, in particular in the familiar space R^3. A fundamental difference between $L_2(\Omega)$ and R^n is that $L_2(\Omega)$ is **infinite dimensional**, whereas the dimension of R^n is $n < \infty$. To see that $L_2(\Omega)$ is indeed infinite dimensional, note that an infinite number of orthonormal functions were introduced in Example 5.2.5, and Theorem 5.2.2 shows that they are linearly independent. Thus, $L_2(\Omega)$ cannot have a basis with a finite number of vectors; i.e., $L_2(\Omega)$ is infinite dimensional. This difference from R^n means that functions in $L_2(\Omega)$ must be represented as linear combinations of an infinite number of basis vectors; i.e., as infinite series. This fact might have been anticipated by the power series representations of functions in Chapter 4.

EXERCISES 5.2

1. Show that the scalar product of Eq. 5.2.2 satisfies the conditions of Eq. 2.4.2.

2. Compute the fourth orthonormalized function in Example 5.2.6, if $u^4(x) = x^3$ on $0 \le x \le 1$.

3. Show that the functions

$$u^i(x) = \cos ix, \quad i = 1, 2, \ldots$$

on $-\pi \le x \le \pi$ are linearly independent.

4. Form an orthonormal sequence from the linearly independent functions of Exercise 3.

5. Find the norm of the function

$$u(x_1, x_2) = \sin \pi x_1 \sin \pi x_2$$

where $\Omega = \{ \mathbf{x} \in R^2 \colon 0 < x_1 < 1, \ 0 < x_2 < 1 \}$

6. Find the distance between two functions

$$u(x_1, x_2) = \sin \pi x_1 \sin \pi x_2 \ \text{ and } \ v(x_1, x_2) = 1$$

where $\Omega = \{ \mathbf{x} \in R^2 \colon 0 < x_1 < 1, \ 0 < x_2 < 1 \}$ (Hint: Use $\| u - v \|_{L_2}$).

7. Find a quadratic polynomial $ax^2 + bx + c$ that is orthogonal to both 1 and x in $L_2(0, 1)$. Is it uniquely determined?

5.3 BASES AND COMPLETENESS OF FUNCTION SPACES

Based on the preceding section, it might be concluded that analysis in a function space is just like analysis in R^n. Unfortunately, life is not this kind. Function spaces are generally not finite dimensional, so bases that contain an infinite number of functions must be dealt with.

Example 5.3.1

The infinite set of functions $1, x, x^2, \ldots$ can be used to construct polynomials of any order; e.g., $a_0 + a_1x + a_2x^2 + \ldots + a_nx^n$. The use of this set of functions to represent an arbitrary function requires a limiting process; i.e., an infinite series. Finally, it is not at all obvious that a sequence of functions in a given function space converges to a function in the space. It is possible that a sequence of continuous functions [i.e., in $C^0(\Omega)$] converges to a discontinuous function [i.e., in $D^0(\Omega)$];

e.g., as shown in Example 4.2.1,

$$\lim_{i \to \infty} x^i = \begin{cases} 0, & 0 \le x < 1 \\ 1, & x = 1 \end{cases} \tag{5.3.1}$$

■

These points of caution should not be taken as a judgment of hopelessness. They are simply warnings that new pitfalls must be faced and care is required to avoid blundering into one and breaking a mathematical leg.

The Space ℓ_2 of Square Summable Sequences

As an introduction to infinite-dimensional vector spaces, the space of n-vectors of real numbers is extended to the space of infinite sequences of real numbers. A vector in this space is defined as

$$\mathbf{a} = [\, a_1, a_2, a_3, \ldots \,]^T \tag{5.3.2}$$

A norm in this space is defined as a direct extension of the definition in the space R^n; i.e.,

$$\| \mathbf{a} \|_{\ell_2} \equiv \left[\sum_{i=1}^{\infty} a_i^2 \right]^{1/2} \tag{5.3.3}$$

Since this involves an infinite sum, not all sequences will define vectors with finite norms. Therefore, the space is limited to only those sequences for which

$$\| \mathbf{a} \|_{\ell_2}^2 \equiv \sum_{i=1}^{\infty} a_i^2 < \infty$$

Example 5.3.2

For the sequence $\{ 1/i \}$; i.e.,

$$\mathbf{a} = [\, 1, 1/2, 1/3, \ldots, 1/i, \ldots \,]^T$$

the ℓ_2 norm is

$$\| \mathbf{a} \|_{\ell_2} = \left[\sum_{i=1}^{\infty} \frac{1}{i^2} \right]^{1/2} < \infty$$

where the finite sum is assured by the result of Example 4.1.6. In contrast, for the sequence $\{ 1/\sqrt{i} \}$; i.e.,

$$\mathbf{b} = [\ 1,\ 1/\sqrt{2},\ 1/\sqrt{3},\ \dots,\ 1/\sqrt{i},\ \dots\ ]^T$$

the ℓ_2 norm is

$$\|\mathbf{b}\|_{\ell_2} = \left[\sum_{i=1}^{\infty} \frac{1}{i} \right]^{1/2} = \infty$$

where divergence of the series to ∞ is assured by the result of Example 4.1.6. ■

Theorem 5.3.1. The space of sequences with finite norm, denoted

$$\ell_2 = \left\{ \mathbf{a} = [\ a_1, a_2, \dots\ ]^T\!:\ a_i,\ i = 1, \dots, \text{real},\ \|\mathbf{a}\|_{\ell_2}^2 \equiv \sum_{i=1}^{\infty} a_i^2 < \infty \right\} \tag{5.3.4}$$

is a vector space. ■

To show that ℓ_2 is a vector space, addition and multiplication by a scalar must be shown to satisfy the postulates of a vector space. If $\mathbf{a} = [\ a_1, a_2, a_3, \dots\ ]^T$ and $\mathbf{b} = [\ b_1, b_2, b_3, \dots\ ]^T$ are vectors such that

$$\sum_{i=1}^{\infty} a_i^2 < \infty \quad \text{and} \quad \sum_{i=1}^{\infty} b_i^2 < \infty$$

then **addition of vectors** in ℓ_2 is defined as

$$\mathbf{a} + \mathbf{b} = [\ a_1 + b_1,\ a_2 + b_2,\ a_3 + b_3,\ \dots\ ]^T \tag{5.3.5}$$

To prove closure under addition, it must be shown that

$$\sum_{i=1}^{\infty} (\ a_i + b_i\ )^2 < \infty$$

To begin the proof, observe that

$$\begin{aligned} (\ a_i + b_i\ )^2 &\le a_i^2 + b_i^2 + 2|\ a_i b_i\ | = a_i^2 + b_i^2 + 2|\ a_i\ ||\ b_i\ | \\ a_i^2 + b_i^2 &\ge 2|\ a_i\ ||\ b_i\ | \end{aligned} \tag{5.3.6}$$

where the second inequality follows from

$$\left(\ |\ a_i\ | - |\ b_i\ |\ \right)^2 = a_i^2 + b_i^2 - 2|\ a_i\ ||\ b_i\ | \ge 0$$

From the inequalities of Eq. 5.3.6,

$$(a_i + b_i)^2 \le 2 (a_i^2 + b_i^2) \tag{5.3.7}$$

so

$$\sum_{i=1}^{n} (a_i + b_i)^2 \le 2 \sum_{i=1}^{n} a_i^2 + 2 \sum_{i=1}^{n} b_i^2$$

In the limit, as n approaches ∞,

$$\sum_{i=1}^{\infty} (a_i + b_i)^2 \le 2 \sum_{i=1}^{\infty} a_i^2 + 2 \sum_{i=1}^{\infty} b_i^2 < \infty$$

Thus, ℓ_2 is closed under the operation of addition.

Multiplication by a scalar γ in ℓ_2 is defined as

$$\gamma \mathbf{a} = [\gamma a_1, \gamma a_2, \gamma a_3, \ldots]^T \tag{5.3.8}$$

Closure under this operation is verified, as follows. Since for finite n,

$$\sum_{i=1}^{n} (\gamma a_i)^2 = \sum_{i=1}^{n} \gamma^2 a_i^2 = \gamma^2 \sum_{i=1}^{n} a_i^2$$

as n approaches ∞,

$$\sum_{i=1}^{\infty} (\gamma a_i)^2 = \sum_{i=1}^{\infty} \gamma^2 a_i^2 = \gamma^2 \sum_{i=1}^{\infty} a_i^2 < \infty$$

The zero vector in ℓ_2 is defined as

$$\mathbf{0} = [0, 0, 0, \ldots]^T \tag{5.3.9}$$

and the negative of a vector is

$$-\mathbf{a} = [-a_1, -a_2, -a_3, \ldots]^T \tag{5.3.10}$$

The other eight postulates of Definition 2.2.1 are easily verified, so ℓ_2 is a vector space. ■

A **scalar product** can now be defined on ℓ_2 as

$$(\mathbf{a}, \mathbf{b})_{\ell_2} \equiv \sum_{i=1}^{\infty} a_i b_i \tag{5.3.11}$$

Note that the **norm** of Eq. 5.3.3 is just

$$\| \mathbf{a} \|_{\ell_2} = (\mathbf{a}, \mathbf{a})_{\ell_2}^{1/2} \tag{5.3.12}$$

To show that for every pair of vectors in ℓ_2 the scalar product of Eq. 5.3.11 exists and is finite, from Eq. 5.3.7, note that

$$a_i b_i \leq | a_i b_i | = | a_i | | b_i | \leq \frac{1}{2}(a_i^2 + b_i^2), \quad \text{no sum on i}$$

The series $\sum_{i=1}^{\infty} a_i b_i$ converges, since

$$\sum_{i=1}^{\infty} a_i b_i \leq \sum_{i=1}^{\infty} | a_i b_i | \leq \frac{1}{2}\left(\sum_{i=1}^{\infty} a_i^2 + \sum_{i=1}^{\infty} b_i^2 \right) < \infty \tag{5.3.13}$$

so Eqs. 5.3.11 and 5.3.12 define a scalar product and the associated norm.

The infinite set of vectors $\psi^1 = [1, 0, 0, \ldots]^T$, $\psi^2 = [0, 1, 0, \ldots]^T$, $\psi^3 = [0, 0, 1, \ldots]^T, \ldots$ is orthonormal in ℓ_2, so they are linearly independent. Thus, ℓ_2 is infinite dimensional. This set of vectors in ℓ_2, each of which is defined by a sequence of real numbers, defines a sequence $\{ \psi^k \}$ in ℓ_2. To distinguish between sequences of vectors and real numbers, a superscript k will be used in the remainder of this subsection to denote vectors and a subscript i will be used to denote elements in a sequence of real numbers. In this case, $\psi^k = [\delta_{ik}]^T$, $k = 1, 2, \ldots$

Convergence of Sequences in Normed Vector Spaces

Definition 5.3.1. A vector space V on which a norm $\| \bullet \|$ is defined is called a **normed vector space.** ■

Thus far, R^n, $L_2(\Omega)$, and ℓ_2 have been shown to be normed vector spaces, the former finite dimensional and the latter two infinite dimensional.

For $\mathbf{a} = [a_1, a_2, \ldots]^T$ in ℓ_2 and $\psi^k = [\delta_{ik}]^T$, $k = 1, 2, \ldots$, define

$$\mathbf{a}^n = \sum_{k=1}^{n} a_k \psi^k = [a_1, a_2, \ldots, a_n, 0, \ldots]^T$$

Then

$$\| \mathbf{a}^n - \mathbf{a} \| = \left(\sum_{i=n+1}^{\infty} a_i^2 \right)^{1/2}$$

which approaches zero as n approaches ∞. The vector **a** may thus be written as

$$\mathbf{a} = \sum_{k=1}^{\infty} a_k \psi^k = \sum_{k=1}^{\infty} (\psi^k, \mathbf{a})\, \psi^k \tag{5.3.14}$$

and the a_i are called components of **a** relative to the orthonormal vectors $\{ \psi^k \}$; i.e., $a_k = (\psi^k, \mathbf{a})$.

In Section 4.2, pointwise and uniform convergence were defined for sequences of real valued functions. In a normed vector space, the notion of convergence can be defined more generally.

Definition 5.3.2. A sequence of vectors $\{ \phi^k \}$ in a normed vector space V is said to **converge** to an element ϕ of the space V if for every $\varepsilon > 0$ there exists an integer N, which may depend on ε, such that

$$\| \phi^n - \phi \| < \varepsilon \tag{5.3.15}$$

whenever $n > N$. This may be denoted

$$\lim_{n\to\infty} \phi^n = \phi$$

■

Another way of viewing convergence of ϕ^n to ϕ is by noting that ϕ^n converges to ϕ if $\lim_{n\to\infty} \| \phi^n - \phi \| = 0$. If Eq. 5.3.15 holds in $L_2(\Omega)$ or ℓ_2, it is said that $\{ \phi^n \}$ **converges in the mean** to ϕ.

Example 5.3.3

Let ϕ^k be the vector $\phi^k = [\, \phi_1^k, \phi_2^k, \ldots \,]^T = [\, \phi_i^k \,]^T$, where

$$\phi_i^k = \frac{1}{(i + 1/k)}$$

For a fixed integer k, the ℓ_2 norm of ϕ^k is

$$\| \phi^k \|_{\ell_2} = \left(\sum_{i=1}^{\infty} \frac{1}{(i + 1/k)^2} \right)^{1/2}$$

$$\leq \left(\sum_{i=1}^{\infty} \frac{1}{i^2} \right)^{1/2} \equiv c < \infty$$

where the fact that $(i + 1/k) \geq i$ and the result of Example 4.1.6 have been used. Thus, each ϕ^k is in ℓ_2 and $\{ \phi^k \}$ is a sequence of vectors in ℓ_2.

Define the scalar limits

$$\phi_i = \lim_{k\to\infty} \phi_i^k = \lim_{k\to\infty} \frac{1}{(i + 1/k)} = \frac{1}{i}, \quad i = 1, 2, \ldots$$

and the vector

$$\phi = [\, 1, 1/2, 1/3, \ldots \,]^T = [\, 1/i \,]^T$$

But Example 5.3.2 showed that this vector is in ℓ_2.

To see that ϕ^k converges to ϕ in ℓ_2, form

$$\| \phi^n - \phi \|_{\ell_2}^2 = \sum_{i=1}^{\infty} \left(\frac{1}{(i + 1/n)} - \frac{1}{i} \right)^2$$

$$= \sum_{i=1}^{\infty} \left(\frac{i - i - 1/n}{(i + 1/n)\, i} \right)^2$$

$$= \frac{1}{n} \sum_{i=1}^{\infty} \left(\frac{1}{(i + 1/n)\, i} \right)^2$$

$$\leq \frac{1}{n} \left(\sum_{i=1}^{\infty} \frac{1}{i^2} \right)^2 = \frac{c^2}{n}$$

Thus,

$$\lim_{n\to\infty} \| \phi^n - \phi \|_{\ell_2} \leq \lim_{n\to\infty} \frac{c}{\sqrt{n}} = 0$$

and indeed $\lim_{n\to\infty} \phi^n = \phi$. ■

Cauchy Sequences and Complete Vector Spaces

Definition 5.3.3. A sequence of vectors $\{ \phi^k \}$ in a normed vector space V is said to be a **Cauchy sequence** if for every $\varepsilon > 0$ there exists an integer N, which may depend on ε, such that

$$\| \phi^n - \phi^m \| < \varepsilon \tag{5.3.16}$$

whenever $n > N$ and $m > N$. ■

Note that if a sequence $\{ \phi^k \}$ is given, the condition of Eq. 5.3.16 can be verified without knowing the limit of the sequence of vectors. It is thus often easier to show that a sequence of vectors is a Cauchy sequence than to find a limit and show convergence.

If a sequence $\{ \phi^k \}$ in a normed vector space V converges to ϕ in V, then it is a Cauchy sequence. To see that this is true, note that for any ε, there exists an N such that

$$\| \phi - \phi^n \| < \frac{\varepsilon}{2} \text{ and } \| \phi - \phi^m \| < \frac{\varepsilon}{2}$$

whenever $n > N$ and $m > N$. By the triangle inequality,

$$\| \phi^n - \phi^m \| \leq \| \phi - \phi^n \| + \| \phi - \phi^m \| < \varepsilon$$

whenever $n > N$ and $m > N$. Thus, $\{ \phi^k \}$ is a Cauchy sequence. In some vector spaces, the converse may not be true; i.e., Cauchy sequences may not have limits in the space.

Definition 5.3.4. A normed vector space V is said to be a **complete vector space** if every Cauchy sequence in V has a limit in V. ■

An important result of real analysis [14] is that the space R of real numbers is complete, when the norm of a real number α is $| \alpha |$. With this fact, a special case of the proof of the theorem that follows shows that the vector space R^n is complete, with the norm

$$\| \mathbf{x} \| = \left(\sum_{i=1}^{n} (x_i)^2 \right)^{1/2}$$

Theorem 5.3.2. The space ℓ_2, with the norm defined in Eq. 5.3.3, is complete. ■

To prove Theorem 5.3.2, let $\{ \phi^k \}$ be a Cauchy sequence in ℓ_2. Then for every $\varepsilon > 0$, there exists an N such that

$$\left(\sum_{i=1}^{\infty} (\phi_i^n - \phi_i^m)^2 \right)^{1/2} < \varepsilon$$

whenever $n > N$ and $m > N$. This implies that for each i,

$$| \phi_i^n - \phi_i^m | < \varepsilon$$

or that the sequence ϕ_i^n with i fixed is a Cauchy sequence of real numbers. Since R is complete, for every i there exists a y_i such that

$$\lim_{n \to \infty} \phi_i^n = y_i$$

Define

$$\mathbf{y} = [\, y_1, y_2, \ldots \,]^T$$

It remains to show that $\lim_{n\to\infty} \| \mathbf{y} - \phi^n \| = 0$ and that $\mathbf{y}$ is in ℓ_2.

For an integer M, consider $\sum_{i=1}^{M} (y_i - \phi_i^n)^2$, which can be rewritten as

$$\sum_{i=1}^{M} (y_i - \phi_i^n)^2 = \sum_{i=1}^{M} (y_i - \phi_i^m + \phi_i^m - \phi_i^n)^2$$

for any integer m. Using the inequality of Eq. 5.3.7,

$$\sum_{i=1}^{M} (y_i - \phi_i^n)^2 \le 2\sum_{i=1}^{M} (y_i - \phi_i^m)^2 + 2\sum_{i=1}^{M} (\phi_i^m - \phi_i^n)^2$$

Since $\{ \phi^i \}$ is a Cauchy sequence in ℓ_2 and since $\lim_{n\to\infty} \phi_i^n = y_i$, for any $\varepsilon > 0$, there exists an N such that

$$\sum_{i=1}^{M} (\phi_i^m - \phi_i^n)^2 < \frac{\varepsilon^2}{4}$$

$$(y_i - \phi_i^m)^2 < \frac{\varepsilon^2}{4M}$$

when $n > N$ and $m > N$. Then,

$$\sum_{i=1}^{M} (y_i - \phi_i^n)^2 < \varepsilon^2$$

for $n > N$. Since M was arbitrary, let M approach ∞, to obtain

$$\| \mathbf{y} - \phi^n \|_{\ell_2}^2 = \sum_{i=1}^{\infty} (y_i - \phi_i^n)^2 < \varepsilon^2$$

for $n > N$ and

$$\sum_{i=1}^{\infty} y_i^2 = \sum_{i=1}^{\infty} [(y_i - \phi_i^n) + \phi_i^n]^2 \leq 2\sum_{i=1}^{\infty} (y_i - \phi_i^n)^2 + 2\sum_{i=1}^{\infty} (\phi_i^n)^2 < \infty$$

Thus, $\mathbf{y}$ is in ℓ_2 and $\lim_{n\to\infty} \| \mathbf{y} - \phi^n \|_{\ell_2} = 0$, which completes the proof. ■

The space ℓ_2 of infinite sequences of real numbers is a complete normed vector space with a scalar product. This collection of properties of a vector space is special enough to deserve a name.

Definition 5.3.5. A complete normed vector space with a scalar product is called a **Hilbert space**. ■

The space ℓ_2 is thus a Hilbert space. Another example of a Hilbert space is R^n, with the usual scalar product

$$(\mathbf{x}, \mathbf{y}) = \sum_{i=1}^{n} x_i y_i$$

Finally, while the proof is beyond the scope of this text, the result that follows is proved in Refs. 14, 15, and 22.

Theorem 5.3.3. For a bounded domain Ω in R^n with a smooth boundary, the space $L_2(\Omega)$, with scalar product and norm of Eqs. 5.2.2 and 5.2.4, is a complete vector space; i.e., it is a Hilbert space.

Least Square Approximation and Complete Sets of Functions

Let a sequence of functions $\{ \phi^i(x) \}$ be orthonormal in $L_2(a, b)$. Consider the problem of approximating a square integrable function $f(x)$, in the **least square** sense, by a linear combination of the $\phi^i(x)$, $i = 1, \ldots, n$. That is, choose $\mathbf{c} = [c_1, c_2, \ldots, c_n]^T$ to minimize

$$F(\mathbf{c}) = \| f - \sum_{i=1}^{n} c_i \phi^i \|^2 = \int_a^b \left(f - \sum_{i=1}^{n} c_i \phi^i \right)^2 dx \tag{5.3.17}$$

Expanding this quantity yields

$$F(\mathbf{c}) = \int_a^b f^2 \, dx + \sum_{i=1}^{n} c_i^2 - 2\sum_{i=1}^{n} c_i \int_a^b f\phi^i \, dx$$

The choice of $\mathbf{c}$ that minimizes $F(\mathbf{c})$ is unaffected by addition of a constant term to

$F(\mathbf{c})$. In particular, minimizing

$$F^*(\mathbf{c}) = \int_a^b f^2\,dx + \sum_{i=1}^{n} c_i^2 - 2\sum_{i=1}^{n} c_i \int_a^b f\phi^i\,dx + \sum_{i=1}^{n}\left(\int_a^b f\phi^i\,dx\right)^2$$

$$= \int_a^b f^2\,dx + \sum_{i=1}^{n}\left(c_i - \int_a^b f\phi^i\,dx\right)^2$$

yields the same $\mathbf{c}$ as minimizing $F(\mathbf{c})$. This proves the following theorem.

Theorem 5.3.4. The functional $F(\mathbf{c})$ of Eq. 5.3.17 is minimized by choosing

$$c_i = \int_a^b f\phi^i\,dx = (\phi^i, f), \quad i = 1, 2, \ldots, n \tag{5.3.18}$$

where $\{\phi^i(x)\}$ is an orthonormal sequence in $L_2(a, b)$. This yields the **least square approximation** of $f(x)$ in $L_2(a, b)$. ■

Example 5.3.4

The set of polynomials $\{\phi^1(x), \phi^2(x)\} = \left\{\frac{1}{\sqrt{2}}, \sqrt{\frac{3}{2}}\,x\right\}$ is orthonormal in $L_2(-1, 1)$. In order to best approximate $f(x) = x^3$ in the least square sense, Eq. 5.3.18 yields

$$c_1 = \int_{-1}^{1} \frac{1}{\sqrt{2}}\,x^3\,dx = \left.\frac{x^4}{4\sqrt{2}}\right|_{-1}^{1} = 0$$

$$c_2 = \int_{-1}^{1} \sqrt{\frac{3}{2}}\,x^4\,dx = \left.\frac{\sqrt{3}}{5\sqrt{2}}\,x^5\right|_{-1}^{1} = \frac{\sqrt{6}}{5}$$

Thus, the function

$$g(x) = c_1\phi^1 + c_2\phi^2 = \frac{3}{5}\,x$$

best approximates $f(x) = x^3$, in the least square sense on the interval $[-1, 1]$, using the $\phi^i(x)$ selected. As shown in Fig. 5.3.1, however, the approximation is not good outside the interval $[-1, 1]$. Since all linear functions can be represented as $c_1\phi^1(x) + c_2\phi^2(x)$, $g(x)$ is the best linear approximation of $f(x) = x^3$, in the least square sense, on the interval $[-1, 1]$.

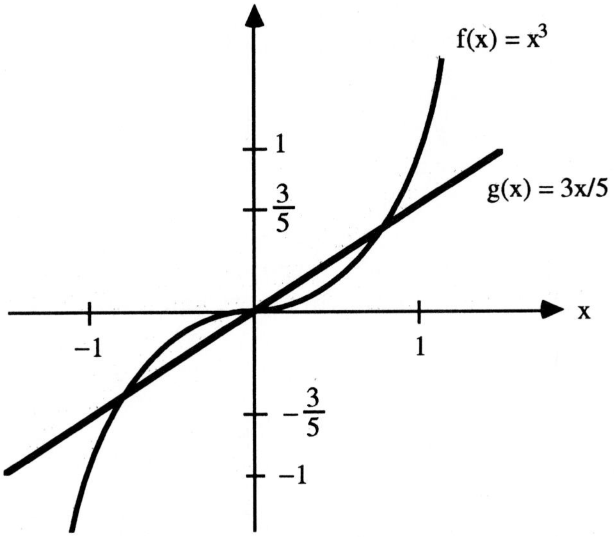

Figure 5.3.1 Linear Least Square Approximation of x^3 ■

The actual minimum value of the function in Eq. 5.3.17 is

$$\| f - \sum_{i=1}^{n} c_i \phi^i \|^2 = \int_a^b f^2\, dx - \sum_{i=1}^{n} c_i^2 \geq 0$$

The quantity on the left is a positive nonincreasing function of n. This does not necessarily imply, however, that as n increases without bound, this quantity necessarily goes to zero. Taking the limit as n approaches ∞ yields **Bessel's inequality**,

$$\infty > \int_a^b f^2\, dx \geq \sum_{i=1}^{\infty} c_i^2 \tag{5.3.19}$$

In terms of $\mathbf{c} = [\, c_1, c_2, \ldots \,]^T$, this is

$$\| \mathbf{c} \|_{\ell_2}^2 \leq \| f \|_{L_2}^2 < \infty$$

so that the series $\sum_{i=1}^{\infty} c_i^2$ converges for any $f(x) \in L_2(a, b)$ and $\mathbf{c} \in \ell_2$. Therefore, corresponding to every function f(x) in $L_2(a, b)$ and any orthonormal sequence $\{ \phi^i(x) \}$ in $L_2(a, b)$, the vector $\mathbf{c} = [\, c_1, c_2, \ldots \,]^T$ defined by Eq. 5.3.18 is in ℓ_2.

Example 5.3.5

Bessel's inequality for the functions in Example 5.3.4 is

$$\frac{6}{25} = c_1^2 + c_2^2 < \int_{-1}^{1} f^2(x)\, dx = \int_{-1}^{1} x^6\, dx = \left.\frac{x^7}{6}\right|_{-1}^{1} = \frac{1}{3}$$

which is a strict inequality. ■

Definition 5.3.6. For a given orthonormal sequence $\{ \phi^i(x) \}$ in $L_2(a, b)$, if Bessel's inequality is an equality for every square integrable function f(x); i.e., if for all $f \in L_2(a, b)$,

$$\lim_{n\to\infty} \| f - \sum_{i=1}^{n} c_i\phi^i \|^2 = \lim_{n\to\infty} \int_a^b \left(f - \sum_{i=1}^{n} c_i\phi^i \right)^2 dx = 0 \qquad (5.3.20)$$

or

$$\| f \|_{L_2} = \| \mathbf{c} \|_{\ell_2}$$

then $\{ \phi^i(x) \}$ is said to be a **complete set of functions**, or a **basis for L_2(a, b)**, and $\sum_{i=1}^{\infty} c_i\phi^i(x)$ converges in the mean to f(x). ■

Since computations with infinite sets of functions are intricate, further examples are delayed until the sections that follow on Fourier series. The remainder of this section is devoted to developing generally applicable properties of the vector space $L_2(a, b)$.

Convergence in the mean in $L_2(a, b)$ does not imply pointwise convergence on (a, b). Consider, for example, a square integrable function g(x) that differs from f(x) on a set of points that does not contribute to the integral; e.g., at a finite number of points. Then,

$$c_i = \int_a^b \phi^i f\, dx = \int_a^b \phi^i g\, dx$$

Thus, $\sum_{i=1}^{\infty} c_i\phi^i(x)$ is the same for f(x) and g(x) and the series converges in the mean to two different functions. In general, stronger conditions on f(x) than merely square integrability are required to guarantee pointwise convergence.

If the orthonormal sequence of functions $\{ \phi^i(x) \}$ is complete, then Bessel's inequality becomes an equality; i.e., if

$$b_i = \int_a^b \phi^i f\,dx = (\phi^i, f)_{L_2}, \quad i = 1, 2, \ldots$$

$$c_i = \int_a^b \phi^i g\,dx = (\phi^i, g)_{L_2}, \quad i = 1, 2, \ldots$$

then, with $\mathbf{b} = [b_1, b_2, \ldots]^T$ and $\mathbf{c} = [c_1, c_2, \ldots]^T$,

$$\| f \|_{L_2}^2 = \int_a^b f^2\,dx = \sum_{i=1}^{\infty} b_i^2 = \| \mathbf{b} \|_{\ell_2}^2$$

$$\| g \|_{L_2}^2 = \int_a^b g^2\,dx = \sum_{i=1}^{\infty} c_i^2 = \| \mathbf{c} \|_{\ell_2}^2$$

Also, using the triangle inequality,

$$\left| (f, g)_{L_2} - \sum_{i=1}^{n} b_i c_i \right| = \left| \int_a^b fg\,dx - \sum_{i=1}^{n} b_i c_i \right|$$

$$= \left| \int_a^b f \left(g - \sum_{i=1}^{n} c_i \phi^i \right) dx \right|$$

$$\leq \| f \|_{L_2} \left(\int_a^b \left(g - \sum_{i=1}^{n} c_i \phi^i \right)^2 dx \right)^{1/2} \tag{5.3.21}$$

approaches zero, since $\{ \phi^i(x) \}$ is complete, and

$$\lim_{n \to \infty} \int_a^b \left(g - \sum_{i=1}^{n} c_i \phi^i \right)^2 dx = 0$$

Taking the limit on both sides of Eq. 5.3.21 as n approaches ∞ proves the **Parseval relation**

$$(f, g)_{L_2} = \int_a^b fg\,dx = \sum_{i=1}^{\infty} b_i c_i = (\mathbf{b}, \mathbf{c})_{\ell_2} \tag{5.3.22}$$

Theorem 5.3.5. A sequence of orthonormal functions $\{ \phi^i(x) \}$ in $L_2(a, b)$ is complete; i.e., it is a basis for $L_2(a, b)$, if and only if 0 is the only function in $L_2(a, b)$ that is orthogonal to all $\phi^i(x)$. ■

For purposes of proof, let $\{ \phi^i(x) \}$ be complete and assume there exists a nonzero normalized function f(x) such that

$$c_i = \int_a^b \phi^i f \, dx = 0$$

for all i. By direct calculation,

$$\lim_{n \to \infty} \int_a^b \left(f - \sum_{i=1}^{n} c_i \phi^i \right)^2 dx = \int_a^b |f|^2 \, dx = 1$$

which contradicts the completeness assumption.

To prove the converse, let 0 be the only function orthogonal to all the $\phi^i(x)$ and assume that $\{ \phi^i(x) \}$ is not complete. If this is so, there exists a function f(x) for which Bessel's inequality is a strict inequality; i.e.,

$$\int_a^b f^2 \, dx - \sum_{i=1}^{\infty} c_i^2 > 0$$

where

$$c_i = (\phi^i, f)$$

However, since $\mathbf{c} \in \ell_2$, the sequence $g_n(x) = \sum_{i=1}^{n} c_i \phi^i(x)$ is a Cauchy sequence in $L_2(a, b)$ so it converges to a function g(x), such that

$$c_i = (\phi^i, g) = (\phi^i, f)$$

Therefore, the function $h(x) = g(x) - f(x)$ is orthogonal to all the $\phi^i(x)$ and

$$\| h \| = \| g - f \| \ge \big| \| f \| - \| g \| \big| = \left| \| f \| - \left(\sum_{i=1}^{\infty} c_i^2 \right)^{1/2} \right| > 0$$

and h(x) is not the zero function. This contradicts the hypothesis and shows that $\{ \phi^i(x) \}$ is complete. ∎

Theorem 5.3.6. The trigonometric functions of Example 5.2.5 form a basis for $L_2(-\pi, \pi)$. Expansion of a function f(x) in $L_2(-\pi, \pi)$ in this basis, using the coefficients of Eq. 5.3.18, yields the **Fourier series** for f(x) that converges to f(x) in the mean; i.e.,

$$\lim_{n \to \infty} \| f(x) - \sum_{i=1}^{n} c_i \phi^i(x) \| = 0$$

■

The proof of Theorem 5.3.6 follows from results established in Section 9.3.

The trigonometric basis for $L_2(\Omega)$ is only one of many bases that exist [15], even though it is the best known. It remains to develop details on how to expand functions in terms of Fourier series and to determine pointwise convergence properties. This is the task of the remainder of this chapter.

EXERCISES 5.3

1. Show that the functions $\phi^i(x) = (1/\sqrt{\pi}) \sin ix$, $i = 1, 2, 3, \ldots$, are not complete in $L_2(0, 2\pi)$.

2. Let f(x) be continuous and have a piecewise continuous derivative on the interval $0 \le x \le 2\pi$, with $f(0) = f(2\pi)$. Show that $\sum_{i=1}^{\infty} i^2 (a_i^2 + b_i^2) < \infty$, where

$$a_n = \frac{1}{\pi} \int_0^{2\pi} f(x) \cos nx \, dx$$

$$b_n = \frac{1}{\pi} \int_0^{2\pi} f(x) \sin nx \, dx$$

[Hint: Apply Bessel's inequality to the function $f'(x)$.]

3. Show that the following sequence of vectors in ℓ_2 is a Cauchy sequence in ℓ_2:

$$\phi^k = \left[\frac{1}{(1 + 1/k)^2}, \frac{1}{(2 + 1/k)^2}, \ldots \right]^T = \left[\frac{1}{(i + 1/k)^2} \right]^T$$

Find its limit $\phi = \lim_{k \to \infty} \phi^k$ and show that $\phi \in \ell_2$.

4. Show that the sequence $\phi^i(x) = x + 1/i$, $i = 1, 2, \ldots$, on the interval $0 < x < 1$ is a Cauchy sequence in $L_2(0, 1)$. Find its limit $\phi(x) = \lim_{i \to \infty} \phi^i(x)$ and show that $\phi(x) \in L_2(0, 1)$.

5.4 CONSTRUCTION OF FOURIER SERIES

Prior to constructing series representations of functions in terms of trigonometric functions, it is helpful to define properties of functions that are to be represented.

Piecewise Continuous Functions

Definition 5.4.1. A function f(x) is said to be **piecewise continuous** in a closed interval [a, b] if there is at most a finite number of points $a = x_1 < x_2 < \ldots < x_n = b$ such that f(x) is continuous in the subintervals $x_j < x < x_{j+1}$ and the **one-sided limits**

$$f(x_j+) = \lim_{\substack{\varepsilon \to 0 \\ \varepsilon > 0}} f(x_j + \varepsilon)$$

$$f(x_j-) = \lim_{\substack{\varepsilon \to 0 \\ \varepsilon > 0}} f(x_j - \varepsilon)$$

exist and are finite for all i = 1, 2, . . . , n − 1, even though they may not be equal. The function space of piecewise continuous functions on [a, b] is denoted as $D^0(a, b)$. ■

Example 5.4.1

Consider the function

$$f(x) = x - j\pi, \quad j\pi \le x < (j+1)\pi, \quad j = -1, 0, 1, \ldots$$

shown in Fig. 5.4.1.

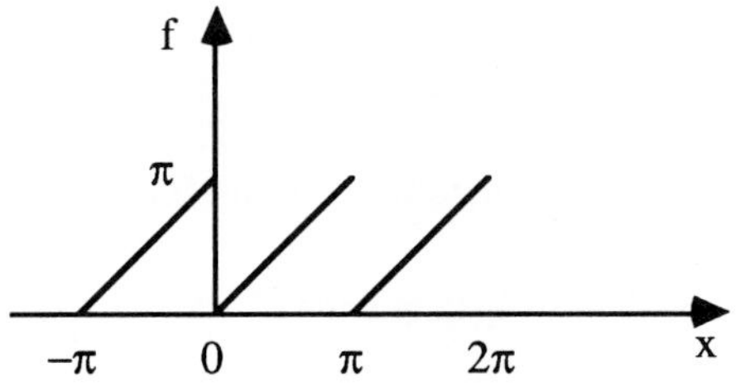

Figure 5.4.1 A Piecewise Continuous Function

This function is continuous, except at $x_j = j\pi$, and the left and right sided limits at $x_j = j\pi$ are

$$f(x_j-) = \lim_{\substack{\varepsilon \to 0 \\ \varepsilon > 0}} f(x_j - \varepsilon)$$

$$= \lim_{\substack{\varepsilon \to 0 \\ \varepsilon > 0}} [\,(j\pi - \varepsilon) - (j-1)\pi\,]$$

$$= j\pi - j\pi + \pi$$

$$= \pi$$

$$f(x_j+) = \lim_{\substack{\varepsilon \to 0 \\ \varepsilon > 0}} f(x_j + \varepsilon)$$

$$= \lim_{\substack{\varepsilon \to 0 \\ \varepsilon > 0}} [\,(j\pi - \varepsilon) - j\pi\,]$$

$$= j\pi - j\pi$$

$$= 0$$

Thus, both limits exist and are finite, so f(x) is piecewise continuous. Note that $f(x_j+) = 0 = f(x_j)$, but $f(x_j-) = \pi \neq f(x_j)$. ■

Functions such as 1/x and sin (1/x) fail to be piecewise continuous in the closed interval [0, 1], because the one-sided limit f(0+) does not exist in both cases.

Definition 5.4.2. The **left-sided derivative** of the function f(x) at a point x_0 is

$$f'(x_0-) = \lim_{\substack{h \to 0 \\ h > 0}} \left[\frac{f(x_0-) - f(x_0 - h)}{h} \right] \tag{5.4.1}$$

and the **right-sided derivative** of the function f(x) at x_0 is

$$f'(x_0+) = \lim_{\substack{h \to 0 \\ h > 0}} \left[\frac{f(x_0 + h) - f(x_0+)}{h} \right] \tag{5.4.2}$$

■

It is clear that if f(x) has a derivative f '(x) at x_0, then the left- and right-sided derivatives exist and have the value $f'(x_0)$. However, left- and right-sided derivatives may exist but not be equal.

Definition 5.4.3. If f(x) is piecewise continuous in an interval [a, b]; i.e., it is continuous in subintervals $x_j < x < x_{j+1}$, $j = 1, \ldots, n < \infty$, and if its first derivative f '(x) is continuous in each of the intervals $x_j < x < x_{j+1}$ and the limits $f'(x_j+)$ and $f'(x_j-)$ exist, then f(x) is called **piecewise smooth**. The space of piecewise smooth functions on [a, b] is denoted $D^1(a, b)$. ■

Example 5.4.2

For $x_j < x < x_{j+1}$ in Example 5.4.1,

$$f'(x) = 1$$

Thus, $f'(x_j-) = 1 = f'(x_j+)$ and the function f(x) in Example 5.4.1 is piecewise smooth. ■

Even and Odd Functions

Definition 5.4.4. A function f(x) is an **even function** if, for all x,

$$f(-x) = f(x) \tag{5.4.3}$$

and an **odd function** if, for all x,

$$f(-x) = -f(x) \tag{5.4.4}$$

■

An even function is one whose graph is symmetric about the ordinate and an odd function has a graph that is symmetric about the origin, as shown in Fig. 5.4.2.

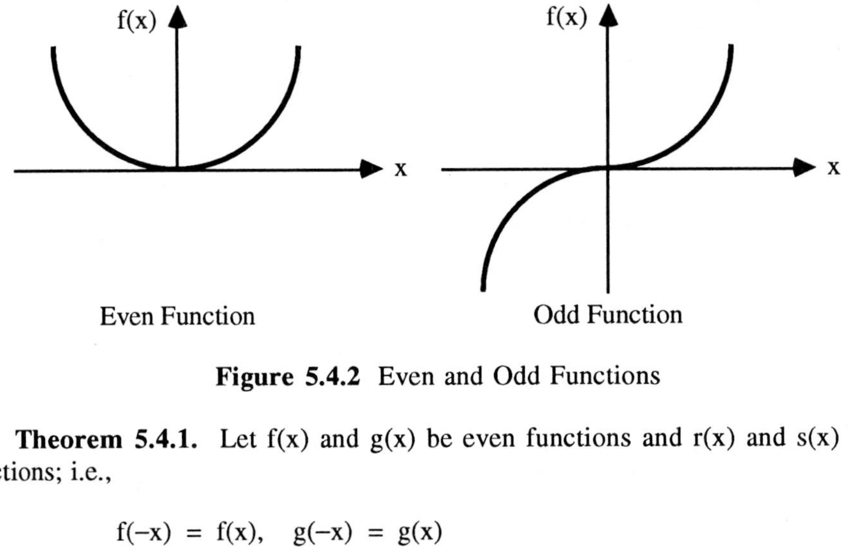

Figure 5.4.2 Even and Odd Functions

Theorem 5.4.1. Let f(x) and g(x) be even functions and r(x) and s(x) be odd functions; i.e.,

$$f(-x) = f(x), \quad g(-x) = g(x)$$

$$r(-x) = -r(x), \quad s(-x) = -s(x)$$

for all x. Then the products $f(x) \times g(x)$ and $r(x) \times s(x)$ are even functions, whereas all other products are odd functions; i.e., products of even functions are even, products of odd functions are even, and products of even and odd functions are odd. ■

To prove Theorem 5.4.1, note first that for products of even functions,

$$f(-x) \times g(-x) = f(x) \times g(x)$$

Similarly, for products of odd functions,

$$r(-x) \times s(-x) = (-r(x)) \times (-s(x))$$

$$= r(x) \times s(x)$$

In contrast, for products of even and odd functions,

$$f(-x) \times r(-x) = f(x) \times (-r(x))$$

$$= (-f(x)) \times r(x)$$

■

Example 5.4.3

The even powers of x; i.e., $f_{2n}(x) = x^{2n}$, are even, since

$$f_{2n}(-x) = (-x)^{2n}$$

$$= (-1)^{2n} x^{2n}$$

$$= x^{2n}$$

$$= f_{2n}(x)$$

The odd powers of x; i.e., $f_{2n+1}(x) = x^{2n+1}$, are odd, since

$$f_{2n+1}(-x) = (-x)^{2n+1}$$

$$= (-1)^{2n+1} x^{2n+1}$$

$$= -x^{2n+1}$$

$$= -f_{2n+1}(x)$$

■

Example 5.4.4

The trigonometric function sin x is odd, since

$$\sin(-x) = -\sin x$$

However, cos x is even, since

$$\cos(-x) = \cos x$$

■

One application in which the symmetry of even and odd functions is significant is in the evaluation of integrals. If f(x) is integrable and even, then for any a,

$$\begin{aligned}\int_{-a}^{a} f(x)\,dx &= \int_{-a}^{0} f(x)\,dx + \int_{0}^{a} f(x)\,dx \\ &= \int_{a}^{0} f(-x)\,d(-x) + \int_{0}^{a} f(x)\,dx \\ &= \int_{0}^{a} f(-x)\,dx + \int_{0}^{a} f(x)\,dx \\ &= 2\int_{0}^{a} f(x)\,dx \end{aligned} \tag{5.4.5}$$

If f(x) is odd, the same argument shows that, for any a,

$$\int_{-a}^{a} f(x)\,dx = 0 \tag{5.4.6}$$

Fourier Series for Periodic Functions

Definition 5.4.5. A function f(x) is said to be a **periodic function** of **period** $p > 0$ if

$$f(x + p) = f(x) \tag{5.4.7}$$

for all x. ■

The best-known examples of periodic functions are the sine and cosine functions. As a special case, a constant function is a periodic function with an arbitrary period p. Thus, the sum $a_0 + a_1 \cos x + a_2 \cos 2x + \ldots + b_1 \sin x + b_2 \sin 2x + \ldots + a_n \cos nx + b_n \sin nx$, has period 2π.

The sequence of functions

$$\{\, 1, \cos x, \sin x, \ldots, \cos nx, \sin nx \,\} \tag{5.4.8}$$

forms an orthogonal sequence on the interval $(-\pi, \pi)$, since

$$\begin{aligned} &\int_{-\pi}^{\pi} (\sin mx)(\sin nx)\,dx = \begin{cases} 0, & \text{if } m \neq n \\ \pi, & \text{if } m = n \end{cases} \\ &\int_{-\pi}^{\pi} (\sin mx)(\cos nx)\,dx = 0, \quad \text{for all } m, n \\ &\int_{-\pi}^{\pi} (\cos mx)(\cos nx)\,dx = \begin{cases} 0, & \text{if } m \neq n \\ \pi, & \text{if } m = n \end{cases} \end{aligned} \tag{5.4.9}$$

for m = 0, 1, . . . and n = 1, 2, . . . To normalize this sequence, multiply the elements of the original orthogonal sequence by the inverse of their norms. Hence, the sequence

$$\left\{ \frac{1}{\sqrt{2\pi}}, \frac{\cos x}{\sqrt{\pi}}, \frac{\sin x}{\sqrt{\pi}}, \ldots, \frac{\cos nx}{\sqrt{\pi}}, \frac{\sin nx}{\sqrt{\pi}} \right\} \tag{5.4.10}$$

is orthonormal, as was shown by direct integration in Example 5.2.5.

Definition 5.4.6. The **Fourier series** representation of f(x) is defined as

$$f(x) \approx \frac{a_0}{2} + \sum_{k=1}^{\infty} (a_k \cos kx + b_k \sin kx) \tag{5.4.11}$$

where the symbol $\approx$ indicates association of the series with f(x), since the series may or may not converge to f(x). The coefficient $a_0/2$ instead of a_0 is used for convenience in later calculations. ■

Let f(x) be an integrable function on the interval $(-\pi, \pi)$ and define the nth partial sum of the series in Eq. 5.4.11 as

$$s_n(x) = \frac{a_0}{2} + \sum_{k=1}^{n} (a_k \cos kx + b_k \sin kx) \tag{5.4.12}$$

or, in terms of the orthonormal functions of Eq. 5.4.10,

$$s_n(x) = \frac{a_0\sqrt{2\pi}}{2}\left(\frac{1}{\sqrt{2\pi}}\right) + \sum_{k=1}^{n}\left(a_k\sqrt{\pi}\,\frac{\cos kx}{\sqrt{\pi}} + b_k\sqrt{\pi}\,\frac{\sin kx}{\sqrt{\pi}}\right)$$

which is to represent f(x) on $(-\pi, \pi)$. The coefficients a_k and b_k in $s_n(x)$ that yield the best approximation to f(x), in the least squares sense; i.e., that minimize

$$\| f - s_n \|^2 = \int_{-\pi}^{\pi} [\, f(x) - s_n(x) \,]^2 \, dx$$

are determined by Eq. 5.3.18 as

$$\frac{a_0\sqrt{2\pi}}{2} = \int_{-\pi}^{\pi} f(x) \frac{1}{\sqrt{2\pi}} \, dx$$

$$a_k\sqrt{\pi} = \int_{-\pi}^{\pi} f(x) \frac{\cos kx}{\sqrt{\pi}} \, dx$$

$$b_k\sqrt{\pi} = \int_{-\pi}^{\pi} f(x) \frac{\sin kx}{\sqrt{\pi}} \, dx$$

More explicitly, the **Fourier coefficients** a_0, a_k, and b_k of f(x) in Eq. 5.4.11 are

$$a_0 = \frac{1}{\pi}\int_{-\pi}^{\pi} f(x)\,dx$$

$$a_k = \frac{1}{\pi}\int_{-\pi}^{\pi} f(x)\cos kx\,dx \tag{5.4.13}$$

$$b_k = \frac{1}{\pi}\int_{-\pi}^{\pi} f(x)\sin kx\,dx$$

Bessel's inequality of Eq. 5.3.19 for Fourier series is

$$\frac{a_0^2}{2} + \sum_{k=1}^{\infty}\left(a_k^2 + b_k^2\right) \le \frac{1}{\pi}\int_{-\pi}^{\pi} f^2(x)\,dx < +\infty \tag{5.4.14}$$

Thus, for any function f(x) in $L_2(-\pi, \pi)$, the Fourier coefficients satisfy the conditions

$$\lim_{k\to\infty} a_k = 0 \quad\text{and}\quad \lim_{k\to\infty} b_k = 0 \tag{5.4.15}$$

The Fourier series is said to **converge in the mean** to f(x) if

$$\lim_{n\to\infty} \| f - s_n \|^2 =$$

$$\lim_{n\to\infty}\int_{-\pi}^{\pi}\left[f(x) - \frac{a_0}{2} - \sum_{k=1}^{n}\left(a_k\cos kx + b_k\sin kx\right)\right]^2 dx = 0 \tag{5.4.16}$$

If the Fourier series converges in the mean to f(x), then Bessel's inequality becomes an equality, i.e.,

$$\frac{a_0^2}{2} + \sum_{k=1}^{\infty}\left(a_k^2 + b_k^2\right) = \frac{1}{\pi}\int_{-\pi}^{\pi} f^2(x)\,dx \tag{5.4.17}$$

Example 5.4.5

Consider the function f(x) = x in the interval $-\pi < x < \pi$. The Fourier coefficients of Eq. 5.4.13 are

$$a_0 = \frac{1}{\pi}\int_{-\pi}^{\pi} f(x)\, dx$$

$$= \frac{1}{\pi}\int_{-\pi}^{\pi} x\, dx$$

$$= 0$$

$$a_k = \frac{1}{\pi}\int_{-\pi}^{\pi} f(x) \cos kx\, dx$$

$$= \frac{1}{\pi}\int_{-\pi}^{\pi} x \cos kx\, dx$$

$$= 0$$

$$b_k = \frac{1}{\pi}\int_{-\pi}^{\pi} f(x) \sin kx\, dx$$

$$= \frac{1}{\pi}\int_{-\pi}^{\pi} x \sin kx\, dx$$

$$= \frac{2}{k}(-1)^{k+1}$$

Hence,

$$x \approx 2\sum_{k=1}^{\infty} (-1)^{k+1} \frac{\sin kx}{k}$$

■

If f(x) is an even function defined on the interval $(-\pi, \pi)$, since cos kx is an even function and sin kx an odd function, Theorem 5.4.1 shows that f(x) cos kx is even and f(x) sin kx is odd. Employing Eqs. 5.4.5 and 5.4.6, the Fourier coefficients of an even function f(x) are

$$a_k = \frac{1}{\pi}\int_{-\pi}^{\pi} f(x) \cos kx\, dx = \frac{2}{\pi}\int_{0}^{\pi} f(x) \cos kx\, dx, \quad k = 0, 1, 2, \ldots \tag{5.4.18}$$

$$b_k = \frac{1}{\pi}\int_{-\pi}^{\pi} f(x) \sin kx\, dx = 0, \quad k = 1, 2, \ldots$$

Hence, the Fourier series of an even function is

$$f(x) \approx \frac{a_0}{2} + \sum_{k=1}^{\infty} a_k \cos kx$$

In a similar manner, if f(x) is an odd function, the function f(x) cos kx is odd and f(x) sin kx is even. As a consequence, the Fourier coefficients of an odd function f(x) are

$$\begin{aligned} a_k &= \frac{1}{\pi}\int_{-\pi}^{\pi} f(x) \cos kx\, dx = 0, \quad k = 0, 1, 2, \ldots \\ b_k &= \frac{1}{\pi}\int_{-\pi}^{\pi} f(x) \sin kx\, dx = \frac{2}{\pi}\int_{0}^{\pi} f(x) \sin kx\, dx, \quad k = 1, 2, \ldots \end{aligned} \tag{5.4.19}$$

and the Fourier series of an odd function is

$$f(x) \approx \sum_{k=1}^{\infty} b_k \sin kx$$

Even and Odd Extensions of Functions

Definition 5.4.7. If a function f(x) is defined in the interval $(0, \pi)$, it may be extended to $(-\pi, \pi)$ in two ways. The first is the **even extension** of f(x), defined as

$$f_e(x) \equiv \begin{cases} f(x), & 0 < x < \pi \\ f(-x), & -\pi < x < 0 \end{cases} \tag{5.4.20}$$

while the second is the **odd extension** of f(x), defined as

$$f_o(x) \equiv \begin{cases} f(x), & 0 < x < \pi \\ -f(-x), & -\pi < x < 0 \end{cases} \tag{5.4.21}$$

■

Since $f_e(x)$ and $f_o(x)$ are even and odd functions with period 2π, respectively, the Fourier series expansions of $f_e(x)$ and $f_o(x)$ are

$$\begin{aligned} f_e(x) &\approx \frac{a_0}{2} + \sum_{k=1}^{\infty} a_k \cos kx \\ a_k &= \frac{2}{\pi}\int_{0}^{\pi} f(x) \cos kx\, dx \end{aligned} \tag{5.4.22}$$

and

$$f_o(x) \approx \sum_{k=1}^{\infty} b_k \sin kx$$

$$b_k = \frac{2}{\pi} \int_0^{\pi} f(x) \sin kx \, dx \tag{5.4.23}$$

Example 5.4.6

The function $f(x) = x$ in the interval $0 \le x \le \pi$ can be extended to an even function on $(-\pi, \pi)$, using Eq. 5.4.20, which yields $f_e(x) = | x |$. By Eq. 5.4.22,

$$a_0 = \frac{2}{\pi} \int_0^{\pi} x \, dx = \pi$$

$$a_k = \frac{2}{\pi} \int_0^{\pi} x \cos kx \, dx = \frac{2}{\pi k^2} (\cos k\pi - 1)$$

Thus,

$$| x | \approx \frac{\pi}{2} + \sum_{k=1}^{\infty} \frac{2}{\pi k^2} (\cos k\pi - 1) \cos kx$$

$$= \frac{\pi}{2} - \sum_{k=1}^{\infty} \frac{4}{\pi (2k - 1)^2} \cos (2k - 1) x$$

■

Thus far, functions have been defined on the interval $(-\pi, \pi)$. In many applications, however, this interval is restrictive and the interval of interest may be arbitrary, say (a, b). Introducing the new variable t, by the transformation

$$x = \frac{1}{2} (b + a) + \frac{(b - a)}{2\pi} t$$

or

$$t = \frac{\pi (2x - b - a)}{b - a} \tag{5.4.24}$$

the interval $a \le x \le b$ is transformed to $-\pi \le t \le \pi$. Using the first of Eq. 5.4.24, the function $f(x(t)) = f [(b + a)/2 + (b - a) t/2\pi] \equiv g(t)$ is defined on the interval $(-\pi, \pi)$. Expanding this function in a Fourier series,

$$g(t) \approx \frac{a_0}{2} + \sum_{k=1}^{\infty} (a_k \cos kt + b_k \sin kt)$$

where

$$a_k = \frac{1}{\pi}\int_{-\pi}^{\pi} g(t)\cos kt\,dt, \quad k = 0, 1, 2, \ldots$$

$$b_k = \frac{1}{\pi}\int_{-\pi}^{\pi} g(t)\sin kt\,dt, \quad k = 1, 2, 3, \ldots$$

Transforming from t back to x, the expansion for f(x) in (a, b) is

$$\begin{aligned} f(x) &= g(t(x)) \\ &\approx \frac{a_0}{2} + \sum_{k=1}^{\infty}\left[a_k \cos\frac{k\pi(2x-b-a)}{(b-a)} + b_k \sin\frac{k\pi(2x-b-a)}{(b-a)}\right] \end{aligned} \tag{5.4.25}$$

where, using the relation $dt = \left(\frac{2\pi}{b-a}\right)dx$,

$$\begin{aligned} a_k &= \frac{2}{(b-a)}\int_a^b f(x)\cos\frac{k\pi(2x-b-a)}{(b-a)}\,dx \\ b_k &= \frac{2}{(b-a)}\int_a^b f(x)\sin\frac{k\pi(2x-b-a)}{(b-a)}\,dx \end{aligned} \tag{5.4.26}$$

EXERCISES 5.4

1. Determine which of the function spaces, D^1 or C^1, the following functions are in:

(a) $f(x) = \begin{cases} x, & 0 \le x \le 1 \\ 0, & 1 < x \le 2 \end{cases}$

(b) $f(x) = \begin{cases} 0, & -1 \le x \le 0 \\ 1/x, & 0 < x \le 1 \end{cases}$

(c) $f(x) = \begin{cases} 2, & 0 \le x < 1 \\ x^2, & 1 < x \le 2 \end{cases}$

(d) $f(x) = \begin{cases} 1 - x, & 1 \le x < 2 \\ x/(x-2), & 2 < x \le 3 \end{cases}$

2. Are the following functions even, odd, or neither even nor odd?

   (a) $x + 2x^2 + 3x^3$

   (b) $x \ln x$

   (c) e^{x^2}

   (d) $x \sin x$

3. Let f(x) be continuously differentiable in the interval (0, a). Let f '(0+) exist and let f(0+) = f(0) = 0. Prove that $f_o(x)$, the odd extension of f(x) to (–a, a), is continuously differentiable in (–a, a).

4. Determine which of the following functions are periodic and find the least positive period of those which are periodic:

   (a) $\sin 3x$

   (b) $x \cos x$

   (c) $\cos x + \cos \pi x$

   (d) $x^2 \sin x$

5. Find the Fourier series of the following functions:

   (a) $f(x) = \begin{cases} x, & -\pi < x < 0 \\ h \text{ (constant)}, & 0 < x < \pi \end{cases}$

   (b) $f(x) = x + \sin x, \quad -\pi \le x \le \pi$

   (c) $f(x) = 1 + x, \quad -\pi \le x \le \pi$

6. Determine the Fourier sine series of the following functions:

   (a) $f(x) = \pi - x, \quad 0 < x < \pi$

   (b) $f(x) = \begin{cases} 1, & 0 < x < \pi/2 \\ 2, & \pi/2 < x < \pi \end{cases}$

7. If f(x) and F(x) are piecewise continuous with periodicity 2π and have Fourier coefficients a_n and b_n and A_n and B_n, respectively, show that

$$\frac{1}{\pi}\int_{-\pi}^{\pi} f(x)\,F(x)\,dx = \frac{1}{2}a_0A_0 + \sum_{n=1}^{\infty}(a_nA_n + b_nB_n)$$

   [Hint: Apply Parseval's relation of Eq. 5.3.22 to f(x) and F(x).]

5.5 POINTWISE CONVERGENCE OF FOURIER SERIES

Pointwise Convergence

Even though the Fourier series associated with a function in $L_2(\Omega)$ converges to the function in the mean, it may fail to converge pointwise. **Pointwise convergence of Fourier series** is investigated in this section. To begin with, consider

$$f(x) \approx \frac{a_0}{2} + \sum_{k=1}^{\infty} (a_k \cos kx + b_k \sin kx)$$

Substituting for a_k and b_k from Eq. 5.4.13 into the partial sum $s_n(x)$ of Eq. 5.4.12,

$$\begin{aligned} s_n(x) &= \frac{1}{2\pi} \int_{-\pi}^{\pi} f(t)\, dt + \frac{1}{\pi} \sum_{k=1}^{n} \left[\left(\int_{-\pi}^{\pi} f(t) \cos kt\, dt \right) \cos kx \right. \\ &\qquad \left. + \left(\int_{-\pi}^{\pi} f(t) \sin kt\, dt \right) \sin kx \right] \\ &= \frac{1}{\pi} \int_{-\pi}^{\pi} f(t) \left[\frac{1}{2} + \sum_{k=1}^{n} (\cos kt \cos kx + \sin kt \sin kx) \right] dt \\ &= \frac{1}{\pi} \int_{-\pi}^{\pi} f(t) \left[\frac{1}{2} + \sum_{k=1}^{n} \cos k (t - x) \right] dt \end{aligned} \tag{5.5.1}$$

Summing the trigonometric identity

$$2 \sin \frac{\alpha}{2} \cos k\alpha = \sin\left(k + \frac{1}{2} \right)\alpha - \sin\left(k - \frac{1}{2} \right)\alpha$$

from k = 1 to k = n yields

$$\begin{aligned} 2 \sin \frac{\alpha}{2} \left(\frac{1}{2} + \sum_{k=1}^{n} \cos k\alpha \right) &= \sin \frac{\alpha}{2} + \left(\sin \frac{3\alpha}{2} - \sin \frac{\alpha}{2} \right) \\ &\quad + \ldots + \left[\sin\left(n + \frac{1}{2} \right)\alpha - \sin\left(n - \frac{1}{2} \right)\alpha \right] \\ &= \sin\left(n + \frac{1}{2} \right)\alpha \end{aligned} \tag{5.5.2}$$

Substituting from Eq. 5.5.2 into Eq. 5.5.1,

$$s_n(x) = \frac{1}{\pi}\int_{-\pi}^{\pi} f(t)\left[\frac{\sin\left(n+\frac{1}{2}\right)(t-x)}{2\sin\left(\frac{t-x}{2}\right)}\right] dt \qquad (5.5.3)$$

Introducing the new variable $s = t - x$, $ds = dt$ and Eq. 5.5.3 is

$$s_n(x) = \frac{1}{\pi}\int_{-\pi-x}^{\pi-x} f(s+x)\,\frac{\sin(n+1/2)s}{2\sin(s/2)}\,ds$$

If f(x) is piecewise continuous and periodic with period 2π, then $s_n(x)$ is also periodic with period 2π. Thus,

$$s_n(x) = \frac{1}{\pi}\int_{-\pi}^{\pi} f(s+x)\,\frac{\sin(n+1/2)s}{2\sin(s/2)}\,ds \qquad (5.5.4)$$

which is known as the **Dirichlet formula** for $s_n(x)$. The coefficient of $f(s+x)$ in the integrand of Eq. 5.5.4 is periodic with period 2π and, from Eq. 5.5.2,

$$\frac{1}{\pi}\int_{-\pi}^{\pi}\frac{\sin(n+1/2)s}{2\sin(s/2)}\,ds = \frac{1}{\pi}\int_{-\pi}^{\pi}\left(\frac{1}{2}+\sum_{k=1}^{n}\cos ks\right)ds = 1 \qquad (5.5.5)$$

Theorem 5.5.1 (Pointwise Convergence Theorem). If f(x) is piecewise smooth in $(-\pi, \pi)$; i.e., f(x) is in $D^1(-\pi, \pi)$, and is periodic with period 2π, then for any x,

$$\frac{a_0}{2} + \sum_{k=1}^{\infty}(a_k \cos kx + b_k \sin kx) = \frac{1}{2}[f(x+) + f(x-)] \qquad (5.5.6)$$

where the Fourier coefficients are given by Eq. 5.4.13 and the notations f(x+) and f(x−) denote limits of f(x) as x is approached from the right and left, respectively. If f(x) is continuous in $(-\pi, \pi)$, the right side of Eq. 5.5.6 reduces to f(x) in $(-\pi, \pi)$. ■

The proof of Theorem 5.5.1 is carried out by showing that

$$\begin{aligned}\lim_{n\to\infty} s_n &= \lim_{n\to\infty}\left[\frac{1}{\pi}\int_{-\pi}^{0} f(x+s)\,\frac{\sin(n+1/2)s}{2\sin(s/2)}\,ds\right.\\ &\qquad\left. + \frac{1}{\pi}\int_{0}^{\pi} f(x+s)\,\frac{\sin(n+1/2)s}{2\sin(s/2)}\,ds\right]\\ &\equiv \lim_{n\to\infty}(I_1 + I_2) = \frac{1}{2}[f(x-) + f(x+)]\end{aligned} \qquad (5.5.7)$$

First, write the integral for I_1 in the form

$$I_1 = \frac{1}{\pi}\int_{-\pi}^{0} [\, f(x+s) - f(x-) + f(x-)\,] \frac{\sin(n + 1/2)s}{2\sin(s/2)}\, ds$$

and note that $\frac{\sin(n + 1/2)s}{2\sin(s/2)}$ is an even function. From Eq. 5.5.5, on the interval $(-\pi, 0)$,

$$\frac{1}{\pi}\int_{-\pi}^{0} f(x-) \frac{\sin(n + 1/2)s}{2\sin(s/2)}\, ds = \frac{f(x-)}{\pi}\int_{-\pi}^{0} \frac{\sin(n + 1/2)s}{2\sin(s/2)}\, ds$$

$$= \frac{f(x-)}{2}$$

Thus,

$$I_1 = \frac{f(x-)}{2} + \frac{1}{\pi}\int_{-\pi}^{0} \frac{f(x+s) - f(x-)}{2\sin(s/2)} \sin(n + 1/2)s\, ds$$

and

$$\lim_{s\to 0, s<0} \left[\frac{f(x+s) - f(x-)}{2\sin(s/2)}\right] = \lim_{s\to 0, s<0} \left[\frac{f(x+s) - f(x-)}{s}\right]\left[\frac{s}{2\sin(s/2)}\right]$$

$$= \lim_{s\to 0, s<0} \left[\frac{f(x+s) - f(x-)}{s}\right]$$

$$= f'(x-)$$

which exists, since $f(x)$ is piecewise smooth. Hence, the function

$$\frac{f(x+s) - f(x-)}{2\sin(s/2)}$$

is piecewise continuous, so it is in $L_2(-\pi, 0)$. By Bessel's inequality (or Eq. 5.4.15), its Fourier sine coefficient must approach zero; i.e.,

$$\lim_{n\to\infty} \int_{-\pi}^{0} \frac{f(x+s) - f(x-)}{2\sin(s/2)} \sin(n + 1/2)s\, ds = 0$$

Therefore,

$$\lim_{n\to\infty} I_1 = \frac{f(x-)}{2}$$

Similarly,

$$\lim_{n\to\infty} I_2 = \frac{f(x+)}{2}$$

Finally, in Eq. 5.5.7,

$$\lim_{n\to\infty} s_n(x) = \lim_{n\to\infty} (I_1 + I_2) = \frac{1}{2} [f(x+) + f(x-)]$$

Thus,

$$\frac{a_0}{2} + \sum_{k=1}^{\infty} (a_k \cos kx + b_k \sin kx) = \frac{1}{2} [f(x+) + f(x-)]$$

which is Eq. 5.5.6.

Note that at a point in $(-\pi, \pi)$ where f(x) is continuous, $f(x+) = f(x-) = f(x)$, in which case

$$f(x) = \frac{a_0}{2} + \sum_{k=1}^{\infty} (a_k \cos kx + b_k \sin kx)$$

which completes the proof of the theorem. ■

Example 5.5.1

Consider the function f(x) defined on the interval $-\pi \le x < \pi$ by

$$f(x) = \begin{cases} 0, & -\pi \le x < 0 \\ x + 1, & 0 \le x < \pi/2 \\ 2x, & \pi/2 \le x < \pi \end{cases}$$

and for all other x by periodic extension; i.e., for all x,

$$f(x + 2\pi) = f(x)$$

To analyze convergence of the Fourier series for f(x) and determine the value to which the series converges at $x = 0$, $x = \pi/2$, and $x = \pi$; first note that f(x) is piecewise continuous in $(-\pi, \pi)$; f'(x) is continuous on the intervals $[-\pi, 0)$, $[0, \pi/2)$, and $[\pi/2, \pi)$; and

$$f'(-\pi +) = 0 = f'(0 -)$$

$$f'(0 +) = 1 = f(\frac{\pi}{2} -)$$

$$f'(\frac{\pi}{2}+) = 2 = f(\pi -)$$

Therefore, f(x) is piecewise smooth in $(-\pi, \pi)$ and is periodic with period 2π. By the pointwise convergence theorem (Theorem 5.5.1),

$$\lim_{n\to\infty} s_n(0) = \frac{1}{2}[\, f(0+) + f(0-)\,] = \frac{1}{2}(1+0) = \frac{1}{2}$$

$$\lim_{n\to\infty} s_n\left(\frac{\pi}{2}\right) = \frac{1}{2}\left[\, f\left(\frac{\pi}{2}+\right) + f\left(\frac{\pi}{2}-\right)\right]$$

$$= \frac{1}{2}\left(\pi + \frac{\pi}{2} + 1\right) = \frac{3\pi+2}{4}$$

$$\lim_{n\to\infty} s_n(\pi) = \frac{1}{2}[\, f(\pi+) + f(\pi-)\,] = \frac{1}{2}(0+2\pi) = \pi$$

■

Uniform Convergence

With the result of Theorem 5.5.1 on convergence of Fourier series for a piecewise smooth function, uniform convergence of Fourier series can be considered.

Theorem 5.5.2 (Uniform and Absolute Convergence Theorem). Let f(x) be a continuous function with period 2π, let f'(x) be piecewise continuous in the interval $(-\pi, \pi)$, and let $f(-\pi) = f(\pi)$. Then, the Fourier series for f(x) is uniformly and absolutely convergent for all x. ■

To prove Theorem 5.5.2, let the Fourier series of f(x) and f'(x) be

$$f(x) = \frac{a_0}{2} + \sum_{k=1}^{\infty} (\, a_k \cos kx + b_k \sin kx\,) \tag{5.5.8}$$

and

$$f'(x) \approx \frac{A_0}{2} + \sum_{k=1}^{\infty} (\, A_k \cos kx + B_k \sin kx\,) \tag{5.5.9}$$

respectively. Since f'(x) is piecewise continuous and $f(-\pi) = f(\pi)$,

$$A_0 = \frac{1}{\pi}\int_{-\pi}^{\pi} f'(x)\, dx = \frac{1}{\pi}[\, f(\pi) - f(-\pi)\,] = 0 \tag{5.5.10}$$

Integrating by parts,

$$A_k = \frac{1}{\pi}\int_{-\pi}^{\pi} f'(x) \cos kx \, dx$$

$$= \frac{1}{\pi}[\, f(x) \cos kx \,] \Big|_{-\pi}^{\pi} + \frac{k}{\pi}\int_{-\pi}^{\pi} f(x) \sin kx \, dx$$

$$= kb_k, \quad k = 1, 2, \ldots \tag{5.5.11}$$

$$B_k = \frac{1}{\pi}\int_{-\pi}^{\pi} f'(x) \sin kx \, dx$$

$$= \frac{1}{\pi}[\, f(x) \sin kx \,] \Big|_{-\pi}^{\pi} - \frac{k}{\pi}\int_{-\pi}^{\pi} f(x) \cos kx \, dx$$

$$= -ka_k, \quad k = 1, 2, \ldots \tag{5.5.12}$$

The coefficients A_k and B_k satisfy Bessel's inequality, so

$$\sum_{k=1}^{\infty} (A_k^2 + B_k^2) \le \frac{1}{\pi}\int_{-\pi}^{\pi} [\, f'(x) \,]^2 \, dx < \infty \tag{5.5.13}$$

From Eqs. 5.5.11 and 5.5.12,

$$\sum_{k=1}^{n} (a_k^2 + b_k^2)^{1/2} = \sum_{k=1}^{n} [\, k^{-2} (A_k^2 + B_k^2) \,]^{1/2} \tag{5.5.14}$$

From Eq. 5.5.13, the series $\sum_{k=1}^{n} (A_k^2 + B_k^2)$ converges to a finite limit. It is shown in Example 4.1.6 that the series $\sum_{k=1}^{n} k^{-1}$ converges to a finite limit. Hence, $\phi = [\, 1, 1/2, \ldots, 1/k, \ldots \,]^T$ and $\psi = [\, (A_1^2 + B_1^2)^{1/2}, \ldots, (A_k^2 + B_k^2)^{1/2}, \ldots \,]^T$ are in ℓ_2. Thus,

$$|\, (\phi, \psi) \,| = \sum_{k=1}^{\infty} [\, k^{-2} (A_k^2 + B_k^2) \,]^{1/2} \le \| \phi \|_{\ell_2} \| \psi \|_{\ell_2}$$

$$= \left[\sum_{k=1}^{\infty} k^{-2} \right]^{1/2} \left[\sum_{j=1}^{\infty} (A_j^2 + B_j^2) \right]^{1/2} \equiv M < \infty$$

Finally, from Eq. 5.5.14, for any n,

$$\sum_{k=1}^{n} (a_k^2 + b_k^2)^{1/2} \leq M \tag{5.5.15}$$

so this series converges.

By the Schwartz inequality in R^2,

$$\begin{aligned} | a_k \cos kx + b_k \sin kx | &= \left| \begin{bmatrix} a_k \\ b_k \end{bmatrix}^T \begin{bmatrix} \cos kx \\ \sin kx \end{bmatrix} \right| \\ &\leq \left\| \begin{bmatrix} a_k \\ b_k \end{bmatrix}^T \right\| \left\| \begin{bmatrix} \cos kx \\ \sin kx \end{bmatrix} \right\| \\ &= (a_k^2 + b_k^2)^{1/2}, \quad k = 1, 2, \ldots \end{aligned} \tag{5.5.16}$$

Application of the Weierstrass M-test of Section 4.2, with the bounds of Eq. 5.5.16 and the convergence result of Eq. 5.5.15, shows that the series

$$\frac{a_0}{2} + \sum_{k=1}^{\infty} (a_k \cos kx + b_k \sin kx)$$

converges uniformly and absolutely. By the pointwise convergence theorem, this series converges to f(x). ■

In Theorem 5.5.2, it is assumed that f(x) is continuous and that f '(x) is piecewise continuous. It can be shown that if f(x) is only piecewise smooth in the interval $(-\pi, \pi)$ and if f(x) is periodic with period 2π, then the Fourier series for f(x) converges uniformly to f(x) in every closed interval that contains no discontinuity of f(x). The partial sums $s_n(x)$ of a Fourier series cannot approach the function f(x) uniformly over any interval that contains a point of discontinuity of f(x), however. The behavior of the deviation of $s_n(x)$ from f(x) in such an interval is known as **Gibbs' phenomenon.**

Example 5.5.2

The Fourier series of

$$f(x) = \begin{cases} -1, & -\pi \leq x \leq 0 \\ 1, & 0 \leq x \leq \pi \end{cases}$$

is

$$f(x) \approx \frac{4}{\pi} \sum_{k=1}^{\infty} \frac{\sin(2k-1)x}{(2k-1)}$$

A typical partial sum $s_n(x)$ is shown in Fig. 5.5.1, showing that $s_n(x)$ oscillates near points of discontinuity of f(x). For large n, the magnitude of oscillation decreases at all points in the interval away from points of discontinuity. Very near points of discontinuity, however, the amplitude of oscillation remains practically independent of n. This illustrates the fact that the Fourier series of a function f(x) does not converge uniformly on any interval that contains a point of discontinuity of f(x).

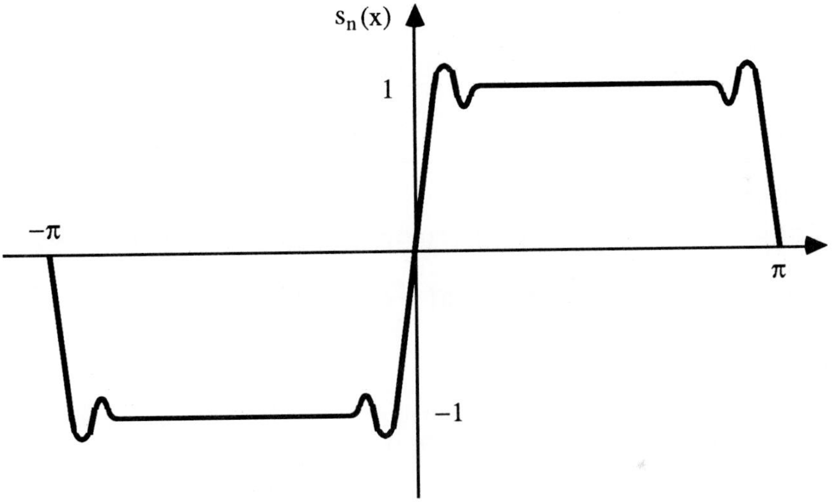

Figure 5.5.1 Gibbs' Phenomenon ■

EXERCISES 5.5

1. Consider the function f(x) defined on the interval $-\pi \le x < \pi$ as

$$f(x) = \begin{cases} (x+\pi)^2, & -\pi \le x < 0 \\ \pi^2, & 0 \le x < \pi/3 \\ 8, & \pi/3 \le x < 2\pi/3 \\ 5, & 2\pi/3 \le x < \pi \end{cases}$$

and for all other x by the periodic extension $f(x + 2\pi) = f(x)$. Analyze the convergence of the trigonometric Fourier series for f(x) and determine those values of x in $-\pi \le x \le \pi$ at which this series does not converge to f(x). Determine the value to which the series converges at each of these points.

2. Find the Fourier series of $f(x) = x^2$, $-\pi \le x \le \pi$, and compute the first four partial sums of the Fourier series and sketch graphs of these partial sums on the same set of axes with a graph of the function. The graphs may be constructed using a computer.

3. Apply Bessel's inequality of Eq. 5.4.14 to show that if f(x) satisfies Theorem 5.5.1, $\sum_{n=1}^{\infty} a_n^2$ and $\sum_{n=1}^{\infty} b_n^2$ converge and that

$$\lim_{n\to\infty} \int_{-\ell}^{\ell} f(x) \cos\left(\frac{n\pi x}{\ell}\right) dx = 0$$

$$\lim_{n\to\infty} \int_{-\ell}^{\ell} f(x) \sin\left(\frac{n\pi x}{\ell}\right) dx = 0$$

This is **Riemann's lemma.**

5.6 DIFFERENTIATION AND INTEGRATION OF FOURIER SERIES

Termwise Differentiation of Fourier Series

Example 5.6.1

To see that termwise differentiation of Fourier series is, in general, not permissible, note that the Fourier series for $f(x) = x$ is

$$f(x) \approx 2\left[\sin x - \frac{\sin 2x}{2} + \frac{\sin 3x}{3} - \dots\right]$$

which converges to $f(x) = x$ for all x in $(-\pi, \pi)$. However, after formal term-by-term differentiation, the series

$$2\,[\cos x - \cos 2x + \cos 3x - \dots]$$

diverges for all x and is certainly not the Fourier series of $f'(x) = 1$. The difficulty arises from the fact that the function $f(x) = x$ in $(-\pi, \pi)$, when extended periodically in the Fourier series expansion, is discontinuous at the points $\pm\pi, \pm 3\pi, \dots$; i.e., $f(-\pi) \ne f(\pi)$. ■

The following theorem shows that uniform convergence, hence continuity of the periodic function, is one of the conditions that must be met if termwise differentiation of a Fourier series is to be permissible.

Theorem 5.6.1 (Fourier Series Differentiation Theorem). Let f(x) be a continuous function in the interval $[-\pi, \pi]$, with $f(-\pi) = f(\pi)$, and let $f'(x)$ be piecewise smooth in that interval. Then the Fourier series for $f'(x)$ can be obtained by termwise differentiation of the series for $f(x)$ and the differentiated series converges to $\frac{1}{2}[f'(x+) + f'(x-)]$. ■

To prove Theorem 5.6.1, let the Fourier series of $f(x)$ and $f'(x)$ be as in Eqs. 5.5.8 and 5.5.9. It was shown, in the proof of Theorem 5.5.2, that if $f(-\pi) = f(\pi)$, the Fourier coefficients of $f'(x)$ are $A_0 = 0$, $A_k = kb_k$, and $B_k = -ka_k$. Thus,

$$f'(x) \approx \sum_{k=1}^{\infty} (-ka_k \sin kx + kb_k \cos kx)$$

This is precisely the result that is obtained by termwise differentiation of Eq. 5.5.8. At a point of discontinuity of $f'(x)$, termwise differentiation is still valid, in the sense that

$$\frac{1}{2}[f'(x+) + f'(x-)] = \sum_{k=1}^{\infty} (-ka_k \sin kx + kb_k \cos kx)$$

Since $f'(x)$ is piecewise smooth, according to the extension noted after the proof of Theorem 5.5.2, the derived series converges pointwise to $f'(x)$ at points of continuity and to $[f'(x+) + f'(x-)]/2$ at points of discontinuity. This completes the proof of the theorem. ■

Example 5.6.2

Determine the Fourier series expansion of $\cos^2 x$ on $0 < x < \pi$ by differentiating an appropriate Fourier series. Let

$$f'(x) = \cos^2 x = \frac{1}{2} + \frac{1}{2}\cos 2x$$

Then,

$$f(x) = \frac{x}{2} + \frac{1}{4}\sin 2x, \quad 0 < x < \pi$$

The even extension of $f(x)$ is continuous, it has a piecewise continuous derivative, and it satisfies $f(-\pi) = f(\pi)$. Its Fourier series representation is thus convergent to

$$f(x) = \frac{a_0}{2} + \sum_{k=1}^{\infty} a_k \cos kx$$

where, by Eq. 5.4.22,

$$a_0 = \frac{2}{\pi}\int_0^{\pi}\left(\frac{x}{2} + \frac{\sin 2x}{4}\right)dx = \frac{\pi}{2}$$

$$a_k = \frac{2}{\pi}\int_0^{\pi}\left(\frac{x}{2} + \frac{\sin 2x}{4}\right)\cos kx\,dx$$

$$= \frac{2}{\pi}\left\{\frac{1}{2k^2}[(-1)^k - 1] + \frac{1}{2(k^2-4)}[(-1)^k - 1]\right\}$$

$$= \frac{2}{\pi}\frac{k^2-2}{k^2(k^2-4)}[(-1)^k - 1], \quad k = 1, 2, \ldots$$

Therefore,

$$f(x) = \frac{\pi}{4} + \sum_{k=1}^{\infty}\frac{2}{\pi}\frac{k^2-2}{k^2(k^2-4)}[(-1)^k - 1]\cos kx$$

Theorem 5.6.1 permits termwise differentiation, so in $0 < x < \pi$,

$$\cos^2 x = f'(x) = \sum_{k=1}^{\infty}\frac{2}{\pi}\frac{k^2-2}{k(k^2-4)}[1 - (-1)^k]\sin kx \qquad \blacksquare$$

Termwise Integration of Fourier Series

Termwise integration of Fourier series is possible under more relaxed conditions than termwise differentiation.

Theorem 5.6.2 (Fourier Series Integration Theorem). Let f(x) be piecewise continuous in $(-\pi, \pi)$ and periodic, with period 2π. Then the Fourier series of f(x), whether convergent or not, can be integrated term by term between any limits; i.e.,

$$\int_a^b f(x)\,dx = \int_a^b \frac{a_0}{2}\,dx + \sum_{k=1}^{\infty}\left[\int_a^b (a_k\cos kx + b_k\sin kx)\,dx\right] \qquad (5.6.1)$$

$\blacksquare$

To prove Theorem 5.6.2, it suffices to show that

$$\int_a^b\left[f(x) - \frac{a_0}{2}\right]dx = \sum_{k=1}^{\infty}\frac{1}{k}[a_k(\sin kb - \sin ka) - b_k(\cos kb - \cos ka)] \qquad (5.6.2)$$

First, define F(x) to be

$$F(x) = \int_0^x \left[f(t) - \frac{a_0}{2} \right] dt$$

Since f(x) is piecewise continuous, F(x) is continuous. Furthermore, $F'(x) = f(x) - \frac{a_0}{2}$ is piecewise continuous. In addition,

$$\begin{aligned} F(x + 2\pi) &= \int_0^x \left[f(t) - \frac{a_0}{2} \right] dt + \int_x^{x+2\pi} \left[f(t) - \frac{a_0}{2} \right] dt \\ &= F(x) + \int_{-\pi}^{\pi} \left[f(t) - \frac{a_0}{2} \right] dt \\ &= F(x) \end{aligned}$$

since f(x) is periodic with period 2π and $a_0 = \frac{1}{\pi} \int_{-\pi}^{\pi} f(x)\, dx$. Since F(x) is a continuous periodic function with a piecewise smooth first derivative, it can be expanded in a Fourier series as

$$F(x) = \frac{A_0}{2} + \sum_{k=1}^{\infty} (A_k \cos kx + B_k \sin kx)$$

which converges uniformly and absolutely, according to Theorem 5.5.2. As in the proof of Theorem 5.5.2, but with a change in sign, $A_k = -b_k/k$ and $B_k = -a_k/k$, for $k > 1$. Hence,

$$F(x) = \frac{A_0}{2} + \sum_{k=1}^{\infty} \left[\frac{1}{k} (a_k \sin kx - b_k \cos kx) \right]$$

From the definition of F(x), it follows that

$$\int_0^x f(t)\, dt = \frac{a_0 x}{2} + \frac{A_0}{2} + \sum_{k=1}^{\infty} \left[\frac{1}{k} (a_k \sin kx - b_k \cos kx) \right]$$

Since $\int_a^b f(t)\, dt = \int_0^b f(t)\, dt - \int_0^a f(t)\, dt$,

$$\int_a^b f(x)\,dx = \frac{a_0}{2}(b-a) + \sum_{k=1}^{\infty}\left[\frac{1}{k}(a_k \sin kb - b_k \cos kb)\right] - \sum_{k=1}^{\infty}\left[\frac{1}{k}(a_k \sin ka - b_k \cos ka)\right]$$

Because these series are absolutely convergent, terms can be rearranged to obtain

$$\int_a^b f(x) = \frac{a_0}{2}(b-a) + \sum_{k=1}^{\infty}\frac{1}{k}[a_k(\sin kb - \sin ka) - b_k(\cos kb - \cos ka)]$$

which is precisely the result obtained by formally integrating the Fourier series for f(x), term by term. ■

Example 5.6.3

Recall from Example 5.4.6 that

$$|x| \approx \frac{\pi}{2} - \sum_{k=1}^{\infty}\frac{4}{\pi(2k-1)^2}\cos(2k-1)x$$

Integrate this Fourier series to obtain the Fourier series for

$$f(x) = \int_0^x |t|\,dt = \begin{cases} \frac{x^2}{2}, & 0 < x < \pi \\ -\frac{x^2}{2}, & -\pi < x < 0 \end{cases}$$

Indefinite integration of each term of the series yields

$$f(x) = \frac{\pi}{2}x - \sum_{k=1}^{\infty}\frac{4}{\pi(2k-1)^3}\sin(2k-1)x + c$$

To find c, note that $f(0) = 0$ and all terms except c on the right are zero at $x = 0$. Thus, $c = 0$. ■

Example 5.6.4

Consider the function $f(x) = x$ on the interval $(-\pi, \pi)$. From Example 5.4.5,

$$x \approx 2\sum_{k=1}^{\infty} (-1)^{k+1} \frac{\sin kx}{k}$$

By Theorem 5.6.2, this series can be integrated term by term to obtain

$$\frac{x^2}{4} = c - \sum_{k=1}^{\infty} \frac{(-1)^{k+1}}{k^2} \cos kx$$

for $-\pi < x < \pi$, where c is a constant of integration. Since the series on the right converges uniformly, it can be integrated term by term from $-\pi$ to π to obtain

$$\int_{-\pi}^{\pi} \frac{x^2}{2}\, dx = 2\left[\int_{-\pi}^{\pi} c\, dx - \sum_{k=1}^{\infty} \frac{(-1)^{k+1}}{k^2} \int_{-\pi}^{\pi} \cos kx\, dx\right]$$

which simplifies to

$$\left.\frac{x^3}{6}\right|_{-\pi}^{\pi} = \frac{\pi^3}{3} = 4\pi c$$

Hence, $c = \frac{\pi^2}{12}$. Therefore, the Fourier series expansion for the function $f(x) = x^2$ is

$$x^2 = 4\left[\frac{\pi^2}{12} - \sum_{k=1}^{\infty} \frac{(-1)^{k+1}}{k^2} \cos kx\right]$$

■

EXERCISES 5.6

1. Determine the Fourier series expansion of each of the following functions by differentiating an appropriate Fourier series:

   (a) $\sin^2 x, \quad 0 < x < \pi$

   (b) $\sin x \cos x, \quad 0 < x < \pi$

2. Find the functions that are represented by the series that are obtained by termwise integration of the following series from 0 to x:

(a) $\sum_{k=1}^{\infty} \frac{(-1)^{k+1}}{k} \sin kx = \frac{x}{2}, \quad -\pi < x < \pi$

(b) $\sum_{k=1}^{\infty} \frac{(-1)^{k+1}}{k} \cos kx = \ln\left(2 \cos \frac{x}{2}\right), \quad -\pi < x < \pi$

(c) $\sum_{k=1}^{\infty} \frac{\sin(2k+1)x}{(2k+1)^3} = \frac{\pi^2 x - \pi x^2}{8}, \quad 0 < x < 2\pi$

3. Find the Fourier series expansion of the function

$$f(x) = \begin{cases} 0, & -\pi < x < 0 \\ x/2, & 0 < x < \pi \end{cases}$$

With the use of this series, show that

$$\frac{\pi^2}{8} = 1 + \frac{1}{3^2} + \frac{1}{5^2} + \frac{1}{7^2} + \ldots$$

5.7 FOURIER SERIES IN TWO VARIABLES

The theory of Fourier series expansion for functions of two variables is analogous to that for functions of one variable. For the sake of simplicity, consider a function $f(\mathbf{x}) = f(x_1, x_2)$ of two variables that is twice continuously differentiable; i.e., that is in $C^2(R^2)$. Let $f(x_1, x_2)$ be periodic with period 2π in both its variables; i.e.,

$$f(x_1 + 2\pi, x_2) = f(x_1, x_2 + 2\pi) = f(x_1, x_2) \tag{5.7.1}$$

If x_2 is held fixed, $f(\bullet, x_2)$ can be expanded into a uniformly convergent Fourier series in the single variable x_1; i.e.,

$$f(x_1, x_2) = \frac{a_0(x_2)}{2} + \sum_{n=1}^{\infty} [\, a_m(x_2) \cos mx_1 + b_m(x_2) \sin mx_1 \,] \tag{5.7.2}$$

in which the coefficients are functions of x_2; i.e.,

$$\begin{aligned} a_m(x_2) &= \frac{1}{\pi} \int_{-\pi}^{\pi} f(x_1, x_2) \cos mx_1 \, dx_1 \\ b_m(x_2) &= \frac{1}{\pi} \int_{-\pi}^{\pi} f(x_1, x_2) \sin mx_1 \, dx_1 \end{aligned} \tag{5.7.3}$$

These coefficients are continuously differentiable functions of x_2 and are periodic of period 2π, so they can be expanded in uniformly convergent Fourier series

$$a_m(x_2) = \frac{a_{m0}}{2} + \sum_{n=1}^{\infty} [\, a_{mn} \cos nx_2 + b_{mn} \sin nx_2 \,]$$
$$b_m(x_2) = \frac{b_{m0}}{2} + \sum_{n=1}^{\infty} [\, c_{mn} \cos nx_2 + d_{mn} \sin nx_2 \,] \tag{5.7.4}$$

where

$$a_{mn} = \frac{1}{\pi^2} \int_{-\pi}^{\pi} \int_{-\pi}^{\pi} f(x_1, x_2) \cos mx_1 \cos nx_2 \, dx_1 \, dx_2$$
$$b_{mn} = \frac{1}{\pi^2} \int_{-\pi}^{\pi} \int_{-\pi}^{\pi} f(x_1, x_2) \cos mx_1 \sin nx_2 \, dx_1 \, dx_2$$
$$c_{mn} = \frac{1}{\pi^2} \int_{-\pi}^{\pi} \int_{-\pi}^{\pi} f(x_1, x_2) \sin mx_1 \cos nx_2 \, dx_1 \, dx_2$$
$$d_{mn} = \frac{1}{\pi^2} \int_{-\pi}^{\pi} \int_{-\pi}^{\pi} f(x_1, x_2) \sin mx_1 \sin nx_2 \, dx_1 \, dx_2 \tag{5.7.5}$$

Substitution of a_m and b_m from Eq. 5.7.4 into Eq. 5.7.2 yields

$$f(x_1, x_2) = \frac{a_{00}}{4} + \frac{1}{2} \sum_{n=1}^{\infty} [\, a_{0n} \cos nx_2 + b_{0n} \sin nx_2 \,]$$
$$+ \frac{1}{2} \sum_{m=1}^{\infty} [\, a_{m0} \cos mx_1 + c_{m0} \sin mx_2 \,]$$
$$+ \frac{1}{2} \sum_{m=1}^{\infty} \sum_{n=1}^{\infty} [\, a_{mn} \cos mx_1 \cos nx_2 + b_{mn} \cos mx_1 \sin nx_2$$
$$+ c_{mn} \sin mx_1 \cos nx_2 + d_{mn} \sin mx_1 \sin nx_2] \tag{5.7.6}$$

which is called the **double Fourier series** for $f(x_1, x_2)$.

Note that when $f(-x_1, x_2) = f(x_1, x_2)$ and $f(x_1, -x_2) = f(x_1, x_2)$, all the coefficients vanish except a_{mn} and the double Fourier series reduces to

$$\begin{aligned} f(x_1, x_2) &= \frac{a_{00}}{4} + \sum_{m=1}^{\infty} \sum_{n=1}^{\infty} a_{mn} \cos mx_1 \cos nx_2 \\ a_{mn} &= \frac{4}{\pi^2} \int_0^{\pi} \int_0^{\pi} f(x_1, x_2) \cos mx_1 \cos nx_2 \, dx_1 \, dx_2 \end{aligned} \tag{5.7.7}$$

When $f(-x_1, x_2) = f(x_1, x_2)$ and $f(x_1, -x_2) = -f(x_1, x_2)$,

$$\begin{aligned} f(x_1, x_2) &= \frac{1}{2} \sum_{n=1}^{\infty} b_{0n} \sin nx_2 + \sum_{m=1}^{\infty} \sum_{n=1}^{\infty} b_{mn} \cos mx_1 \sin nx_2 \\ b_{mn} &= \frac{4}{\pi^2} \int_0^{\pi} \int_0^{\pi} f(x_1, x_2) \cos mx_1 \sin nx_2 \, dx_1 \, dx_2 \end{aligned} \tag{5.7.8}$$

Finally, when $f(-x_1, x_2) = -f(x_1, x_2)$ and $f(x_1, -x_2) = -f(x_1, x_2)$,

$$\begin{aligned} f(x_1, x_2) &= \sum_{m=1}^{\infty} \sum_{n=1}^{\infty} d_{mn} \sin mx_1 \sin nx_2 \\ d_{mn} &= \frac{4}{\pi^2} \int_0^{\pi} \int_0^{\pi} f(x_1, x_2) \sin mx_1 \sin nx_2 \, dx_1 \, dx_2 \end{aligned} \tag{5.7.9}$$

Example 5.7.1

Consider expansion of the function $f(x_1, x_2) = x_1 x_2$ in a double Fourier series. Since $f(-x_1, x_2) = -x_1 x_2 = -f(x_1, x_2)$ and $f(x_1, -x_2) = -x_1 x_2 = -f(x_1, x_2)$, from Eq. 5.7.9,

$$\begin{aligned} d_{mn} &= \frac{4}{\pi^2} \int_0^{\pi} \int_0^{\pi} x_1 x_2 \sin mx_1 \sin nx_2 \, dx_1 \, dx_2 \\ &= (-1)^{m+n} \frac{4}{mn} \end{aligned}$$

Thus, the double Fourier series for $f(x_1, x_2)$, in $-\pi < x_1 < \pi$ and $-\pi < x_2 < \pi$, is

$$f(x_1, x_2) = 4 \sum_{m=1}^{\infty} \sum_{n=1}^{\infty} (-1)^{m+n} \frac{\sin mx_1 \sin nx_2}{mn}$$

■

EXERCISES 5.7

1. Determine the double Fourier series of the following functions:

   (a) $f(x_1, x_2) = 1, \quad 0 \le x_1 \le \pi, \quad 0 \le x_2 \le \pi$

   (b) $f(x_1, x_2) = x_1^2 \sin x_2, \quad -\pi \le x_1 \le \pi, \quad -\pi \le x_2 \le \pi$

2. Develop formulas for the general double Fourier series expansion of a function in the rectangle $-a \le x_1 \le a$, $-b \le x_2 \le b$.

6

MATHEMATICAL MODELS IN MECHANICS

As an introduction to the study of partial differential equations that arise in engineering applications, it is helpful to define a set of model problems that contain many of the features of more complex applications. The differential equations that govern these model problems are derived using techniques that may be employed in more general applications. The resulting equations represent typical models of more general field equations that are encountered in fluid dynamics, elasticity, and heat transfer. The purpose of this chapter is to illustrate methods of obtaining the governing equations for such problems. The differential equations derived here are studied in Chapters 7 and 8, where methods for solving specific classes of equations are developed and applied.

6.1 MULTIPLE INTEGRAL THEOREMS

Since the study of mechanics often deals with static and dynamic behavior of material or energy that is continuously distributed over one or more space dimensions, a method is required to quantify the properties of the material state, such as mass, momentum, volume, and other quantities that are associated with an infinite number of points that make up the **continuum**. Such quantities can be described in terms of **multiple integrals.** A few **multiple integral theorems** play a critical role in quantitatively expressing physical laws that govern the state of a continuum. It is presumed that the reader is familiar with definitions and basic ideas of integration. The purpose of this section is to summarize multiple integral properties and theorems and derive results that will be used in the study of techniques for modeling the behavior of continua. A minimal number of proofs are included in this section. The interested reader is referred to Refs. 8 and 9 for details.

Properties of Multiple Integrals

Consider an area A in R^2 and a volume V in R^3, both of which represent physical domains that are bounded by smooth curves and surfaces, respectively. A function $f(\mathbf{x})$ denotes a scalar valued function of the vector variable $\mathbf{x}$ in either R^2 or R^3. The integral of such scalar valued functions over an area A or volume V is denoted as

$$\iint_A f(\mathbf{x})\, dx_1\, dx_2, \qquad \iiint_V f(\mathbf{x})\, dx_1\, dx_2\, dx_3 \tag{6.1.1}$$

If c is a scalar and g(**x**) is a second function, then the following linear properties of multiple integrals hold:

$$\iint_A c\, f(\mathbf{x})\, dx_1\, dx_2 = c \iint_A f(\mathbf{x})\, dx_1\, dx_2$$

$$\iiint_V c\, f(\mathbf{x})\, dx_1\, dx_2\, dx_3 = c \iiint_V f(\mathbf{x})\, dx_1\, dx_2\, dx_3$$

$$\iint_A [\, f(\mathbf{x}) + g(\mathbf{x})\,]\, dx_1\, dx_2 = \iint_A f(\mathbf{x})\, dx_1\, dx_2 + \iint_A g(\mathbf{x})\, dx_1\, dx_2 \tag{6.1.2}$$

$$\iiint_V [\, f(\mathbf{x}) + g(\mathbf{x})\,]\, dx_1\, dx_2 = \iiint_V f(\mathbf{x})\, dx_1\, dx_2\, dx_3 + \iiint_V g(\mathbf{x})\, dx_1\, dx_2\, dx_3$$

If an area A is subdivided into areas A_1 and A_2, as shown in Fig. 6.1.1, and a volume V is likewise subdivided into volumes V_1 and V_2, then

$$\iint_A f(\mathbf{x})\, dx_1\, dx_2 = \iint_{A_1} f(\mathbf{x})\, dx_1\, dx_2 + \iint_{A_2} f(\mathbf{x})\, dx_1\, dx_2$$

$$\iiint_V f(\mathbf{x})\, dx_1\, dx_2\, dx_3 = \iiint_{V_1} f(\mathbf{x})\, dx_1\, dx_2\, dx_3 + \iiint_{V_2} f(\mathbf{x})\, dx_1\, dx_2\, dx_3 \tag{6.1.3}$$

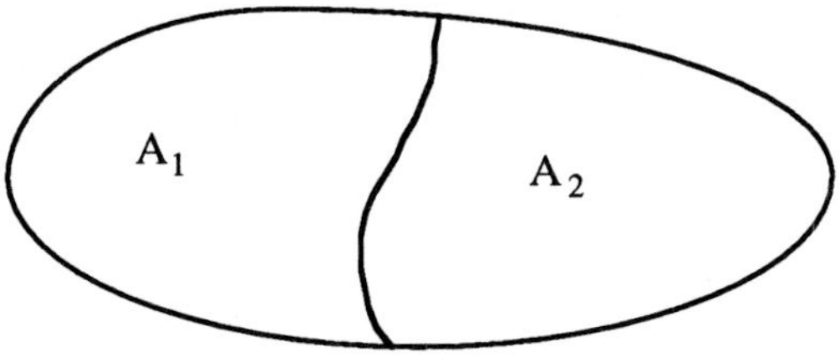

Figure 6.1.1 Subdivision of Area A

If f(**x**) ≤ g(**x**) throughout an area A or volume V, then

$$\iint_A f(\mathbf{x})\, dx_1\, dx_2 \le \iint_A g(\mathbf{x})\, dx_1\, dx_2$$

$$\iiint_V f(\mathbf{x})\, dx_1\, dx_2\, dx_3 \le \iiint_V g(\mathbf{x})\, dx_1\, dx_2\, dx_3 \tag{6.1.4}$$

It follows that, since $-\mid f(\mathbf{x}) \mid \le f(\mathbf{x}) \le \mid f(\mathbf{x}) \mid$, Eq. 6.1.4 may be applied to obtain

$$\left| \iint_A f(\mathbf{x})\, dx_1\, dx_2 \right| \le \iint_A \mid f(\mathbf{x}) \mid dx_1\, dx_2$$

$$\left| \iiint_V f(\mathbf{x})\, dx_1\, dx_2\, dx_3 \right| \le \iiint_V \mid f(\mathbf{x}) \mid dx_1\, dx_2\, dx_3 \tag{6.1.5}$$

The Mean Value Theorem and Leibniz's Rule

An important result for derivation of differential equations of continuum mechanics is the **mean value theorem**, which holds for a **simply connected region**; i.e., a region for which each closed curve or surface in the region encloses only points in the region. There are, thus, no holes in simply connected regions.

Theorem 6.1.1 (Mean Value Theorem). If the function $f(\mathbf{x})$ is continuous in a simply connected area A or volume V, then there is a point $\bar{\mathbf{x}}$ in A or V such that

$$\iint_A f(\mathbf{x})\, dx_1\, dx_2 = A\, f(\bar{\mathbf{x}})$$

$$\iiint_V f(\mathbf{x})\, dx_1\, dx_2\, dx_3 = V\, f(\bar{\mathbf{x}}) \tag{6.1.6}$$

i.e., the function $f(\mathbf{x})$ achieves its **mean value** at $\bar{\mathbf{x}}$. ■

Example 6.1.1

The average value of $f(x_1, x_2) = ax_1 - x_1^2$ on the rectangle $-a \le x_1 \le a$, $0 \le x_2 \le 1$, can be found from

$$\int_0^1 \int_{-a}^{a} (ax_1 - x_1^2)\, dx_1\, dx_2 = \frac{1}{2} ax_1^2 - \frac{1}{3} x_1^3 \Big|_{-a}^{a} = -\frac{2}{3} a^3$$

$$A = a - (-a) = 2a$$

and Eq. 6.1.6 may be used to obtain the mean value $f(\bar{\mathbf{x}}) = -(2/3)(a^3/2a) = -a^2/3$, which gives $\bar{x}_1 = (1 - \sqrt{7/3})(a/2)$ and $\bar{x}_2$ is any value in (0, 1). ■

While Eq. 6.1.6 is not particularly useful from a computational point of view, it is often important to know that there is a point in a region of a continuum such that the integral can be expressed as the value of the integrand at that point times the area or volume of the region.

To illustrate use of the mean value theorem, and as a precursor to development of derivative formulas associated with integrals, consider the integral of a function over an interval, where both the function and the end points depend on a scalar parameter α,

$$J(\alpha) = \int_{a(\alpha)}^{b(\alpha)} f(x, \alpha)\, dx \tag{6.1.7}$$

It is clear that the value of this integral is a function of the parameter α, as indicated by the notation $J(\alpha)$. To differentiate this expression with respect to α, form the **difference quotient**

$$\begin{aligned}\frac{J(\alpha+\delta) - J(\alpha)}{\delta} &= \frac{1}{\delta}\left\{ \int_{a(\alpha+\delta)}^{b(\alpha+\delta)} f(x, \alpha+\delta)\, dx - \int_{a(\alpha)}^{b(\alpha)} f(x, \alpha)\, dx \right\} \\ &= \frac{1}{\delta}\left\{ \int_{b(\alpha)}^{b(\alpha+\delta)} f(x, \alpha+\delta)\, dx - \int_{a(\alpha)}^{a(\alpha+\delta)} f(x, \alpha+\delta)\, dx \right. \\ &\qquad \left. + \int_{a(\alpha)}^{b(\alpha)} [\, f(x, \alpha+\delta) - f(x, \alpha)\,]\, dx \right\}\end{aligned}$$

If all functions involved are continuous, the mean value theorem may be applied to each integral, to obtain

$$\begin{aligned}\frac{J(\alpha+\delta) - J(\alpha)}{\delta} &= \left[\frac{b(\alpha+\delta) - b(\alpha)}{\delta} \right] f(\bar{x}, \alpha+\delta) \\ &\quad - \left[\frac{a(\alpha+\delta) - a(\alpha)}{\delta} \right] f(\bar{\bar{x}}, \alpha+\delta) \\ &\quad + \int_{a(\alpha)}^{b(\alpha)} \left[\frac{f(x, \alpha+\delta) - f(x, \alpha)}{\delta} \right] dx\end{aligned}$$

where $\bar{x}$ is between $b(\alpha)$ and $b(\alpha+\delta)$ and $\bar{\bar{x}}$ is between $a(\alpha)$ and $a(\alpha+\delta)$. Assume now that the functions $a(\alpha)$, $b(\alpha)$, and $f(x, \alpha)$ are differentiable with respect to α. Taking the limit of both sides of the above expressions as δ approaches 0, **Leibniz's Rule of Differentiation** is obtained as

$$\frac{dJ}{d\alpha} = \frac{db}{d\alpha} f(b(\alpha), \alpha) - \frac{da}{d\alpha} f(a(\alpha), \alpha) + \int_{a(\alpha)}^{b(\alpha)} \frac{\partial f(x, \alpha)}{\partial \alpha}\, dx \tag{6.1.8}$$

Example 6.1.2

Consider the following integral, which depends on α:

$$F(\alpha) = \int_{\alpha^2}^{\alpha^3} \tan(\alpha x^2)\, dx$$

Using Leibniz's Rule of Differentiation, the derivative of $F(\alpha)$ is

$$F'(\alpha) = 3\alpha^2 \tan(\alpha^7) - 2\alpha \tan(\alpha^5) + \int_{\alpha^2}^{\alpha^3} x^2 \sec^2(\alpha x^2)\, dx \quad \blacksquare$$

Line and Surface Integrals

If C is a continuous curve in R^2 and S is a connected surface in R^3, then **line integrals** and **surface integrals** of a function $f(\mathbf{x})$ on R^2 and R^3, defined on C and S, respectively, are denoted as

$$\int_C f(\mathbf{x})\, dC, \qquad \iint_S f(\mathbf{x})\, dS \tag{6.1.9}$$

where dC and dS are **differential arc length** and **differential surface area** on the curve C and the surface S, respectively. These integrals obey the same linearity, inequality, and mean value properties of Eqs. 6.1.2–6.1.6. Such line and surface integrals play a key role in describing motion of material and energy across curves or surfaces in a continuum and are used frequently in developing governing equations for continuum behavior.

Let $P(\mathbf{x})$ and $Q(\mathbf{x})$ be scalar valued functions defined in R^2. It is assumed that these functions are continuously differentiable on an open set in R^2 that contains a region A that is bounded by a closed curve C, as shown in Fig. 6.1.2. Note that the region A has been selected so that a vertical line intersects curve C in only two points. A more complex region may be subdivided into several regions that satisfy such a condition.

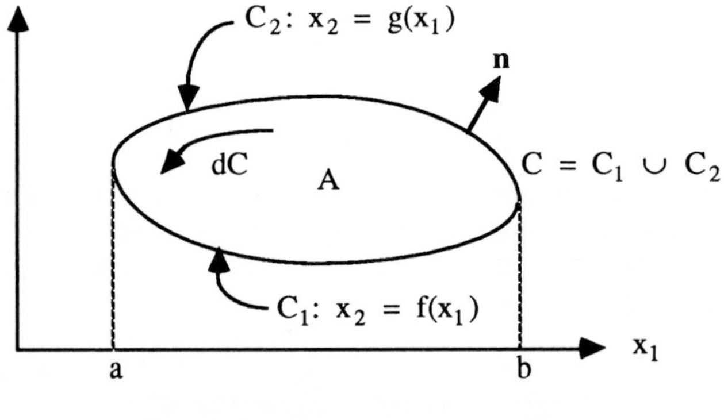

Figure 6.1.2 Domain of Integration

Due to the simple nature of this region, the integral of a quantity over A can be written as an **iterated integral**; i.e.,

$$\iint_A \frac{\partial P}{\partial x_2}\, dx_1\, dx_2 = \int_a^b \int_{f(x_1)}^{g(x_1)} \frac{\partial P}{\partial x_2}\, dx_2\, dx_1$$

$$= \int_a^b P(x_1, g(x_1))\, dx_1 - \int_a^b P(x_1, f(x_1))\, dx_1$$

$$= -\int_b^a P(x_1, g(x_1))\, dx_1 - \int_a^b P(x_1, f(x_1))\, dx_1 \tag{6.1.10}$$

Taking counterclockwise as positive for integration around the closed curve C,

$$\iint_A \frac{\partial P}{\partial x_2}\, dx_1\, dx_2 = -\int_C P\, dx_1 \tag{6.1.11}$$

Performing a similar partitioning of the multiple integral, but using integration with respect to x_1 to reduce the integral to boundary form,

$$\iint_A \frac{\partial Q}{\partial x_1}\, dx_1\, dx_2 = \int_C Q\, dx_2 \tag{6.1.12}$$

where the change in sign on the right side is due to the orientation of the differential as a counterclockwise sweep around the curve C is made. Adding Eqs. 6.1.11 and 6.1.12, the theorem that follows is obtained.

Green and Divergence Theorems

Theorem 6.1.2 (Green's Theorem). If real valued functions $P(x_1, x_2)$ and $Q(x_1, x_2)$ are continuously differentiable in an open region that contains the region A, which is bounded by a continuous curve C, then

$$\iint_A \left[\frac{\partial Q}{\partial x_1} - \frac{\partial P}{\partial x_2} \right] dx_1\, dx_2 = \int_C [\, P\, dx_1 + Q\, dx_2\,] \tag{6.1.13}$$

where integration around the boundary C is counterclockwise. ■

Example 6.1.3

With $P(x_1, x_2) = x_1 x_2$ and $Q(x_1, x_2) = x_1^2 + x_2^2$ in the integral of Eq. 6.1.13 over the area $A = \{\, \mathbf{x} \in R^2 \text{: } 0 \le x_1 \le 1, 0 \le x_2 \le 1 \,\}$, Green's theorem can be used to obtain

$$\int_C x_1 x_2\, dx_1 + \int_C (x_1^2 + x_2^2)\, dx_2 = \int_0^1 \int_0^1 (2x_1 - x_1)\, dx_1\, dx_2$$

$$= \frac{1}{2} x_1^2 \Big|_0^1 = \frac{1}{2}$$

■

From the diagram of Fig. 6.1.3, the **unit normal** and **unit tangent** vectors **n** and **s**, respectively, may be written as

$$\mathbf{n} = \begin{bmatrix} n_1 \\ n_2 \end{bmatrix}, \quad \mathbf{s} = \begin{bmatrix} -n_2 \\ n_1 \end{bmatrix} \tag{6.1.14}$$

where n_1 and n_2 are the direction cosines of the normal **n**. A vector valued function $\mathbf{F}(\mathbf{x}) = [\, F_1(\mathbf{x}), F_2(\mathbf{x})\,]^T$ may be integrated over the boundary C, in terms of either differential coordinates or differential arc length, as

$$\begin{aligned} \int_C \mathbf{F}^T\mathbf{s}\, dC &= \int_C [\, F_1, F_2\,] \begin{bmatrix} -n_2 \\ n_1 \end{bmatrix} dC \\ &= \int_C (- F_1 \cos\beta + F_2 \cos\alpha)\, dC \\ &= \int_C (F_1\, dx_1 + F_2\, dx_2) \end{aligned} \tag{6.1.15}$$

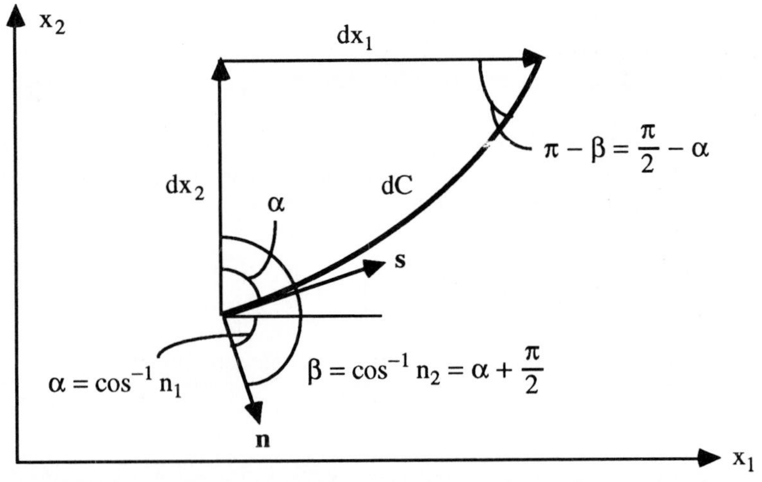

Figure 6.1.3 Differential Arc dC

The geometric relations of Fig. 6.1.3 and Green's theorem can be used to obtain

$$\int_C \mathbf{F}^T\mathbf{n}\, dC = \int_C [\, F_1 n_1 + F_2 n_2 \,]\, dC = \int_C [\, F_1\, dx_2 - F_2\, dx_1 \,]$$

$$= \iint_A \left[\frac{\partial F_1}{\partial x_1} + \frac{\partial F_2}{\partial x_2} \right] dx_1\, dx_2 \tag{6.1.16}$$

Defining the **gradient operator** as $\nabla \equiv \left[\frac{\partial}{\partial x_1}, \frac{\partial}{\partial x_2} \right]^T$ in R^2 and $\nabla \equiv \left[\frac{\partial}{\partial x_1}, \frac{\partial}{\partial x_2}, \frac{\partial}{\partial x_3} \right]^T$ in R^3, $\nabla^T\mathbf{F}(\mathbf{x}) = \text{div}\, \mathbf{F}(\mathbf{x})$ is called the **divergence of a vector valued function** $\mathbf{F}(\mathbf{x})$, and the following theorem is obtained.

Theorem 6.1.3 (Divergence Theorem). If the vector valued function $\mathbf{F}(\mathbf{x}) = [\, F_1(\mathbf{x}), F_2(\mathbf{x}) \,]^T$ is continuously differentiable in an open region of R^2 that contains the domain A, which is bounded by the smooth curve C, then

$$\iint_A \text{div}\, \mathbf{F}\, dx_1\, dx_2 \equiv \iint_A \nabla^T\mathbf{F}\, dx_1\, dx_2 = \int_C \mathbf{F}^T\mathbf{n}\, dC \tag{6.1.17}$$

In a three-dimensional space, an analogous formula holds; i.e.,

$$\iiint_V \text{div}\, \mathbf{F}\, dx_1\, dx_2\, dx_3 \equiv \iiint_V \nabla^T\mathbf{F}\, dx_1\, dx_2\, dx_3$$

$$= \iint_S \mathbf{F}^T\mathbf{n}\, dS \tag{6.1.18}$$

where $\mathbf{F}(\mathbf{x}) = [\, F_1(\mathbf{x}), F_2(\mathbf{x}), F_3(\mathbf{x}) \,]^T$ is a continuously differentiable vector valued function in an open region of R^3 that contains V, which is bounded by a smooth surface S, and $\mathbf{n}(\mathbf{x})$ is the unit outward normal to surface S. ■

For proof of the three-dimensional divergence theorem, the reader is referred to Ref. 8.

Example 6.1.4

Let S be the surface of a region V of R^3, as shown in Fig. 6.1.4, for which the divergence theorem is applicable. Let O be the origin of the coordinate system and let $P(x_1, x_2, x_3)$ be a variable point on S. Then, the volume of the region is

$$\text{Vol} = \frac{1}{3} \iint_S r \cos\theta\, dS$$

where $r = |\, \mathbf{r} \,|$ is the distance $\overline{OP}$ and θ is the angle between the directed line OP and the outward normal to S at P.

To show this, let $F_1 = x_1$, $F_2 = x_2$, and $F_3 = x_3$. Then, $\text{div}\,\mathbf{F}(\mathbf{x}) = 1 + 1 + 1 = 3$ and $\mathbf{F}^T\mathbf{n} = r\cos\theta$. From Eq. 6.1.18,

$$\iiint_\Omega 3\, dV = \iint_S r\cos\theta\, dS$$

Therefore,

$$V = \iiint_\Omega dV = \frac{1}{3}\iint_S r\cos\theta\, dS$$

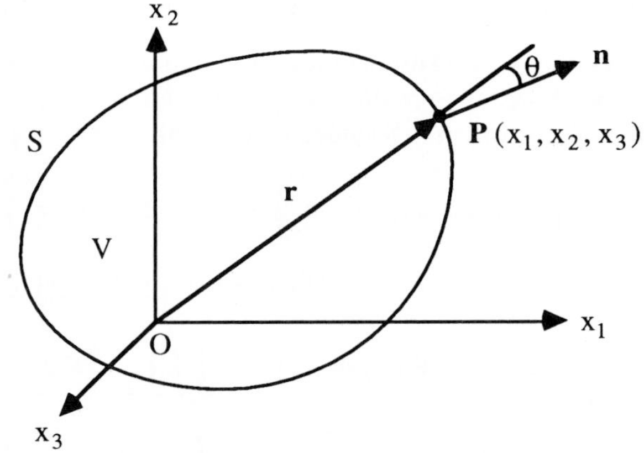

Figure 6.1.4 A region V With Boundary S ■

In subsequent developments, interpretation of certain derivative expressions on the boundary of an area or volume as **directional derivatives** will be helpful. More specifically, the scalar product of the gradient of a function and a unit vector provides the derivative of that function with respect to a coordinate in the direction of the unit vector. In the case of the unit outward normal **n** to a surface, the **normal derivative** of a function $g(\mathbf{x})$ is

$$\frac{\partial g}{\partial n} \equiv (\nabla g)^T\mathbf{n} = \frac{\partial g}{\partial x_1} n_1 + \frac{\partial g}{\partial x_2} n_2 + \frac{\partial g}{\partial x_3} n_3$$

where $\nabla g = \left[\frac{\partial g}{\partial x_1}, \frac{\partial g}{\partial x_2}, \frac{\partial g}{\partial x_3} \right]^T$.

The gradient operator may be employed twice to a scalar valued function $g(\mathbf{x})$ in R^2 or R^3 to obtain the **Laplace operator**, defined as

$$\nabla^2 g \equiv \operatorname{div}(\nabla g) = \nabla^T(\nabla g) = \frac{\partial^2 g}{\partial x_1^2} + \frac{\partial^2 g}{\partial x_2^2}, \text{ in } R^2$$

$$\nabla^2 g \equiv \operatorname{div}(\nabla g) = \nabla^T(\nabla g) = \frac{\partial^2 g}{\partial x_1^2} + \frac{\partial^2 g}{\partial x_2^2} + \frac{\partial^2 g}{\partial x_3^2}, \text{ in } R^3 \tag{6.1.19}$$

Employing Eqs. 6.1.17 and 6.1.18, with $\mathbf{F} = \nabla g$ and Eq. 6.1.19, **Green's identities** are obtained as

$$\iint_A \nabla^2 g \, dx_1 \, dx_2 = \int_C (\nabla g)^T \mathbf{n} \, dC = \int_C \frac{\partial g}{\partial n} \, dC$$

$$\iiint_V \nabla^2 g \, dx_1 \, dx_2 \, dx_3 = \iint_S (\nabla g)^T \mathbf{n} \, dS = \iint_S \frac{\partial g}{\partial n} \, dS \tag{6.1.20}$$

for R^2 and R^3, respectively.

From Eq. 6.1.16, with $F_i = uv$ and $F_j = 0$, $i \neq j$, the following identity is obtained:

$$\int_C uvn_i \, dC = \iint_A \frac{\partial (uv)}{\partial x_i} \, dx_1 \, dx_2$$

$$= \iint_A u \frac{\partial v}{\partial x_i} \, dx_1 \, dx_2 + \iint_A v \frac{\partial u}{\partial x_i} \, dx_1 \, dx_2$$

This important result is summarized in the form of a theorem.

Theorem 6.1.4 (Integration by Parts). Let $u(\mathbf{x})$ and $v(\mathbf{x})$ be in $C^1(A)$, where A is a region in R^2 with a smooth boundary C. Then

$$\iint_A u \frac{\partial v}{\partial x_i} \, dx_1 \, dx_2 = \int_C uvn_i \, dC - \iint_A v \frac{\partial u}{\partial x_i} \, dx_1 \, dx_2 \tag{6.1.21}$$

Similarly, if $u(\mathbf{x})$ and $v(\mathbf{x})$ are in $C^1(V)$, where V is a region in R^3 with a smooth boundary, then

$$\iiint_V u \frac{\partial v}{\partial x_i} \, dx_1 \, dx_2 \, dx_3 = \iint_S uvn_i \, dS - \iiint_V v \frac{\partial u}{\partial x_i} \, dx_1 \, dx_2 \, dx_3 \tag{6.1.22}$$

■

Note the similarity between Eqs. 6.1.21 and 6.1.22 and the elementary integration by parts formula [8, 9],

$$\int_a^b v \frac{du}{dx} \, dx = v(x) \, u(x) \Big|_a^b - \int_a^b u \frac{dv}{dx} \, dx$$

The Implicit Function Theorem

A final topic that is important in studying continuum behavior is the transformation of an integral quantity associated with the continuum. First, recall the standard implicit function theorem from advanced calculus [8].

Theorem 6.1.5 (Implicit Function Theorem). Let y_i, $i = 1, \ldots, n$, denote a set of coordinates in R^n, with the usual Cartesian system denoted x_i, $i = 1, \ldots, n$. Let the coordinate systems be related by the **transformation**

$$x_i = g_i(\mathbf{y}), \quad i = 1, \ldots, n \tag{6.1.23}$$

where the functions $g_i(\mathbf{y})$ are differentiable. Equation 6.1.23 can be solved for $\mathbf{y}$ as a differentiable function of $\mathbf{x}$ in a neighborhood of each point $\mathbf{y}$ in R^n if the **Jacobian matrix** of the functions $g_i(\mathbf{y})$ is nonsingular; i.e.,

$$\left\| \left[\frac{\partial g_i(\mathbf{y})}{\partial y_j} \right] \right\| \neq 0 \tag{6.1.24}$$

■

Example 6.1.5

Consider the coordinate transformation

$$x_1 = g_1(\mathbf{y}) = y_1 + 2y_2$$

$$x_2 = g_2(\mathbf{y}) = 3y_1 + y_2$$

Since

$$\left\| \left[\frac{\partial g_i(\mathbf{y})}{\partial y_j} \right] \right\| = \begin{vmatrix} 1 & 2 \\ 3 & 1 \end{vmatrix} = -5 \neq 0$$

by the implicit function theorem, the above equation can be solved to obtain

$$y_1 = f_1(\mathbf{x}) = (-x_1 + 2x_2)/5$$

$$y_2 = f_2(\mathbf{x}) = (3x_1 - x_2)/5$$

■

Example 6.1.6

Consider the transformation between x-y Cartesian coordinates and r-θ polar coordinates, given as

$$x = g_1(r, \theta) = r \cos \theta$$

$$y = g_2(r, \theta) = r \sin \theta$$

The determinant of the Jacobian matrix of this transformation is

$$\begin{vmatrix} \dfrac{\partial g_1}{\partial r} & \dfrac{\partial g_1}{\partial \theta} \\ \dfrac{\partial g_2}{\partial r} & \dfrac{\partial g_2}{\partial \theta} \end{vmatrix} = \left| \begin{bmatrix} \cos\theta & -r\sin\theta \\ \sin\theta & r\cos\theta \end{bmatrix} \right| = r\cos^2\theta + r\sin^2\theta = r$$

As long as $r \neq 0$, by the implicit function theorem, this coordinate transformation can be solved to yield

$$r = \sqrt{x^2 + y^2}$$

$$\theta = \tan^{-1}\frac{y}{x}$$

■

Often, variables will be changed to describe an area or volume in R^2 or R^3, over which a multiple integral is to be evaluated. Denote by A' and V' the transformed area and volume A and V, respectively, where the equations of transformation are as in Eq. 6.1.23.

Theorem 6.1.6. If the transformation of coordinates of Eq. 6.1.23 is continuously differentiable and if Eq. 6.1.24 holds, then for domains A and V in R^2 and R^3, respectively, with smooth boundaries,

$$\iint_A F(\mathbf{x})\, dx_1\, dx_2 = \iint_{A'} F(\mathbf{g}(\mathbf{y})) \left| \left[\frac{\partial g_i(\mathbf{y})}{\partial y_j} \right] \right| dy_1\, dy_2$$

$$\iiint_V F(\mathbf{x})\, dx_1\, dx_2\, dx_3 \tag{6.1.25}$$

$$= \iiint_{V'} F(\mathbf{g}(\mathbf{y})) \left| \left[\frac{\partial g_i(\mathbf{y})}{\partial y_j} \right] \right| dy_1\, dy_2\, dy_3$$

■

For proof, the reader is referred to Ref. 8.

EXERCISES 6.1

1. Show that if two functions $f(\mathbf{x})$ and $g(\mathbf{x})$ are continuous and if

$$\iint_A f\, dx_1\, dx_2 = \iint_A g\, dx_1\, dx_2$$

or

$$\iiint_V f\, dx_1\, dx_2\, dx_3 = \iiint_V g\, dx_1\, dx_2\, dx_3$$

for all regions A and V in R^2 or R^3, respectively, then $f(\mathbf{x}) = g(\mathbf{x})$ throughout R^2 and R^3, respectively.

2. Let $J(\theta) = \int_0^{\theta} \sin(x+\theta)\, dx$. Compute $\frac{d J(\theta)}{d\theta}$ by

   (a) using Leibniz's rule.

   (b) evaluating the definite integral and then differentiating with respect to θ.

3. Let $u(\mathbf{x})$ and $v(\mathbf{x})$ be scalar valued functions in $C^2(A)$, where A is a domain in R^2 with a continuous boundary C.

   (a) Show that $\nabla^T(u\nabla v) = (\nabla u)^T(\nabla v) + u\nabla^2 v$; i.e.,

$$\frac{\partial}{\partial x_1}\left(u\frac{\partial v}{\partial x_1}\right) + \frac{\partial}{\partial x_2}\left(u\frac{\partial v}{\partial x_2}\right) = \left[\frac{\partial u}{\partial x_1}, \frac{\partial u}{\partial x_2}\right]\begin{bmatrix}\frac{\partial v}{\partial x_1}\\ \frac{\partial v}{\partial x_2}\end{bmatrix} + u\left[\frac{\partial^2 v}{\partial x_1^2} + \frac{\partial^2 v}{\partial x_2^2}\right]$$

   (b) Apply the divergence theorem to prove **Green's formulas**

$$\int_C u\frac{\partial v}{\partial n}\, dC = \iint_A (\nabla u)^T(\nabla v)\, dx_1\, dx_2 + \iint_A u\,\nabla^2 v\, dx_1\, dx_2$$

   and

$$\int_C \left[u\frac{\partial v}{\partial n} - v\frac{\partial u}{\partial n}\right] dC = \iint_A [\,u\,\nabla^2 v - v\,\nabla^2 u\,]\, dx_1\, dx_2$$

4. Show that the results of Exercise 3 hold in R^3, with C replaced by S and A replaced by V.

5. Let $u(\mathbf{x})$ and $v(\mathbf{x})$ be in $C^2(A)$, with $u(\mathbf{x}) = v(\mathbf{x}) = 0$ on the boundary C of A. Show that, in this case,

$$\iint_A v\,\nabla^2 u\, dx_1\, dx_2 = \iint_A u\,\nabla^2 v\, dx_1\, dx_2$$

6. Show that if $u(\mathbf{x})$ is in $C^2(A)$ and $u(\mathbf{x}) = 0$ on the boundary C of A, then

$$-\iint_A u\,\nabla^2 u\, dx_1\, dx_2 \geq 0$$

7. Extend Exercise 6 to R^3.

8. Let $p(\mathbf{x})$ denote the pressure in a three-dimensional flow field. Let $p(\mathbf{x})$ be differentiable and let V be a volume element of fluid with boundary S. Use the divergence theorem to transform the net force on S due to pressure p in the x_i direction; i.e., $-\iint_S p n_i\, dS$, to an integral over V, namely, $-\iiint_V \frac{\partial p}{\partial x_i}\, dx_1\, dx_2\, dx_3$.

6.2 MATERIAL DERIVATIVE AND DIFFERENTIAL FORMS

Certain key concepts that are not apparently associated with multiple integral theorems play an important role in development of the governing equations of continuum behavior. Specifically, material derivatives are important in deriving the differential equations of motion for a continuum. The theory of differential forms also plays a key role in conservation laws of mechanics, as a direct result of multiple integral theorems.

Description of a Continuum

In this section, a **continuum** will be considered as a region of space that is composed of an infinite number of **material particles**. To keep track of these particles, a coordinate system is chosen and the coordinates of a particle at time $t = 0$ are $\mathbf{a} = [\, a_1, a_2, a_3 \,]^T$. At a later time, the particle has moved to physical coordinates $\mathbf{x} = [\, x_1, x_2, x_3 \,]^T$. Thus, motion of the particles is described by the equations

$$x_i = x_i(\mathbf{a}, t), \quad i = 1, 2, 3 \tag{6.2.1}$$

where $\mathbf{a}$ locates the particle at $t = 0$. It is presumed that the functions $x_i(\mathbf{a}, t)$ are continuously differentiable and that the determinant of the Jacobian matrix of their derivatives with respect to the a_i does not vanish. Thus, $\mathbf{a}$ can be determined as a function of $\mathbf{x}$ and t. Consider a volume V that is made up of a collection of material particles at time $t = 0$. As time progresses, this volume will deform as the particles move. The deformed volume at time t is denoted V(t).

Example 6.2.1

A basic quantity in the study of motion is the mass of a collection of material points, which must be preserved. Thus, the **law of conservation of mass** may be written, in terms of material density $\rho_0(\mathbf{a})$ at time $t = 0$ and $\rho(\mathbf{x}(\mathbf{a}, t))$ at time t, as

$$\iiint_{V(t)} \rho(\mathbf{x}(\mathbf{a}, t))\, dx_1\, dx_2\, dx_3 = \iiint_{V(0)} \rho_0(\mathbf{a})\, da_1\, da_2\, da_3 \tag{6.2.2}$$

where the integral on the left is taken over the deformed volume at time t and the integral on the right is taken over the initial volume of material at time $t = 0$. Employing Eq. 6.1.25, variables in the integral on the left side of Eq. 6.2.2 can be changed, to obtain

$$\iiint_{V(0)} \rho(\mathbf{x}(\mathbf{a}, t)) \left| \left[\frac{\partial x_i}{\partial a_j} \right] \right| da_1\, da_2\, da_3$$

$$= \iiint_{V(0)} \rho_0(\mathbf{a})\, da_1\, da_2\, da_3 \tag{6.2.3}$$

Since Eq. 6.2.3 must hold for any volume element in the continuum, V(0) can be taken as a very small neighborhood of point $\mathbf{a}$. The mean value theorem, as expressed by the second of Eq. 6.1.6, yields

$$\mathrm{Vol}(0)\left[\rho(\mathbf{x}(\bar{\mathbf{a}}, t))\left|\left[\frac{\partial x_i}{\partial a_j}\right]\right| - \rho_0(\mathbf{a})\right] = 0$$

for some point $\bar{\mathbf{a}}$ in V(0), where Vol(0) is the volume of V(0). Since Vol(0) ≠ 0,

$$\rho(\mathbf{x}(\bar{\mathbf{a}}, t))\left|\left[\frac{\partial x_i}{\partial a_j}\right]\right| - \rho_0(\bar{\mathbf{a}}) = 0$$

Letting V(0) shrink to the point **a**, since all terms appearing are continuous in **a** and $\bar{\mathbf{a}}$ must converge to **a**,

$$\rho(\mathbf{x}(\mathbf{a}, t))\left|\left[\frac{\partial x_i}{\partial a_j}\right]\right| = \rho_0(\mathbf{a}) \tag{6.2.4}$$

This provides a relationship between the initial density $\rho_0(\mathbf{a})$ and the density $\rho(\mathbf{x}(\mathbf{a}, t))$ at any time during the continuum motion. ■

There are many integral quantities in continuum behavior that vary with time. Typically, consider the integral

$$J(t) = \iiint_{V(t)} f(\mathbf{x}, t)\, dx_1\, dx_2\, dx_3 \tag{6.2.5}$$

where f(**x**, t) denotes a property of the continuum, such as the density in Eq. 6.2.2. It is important to be able to determine the rate of change of the quantity J with time, for the set of particles that occupy V(t). Any given particle at time t is located at a point **x**. After a differential time Δt, it will move to the new position $\mathbf{x}' = \mathbf{x} + \mathbf{v}\Delta t$, where **v** is its velocity. At time t + Δt, the boundary S of V will deform to a neighboring surface S', which bounds the deformed volume V' = V(t + Δt) of Fig. 6.2.1.

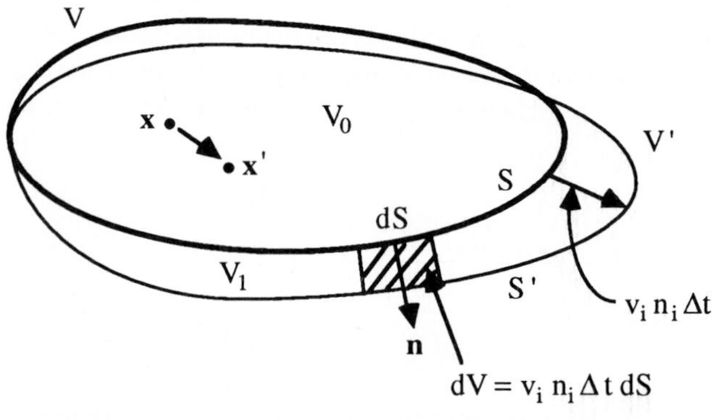

Figure 6.2.1 Deformation of a Moving Volume

Material Derivative

Define the **material derivative** of J(t) in Eq. 6.2.5 as

$$\frac{dJ}{dt} = \lim_{\Delta t \to 0} \frac{1}{\Delta t}\left[\iiint_{V'} f(\mathbf{x}', t + \Delta t)\ dx_1'\ dx_2'\ dx_3' - \iiint_{V} f(\mathbf{x}, t)\ dx_1\ dx_2\ dx_3 \right] \qquad (6.2.6)$$

There are two integral contributions to the difference on the right side of Eq. 6.2.6; one over the common domain V_0 of V and V' and the other over the region V_1, where V and V' differ. The former contribution to the limit in Eq. 6.2.6 is simply

$$\lim_{\Delta t \to 0} \iiint_{V_0} \left[\frac{f(\mathbf{x}', t + \Delta t) - f(\mathbf{x}, t)}{\Delta t} \right] dx_1\ dx_2\ dx_3 = \iiint_{V} \frac{\partial f(\mathbf{x}, t)}{\partial t}\ dx_1\ dx_2\ dx_3 \qquad (6.2.7)$$

since in the limit as Δt approaches 0, V_0 approaches V.

The contribution due to motion normal to surface S is determined by the normal displacement of particles on the boundary S, which is $v_i n_i \Delta t$, where n_i are direction cosines of the outer normal **n** of S and summation notation is used. The volume within V_1 that is swept out by particles that occupy an element of area dS on S is $dV = v_i n_i \Delta t\, dS$. Thus, the contribution to the integral of Eq. 6.2.6 from V_1 is

$$\lim_{\Delta t \to 0} \frac{1}{\Delta t} \iiint_{V_1} f(\mathbf{x}', t + \Delta t)\ v_i n_i\ \Delta t\ ds = \iint_{S(t)} f(\mathbf{x}, t)\ v_i n_i\ ds$$

Adding this to the contribution over V_0 from Eq. 6.2.7, the material derivative of J(t) is

$$\frac{dJ}{dt} \equiv \frac{d}{dt} \iiint_{V(t)} f(\mathbf{x}, t)\ dx_1\ dx_2\ dx_3 = \iiint_{V(t)} \frac{\partial f(\mathbf{x}, t)}{\partial t}\ dx_1\ dx_2\ dx_3 + \iint_{S(t)} f(\mathbf{x}, t)\ v_i n_i\ dS \qquad (6.2.8)$$

It is interesting to note that this result is an extension of Leibniz's rule of differentiation of Eq. 6.1.8.

The surface integral in Eq. 6.2.8 is precisely the right side of Eq. 6.1.18, if the vector **F** is interpreted as having components $F_i = f v_i$. Equation 6.1.18 then yields (note that summation notation is used)

$$\frac{dJ}{dt} = \iiint_{V(t)} \left[\frac{\partial f}{\partial t} + \frac{\partial}{\partial x_i}(f v_i) \right] dx_1\, dx_2\, dx_3 \tag{6.2.9}$$

Noting that $\frac{\partial}{\partial x_i}(f v_i) = \frac{\partial f}{\partial x_i}\frac{dx_i}{dt} + f\frac{\partial v_i}{\partial x_i}$, Eq. 6.2.9 can be rewritten as

$$\begin{aligned}\frac{dJ}{dt} &= \frac{d}{dt}\iiint_{V(t)} f(\mathbf{x}, t)\, dx_1\, dx_2\, dx_3 \\ &= \iiint_{V(t)} \left[\frac{d\,f(\mathbf{x}, t)}{dt} + f\frac{\partial\, v_i(\mathbf{x}, t)}{\partial x_i} \right] dx_1\, dx_2\, dx_3 \end{aligned}\tag{6.2.10}$$

which is a useful alternate form of the material derivative of J(t).

Equations 6.2.8 and 6.2.10 are alternate forms of the **material derivative** of the spatial integral of the function f(**x**, t) over the volume V(t), which is moving with time. They play an important role in deriving the equations of fluid mechanics and other applications in which movement of the continuum occurs over time.

Example 6.2.2

Continuing with Example 6.2.1 involving the total mass in a given volume, as in Eq. 6.2.2, note that if mass is neither created nor destroyed in a given flow field, then for a fixed set of particles that occupy a continuously deforming volume V(t),

$$\frac{d}{dt}\iiint_{V(t)} \rho(\mathbf{x}, t)\, dx_1\, dx_2\, dx_3 = 0 \tag{6.2.11}$$

Employing Eq. 6.2.10, this results in the integral form of the **law of conservation of mass**

$$\iiint_{V(t)} \left[\frac{d\,\rho(\mathbf{x}, t)}{dt} + \rho\frac{\partial\, v_i(\mathbf{x}, t)}{\partial x_i} \right] dx_1\, dx_2\, dx_3 = 0 \tag{6.2.12}$$

or, using Eq. 6.2.8,

$$\iiint_{V(t)} \frac{\partial\, \rho(\mathbf{x}, t)}{\partial t}\, dx_1\, dx_2\, dx_3 + \iint_{S(t)} \rho(\mathbf{x}, t)\, v_i(\mathbf{x}, t)\, n_i\, dS = 0 \tag{6.2.13}$$

If the integrand in Eq. 6.2.12 is continuous, then by the mean value theorem, there is a point $\bar{\mathbf{x}}$ in V(t) where

$$\frac{d\,\rho(\bar{\mathbf{x}}, t)}{dt} + \rho(\bar{\mathbf{x}}, t)\frac{\partial\, v_i(\bar{\mathbf{x}}, t)}{\partial x_i} = 0$$

Since the volume element V(t) is arbitrary, it can be shrunk to any point to yield the differential equation form of the **law of conservation of mass** (often called the **continuity equations**) as

$$\frac{d\rho}{dt} + \rho \frac{\partial v_i}{\partial x_i} = 0$$

or (6.2.14)

$$\frac{\partial \rho}{\partial t} + \frac{\partial (\rho v_i)}{\partial x_i} = 0$$

These continuity equations will be used in a later section to study fluid flow. Note that Eq. 6.2.14 is valid only if ρ and v_i have continuous first derivatives. Even if these functions are not differentiable, Eq. 6.2.12 and Eq. 6.2.13 still have physical significance, as laws of conservation of mass. ■

Differential Forms

A differential expression

$$f_i(\mathbf{x})\, dx_i \tag{6.2.15}$$

where summation notation is used, is called a **differential form** in n variables. Such differential forms arise often in the study of problems of mechanics.

Example 6.2.3

For $\mathbf{x} \in R^3$, the following expressions are differential forms:

$$f_i(\mathbf{x})\, dx_i = 2x_2x_3\, dx_1 + x_1x_3\, dx_2 + x_1x_2\, dx_3$$

$$g_i(\mathbf{x})\, dx_i = 2x_1x_2x_3\, dx_1 + x_1^2x_3\, dx_2 + x_1^2x_2\, dx_3$$

Note that the second differential form is simply the product of x_1 and the first differential form. ■

Definition 6.2.1. The differential form of Eq. 6.2.15 is said to be an **exact differential form** at point **x** if there is a differentiable function F(**x**), defined in a neighborhood of **x**, such that

$$dF(\mathbf{x}) = f_i(\mathbf{x})\, dx_i \tag{6.2.16}$$

at all points in this neighborhood; i.e., if $f_i(\mathbf{x}) = \dfrac{\partial F(\mathbf{x})}{\partial x_i}$. ■

Theorem 6.2.1. Let D be a simply connected open region in R^n and let $f_i(\mathbf{x})$ have continuous first partial derivatives in D. Then the differential form of Eq. 6.2.15 is exact in D if and only if

$$\frac{\partial f_i(\mathbf{x})}{\partial x_j} = \frac{\partial f_j(\mathbf{x})}{\partial x_i}, \quad i, j = 1, \ldots, n \tag{6.2.17}$$

at each point in D. Further, if the form is exact in D, the function $F(\mathbf{x})$ in Eq. 6.2.16 may be written as a line integral

$$F(\mathbf{x}) - F(\mathbf{a}) = \int_C f_i(\mathbf{x})\, dx_i \tag{6.2.18}$$

where C is any smooth curve in the domain D from point **a** to point **x**. Finally, the integral in Eq. 6.2.18 is independent of the path C. ■

This theorem is proved here for n = 2. The proof for $n \geq 3$, the interested reader is referred to Refs. 3 and 8.

To show that the conditions of Eq. 6.2.17 are necessary, it will be shown that Eq. 6.2.17 holds when the function $F(\mathbf{x})$ satisfies

$$f_i(\mathbf{x}) = \frac{\partial F(\mathbf{x})}{\partial x_i} \tag{6.2.19}$$

Since the functions $f_i(\mathbf{x})$ are continuously differentiable, the function $F(\mathbf{x})$ is twice continuously differentiable. Differentiating both sides of Eq. 6.2.19 with respect to x_j yields

$$\frac{\partial f_i(\mathbf{x})}{\partial x_j} = \frac{\partial^2 F(\mathbf{x})}{\partial x_j\, \partial x_i} = \frac{\partial^2 F(\mathbf{x})}{\partial x_i\, \partial x_j} = \frac{\partial f_j(\mathbf{x})}{\partial x_i} \tag{6.2.20}$$

which is Eq. 6.2.17, as was to be shown.

Conversely, beginning with the assumption that Eq. 6.2.17 holds, it must now be shown that a function $F(\mathbf{x})$ satisfying Eq. 6.2.16 exists and is given by Eq. 6.2.18. Choosing any fixed point **a** in D, let $\mathbf{x}^1$ be any other point in D. Let C be a smooth curve from **a** to $\mathbf{x}^1$ and define

$$G(\mathbf{x}^1) = \int_C f_i(\mathbf{x})\, dx_i \tag{6.2.21}$$

Let C_1 be any other smooth path to $\mathbf{x}^1$ in the domain D, as shown in Fig. 6.2.2.

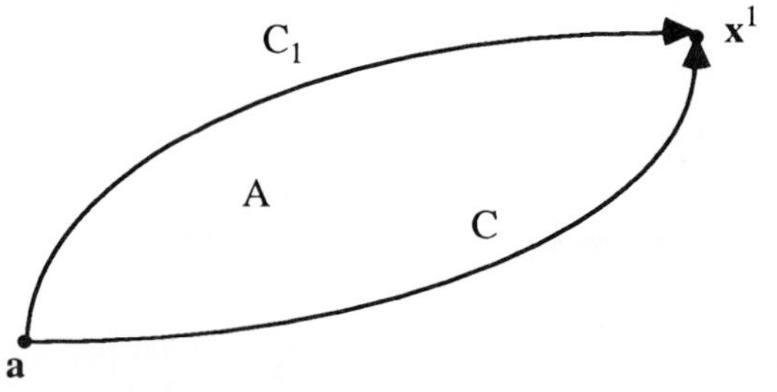

Figure 6.2.2 Paths From **a** to $\mathbf{x}^1$

It will now be shown that

$$\int_C f_i(\mathbf{x})\, dx_i = \int_{C_1} f_i(\mathbf{x})\, dx_i \tag{6.2.22}$$

which will show that the integral of Eq. 6.2.18 is independent of path from **a** to $\mathbf{x}^1$. This is equivalent to showing that the line integral of $f_i(\mathbf{x})\, dx_i$, around the closed curve in Fig. 6.2.2, is zero. Denoting this curve as Γ and the interior of the curve as region A, Green's theorem (Theorem 6.2.1) can be used to obtain

$$\int_\Gamma f_i(\mathbf{x})\, dx_i = \iint_A \left[-\frac{\partial f_1(\mathbf{x})}{\partial x_2} + \frac{\partial f_2(\mathbf{x})}{\partial x_1} \right] dx_1\, dx_2 = 0 \tag{6.2.23}$$

where the integral over A is zero, by virtue of Eq. 6.2.17. Thus, the integral identity of Eq. 6.2.22 holds. This completes the proof of the theorem for n = 2. ■

Example 6.2.4

To see if the differential forms of Example 6.2.3 are exact, note that

$$\frac{\partial f_1(\mathbf{x})}{\partial x_2} = 2x_3 \neq x_3 = \frac{\partial f_2(\mathbf{x})}{\partial x_1}$$

so the first differential form is not exact. Thus, there is no function $F(\mathbf{x})$ for which $d\,F(\mathbf{x}) = f_i(\mathbf{x})\, dx_i$.

For the second differential form,

$$\frac{\partial g_1(\mathbf{x})}{\partial x_2} = 2x_1x_3 = \frac{\partial g_2(\mathbf{x})}{\partial x_1}$$

$$\frac{\partial g_1(\mathbf{x})}{\partial x_3} = 2x_1x_2 = \frac{\partial g_3(\mathbf{x})}{\partial x_1}$$

$$\frac{\partial g_2(\mathbf{x})}{\partial x_3} = x_1^2 = \frac{\partial g_3(\mathbf{x})}{\partial x_2}$$

so the second differential form is exact. In fact, a simple calculation shows that for $G(\mathbf{x}) = x_1{}^2x_2x_3$, $d\,G(\mathbf{x}) = g_i(\mathbf{x})\, dx_i$. The factor $u(\mathbf{x}) = x_1$ that multiplies the first differential form, which is not exact, to obtain the second differential form, which is exact, is called an **integrating factor**. ■

As an application of the theory of exact differential forms in mechanics, consider the work done on an n-dimensional mechanical system, with generalized coordinates **x** and generalized force **f**(**x**) in R^n. The work done on the system in undergoing a differential displacement d**x** is

$$dW(\mathbf{x}) = f_i(\mathbf{x})\, dx_i \tag{6.2.24}$$

The total work done in displacing the system from point $\mathbf{x}^1$ to point $\mathbf{x}^2$, over a curve C, may be written as the integral

$$W(\mathbf{x}) = \int_C f_i(\mathbf{x})\, dx_i \tag{6.2.25}$$

According to Theorem 6.2.1, this work is independent of the path C from $\mathbf{x}^1$ to $\mathbf{x}^2$, if and only if the differential form $f_i(\mathbf{x})\, dx_i$ is exact; i.e., if and only if Eq. 6.2.17 holds. Mechanical systems for which work done by a force field is independent of path traversed are called **conservative systems.**

EXERCISES 6.2

1. Let V be a volume element in a flow field, with boundaries as in Exercise 8 of Section 6.1. Compute the time derivative of the momentum in the x_i direction; i.e.,

$$\frac{d}{dt} \iiint_{V(t)} \rho(\mathbf{x}, t)\, v_i(\mathbf{x}, t)\, dx_1\, dx_2\, dx_3$$

2. Determine whether the following differential forms are exact:

(a) $f_i(\mathbf{x})\, dx_i = 2x_1\, dx_1 + x_4^3\, dx_2 + 2\, dx_3 + 3x_2 x_4^2\, dx_4$

(b) $g_i(\mathbf{x})\, dx_i = (2 + 6x_1 x_2 + 2x_3^2)\, dx_1 + 2x_1^2\, dx_2 + 2x_1 x_3\, dx_3$

3. If the factor $u(\mathbf{x}) = x_1$ is multiplied through the differential form of Exercise 2 (b) above, show that $u(\mathbf{x})\, g_i(\mathbf{x})\, dx_i$ is exact.

4. Find functions $F(\mathbf{x})$ and $G(\mathbf{x})$ such that, for the differential forms $f_i(\mathbf{x})\, dx_i$ and $u(\mathbf{x})\, g_i(\mathbf{x})\, dx_i$ of Exercises 2 and 3,

$$dF(\mathbf{x}) = \frac{\partial F(\mathbf{x})}{\partial x_i}\, dx_i = f_i(\mathbf{x})\, dx_i$$

$$dG(\mathbf{x}) = \frac{\partial G(\mathbf{x})}{\partial x_i}\, dx_i = u(\mathbf{x})\, g_i(\mathbf{x})\, dx_i$$

5. Evaluate the integral

$$\int_S \left(\frac{x}{x^2 - y^2}\, dx + \frac{y}{y^2 - x^2}\, dy \right)$$

where S is a curve from (1, 0) to (5, 4) and lies between the lines $y = x$ and $y = -x$.

6.3 VIBRATING STRINGS AND MEMBRANES

The simplest examples of continuum mechanics problems that contain the basic elements found in more complex applications are strings and membranes [11, 17, 18]. These bodies represent the effect of distributed mass over one and two space dimensions, while avoiding the technical complexity of distributed systems that require higher-order differential operators to represent force-displacement relations.

The Transversely Vibrating String

A string is fixed at both ends, as shown in Fig. 6.3.1(a). The abscissa x in this graph is a physical coordinate that is measured along the undeformed string and the ordinate u represents displacement of the string. For a given set of initial conditions (position and velocity) and applied forces, the string will move, or **vibrate**, in the plane of the applied loads. Denoting the time coordinate by t, the displacement function of the string will be a function of both the space and time variables; i.e., u = u(x, t).

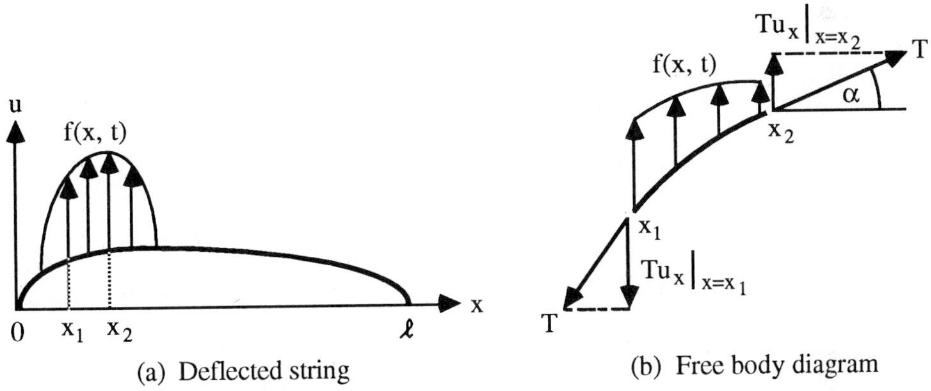

Figure 6.3.1 Vibrating String

The string is initially stretched taut with tension T. It is presumed that deflection of the string occurs only in the u direction; i.e., normal to the x-axis. It is further assumed that only small amplitude oscillations occur, so that arc length, which is measured along the string, is approximately equal to distance measured along the x-axis. In this case, there is no significant elongation of the string. Thus, the tension in the string is constant and is directed tangent to the string at every point, as shown in Fig. 6.3.1(b). The projections of this tensile force in the x and u directions are

$$T_x(x) = T\cos\alpha \approx T \tag{6.3.1}$$

$$T_u(x) = T\sin\alpha \approx T\tan\alpha = Tu_x \tag{6.3.2}$$

where small angle approximations for cos α and sin α are employed; i.e., $\cos\alpha \approx 1$, $\sin\alpha \approx \tan\alpha = \partial u/\partial x = u_x$.

Let x be a typical point along the string and select points x_1 and x_2 on the string so that $x_1 < x < x_2$. The **momentum** of the segment $[x_1, x_2]$ of the string, in the direction of the u-axis, is

$$M(t) = \int_{x_1}^{x_2} \rho(\xi)\, u_t(\xi, t)\, d\xi \tag{6.3.3}$$

where ρ is the mass density per unit length of the string and u_t is velocity.

For any time t, select t_1 and t_2 so that $t_1 < t < t_2$. The change in momentum of the segment $[x_1, x_2]$ of string in the time interval $[t_1, t_2]$ is

$$\begin{aligned} \Delta M &= \int_{x_1}^{x_2} \rho(\xi)\,[\, u_t(\xi, t_2) - u_t(\xi, t_1)\,]\, d\xi \\ &= \int_{t_1}^{t_2}\int_{x_1}^{x_2} \rho(\xi)\left[\frac{u_t(\xi, t_2) - u_t(\xi, t_1)}{t_2 - t_1}\right] d\xi\, dt \end{aligned} \tag{6.3.4}$$

Similarly, the **impulse** of the force applied to this segment of the string, over the same time interval, is the time integral of the applied force, which is

$$\begin{aligned} I &= \int_{t_1}^{t_2} T\,[\, u_x(x_2, \tau) - u_x(x_1, \tau)\,]\, d\tau + \int_{x_1}^{x_2}\int_{t_1}^{t_2} f(\xi, \tau)\, d\tau\, d\xi \\ &= \int_{x_1}^{x_2}\int_{t_1}^{t_2} \left\{ T\left[\frac{u_x(x_2, \tau) - u_x(x_1, \tau)}{x_2 - x_1}\right] + f(\xi, \tau)\right\} d\tau\, d\xi \end{aligned} \tag{6.3.5}$$

where $f(\xi, \tau)$ is the external force applied at point ξ and time τ.

Equating the impulse to the change in momentum of this segment of the string, the **integral equation of motion** is obtained as

$$\begin{aligned} &\int_{x_1}^{x_2}\int_{t_1}^{t_2} \rho(\xi)\left[\frac{u_t(\xi, t_2) - u_t(\xi, t_1)}{t_2 - t_1}\right] d\tau\, d\xi \\ &\quad = \int_{x_1}^{x_2}\int_{t_1}^{t_2} T\left[\frac{u_x(x_2, \tau) - u_x(x_1, \tau)}{x_2 - x_1}\right] d\tau\, d\xi + \int_{x_1}^{x_2}\int_{t_1}^{t_2} f(\xi, \tau)\, d\tau\, d\xi \end{aligned} \tag{6.3.6}$$

If the displacement u(x, t) is presumed to have two continuous time and space derivatives with respect to x and t, then applying the mean value theorem yields

$$\rho(\bar{x})\left[\frac{u_t(\bar{x}, t_2) - u_t(\bar{x}, t_1)}{t_2 - t_1}\right] = T\left[\frac{u_x(x_2, \bar{t}) - u_x(x_1, \bar{t})}{x_2 - x_1}\right] + f(\bar{x}, \bar{t})$$

where $x_1 \le \bar{x} \le x_2$ and $t_1 \le \bar{t} \le t_2$. Taking the limit as t_1 approaches t_2 and x_1 approaches x_2, $\bar{x}$ approaches x_1, $\bar{t}$ approaches t_1, and letting $t_1 = t$ and $x_1 = x$, the **differential equation of motion** is obtained as

$$\rho(x)\, u_{tt} = T u_{xx} + f(x, t) \tag{6.3.7}$$

which must hold at each point along the string, for all time. In the case of constant mass density, this equation can be written in the form

$$u_{tt} - a^2 u_{xx} = F(x, t) \tag{6.3.8}$$

where $a^2 = T/\rho$ and $F(x, t) = (1/\rho)\, f(x, t)$. This is the second-order **partial differential equation** for a **vibrating string**. It is called the **wave equation**. If $F(x, t) = 0$, it is called the **homogeneous wave equation**.

As in the case of dynamics of particles, **initial conditions** on displacement and velocity must be specified in order to have a well-posed dynamics problem. In the case of a string, the initial displacement and velocity of each particle in the string must be specified; i.e.,

$$\begin{aligned} u(x, 0) &= g(x) \\ u_t(x, 0) &= h(x) \end{aligned} \tag{6.3.9}$$

for all x in the open interval $(0, \ell)$. It is clear from Fig. 6.3.1(a) that the lateral displacement of the string at each end must be 0. This leads to the **boundary conditions**

$$u(0, t) = u(\ell, t) = 0 \tag{6.3.10}$$

for all $t \ge 0$.

The differential equation of Eq. 6.3.8, initial conditions of Eq. 6.3.9, and boundary conditions of Eq. 6.3.10 form what is commonly known as an **initial-boundary-value problem** of partial differential equations. This problem for the vibrating string is a prototype that is studied in the literature on **mathematical physics**. This prototype equation exhibits many of the characteristics of higher-order partial differential equations and systems of partial differential equations that arise in the study of continuum mechanics. Most of the solution techniques studied in this text for this problem may be used to solve more complex problems of mechanics.

Example 6.3.1

An important assumption was made in the foregoing development in order to obtain the second-order differential equation of Eq. 6.3.8. It was assumed that continuous second derivatives of the deflection function exist. Such assumptions are common in derivation of the equations of continuum mechanics and must be recognized as limitations on the ability of the differential equation to describe all situations that may be of interest in applications. For example, consider a string that is loaded at a single point, as shown in Fig. 6.3.2. Such a situation is of real concern, not only in this study of vibrating strings, but in related problems of solid and fluid mechanics. From Fig. 6.3.2, it is clear that no second derivative of u(x, t) with respect to x exists at the point x_0 where the load $f_0(t)$ is applied.

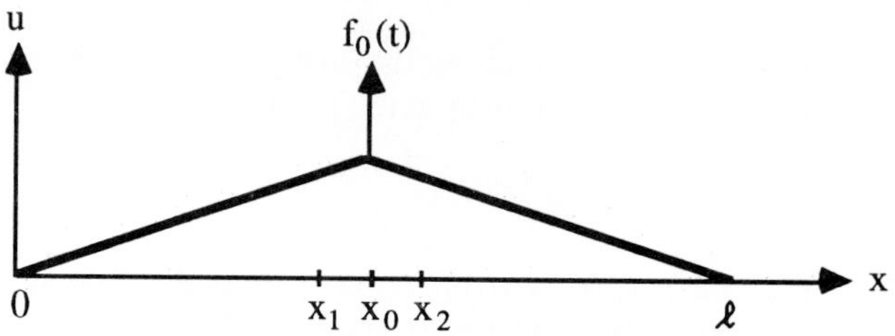

Figure 6.3.2 String With Point Load

Consider two points x_1 and x_2 on either side of the point of application of the load. As in the preceding development, equating the change in momentum to the impulse applied to the segment of the string between x_1 and x_2, over the period t_1 to t_2,

$$\int_{x_1}^{x_2} \rho(\xi) [u_t(\xi, t_2) - u_t(\xi, t_1)] d\xi$$

$$= \int_{t_1}^{t_2} T [u_x(x_2, \tau) - u_x(x_1, \tau)] d\tau + \int_{t_1}^{t_2} f_0(\tau) d\tau \quad (6.3.11)$$

Since the integrand on the left is a bounded and continuous function, as the limit is taken with x_1 approaching x_0 from the left and x_2 approaching x_0 from the right, the integral on the left vanishes and Eq. 6.3.11 becomes

$$0 = \int_{t_1}^{t_2} T [u_x(x_0+, \tau) - u_x(x_0-, \tau)] d\tau + \int_{t_1}^{t_2} f_0(\tau) d\tau \quad (6.3.12)$$

where x_0+ and x_0- denote limits from the right and left, respectively. Applying the mean value theorem and letting t_2 approach t_1,

$$u_x(x_0+, t) - u_x(x_0-, t) = -\frac{1}{T} f_0(t) \quad (6.3.13)$$

The slope of the string at the point x_0 of applied load has a discontinuity that is given by Eq. 6.3.13. It is clear that no second derivative of u(x, t) with respect to the space variable exists at this point, so the second-order differential equation of Eq. 6.3.8

cannot apply at point x_0. This illustrates that assumptions made for mathematical convenience in deriving equations will often preclude the applicability of the resulting equations to situations that may be of practical interest. This example is provided simply to illustrate that assumptions concerning the regularity properties of functions involved in the problem being studied must be very carefully stated and interpreted. ■

In summary, consider a smoothly loaded string, so that candidate deflection functions have two continuous derivatives. Also, consider the case in which the string is initially undeformed and at rest; i.e., $g(x) = h(x) = 0$ in Eq. 6.3.9. The set of admissible deflections can now be described as

$$D = \{ u(x, t) \in C^2 : \ u(x, 0) = u_t(x, 0) = 0 \ \text{for all } x \text{ in } (0, \ell) \text{ and}$$
$$u(0, t) = u(\ell, t) = 0 \ \text{for all } t \geq 0 \} \quad (6.3.14)$$

Note that functions in D satisfy the differentiability characteristics that are required of the solution, as well as the initial and boundary conditions of Eqs. 6.3.9 and 6.3.10.

A spatial **linear operator** A may now be associated with the differential equation of Eq. 6.3.8 and the collection of admissible displacements of Eq. 6.3.14,

$$Au \equiv -a^2 u_{xx} \quad (6.3.15)$$

This is called a linear operator, since it satisfies the conditions

$$A(u+v) = -a^2 (u_{xx} + v_{xx}) = Au + Av$$

$$A(\alpha u) = -a^2 \alpha u_{xx} = \alpha Au$$

for all $u(x, t)$ and $v(x, t)$ that have two continuous derivatives with respect to x and for all constants α. The initial-boundary-value problem for motion of a string may now be stated as follows: Find a deflection function $u(x, t)$ in D of Eq. 6.3.14 that satisfies the **operator equation**

$$u_{tt} + Au = F(x, t) \quad (6.3.16)$$

The Transversely Vibrating Membrane

A thin elastic film, called a **membrane**, is stretched over a closed plane curve Γ, as shown in Fig. 6.3.3. It offers no in-plane resistance to stretching and distortion. Due to transverse excitation, the membrane may be set into vibratory motion, as if the head of a drum were struck. The objective is to derive the differential equation that governs transverse vibration of the membrane.

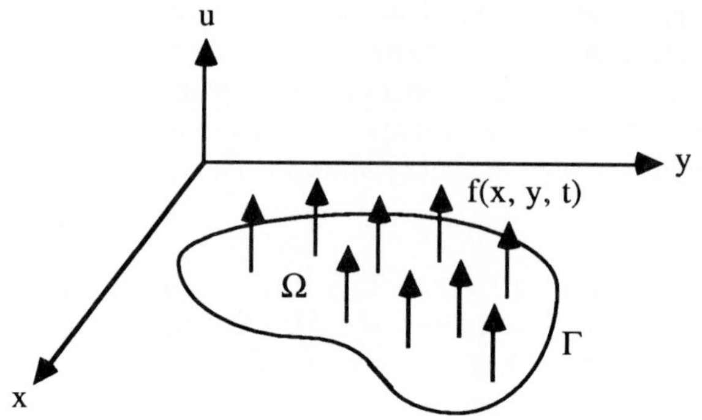

Figure 6.3.3 Transversely Vibrating Membrane

Consider a deformed element of the membrane shown in Fig. 6.3.4 that contains a typical point (x, y) at time t. A tension T per unit length acts in the membrane, on the curve C that bounds the element that is imagined cut from the membrane. The tension T is tangent to the membrane and normal to C, since the membrane can support no bending moment and no shear can be supported by the membrane material. In general, the tension can be a function of both space variables and time; i.e., the tension is T(x, y, t).

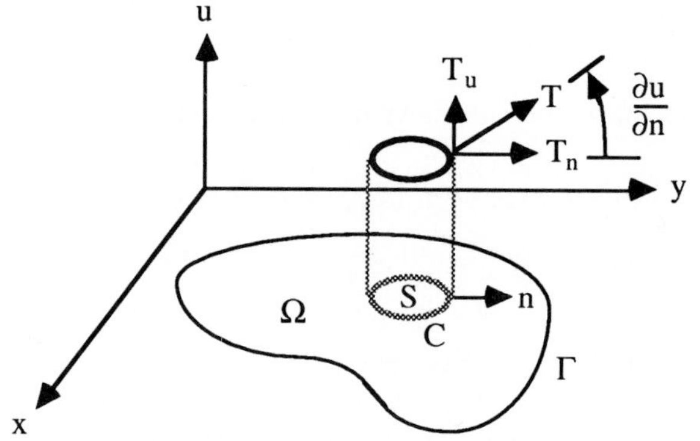

Figure 6.3.4 Element of Vibrating Membrane

In this study of small amplitude vibration of the membrane, the angle made by the tangent plane with the x-y plane is small, so squared terms in first derivatives can be ignored. It can be concluded, therefore, that projections of the surface tension onto the x-y plane and the u-axis are

$$\begin{aligned} T_n &\approx T \\ T_u &\approx T \frac{\partial u}{\partial n} \end{aligned} \tag{6.3.17}$$

As noted in the preceding, the tension T(x, y, t) can vary over the domain that is bounded by the curve Γ. Since only transverse vibration of the membrane occurs, there is no movement of the membrane in the x-y plane. This fact implies certain behavior of the tension, which may be investigated by considering the rectangular element of membrane shown in Fig. 6.3.5.

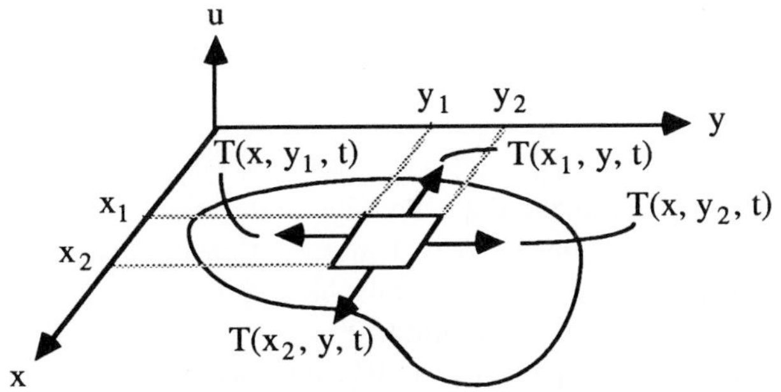

Figure 6.3.5 Rectangular Element of Membrane

Since there is no motion of the membrane in the x-y plane, the net force on the rectangular element in both the x and y directions must be 0; i.e.,

$$\int_{y_1}^{y_2} [\, T(x_2, y, t) - T(x_1, y, t) \,]\, dy = 0$$

$$\int_{x_1}^{x_2} [\, T(x, y_2, t) - T(x, y_1, t) \,]\, dx = 0 \qquad (6.3.18)$$

Employing the mean value theorem and taking limits as y_1 approaches y_2 and x_1 approaches x_2 in Eq. 6.3.18,

$$T(x_2, y, t) = T(x_1, y, t)$$

$$T(x, y_2, t) = T(x, y_1, t) \qquad (6.3.19)$$

This implies that tension is independent of both x and y. Hence, it is a function of only t. Since there is no spatial variation of tension in the membrane and no in-plane motion of the membrane, there is no impetus for temporal variation of the tension. Thus, the tension is constant; i.e.,

$$T(x, y, t) = T \qquad (6.3.20)$$

Equations of motion can now be written for the typical element of membrane shown in Fig. 6.3.4, whose projection onto the x-y plane is a domain S with boundary C. Setting the change in the z-component of momentum of the element from time t_1 to t_2 equal to the impulse of the z-components of force that act on the element during that period, the **integral equation of motion** is obtained as

$$\iint_S [\, u_t(x, y, t_2) - u_t(x, y, t_1) \,]\, \rho(x, y)\, dx\, dy$$

$$= \int_{t_1}^{t_2} \int_C T \frac{\partial u}{\partial n}\, dC\, dt + \int_{t_1}^{t_2} \iint_S f(x, y, t)\, dx\, dy\, dt \qquad (6.3.21)$$

where $\rho(x, y)$ is the mass per unit area of the membrane and $f(x, y, t)$ is the lateral force per unit area that is applied in the u direction to the membrane.

In order to transform Eq. 6.3.21 into a differential equation, it is assumed that two derivatives of u exist and are continuous. Note that this eliminates the possibility of point or line loads on the membrane. Putting $\mathbf{F} = \nabla u$ in the divergence theorem, the boundary integral on the right of Eq. 6.3.21 can be transformed to an integral over the domain S. A manipulation of the integral on the left side of Eq. 6.3.21 yields

$$\int_{t_1}^{t_2} \iint_S \left[\frac{u_t(x, y, t_2) - u_t(x, y, t_1)}{t_2 - t_1} \right] \rho(x, y)\, dx\, dy\, dt$$

$$= \int_{t_1}^{t_2} \iint_S T\,(\, u_{xx} + u_{yy} \,)\, dx\, dy\, dt + \int_{t_1}^{t_2} \iint_S f\, dx\, dy\, dt \qquad (6.3.22)$$

Employing the mean value theorem yields

$$\left[\frac{u_t(\bar{x}, \bar{y}, t_2) - u_t(\bar{x}, \bar{y}, t_1)}{t_2 - t_1} \right] \rho(\bar{x}, \bar{y})$$

$$= T\,[\, u_{xx}(\bar{x}, \bar{y}, \bar{t}) + u_{yy}(\bar{x}, \bar{y}, \bar{t}) \,] + f(\bar{x}, \bar{y}, \bar{t}) \qquad (6.3.23)$$

where $(\bar{x}, \bar{y})$ is in S and $t_1 < \bar{t} < t_2$. Taking the limit as the domain S shrinks to the point (x, y) and the time interval $[t_1, t_2]$ shrinks to t, the **differential equation of motion** is obtained as

$$\rho(x, y)\, u_{tt} - T\,(\, u_{xx} + u_{yy} \,) = f(x, y, t) \qquad (6.3.24)$$

For a membrane of constant mass distribution over the domain Ω, $\rho(x, y)$ is constant and this differential equation reduces to

$$u_{tt} - a^2\,(\, u_{xx} + u_{yy} \,) = F(x, y, t) \qquad (6.3.25)$$

where $a^2 = T/\rho$ and $F(x, y, t) = f(x, y, t)/\rho$. This is called the **wave equation** in two space dimensions for the **vibrating membrane**.

As in particle equations of motion, it is essential to specify the initial value of the displacement of the membrane and its initial velocity, throughout the domain Ω. This results in the **initial conditions**

$$\begin{aligned} u(x, y, 0) &= g(x, y) \\ u_t(x, y, 0) &= h(x, y) \end{aligned} \tag{6.3.26}$$

in Ω. Further, it is required that the displacement of the membrane at the boundary Γ is 0. This results in the **boundary condition**

$$u(x, y, t) = 0 \tag{6.3.27}$$

on Γ, for all $t \geq 0$.

It may be noted that the differential equation of Eq. 6.3.24, the initial conditions of Eq. 6.3.26, and the boundary condition of Eq. 6.3.27 form an **initial-boundary-value problem** similar to that obtained for the vibrating string. The only difference between these problems is the fact that the membrane equation has one additional space variable. As will be noted in later studies of these differential equations, there is considerable similarity in behavior of their solutions. Techniques that can be used to solve the vibrating string problem can also be used to solve the vibrating membrane problem.

Considering the case of motion of a membrane that is initially undeformed and at rest; i.e., $g(x, y) = h(x, y) = 0$ in Eq. 6.3.26, the collection of admissible membrane deflections for a continuous load can be defined as

$$D = \{ u(x, y, t) \in C^2(\Omega): \; u(x, y, 0) = u_t(x, y, 0) = 0 \text{ in } \Omega \text{ and } u(x, y, t) = 0 \text{ on } \Gamma, \text{ for all } t \} \tag{6.3.28}$$

A **linear operator** may be associated with Eq. 6.3.25 and the collection of admissible displacement functions D of Eq. 6.3.28,

$$Au \equiv -a^2\nabla^2 u \tag{6.3.29}$$

The initial-boundary-value problem for motion of a membrane may now be stated as follows: Find the deflection function in D that satisfies the operator equation

$$u_{tt} + Au = F(x, y, t) \tag{6.3.30}$$

Steady State String and Membrane Problems

Steady state vibration, or **harmonic oscillation**, and **static displacement** of strings and membranes represent important classes of applications that arise in mechanics. Consider first harmonic oscillation of the string, in which the time dependence of the displacement function is harmonic; i.e.,

$$u(x, t) = v(x) \sin \omega t \tag{6.3.31}$$

where ω is the frequency of harmonic oscillation, or **natural frequency** of vibration. Substitution of u(x, t) from Eq. 6.3.31 into the homogeneous equation of Eq. 6.3.7; i.e., with f(x, t) = 0, yields

$$(T v'' + \rho(x)\, \omega^2 v) \sin \omega t = 0 \tag{6.3.32}$$

The boundary condition of Eq. 6.3.10, for all t, is

$$v(0) \sin \omega t = v(\ell) \sin \omega t = 0 \tag{6.3.33}$$

Since Eqs. 6.3.32 and 6.3.33 must hold for all t, the coefficients of sin ωt must be equal to zero. Defining $\omega^2/T = \lambda$, an **eigenvalue problem** is obtained as

$$\begin{aligned} - v'' &= \lambda\, \rho(x)\, v \\ v(0) &= v(\ell) = 0 \end{aligned} \tag{6.3.34}$$

In case ρ is constant and a is as defined in Eq. 6.3.8, this problem can be solved by noting that the solution of Eq. 6.3.34 is $v(x) = \sin(\omega x/a)$. In order that $v(\ell) = 0$, it is necessary that $\omega\ell/a = n\pi$, where n is an integer. Thus, $\omega_n = n\pi a/\ell$. The **eigenvalues** and their associated **eigenfunctions** of Eq. 6.3.34 are thus

$$\begin{aligned} \frac{\omega_n}{a} &= \frac{n\pi}{\ell} \\ v_n(x) &= \sin \frac{n\pi}{\ell} x \end{aligned} \tag{6.3.35}$$

Similar to the case of a harmonically vibrating string, consider harmonic vibration of a membrane; e.g., vibration of a drum head. As in the case of the string, the displacement is of the form

$$u(x, y, t) = v(x, y) \sin \omega t \tag{6.3.36}$$

In order that the boundary condition in Eq. 6.3.27 holds for all t, v(x, y) must satisfy

$$v(x, y) = 0 \tag{6.3.37}$$

on Γ. The homogeneous form of the operator equation of Eq. 6.3.30 now becomes $- \omega^2 v(x, y) \sin \omega t - a^2 \nabla^2 v(x, y) \sin \omega t = 0$, for all t. Defining $\lambda = \omega^2/a^2$, the eigenfunction v(x, y) must satisfy the operator **eigenvalue equation**

$$Av \equiv - \nabla^2 v = \lambda v \tag{6.3.38}$$

where v(x, y) belongs to the set

$$D_A = \{ v(x, y) \in C^2(\Omega) : v(x, y) = 0 \text{ on } \Gamma \} \tag{6.3.39}$$

Note the similarity between the eigenvalue equation of Eq. 6.3.38 for a function v(x, y) and the matrix eigenvalue problem of Eq. 3.1.1. While there are significant technical differences between matrix and differential operator eigenvalue problems, there are also striking similarities.

Having the equations of motion of a membrane, the differential equations that govern **equilibrium of a membrane**, under a given load f(x, y), may be obtained by noting that the lateral deflection of the membrane is not a function of time; i.e., u = u(x, y). Further, the initial conditions of Eq. 6.3.26 are not meaningful and the boundary condition of Eq. 6.3.27 is independent of time. Therefore, the differential equation

$$u_{xx} + u_{yy} = (1/T)\, f(x, y) \tag{6.3.40}$$

in Ω and the boundary condition

$$u(x, y) = 0 \tag{6.3.41}$$

on Γ govern equilibrium of the membrane. These equations form a **boundary-value problem** of partial differential equations.

The differential equation of Eq. 6.3.40 and the boundary condition of Eq. 6.3.41 can be characterized by the **operator equation**

$$\begin{aligned} Au &\equiv -\nabla^2 u = -(1/T)\, f(x, y) \\ D_A &= \{ u \in C^2(\Omega):\ u = 0 \text{ on } \Gamma \} \end{aligned} \tag{6.3.42}$$

for a continuous load f(x, y).

EXERCISES 6.3

1. Show that the set of admissible displacements D of Eq. 6.3.14 is a vector space.

2. Define a collection of admissible deflections $\bar{D}$ for the case $g(x) \neq 0 \neq h(x)$ in Eq. 6.3.9. Is this a vector space?

3. Given a function $\bar{u}(x, t)$ that satisfies Eq. 6.3.9, with $g(x) \neq 0 \neq h(x)$, show that the change in variables $\tilde{u}(x, t) = u(x, t) - \bar{u}(x, t)$ can be employed to reduce the initial conditions to homogeneous form. Obtain the modified equation of Eq. 6.3.8 for $\tilde{u}(x, t)$.

4. Derive the boundary condition at the right end of the string shown, where the mass of the pin that slides freely in the vertical slot is negligible.

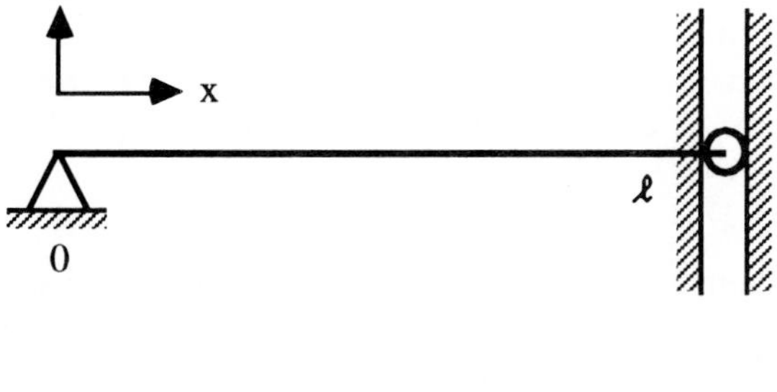

5. Derive the equation of motion of a vibrating string in a viscous medium; i.e., with a resisting force

$$\mu \frac{\partial u(x, t)}{\partial t}$$

in the negative u direction, where μ is a **coefficient of viscosity** per unit length of the string.

6. Show that D of Eq. 6.3.28 is a vector space.

7. Define the set of admissible membrane deflections $\bar{D}$, in case $g(x, y) \neq 0 \neq h(x, y)$, in Eq. 6.3.26.

8. Given a function $\bar{u}(x, y, t)$ that satisfies Eq. 6.3.26 with $g(x, y) \neq 0 \neq h(x, y)$, show that the transformation $u(x, y, t) = \bar{u}(x, y, t) + \tilde{u}(x, y, t)$ can be employed to put the problem in the form of Eq. 6.3.25, with a modified right side and with the domain given in Eq. 6.3.28.

9. For a circular membrane, rewrite Eq. 6.3.25 in terms of polar coordinates.

10. Show that D_A of Eq. 6.3.39 is a vector space.

11. Derive a boundary-value problem for equilibrium of a string under the action of a continuous load f(x) and boundary conditions of Eq. 6.3.10.

6.4 HEAT AND DIFFUSION PROBLEMS

In this section, problems of the flow of heat in solids and the flow of gas in porous materials are studied. While these problems have fundamentally different physical properties, it is interesting to note that their mathematical properties are virtually identical.

Conduction of Heat in Solids

Consider a domain Ω in R^3 that is bounded by a closed surface Γ. Let u(x, y, z, t) be the **temperature** at a point (x, y, z) at time t. The following important facts govern temperature distribution and heat flow:

(1) If the temperature is not constant, heat flows from regions of higher temperature to regions of lower temperature.

(2) **Fourier's law** states that the rate of **heat flow** is proportional to the gradient of the temperature. Thus, the **rate of heat flow** $\mathbf{v}(x, y, z, t) = [\, v_1(x, y, z, t), v_2(x, y, z, t), v_3(x, y, z, t)\,]^T$ in an **isotropic body** is

$$\mathbf{v} = -K\nabla u \tag{6.4.1}$$

where $K > 0$ is a constant, called the **thermal conductivity** of the body.

Let V be an arbitrary region that is bounded by a closed surface S in Ω. Then the rate of heat loss from V is

$$\iint_S \mathbf{v}^T\mathbf{n}\, dS = -K\iint_S \nabla u^T\mathbf{n}\, dS = -K\iint_S \frac{\partial u}{\partial n}\, dS$$

where **n** is the outward unit normal of S. By the divergence theorem, if u(x, y, z, t) is twice continuously differentiable with respect to x, y, and z,

$$\iint_S \mathbf{v}^T\mathbf{n}\, dS = -K\iiint_V \nabla^2 u\, dx\, dy\, dz \tag{6.4.2}$$

The amount of heat in V at a given time is

$$\iiint_V \sigma\rho u\, dx\, dy\, dz \tag{6.4.3}$$

where ρ is the **density** of the material of the body and σ is its **specific heat**, both assumed to be constant. Assuming that the temperature u(x, y, z, t) is differentiable with respect to t and using Eq. 6.4.1, the rate of decrease in heat content of V is

$$-\iiint_V \sigma\rho \frac{\partial u}{\partial t}\, dx\, dy\, dz \tag{6.4.4}$$

Since the rate of decrease of heat contained in V must be equal to the amount of heat leaving V per unit of time, from Eqs. 6.4.2 and 6.4.4,

$$-\iiint_V \sigma\rho u_t\, dx\, dy\, dz = -K\iiint_V \nabla^2 u\, dx\, dy\, dz$$

or

$$\iiint_V \left[\sigma\rho \frac{\partial u}{\partial t} - K\nabla^2 u\right] dx\, dy\, dz = 0 \tag{6.4.5}$$

for an arbitrary region V in Ω. Using the mean value theorem,

$$\sigma\rho u_t(\bar{x}, \bar{y}, \bar{z}, t) - K\nabla^2 u(\bar{x}, \bar{y}, \bar{z}, t) = 0$$

where $(\bar{x}, \bar{y}, \bar{z})$ is in V. Letting the volume element V shrink to a typical point (x, y, z), the following differential equation is obtained:

$$u_t - k^2\nabla^2 u = 0 \tag{6.4.6}$$

where $k^2 = K/\sigma\rho$. This is known as the **heat equation**.

In addition to the differential equation of Eq. 6.4.6, the temperature distribution u(x, y, z, t) is determined by the **initial temperature distribution** within the body and by the temperature maintained at the boundary Γ of Ω. These conditions result in the **initial conditions**

$$u(x, y, z, 0) = p(x, y, z) \tag{6.4.7}$$

in Ω and the **boundary conditions**

$$u(x, y, z, t) = q(x, y, z, t) \tag{6.4.8}$$

on Γ, and for all $t \geq 0$.

It may be noted that the differential equation of Eq. 6.4.6, the initial condition of Eq. 6.4.7, and the boundary condition of Eq. 6.4.8 form an **initial-boundary-value problem** that is similar to the problem defined for the vibrating string and vibrating membrane in Section 6.3. The essential difference is that only one time derivative occurs in the differential equation and only the value of the dependent variable u(x, y, z, t) is specified as an initial condition. The fact that only one time derivative of the dependent variable occurs in the differential equation leads to considerably different mathematical properties of the initial-boundary-value problem associated with heat conduction, as compared with the equations of the vibrating string and vibrating membrane. The reasons for these differences in behavior will become more apparent in Chapters 7 and 8.

It is worthy of note that for the study of heat transfer in a bar with one space dimension, or in a plate with two space dimensions, the differential equation of Eq. 6.4.6 maintains precisely the same form, with the spatial differential operator on the right side containing only those space derivatives that are appropriate.

It may be further noted that a **steady state temperature distribution** in a two-dimensional body is independent of time, so the time derivative on the left side of Eq 6.4.6 vanishes and the initial condition of Eq. 6.4.7 is meaningless. The differential equation for steady state temperature u(x, y) is thus

$$\nabla^2 u = 0 \tag{6.4.9}$$

in Ω, and the boundary conditions are

$$u(x, y) = q(x, y) \tag{6.4.10}$$

on Γ.

Note that the **boundary-value problem** formed by Eqs 6.4.9 and 6.4.10 is of the same form as the boundary-value problem for static deflection of a membrane in Section 6.3. The difference is that the membrane differential equation is nonhomogeneous and it has homogeneous boundary conditions. This suggests that there is a relation between the steady state temperature distribution in a plane body and the static deflection of a membrane that is stretched over a boundary of the same shape.

To make the analogy between thermal and membrane deflection problems sharper, presume that a $C^2(\Omega)$ function $\bar{u}(x, y)$ that satisfies Eq. 6.4.10 is known. Defining $\tilde{u}(x, y) = u(x, y) - \bar{u}(x, y)$, a set of admissible temperatures is obtained as

$$D = \{ \tilde{u}(x, y) \text{ in } C^2(\Omega): \ \tilde{u}(x, y) = 0 \ \text{ on } \Gamma \} \tag{6.4.11}$$

Then the operator equation of Eq. 6.4.9 takes on the nonhomogeneous form

$$\nabla^2 \tilde{u} = -\nabla^2 \bar{u} \tag{6.4.12}$$

in Ω, where the right side of Eq. 6.4.12 is known.

Diffusion of Gas in Porous Materials

If a **porous material** is nonuniformly filled with a **gas**, then **diffusion of gas** occurs from a region of higher concentration to a region of lower concentration. For illustrative purposes, diffusion of a gas along a one-dimensional porous material is depicted in Fig. 6.4.1. At any instant of time, the mass per unit volume of gas in a plane normal to the axis of the porous solid is u(x, t).

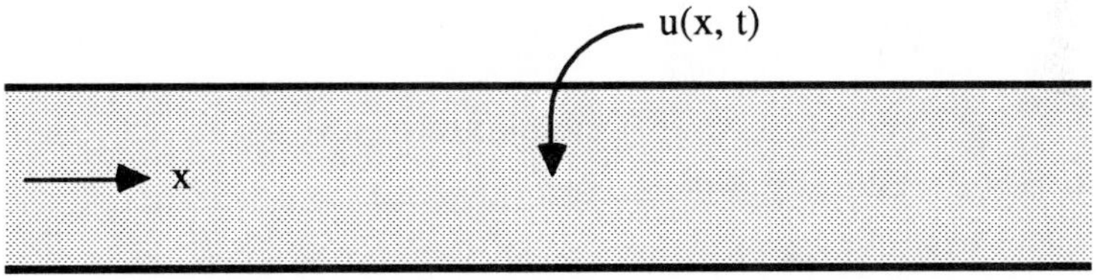

Figure 6.4.1 Diffusion of Gas in a Porous Medium

For a general porous body Ω in R^3, with boundary Γ, the physical relation that governs motion of gas is **Fick's law**,

$$\mathbf{q} = -D\nabla u \tag{6.4.13}$$

where u(x, y, z, t) is the **mass density of gas**, $\mathbf{q}$(x, y, z, t) is the **rate of mass flow** of gas across a unit area, and the scalar $D > 0$ is the **diffusion coefficient** associated with the porous material. The rate of **gas flow** from a typical volume V in Ω, with boundary S is

$$\iint_S \mathbf{q}^T\mathbf{n}\, dS = -D \iint_S \nabla u^T \mathbf{n}\, dS = -D \iint_S \frac{\partial u}{\partial n}\, dS$$

where $\mathbf{n}$(x, y, z) is the outward unit normal of S. By the divergence theorem,

$$\iint_S \mathbf{q}^T\mathbf{n}\, dS = -D \iiint_V \nabla^2 u\, dx\, dy\, dz \tag{6.4.14}$$

The mass of gas in volume V at a given time is

$$Q = \iiint_V Cu\, dx\, dy\, dz \tag{6.4.15}$$

where the constant $C > 0$ is the **coefficient of porosity**, which is the volume of pores (voids) per unit of volume occupied by the porous material. Thus, the rate of decrease of the mass of gas in V is

$$-\frac{dQ}{dt} = -C\iiint_V \frac{\partial u}{\partial t}\, dx\, dy\, dz \tag{6.4.16}$$

Equating the rate of gas flow out of V, from Eq. 6.4.14, to the rate of decrease in gas content in V, from Eq. 6.4.16,

$$\iiint_V [D\nabla^2 u - Cu_t]\, dx\, dy\, dz = 0 \tag{6.4.17}$$

Applying the mean value theorem,

$$D\nabla^2 u(\bar{x}, \bar{y}, \bar{z}, t) - Cu_t(\bar{x}, \bar{y}, \bar{z}, t) = 0$$

where $(\bar{x}, \bar{y}, \bar{z})$ is in V. Letting V shrink to a typical point (x, y, z) and defining $k^2 = D/C$, the **diffusion equation** is obtained as

$$u_t = k^2\nabla^2 u \tag{6.4.18}$$

If the concentration of gas at the boundary Γ of Ω and the initial concentration of gas in Ω are known, then the **initial conditions** and **boundary conditions** that must be satisfied by the solution of the diffusion problem are

$$\begin{aligned} u(x, y, z, 0) &= f(x, y, z), \quad \text{in } \Omega \\ u(x, y, z, t) &= g(x, y, z), \quad \text{on } \Gamma, \text{ for all } t \geq 0 \end{aligned} \tag{6.4.19}$$

Comparing these equations with Eqs. 6.4.6, 6.4.7, and 6.4.8, it is noted that the differential equation and the initial and boundary conditions for the diffusion problem are of precisely the same form as those for the heat conduction problem. Therefore, it can be expected that the mathematical characteristics of gas flow in a porous medium will be identical to those of heat propagation in a solid.

EXERCISES 6.4

1. Formulate an operator equation, with a set of admissible solutions, to represent the initial-boundary-value-problem of Eqs. 6.4.6 through 6.4.8. Is this operator linear?

2. Define a change of variable in the operator equation of Exercise 1 that makes the set of admissible temperature distributions a linear space. [Hint: Assume that a function that satisfies the initial and boundary conditions of Eqs. 6.4.7 and 6.4.8 is known.]

3. Write the initial-boundary-value problem for the temperature in a rod of length ℓ that is thermally insulated along its lateral surface, has an initial temperature distribution $u(x, 0) = f(x)$, and for which the ends of the rod are maintained at given temperatures $u(0, t) = g(t)$ and $u(\ell, t) = h(t)$.

4. Repeat Exercise 3, but with the ends of the rod insulated; i.e., $u_x(0, t) = u_x(\ell, t) = 0$.

5. On the lateral surface of a rod, heat is conducted into the surrounding medium at a rate $v(x, t) = c\,[\,u(x, t) - T(x, t)\,]$ per unit length of rod, where c is the **coefficient of convectivity**, $T(x, t)$ is the temperature of the surrounding gas, temperature variation across the cross section of the rod is negligible, and the ends of the rod are insulated. Write the governing initial-boundary-value equations for the system.

6. Derive a heat equation for a medium in which a chemical reaction generates heat at a rate $Q(x, t)$ heat units per unit of mass of material.

7. Derive a heat transfer equation in a cylindrical body, where $u = u(r, z, t)$, $u(r_0, z, t) = u(r, 0, t) = u(r, \ell, t) = 0$, and $u(r, z, 0) = f(r, z)$.

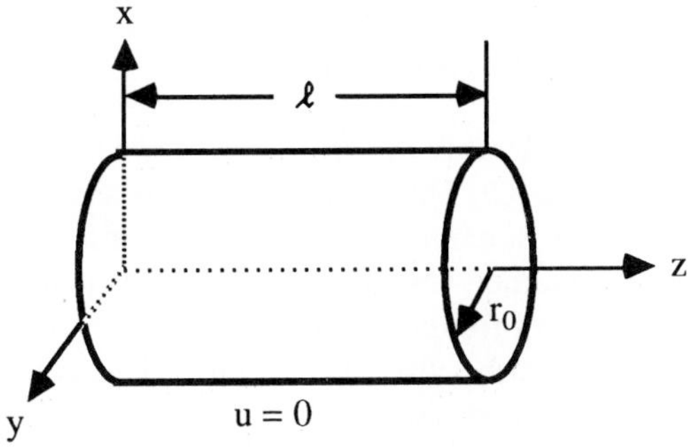

8. Write the equation for diffusion of a gas in a cylindrical volume of porous material, as shown in Fig. 6.4.1, where uniform concentration of gas across each cross section is assumed. The initial concentration is $u(x, 0) = f(x)$ and the ends of the volume are sealed; i.e., $u_x(0, t) = u_x(\ell, t) = 0$.

9. Repeat Exercise 8, but with the ends of the column semipermeable; i.e., the rates of gas flow out of the tube are

$$q(0, t) = K\,[\,u(0, t) - v(t)\,]$$

$$q(\ell, t) = K\,[\,u(\ell, t) - v(t)\,]$$

where $v(t)$ is the gas concentration outside the porous medium.

10. Write the diffusion equation for a column of porous material in which a chemical reaction generates gas at a rate Q(x, t) and gas flows out of the ends according to the law

$$q(0, t) = -Du(0, t)$$

$$q(\ell, t) = Du(\ell, t)$$

11. In case Q(x, t) does not depend on time, write a steady state equation of flow for the problem in Exercise 10.

12. Define a relationship between the thermal and diffusion constants and the initial conditions so that the temperature in Exercise 4 is identical to the gas density in Exercise 8.

6.5 FLUID DYNAMICS

A major topic in the field of continuum mechanics is fluid and gas dynamics. The basic equations that govern flow of fluids and gases are derived in this section, to include special cases that are of interest in applications.

Dynamics of Compressible Fluids

Fluid dynamics is considerably different, in a physical sense, from the dynamics problems discussed in Section 6.3. The fundamental difference is that an ideal fluid cannot resist shearing stress. That is, continued deformation of a fluid can occur such that rates of shearing strain are maintained and large deformations occur in the continuum. Thus, it is natural to study the flow of fluids from the point of view of determining a velocity distribution in the continuum, rather than determining displacements of individual fluid particles, as is done in vibration problems.

The study of fluid dynamics in this section is limited to **plane flow**. That is, only two components of velocity at each point in the **flow field** Ω occur, as shown in Fig. 6.5.1. The x and y **components of velocity** are denoted u(x, y, t) and v(x, y, t), respectively. They vary with location in the flow field and with time.

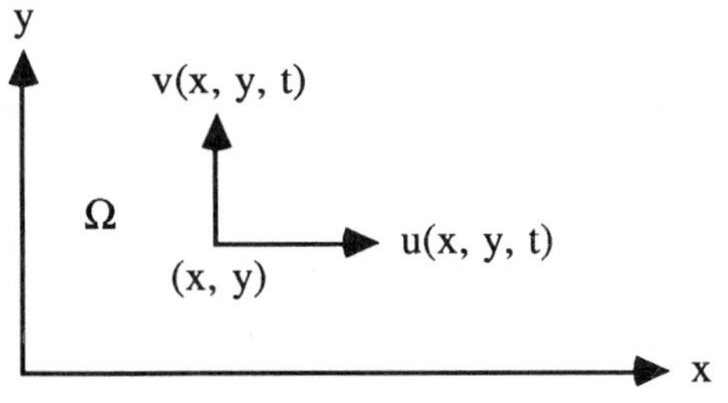

Figure 6.5.1 Two-dimensional Flow Field

In addition to determining the velocity of the flow field, **fluid density** and **pressure** within the flow field must also be determined. Density of the fluid will be denoted by $\rho(x, y, t)$ and pressure in the fluid will be denoted by $p(x, y, t)$. Finally, an **external force** may act on each unit of mass in the flow field. The components of this force will be denoted $F_x(x, y, t)$ and $F_y(x, y, t)$. Such forces may arise from gravitational effects or from interaction of the fluid with some other field of influence, such as magnetic or electrostatic fields that may have some effect on the fluid.

Consider a typical element of fluid in the flow field, as shown in Fig. 6.5.2, that occupies an area $S(t) \subset \Omega$ and has boundary $C(t)$. Consideration is limited here to **ideal fluid**, in which no viscosity exists. That is, only a pure pressure exists in the fluid and no shear is supported along any boundary in the flow field. Assuming that the pressure p is differentiable, the components of the resultant force that act on the element of fluid in S in the x and y directions are

$$
\begin{aligned}
-\int_C p n_x \, d\Gamma + \iint_S \rho F_x \, dx\, dy &= -\iint_S \frac{\partial p}{\partial x} \, dx\, dy + \iint_S \rho F_x \, dx\, dy \\
-\int_C p n_y \, d\Gamma + \iint_S \rho F_y \, dx\, dy &= -\iint_S \frac{\partial p}{\partial y} \, dx\, dy + \iint_S \rho F_y \, dx\, dy
\end{aligned}
\tag{6.5.1}
$$

where the divergence theorem has been used to transform boundary integrals over C to integrals over S.

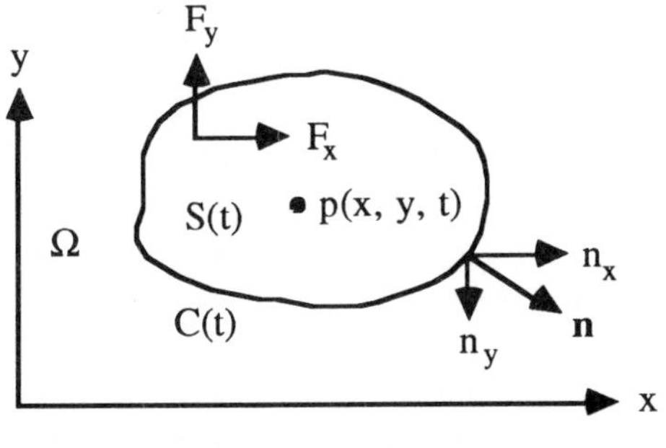

Figure 6.5.2 Element in Flow Field

For the set of mass points in S of Fig. 6.5.2, the components of the resultant applied force in Eq. 6.5.1 must equal the time derivatives of the corresponding components of momentum. The components of **momentum of fluid** in S in the x and y directions are

$$
\begin{aligned}
&\iint_{S(t)} \rho u \, dx\, dy \\
&\iint_{S(t)} \rho v \, dx\, dy
\end{aligned}
\tag{6.5.2}
$$

Using the material derivative of Section 6.2 to compute the time rate of change of momentum and equating the results to the corresponding forces in Eq. 6.5.1,

$$\iint_{S(t)} [(\rho u)_t + (\rho u^2)_x + (\rho uv)_y + p_x - \rho F_x]\, dx\, dy = 0$$

$$\iint_{S(t)} [(\rho v)_t + (\rho uv)_x + (\rho v^2)_y + p_y - \rho F_y]\, dx\, dy = 0$$

for all regions S(t) in the flow field. Applying the mean value theorem and letting S shrink to a point, the **equations of fluid motion** are obtained in the form

$$\begin{aligned} (\rho u)_t + (\rho u^2 + p)_x + (\rho uv)_y &= \rho F_x \\ (\rho v)_t + (\rho uv)_x + (\rho v^2 + p)_y &= \rho F_y \end{aligned} \tag{6.5.3}$$

Recall from Eq. 6.2.14 of Section 6.2, the condition of **conservation of mass** is

$$\rho_t + (\rho u)_x + (\rho v)_y = 0 \tag{6.5.4}$$

This is called the **continuity equation** and states that mass is neither created nor destroyed in the flow field. The **equations of fluid dynamics** of Eqs. 6.5.3 and 6.5.4 can be written in **divergence form**; i.e.,

$$\begin{bmatrix} \rho u \\ \rho v \\ \rho \end{bmatrix}_t + \begin{bmatrix} \rho u^2 + p \\ \rho uv \\ \rho u \end{bmatrix}_x + \begin{bmatrix} \rho uv \\ \rho v^2 + p \\ \rho v \end{bmatrix}_y = \begin{bmatrix} \rho F_x \\ \rho F_y \\ 0 \end{bmatrix} \tag{6.5.5}$$

where u(x, y, t), v(x, y, t), p(x, y, t), and ρ(x, y, t) are unknowns. This form of the nonlinear equations of fluid motion is the direct result of conservation laws used in their derivation.

In order to extract physically meaningful conclusions from the two-dimensional equations of fluid dynamics, it is occasionally helpful to write Eq. 6.5.3 in a slightly different form. Carrying out the differentiation in Eq. 6.5.3, employing the continuity equation of Eq. 6.5.4, and multiplying by 1/ρ(x, y, t),

$$\begin{aligned} u_t + uu_x + vu_y + \frac{1}{\rho} p_x &= F_x \\ v_t + uv_x + vv_y + \frac{1}{\rho} p_y &= F_y \end{aligned} \tag{6.5.6}$$

In order to proceed further, it is necessary to introduce a relation between ρ and p. Without delving into the theory of **thermodynamics**, the case of **constant entropy** (**isentropic flow**) is treated, for which there exists an algebraic relation $p = f(\rho)$, called the **equation of state**. An important special case that is valid for gases is the isentropic gas law

$$p = A\rho^{\gamma} \tag{6.5.7}$$

where A is a constant and γ is the **ratio of specific heats** of the ideal gas being studied. The condition A = constant along a particle path can be enforced by writing

$$\frac{dA}{dt} = A_t + uA_x + vA_y = 0 \tag{6.5.8}$$

Using Eq. 6.5.7 to eliminate A from Eq. 6.5.8, a partial differential equation is obtained as

$$\left(\frac{p}{\rho^{\gamma}}\right)_t + u\left(\frac{p}{\rho^{\gamma}}\right)_x + v\left(\frac{p}{\rho^{\gamma}}\right)_y = 0 \tag{6.5.9}$$

For gases, the ratio of specific heats γ is not much greater than 1. In particular, $\gamma = 7/5 = 1.4$ for air. Consequently, Eq. 6.5.9 plays a significant role in most problems of **aerodynamics.** On the other hand, the density for **liquids** is often virtually independent of pressure. Thus, in **hydrodynamics**, Eq. 6.5.7 can be replaced by the more elementary equation of **incompressibility**

$$\rho = \text{constant} \tag{6.5.10}$$

and Eq. 6.5.9 can be omitted.

A common condition that must be satisfied at a **solid-fluid boundary** Γ of the flow field Ω is that the fluid does not penetrate the boundary and that no gaps occur between the boundary and the fluid. If the boundary is stationary, this condition may be stated as

$$un_x + vn_y = 0$$

on Γ; i.e., that the component of the fluid velocity normal to the boundary must be zero, where $n_x(x, y)$ and $n_y(x, y)$ are direction cosines of the normal to the boundary.

When the boundary is in motion, the fluid velocity component normal to the boundary must equal the normal component of velocity of the boundary. Letting $\dot{\upsilon}(x, y)$ represent the **normal velocity of the boundary**, the boundary condition becomes

$$\dot{\upsilon} = un_x + vn_y$$

on Γ.

Irrotational, Incompressible, and Steady Isentropic Flows

Note that the equations of motion for a fluid are nonlinear in the variables of the problem. This fact tends to make the study of fluid dynamics considerably more complicated than that of vibrating strings and membranes, whose governing equations are linear. There are, however, special cases of fluid dynamics that can be treated by specialized techniques, which is the focus of the remainder of this section.

Circulation (Potential Flow): For any closed curve C in an **isentropic flow field** Ω that encloses a region S, as in Fig. 6.5.2, a quantity called the **circulation** associated with the curve is defined as the integral of the projection of the velocity vector onto the tangent to the curve,

$$\xi = \int_C (u\, dx + v\, dy) = \iint_S \left(\frac{\partial v}{\partial x} - \frac{\partial u}{\partial y} \right) dx\, dy \tag{6.5.11}$$

where the integral over S is obtained by applying Green's theorem. Using the material derivative to compute the time derivative of ξ, after some manipulation,

$$\begin{aligned} \frac{d\xi}{dt} = \iint_S \Big\{ & \frac{\partial}{\partial x} [v_t + u v_x - u u_y] \\ & + \frac{\partial}{\partial y} [- u_t + v v_x - v u_y] \Big\} dx\, dy \end{aligned} \tag{6.5.12}$$

Substituting from Eq. 6.5.6, with external forces equal to zero, yields

$$\begin{aligned} \frac{d\xi}{dt} = \iint_S \Big\{ & \frac{\partial}{\partial x} \left[- \frac{1}{\rho} p_y - v v_y - u u_y \right] \\ & + \frac{\partial}{\partial y} \left[\frac{1}{\rho} p_x + u u_x + v v_x \right] \Big\} dx\, dy \end{aligned}$$

Using Green's theorem to transform this to a boundary integral and manipulating,

$$\begin{aligned} \frac{d\xi}{dt} &= - \frac{1}{2} \int_C [d(u^2) + d(v^2)] - \int_C \frac{1}{\rho} (p_x\, dx + p_y\, dy) \\ &= - \int_C \frac{dp}{\rho} = 0 \end{aligned} \tag{6.5.13}$$

which follows, since the first integrand is a total differential and since an equation of state such as $p = f(\rho)$ determines ρ as a function of p. Thus, in the absence of external forces, the circulation in the flow field is invariant with time. If the flow starts without rotation, then $\xi = 0$ for all time; i.e., **irrotational flow** occurs throughout its time history.

Let S in Fig. 6.5.2 shrink to a typical point (x, y) in Eq. 6.5.11 and use the mean value theorem to conclude that $\partial v(x, y, t)/\partial x = \partial u(x, y, t)/\partial y$. Thus, the differential form $u(x, y, t)\,dx + v(x, y, t)\,dy$ is exact, at each time t. Theorem 6.2.1 then implies that there exists a function $\phi(x, y, t)$, called a **velocity potential**, such that

$$\frac{\partial \phi}{\partial x} = u$$
$$\frac{\partial \phi}{\partial y} = v \tag{6.5.14}$$

Inserting Eq. 6.5.14 into Eq. 6.5.6, premultiplying the first equation by dx and the second by dy, adding, and integrating over a curve C that is not necessarily closed yields

$$\int_C \left[\left(\phi_{tx} + uu_x + vv_x + \frac{1}{\rho} p_x \right) dx + \left(\phi_{ty} + uu_y + vv_y + \frac{1}{\rho} p_y \right) dy \right] = 0$$

providing there is no external force. With some rearrangement, this equation becomes

$$\phi_t + \frac{1}{2}(u^2 + v^2) + \int_C \frac{dp}{\rho} = H(t) \tag{6.5.15}$$

where H is a function of time alone. This equation is called **Bernoulli's law**, or **Bernoulli's equation.** Along with a state equation $p = f(\rho)$, it allows computation of the pressure field, once the velocity potential is known. In this sense, irrotational isentropic fluid dynamics problems are reduced to determination of the velocity potential that satisfies the continuity equation of Eq. 6.5.4, since the equations of motion of Eq. 6.5.6 are identically satisfied.

Incompressible Potential Flow: In case an isentropic, irrotational flow is also an **incompressible flow**, which is a good approximation for liquids, the density ρ is constant and Eq. 6.5.4 reduces to the familiar **Laplace's equation** for the velocity potential $\phi(x, y)$; i.e.,

$$\nabla^2 \phi \equiv \phi_{xx} + \phi_{yy} = 0 \tag{6.5.16}$$

Example 6.5.1

Consider the steady (i.e., time independent) flow around a cylinder (such as a bridge pile) in a flow field Ω that is rectilinear at infinity, as shown in Fig. 6.5.3. The flow velocity at infinity is taken as $u|_\infty = 1$ and $v|_\infty = 0$, so that at infinity, $\phi(x, y) = x$. At the boundary Γ of the cylinder, the normal velocity of fluid flow must be zero. This simply reflects the fact that the normal derivative of the velocity potential on Γ is zero. This steady flow field is governed by the following **boundary-value problem** for the velocity potential $\phi(x, y)$:

$$\begin{aligned} \nabla^2\phi &= 0, \quad \text{in } \Omega \\ \frac{\partial\phi}{\partial n} &= 0, \quad \text{on } \Gamma \\ \phi\Big|_{\infty} &= x \end{aligned} \tag{6.5.17}$$

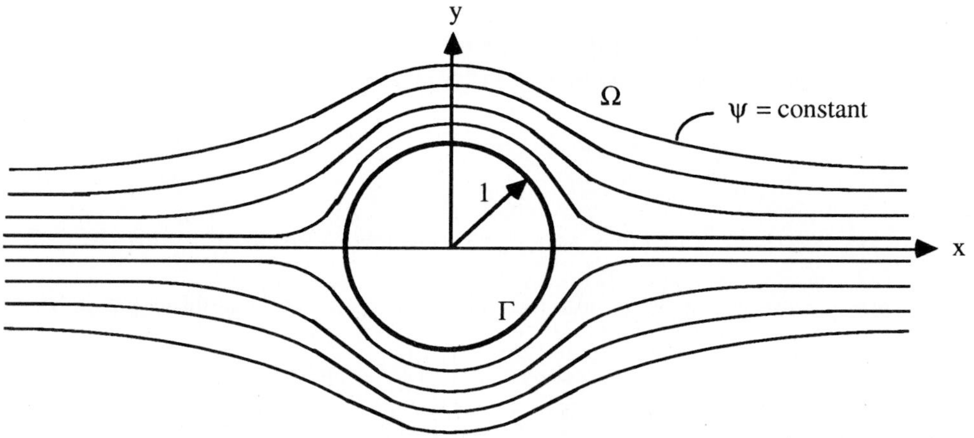

Figure 6.5.3 Steady Flow Around a Cylinder ■

Steady Isentropic Flow: Regardless of whether the flow is irrotational, all time derivatives in the equations of motion vanish in a case of **steady flow**. In this case, the continuity equation of Eq. 6.5.4 implies the existence of another potential function, called a **stream function** $\psi = \psi(x, y)$, such that

$$\begin{aligned} \rho u &= \frac{\partial\psi}{\partial y} \\ \rho v &= -\frac{\partial\psi}{\partial x} \end{aligned} \tag{6.5.18}$$

in Ω. The time rate of change of the stream function along the path swept out by a particle is

$$\dot{\psi} = \psi_x u + \psi_y v = \frac{1}{\rho}(\psi_x\psi_y - \psi_y\psi_x) = 0 \tag{6.5.19}$$

Thus, curves $\psi(x, y)$ = constant follow fluid particle paths, so they are called **stream lines**. A set of stream lines is shown in Fig. 6.5.3.

One-dimensional Isentropic Flow: In case of flow only in the x direction, without external forces, the equations of motion and continuity reduce to

$$u_t + uu_x + \frac{1}{\rho} p_x = 0 \tag{6.5.20}$$

$$\rho_t + \rho_x u + \rho u_x = 0 \tag{6.5.21}$$

The isentropic equation of state for the fluid is

$$p = A\rho^{\gamma} \tag{6.5.22}$$

where A is a constant. Defining a new parameter c as

$$c^2 = \frac{dp}{d\rho} = A\gamma\rho^{\gamma-1} \tag{6.5.23}$$

which depends only on ρ, p may be eliminated from Eq. 6.5.20, to obtain

$$u_t + uu_x + \frac{c^2}{\rho} \rho_x = 0 \tag{6.5.24}$$

Steady, Isentropic, Irrotational Flow: For the steady isentropic flow case, Eq. 6.5.22 provides the equation of state, which may be substituted directly into Bernoulli's equation of Eq. 6.5.15, with $\phi_t = 0$, to obtain

$$\frac{1}{2}(u^2 + v^2) + \frac{A\gamma\rho^{\gamma-1}}{\gamma - 1} = H \tag{6.5.25}$$

where H is a constant. Differentiating this equation first with respect to x and then with respect to y allows development of the formulas

$$\rho_x = -\frac{\rho}{c^2}(u\phi_{xx} + v\phi_{yx})$$

$$\rho_y = -\frac{\rho}{c^2}(u\phi_{xy} + v\phi_{yy})$$

Writing Eq. 6.5.4 in terms of the velocity potential yields

$$(\rho\phi_x)_x + (\rho\phi_y)_y = 0$$

Expanding the derivatives and substituting from the preceding equations, a second-order partial differential equation is obtained as

$$\left(1 - \frac{u^2}{c^2}\right)\phi_{xx} - \frac{2uv}{c^2}\phi_{xy} + \left(1 - \frac{v^2}{c^2}\right)\phi_{yy} = 0 \qquad (6.5.26)$$

Equation 6.5.26 is a special case of the nonlinear equations of fluid flow. This equation will be important later in demonstrating flow characteristics as a function of the velocity of flow of the fluid.

Propagation of Sound in Gas

Consider a **gas** at rest that is subjected to a small initial excitation. Since only small motion of particles in the gas occurs, the process proceeds isentropically and the equation of state of Eq. 6.5.22 holds. Denoting the pressure and density in still gas as p_0 and ρ_0, a new quantity s(x, t) is defined, called the **compression**,

$$s = (\rho - \rho_0)/\rho_0$$

or

$$\rho = \rho_0(1 + s) \qquad (6.5.27)$$

For small amplitude oscillations (sound waves) in the gas, the compression should be small, and second-order terms involving this variable can be neglected.

Employing the approximations

$$(\rho u)_x = \rho_x u + \rho u_x \approx \rho_0 u_x$$

$$\frac{1}{\rho} p_x \approx \frac{1}{\rho_0} p_x \qquad (6.5.28)$$

and ignoring product terms that involve gas flow velocities and their derivatives, Eqs. 6.5.4 and 6.5.6 become

$$u_t + \frac{1}{\rho_0} p_x = 0$$

$$\rho_t + \rho_0 u_x = 0 \qquad (6.5.29)$$

where no external force is applied. Employing Eq. 6.5.27, the equation of state of Eq. 6.5.22, and Eq. 6.5.23, the reduced equations are obtained as

$$u_t + c_0^2 s_x = 0$$

$$s_t + u_x = 0 \qquad (6.5.30)$$

where c_0 denotes the value of c at the air density ρ_0 in still gas.

Differentiating the first of Eq. 6.5.30 with respect to x and the second with respect to t and substituting the second into the first, the second-order differential equation for the compression s(x, t) is

$$s_{tt} - c_0^2 s_{xx} = 0 \tag{6.5.31}$$

It should be recalled that Eq. 6.5.31 holds only for small vibratory motion; i.e., **propagation of sound** in a gas, commonly known as the **equation of acoustics**. Note also that this is the same differential equation that governed the vibration of a string, so it can be expected that the same type of physical behavior will be associated with string vibration as occurs in propagation of sound in gas.

EXERCISES 6.5

1. For a static fluid in which x is horizontal and y is vertical, $u = v = 0$, $F_x = 0$, and $F_y = -g$. Reduce the general equations of motion to represent this static case, called **hydrostatics**.
2. If the static fluid in Exercise 1 is of constant density (e.g., water), derive an expression for pressure as a function of depth.
3. Extend the equations of motion and continuity to three space dimensions and time.
4. Which of the following are velocity potentials for incompressible irrotational flow?

   (a) $\phi = x + y$

   (b) $\phi = x^2 + y^2$

   (c) $\phi = \sin(x + y)$

   (d) $\phi = \ln x$
5. Use the results of Exercise 3 to define the flow field of an incompressible fluid that is spinning steadily counterclockwise with a vertical cylinder, about its axis, at angular speed ω rad/sec. Use the fact that $u = -\omega y$, $v = \omega x$, and $w = 0$, where w is the z component of velocity. Use the equations of motion to find the pressure p as a function of x, y, and z. Finally, put p = constant to find the shape of the free surface of the fluid.

6.6 ANALOGIES (PHYSICAL CLASSIFICATION)

As may be noted for the very different physical problems treated in the preceding sections, the same form of partial differential equations arise; with only variations in the form of the nonhomogeneous terms in the differential equations, boundary conditions, and initial conditions. In some cases, the differential equations, boundary conditions, and initial conditions are identical, even for problems that have considerably different physical significance. Therefore, **analogies** can be considered in a mathematical sense, and even in a physical sense when experimentation in one problem may be readily performed but may be difficult in an analogous problem. To illustrate this point, four different forms of second-order partial differential equations are shown in Table 6.6.1 and the physical problems that are described by each form of differential equation are tabulated.

Table 6.6.1 Analogous Physical Problems

$u_{tt} - a^2\nabla^2 u = f$	$u_t - k^2\nabla^2 u = f$	$\nabla^2 u = f$	$-\nabla^2 u = \frac{\omega^2}{a^2} u$
Vibrating String	Conduction of Heat in Solids	Static Deflection of String	Harmonic Vibrating String
Vibrating Membrane		Static Deflection of Membrane	
Sound Propagation in Gas	Diffusion of Gas in a Porous Medium	Steady State Temperature Distribution in Solids	Harmonic Vibrating Membrane
		Steady Incompressible, Irrotational Flow of Fluids	

First, note a fundamental difference between problems governed by **Laplace's equation**, in the third column, and those governed by the wave and heat equations in the first and second columns. Laplace's equation is associated with static or steady state (time independent) problems. It will be noted in Chapters 7 and 8 that very different mathematical and physical properties are associated with the solution of static problems, as compared with dynamics problems. Finally, as noted in the fourth column, harmonic vibration is governed by an eigenvalue problem.

Within the class of dynamics problems, the only essential difference between the wave and heat equations is the order of the highest time derivative that appears in the differential equation. While in some ways the solutions of these time dependent problems are similar, fundamental differences in the two classes will be noted in Chapters 7 and 8.

The model problems selected from different fields of mechanics in this chapter illustrate techniques that may be used to derive the partial differential equations of continuum behavior and to show that there are a limited number of basic types of differential equations that arise in the study of second-order partial differential equations. Such problems and associated equations are commonly called **equations of mathematical physics.** They are used to study the fundamental properties of differential equations, to provide insight into techniques that can be used for the study of more complex problems. When more complex problems arise, algebraic messiness of the governing equations involved will occur. It will be found, however, that many of the fundamental features found in the model problems derived in this chapter are present in more general and complex problems. It is essential, therefore, to study the second-order differential equations associated with these model problems, for the purposes of developing solution techniques and physical and mathematical insights into techniques for solving partial differential equations of mechanics.

A second key point that must be noted in this discussion is that the problems highlighted in Table 6.6.1 are linear and, as was noted in Section 6.5, problems such as those in fluid dynamics are governed by nonlinear differential equations. This is also true of the field of finite elasticity and other fields of nonlinear mechanics. While nonlinear differential equations arise naturally in general continua, no uniformly applicable solution technique is known. Solution techniques for the various types of nonlinear continua must, therefore, be studied on an ad hoc basis. No attempt is made in this text to be complete in treatment of such nonlinear problems. Rather, concentration is on development of broadly applicable solution techniques and analysis methods that can be employed in studying problems that arise in mechanics and to provide experience and a stepping-off point for the study of methods for solving the more complex problems of mechanics.

7

INTRODUCTION TO THE THEORY OF LINEAR PARTIAL DIFFERENTIAL EQUATIONS

Thus far, analytical foundations and the equations of mechanics that govern broad classes of engineering applications have been treated. As seen in Chapter 6, many of the partial differential equations of mechanics have common mathematical forms, even though the physical phenomena they describe are radically different. These equations serve as prototypes for the study of partial differential equations. It is shown in this chapter that when the subject of linear partial differential equations is approached from a purely theoretical point of view, precisely the same forms of equations arise as were derived from physical considerations. Linear second-order partial differential equations are thus naturally classified into three distinct types, from a mathematical point of view.

7.1 FIRST-ORDER PARTIAL DIFFERENTIAL EQUATIONS

Before launching into the subject of second-order partial differential equations, it is appropriate to say a few words about **first-order partial differential equations**. Many treatments of partial differential equations either neglect the first-order case entirely, or treat it only briefly. The reason is twofold. First, most important equations of mechanics involve second derivatives. The second reason is mathematical in nature, rather than physical. The problem of solving a first-order partial differential equation is closely related to solving ordinary differential equations. Once this fact is understood, there is little point in belaboring the study of first-order partial differential equations, except for explaining some of the techniques involved. Since one of these techniques will be used later, in analysis of second-order partial differential equations, it is appropriate to illustrate the technique for a special case.

Reduction to Ordinary Differential Equations

Consider the **homogeneous first-order partial differential equation**

$$a(x, y)\, u_x + b(x, y)\, u_y = 0 \tag{7.1.1}$$

where $u_x \equiv \partial u/\partial x$ and $u_y \equiv \partial u/\partial y$ are partial derivatives of the function u(x, y) that is to be found and a(x, y) and b(x, y) are given functions of x and y. The goal is to find a function u(x, y) such that u = u(x, y) is a solution of Eq. 7.1.1. The relation u = u(x, y) describes a surface in x-y-u space. Writing the equation of the **solution surface** as $G(x, y, u) \equiv u(x, y) - u = 0$, its differential is $u_x\, dx + u_y\, dy - du = [\, u_x, u_y, -1\,]\, [\, dx, dy, du\,]^T = 0$, for all vectors $[\, dx, dy, du\,]^T$ in the tangent plane of the solution surface. Thus, the vector $[\, G_x, G_y, G_u\,]^T = [\, u_x, u_y, -1\,]^T$ is normal to the solution surface, as shown in Fig. 7.1.1. The partial differential equation of Eq. 7.1.1 may thus be interpreted as stating that the scalar product of the vectors $[\, u_x, u_y, -1\,]^T$ and $[\, a, b, 0\,]^T$ is zero. This means that the vector $[\, a, b, 0\,]^T$ must be tangent to the solution surface, as illustrated in Fig. 7.1.1.

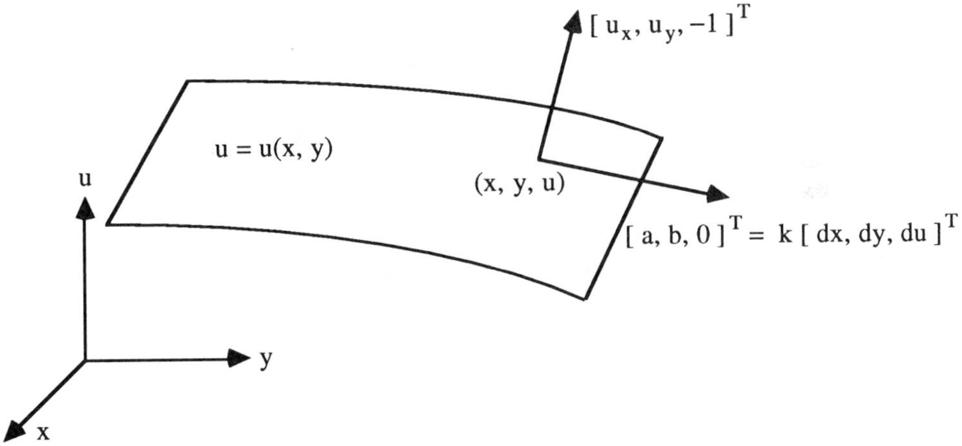

Figure 7.1.1 Solution Surface

The solution surface may be obtained by finding a family of **solution curves** that have $[\, a, b, 0\,]^T$ as their tangent vector at each point (x, y, u), since such curves must be in the solution surface. To find such curves, note that a differential $[\, dx, dy, du\,]^T$ that is tangent to a solution curve must be parallel to the vector $[\, a, b, 0\,]^T$. This condition defines the following ordinary differential equations for solution curves:

$$\frac{dx}{a(x, y)} = \frac{dy}{b(x, y)} \qquad du = 0 \tag{7.1.2}$$

General Solutions

The general solution of the second of Eqs. 7.1.2 is simply

$$u = c_1 \tag{7.1.3}$$

which means that each solution curve being sought is the intersection of a plane $u = c_1$ that is parallel to the xy-plane and the surface defined by the solution of

$$\frac{dy}{dx} = \frac{b(x, y)}{a(x, y)} \tag{7.1.4}$$

The general solution of Eq. 7.1.4 is

$$f(x, y) = c_2 \tag{7.1.5}$$

In x-y-u space, the surface defined by Eq. 7.1.5 is a cylinder (not necessarily circular) that extends to infinity, parallel to the u-axis. A typical solution curve defined by Eqs. 7.1.3 and 7.1.5 is shown in Fig. 7.1.2.

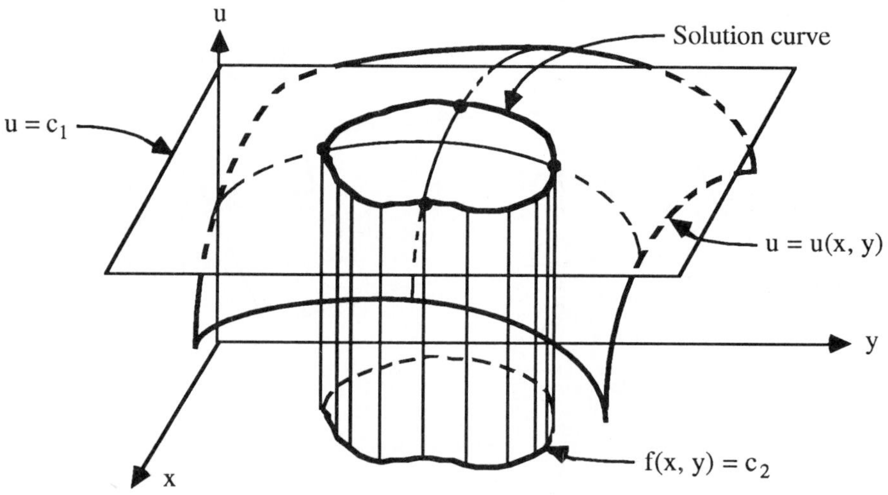

Figure 7.1.2 Geometry of Solution Curves

The two-parameter family of curves that is represented by Eqs. 7.1.3 and 7.1.5 has the property that at any point (x, y, u) of a curve in this family, the vector $[\ a, b, 0\ ]^T$ is in the direction of the tangent to the curve. These curves may be used to generate a solution surface of the original partial differential equation, by establishing an arbitrary relationship between the parameters; i.e.,

$$c_1 = W(c_2) \tag{7.1.6}$$

where W is an arbitrary continuously differentiable function of a single variable. From Eqs. 7.1.3 and 7.1.5, this leads to

$$u = W(\ f(x, y)\ ) \tag{7.1.7}$$

as the **general solution** of Eq. 7.1.1. To verify that Eq. 7.1.7 yields a solution of Eq. 7.1.1, for an **arbitrary function** $W(\bullet)$, note that $u_x = W'(f)\ f_x$ and $u_y = W'(f)\ f_y$. Thus,

$$a(x, y)\ u_x + b(x, y)\ u_y = W'(f)\ [\ a(x, y)\ f_x + b(x, y)\ f_y\ ] \tag{7.1.8}$$

From Eq. 7.1.5,

$$f_x\,dx + f_y\,dy = 0$$

or

$$\frac{dy}{dx} = -\frac{f_x}{f_y} \tag{7.1.9}$$

From Eqs. 7.1.4 and 7.1.9,

$$\frac{b}{a} = -\frac{f_x}{f_y} \tag{7.1.10}$$

or

$$a f_x + b f_y = 0 \tag{7.1.11}$$

Thus, the right side of Eq. 7.1.8 is zero and Eq. 7.1.7 provides a solution of Eq. 7.1.1, with an arbitrary function W(•).

Example 7.1.1

Consider the equation

$$e^x u_x + e^y u_y = 0 \tag{7.1.12}$$

The solution of this equation is given by Eq. 7.1.7, where f(x, y) = c is the solution of

$$\frac{dy}{dx} = \frac{e^y}{e^x}$$

or

$$e^{-y}\,dy = e^{-x}\,dx$$

The indefinite integral of this equation is

$$-e^{-y} = -e^{-x} + c$$

or

$$f(x, y) = e^{-x} - e^{-y}$$

From Eq. 7.1.7, the general solution of Eq. 7.1.12 is

$$u = W(e^{-x} - e^{-y})$$

where W is an arbitrary function. ■

It is interesting to note that, whereas the general solution of a first-order ordinary differential equation contains an arbitrary constant, the general solution of a first-order partial differential equation contains an arbitrary function. This fundamental difference illustrates that there are infinitely many more solutions of partial differential equations than of ordinary differential equations. Further, since computations with functions are much more complicated than with real numbers, the usefulness of a general solution of a partial differential equation is limited. For example, the technique of solving boundary-value problems of ordinary differential equations by finding a general solution and then finding the constants of integration to satisfy boundary conditions does not carry over to partial differential equations.

Although this introduction by no means exhausts the subject of first-order partial differential equations, it illustrates the fact that solving first-order partial differential equations is related to solving ordinary differential equations. The technique demonstrated here will be encountered later, in the study of second-order partial differential equations. For more extensive treatments of the first-order case, the reader is referred to Refs. 11, 12, 18, and 19.

EXERCISES 7.1

1. Find the general solutions of

   (a) $3u_x + 4u_y = 0$

   (b) $xu_x + u_y = 1$

   (c) $\cos y\, u_x + \sin x\, u_y = 0$

   (d) $e^x u_x + e^y u_y = u$

2. Derive a set of ordinary differential equations analogous to Eq. 7.1.2 for a general solution of the linear first-order partial differential equation in three independent variables

$$a(x, y, z)\, u_x + b(x, y, z)\, u_y + c(x, y, z)\, u_z = 0$$

3. Find the general solutions of

   (a) $(\cos x \cos y)\, u_x - (\sin x \sin y)\, u_y + (\sin x \cos y)\, u_z = 0$

   (b) $yzu_x + zxu_y + xyu_z = -xyz$

7.2 GENERAL THEORY OF SECOND-ORDER PARTIAL DIFFERENTIAL EQUATIONS

Second-order Equations in Two Variables

In considering second-order partial differential equations, attention is initially restricted to those that involve just two independent variables. A **second-order partial differential equation** (linear or nonlinear) can be expressed in the general form

$$F(u_{xx}, u_{xy}, u_{yy}, u_x, u_y, u, x, y) = 0 \tag{7.2.1}$$

where u = u(x, y) is an unknown function of x and y and

$$u_{xx} \equiv \frac{\partial^2 u}{\partial x^2}, \quad u_{xy} \equiv \frac{\partial^2 u}{\partial x \partial y}, \quad u_{yy} \equiv \frac{\partial^2 u}{\partial y^2}$$

The first question to be considered is, Does there exist a general method of solving Eq. 7.2.1? If the answer is yes, the second question is, In what form will the solution be expressed? Finally, is the method useful in practice? Briefly stated, the answers to these three questions are as follows:

(1) Yes, there is a general method of solving Eq. 7.2.1, which is stated in the Cauchy–Kowalewski theorem that follows.
(2) The solution may be expressed in the form of power series.
(3) No, the method is not useful in constructing solutions, but it provides valuable insights into the nature of solutions.

These answers may seem either surprising, disappointing, or both. It is useful to consider the reasons for them because of the light they shed on the nature of partial differential equations. Since analytic functions can be written in power series form, it seems natural to assume a **power series solution** of the form

$$u(x, y) = \sum_{m,n=0}^{\infty} a_{mn} (x - x_0)^m (y - y_0)^n \tag{7.2.2}$$

From Taylor's series in more than one variable, if such a series representation is valid, the coefficients a_{mn} are expressible in terms of the partial derivatives of u(x, y) at some point (x_0, y_0), specifically

$$a_{mn} = \frac{1}{m!\, n!} \frac{\partial^{m+n} u}{\partial x^m \partial y^n} \bigg|_{x = x_0,\, y = y_0} \tag{7.2.3}$$

Since the differential equation of Eq. 7.2.1 expresses a relation between u(x, y) and its derivatives, it might be expected that, given certain initial values of the function u(x, y) and its lower-order derivatives, the values of all higher-order derivatives could be obtained by successive differentiation of the differential equation. Then it would be necessary only to show that the series that is constructed according to this method converges. This is precisely what the Cauchy–Kowalewski theorem does. An outline of the process is presented in sufficient detail to observe its strengths and weaknesses, but the theorem is not proved in detail, because of the technical complexities involved [11, 19].

First, assume that Eq. 7.2.1 can be solved, algebraically, for one of the second partial derivatives of u(x, y); e.g.,

$$u_{xx} = G(x, y, u, u_x, u_y, u_{xy}, u_{yy}) \tag{7.2.4}$$

For the method to work, the function G must be analytic; i.e., it must be expressible as a convergent power series in some neighborhood of an initial point. Assume next, without loss of generality, that $x_0 = y_0 = 0$ denotes the initial point about which G is analytic at $(0, 0, u_0, p_0, q_0, s_0, t_0)$, where $u(0, 0) \equiv u_0$, $u_x(0, 0) \equiv p_0$, $u_y(0, 0) \equiv q_0$, $u_{xy}(0, 0) \equiv s_0$, and $u_{yy}(0, 0) \equiv t_0$.

Cauchy Data

Note that to determine the coefficients a_{mn} in Eq. 7.2.3 for $m < 2$ and for all n, using Eq. 7.2.3, it is sufficient to have **initial conditions**, called **Cauchy data,**

$$\begin{aligned} u(0, y) &= f(y) \\ u_x(0, y) &= g(y) \end{aligned} \tag{7.2.5}$$

where f(y) and g(y) are given analytic functions of y, in some neighborhood of the origin. By differentiating Eq. 7.2.4 and substituting $x = 0$ and $y = 0$, the value of any coefficient a_{mn} with $m \geq 2$ may now be found from Eq. 7.2.3. To see why this is so, construct the following matrix of u(x, y) and its derivatives at the point (0, 0):

m \ n	0	1	2	3	...
0	$u(0, 0)$	$u_y(0, 0)$	$u_{yy}(0, 0)$	$u_{yyy}(0, 0)$	...
1	$u_x(0, 0)$	$u_{xy}(0, 0)$	$u_{xyy}(0, 0)$	...	
2	$u_{xx}(0, 0)$	$u_{xxy}(0, 0)$	...		
3	$u_{xxx}(0, 0)$	...			
.	...				

(7.2.6)

The first row of this matrix is computable from the first of Eq. 7.2.5 as

$$u(0, 0) = f(0), \quad u_y(0, 0) = f'(0), \quad u_{yy}(0, 0) = f''(0),$$
$$u_{yyy}(0, 0) = f'''(0), \ldots$$

Likewise, the second of Eq. 7.2.5 yields the second row of the matrix as

$$u_x(0, 0) = g(0), \quad u_{xy}(0, 0) = g'(0), \quad u_{xyy}(0, 0) = g''(0), \ldots$$

To find other rows in the matrix, which will then permit construction of the power series solution, Eq. 7.2.4 is differentiated the appropriate number of times with respect to x and y and values in previously computed rows are substituted.

Example 7.2.1

Let u(x, y) satisfy the differential equation and initial conditions

$$\begin{aligned} 6u_{xx} + u_{yy} + u_y &= 0 \\ u(0, y) = 0, \quad u_x(0, y) &= e^{2y} \end{aligned} \tag{7.2.7}$$

Assume a solution of the form

$$u(x, y) = \sum_{m=0,n=0}^{\infty} a_{mn} x^m y^n \tag{7.2.8}$$

where

$$a_{mn} = \frac{1}{m!\, n!} \left[\frac{\partial^{m+n} u}{\partial x^m \, \partial y^n} \right] \Bigg|_{x=0,\, y=0} \tag{7.2.9}$$

The initial conditions of Eq. 7.2.7 imply that the first row (m = 0) of the matrix in Eq. 7.2.6 is made up of zeros and that the second row (m = 1) is

$$1, 2, 4, \ldots, 2^n, \ldots$$

From the differential equation of Eq. 7.2.7,

$$u_{xx} = -\frac{1}{6}(u_y + u_{yy}) \tag{7.2.10}$$

Differentiating the first of the initial conditions of Eq. 7.2.7 with respect to y yields $u_y(0, 0) = u_{yy}(0, 0) = 0$, so $u_{xx}(0, 0) = 0$. By differentiating Eq. 7.2.10 any number of times with respect to y and then letting x = y = 0, it is found that every entry in the third row of Eq. 7.2.6 is zero. To find values in the fourth row (m = 3), differentiate Eq. 7.2.10 with respect to x, to obtain

$$u_{xxx} = -\frac{1}{6}(u_{xy} + u_{xyy}) \tag{7.2.11}$$

Using the second initial condition of Eq. 7.2.7, this yields $u_{xxx}(0, y) = -(1/6)(2e^{2y} + 4e^{2y}) = -e^{2y}$ and $u_{xxx}(0, 0) = -1$. By successively differentiating Eq. 7.2.11 with respect to y and letting x = y = 0, the fourth row of Eq. 7.2.6 is found to be

$$-1, -2, -4, \ldots, -2^n, \ldots$$

Continuing in this manner, it is found that matrix entries in every row with m even are 0 and that entries in rows with m odd are either

$$1, 2, 4, \ldots, 2^n, \ldots$$

or

$$-1, -2, -4, \ldots, -2^n, \ldots$$

if $(1/2)(m-1)$ is even or odd, respectively.

From Eq. 7.2.9, the power series coefficients are

$$a_{mn} = \begin{cases} 0, & m \text{ even} \\ \dfrac{(-1)^{(m-1)/2}\, 2^n}{m!\, n!}, & m \text{ odd} \end{cases}$$

and the power series of Eq. 7.2.2 for u(x, y) is

$$u(x, y) = \sum_{m=1,3,5,\ldots} \sum_{n=0}^{\infty} \frac{(-1)^{(m-1)/2}\, 2^n}{m!\, n!} x^m y^n \tag{7.2.12}$$

In this example, the series of Eq. 7.2.12 represents the solution of the initial-value problem everywhere, since by formal manipulation, u(x, y) may be expressed in closed form as

$$u(x, y) = \sum_{n=0}^{\infty} \frac{(2y)^n}{n!} \sum_{m=1,3,5,\ldots} \frac{(-1)^{(m-1)/2}\, x^m}{m!}$$

which is an expression that converges everywhere to

$$u(x, y) = e^{2y} \sin x$$

This function satisfies the differential equation and the initial conditions of Eq. 7.2.8. ∎

The differential equation of Eq. 7.2.4 and the initial conditions of Eq. 7.2.5 comprise an **initial-value problem**, called a **Cauchy problem**. Since any analytic function u(x, y) is the solution of some Cauchy problem, the matrix of Eq. 7.2.6 implies that the general solution of a second-order partial differential equation in two independent variables contains two arbitrary functions f and g of a single variable. This is an example of a more general rule [19].

Theorem 7.2.1. The general solution of an nth order partial differential equation in m independent variables contains n arbitrary functions of $m-1$ independent variables. ∎

The well-known situation for ordinary differential equations is a special case of Theorem 7.2.1, where $m = 1$ and a function of $m - 1 = 0$ independent variables is a constant.

Thus, the general solution of an nth order differential equation in one independent variable (an ordinary differential equation) contains n arbitrary constants. This result is also consistent with the first-order partial differential equation in two independent variables ($n = 1$, $m = 2$) in Example 7.1.1, whose general solution contained an arbitrary function $W(\bullet)$ of one variable.

The Cauchy–Kowalewski Existence Theorem

The foregoing may be summarized by giving a statement of the Cauchy–Kowalewski theorem. This statement has been deferred until now, with the hope that the preceding discussion and examples will make it more meaningful.

Theorem 7.2.2 (Cauchy–Kowalewski Theorem). Let f(y) and g(y) be analytic functions in a neighborhood of y_0; let $u_0 = f(y_0)$, $p_0 = g(y_0)$, $q_0 = f'(y_0)$, $s_0 = g'(y_0)$, and $t_0 = f''(y_0)$; and let the function G(x, y, u, p, q, s, t) be analytic in a neighborhood of $(x_0, y_0, u_0, p_0, q_0, s_0, t_0)$. Then, in some neighborhood of (x_0, y_0), there is one and only one solution u(x, y) of the partial differential equation

$$u_{xx} = G(x, y, u, u_x, u_y, u_{xy}, u_{yy}) \tag{7.2.13}$$

that satisfies the initial conditions

$$\begin{aligned} u(x_0, y) &= f(y) \\ u_x(x_0, y) &= g(y) \end{aligned} \tag{7.2.14}$$

This solution is analytic; i.e., it is representable by a power series

$$u(x, y) = \sum_{m,n=0}^{\infty} a_{mn} (x - x_0)^m (y - y_0)^n \tag{7.2.15}$$

that converges in a neighborhood of (x_0, y_0), where

$$a_{mn} = \frac{1}{m!\,n!} \left[\frac{\partial^{m+n} u}{\partial x^m \partial y^n} \right] \Bigg|_{x = x_0,\, y = y_0} \qquad m, n = 0, 1, 2, \ldots \tag{7.2.16}$$

■

The Cauchy–Kowalewski theorem has been stated for a second-order partial differential equation in two independent variables. Similar results are also true for partial differential equations of arbitrary order in more than two independent variables [19]. The theorem provides valuable information about solutions of partial differential equations, as follows:

(1) The Cauchy–Kowalewski theorem provides a general method of solution (power series) that is valid when the partial differential equation and initial

conditions satisfy analyticity conditions. Unfortunately, the method is inapplicable in most practical situations.

(2) The general solution of a second-order partial differential equation in two independent variables contains two arbitrary functions of a single variable.

(3) The arbitrary functions that appear in the general solution correspond to the initial conditions and indicate the number of initial conditions that are appropriate in an initial-value problem, for which a solution exists and is unique.

One final remark about what the Cauchy–Kowalewski theorem does not say is in order. Since it addresses only initial-value problems, specifically Cauchy problems, it does not say anything about the solution of **boundary-value problems** that arise frequently in applications. For example, suppose a solution of Eq. 7.2.4 is sought that satisfies, instead of the initial conditions of Eq. 7.2.5, **boundary conditions** such as

$$\begin{aligned} u(x_1, y) &= f(y) \\ u(x_2, y) &= g(y) \end{aligned} \tag{7.2.17}$$

Instead of specifying $u(x, y)$ and one of its first partial derivatives along the line $x = x_0$, values of $u(x, y)$ only have been specified along two different lines $x = x_1$ and $x = x_2$. It is apparent that a power series approach will be of little help here, since neither condition alone is sufficient to determine the coefficients of a power series for the solution. One condition (say, that along $x = x_1$) might be applied to obtain some information about the coefficients. There is no assurance, however, that the assumed power series about (x_1, y_0), with partially determined coefficients, would converge in a sufficiently large neighborhood of (x_1, y_0) to include the point (x_2, y_0), where the other condition would need to be applied. Thus, approaches other than power series methods must be sought for the study of boundary-value problems.

EXERCISES 7.2

1. Consider a steady, isentropic, irrotational flow of air that is governed by Eq. 6.5.26; i.e.,

$$\left(1 - \frac{u^2}{c^2}\right)\phi_{xx} - \left(\frac{2uv}{c^2}\right)\phi_{xy} + \left(1 - \frac{v^2}{c^2}\right)\phi_{yy} = 0$$

The initial condition is uniform flow in the x direction, across the line $x = 0$; i.e.,

$$\begin{aligned} \phi(0, y) &= 0 \\ \phi_x(0, y) &= u_0 \end{aligned}$$

(a) Under what condition does the Cauchy–Kowalewski theorem apply to this problem?

(b) If the parameter c was known to be the speed of sound in the fluid, what mathematical significance does $u = c$ have?

2. If $F(x, y, t)$, $g(x, y)$, and $h(x, y)$ in Eqs. 6.3.25 and 6.3.26 are analytic, what does the Cauchy–Kowalewski theorem say about the vibrating membrane problem on an infinite physical region Ω?

3. If $p(x, y, z)$ in Eq. 6.4.7 is analytic, what does the Cauchy–Kowalewski theorem say about the heat conduction problem of Eqs. 6.4.6 and 6.4.7 in an infinite physical region Ω?

7.3 CHARACTERISTIC CURVES

Initial Curves

Application of the Cauchy–Kowalewski theorem is now extended to initial conditions other than those of the form of Eq. 7.2.5; i.e., conditions specified along the y-axis. Let a curve C be given in the x-y plane, as shown in Fig. 7.3.1(a), with unit vector $\nu = [\nu_x, \nu_y]^T$ that is not tangent to the curve C. A generalization of the initial conditions of Eq. 7.2.14, which may be viewed in Fig. 7.3.1(b) as values of $u(x, y)$ on the vertical line $x = x_0$ and its first directional derivative u_ν, is

$$\begin{aligned} u(x, y) &= f(x, y) \\ u_\nu(x, y) &= g(x, y) \end{aligned} \tag{7.3.1}$$

for (x, y) on C, where $u_\nu = u_x\nu_x + u_y\nu_y$. The curve C is called an **initial curve**.

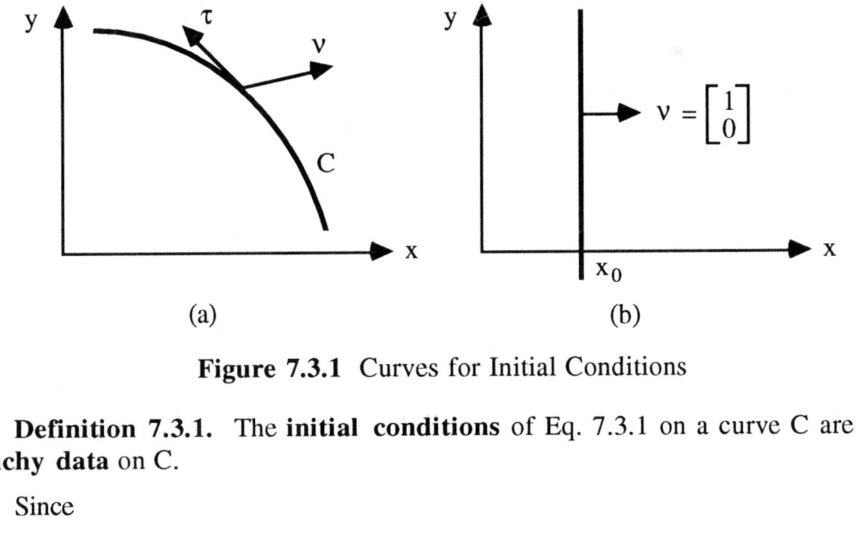

(a) (b)

Figure 7.3.1 Curves for Initial Conditions

Definition 7.3.1. The **initial conditions** of Eq. 7.3.1 on a curve C are called **Cauchy data** on C. ■

Since

$$u_\nu = u_x\nu_x + u_y\nu_y$$

and

$$u_\tau = u_x\tau_x + u_y\tau_y$$

where $\tau = [\ \tau_x, \tau_y\ ]^T$ is the unit tangent vector to the curve C and $\nu = [\ \nu_x, \nu_y\ ]^T$ is not tangent to the curve C, using the Cauchy data of Eq. 7.3.1, the first derivatives u_x and u_y can be found. In general, for multivariable problems, it is shown in Ref. 18 (p. 193) that Cauchy data on the surface C can be used to find all the first derivatives of the unknown function u on the surface C.

For certain types of second-order equations that will be studied in more detail later, it will be found that there exist exceptional curves, along which initial values of u and its first directional derivative cannot be specified, without destroying either existence or uniqueness of the solution.

Definition 7.3.2. An initial curve in the x-y plane along which the differential equation of Eq. 7.2.1 and Cauchy data (function and first directional derivative) of Eq. 7.3.1 fail to determine all derivatives of the dependent variable u(x, y) is called a **characteristic curve.** ■

At this point, consideration is limited to how to determine whether a given initial curve is a characteristic curve. In a more general analysis of second-order linear equations in Section 7.5, a method is given to find all possible characteristic curves. Note that if the initial curve is the y-axis, as in Fig. 7.3.1(b) with $x_0 = 0$, it should be possible to solve the differential equation algebraically for u_{xx}, in terms of the other variables that appear in the equation. If not, then the y-axis is a characteristic curve. The matrix of Eq. 7.2.6 makes this clear, since if initial conditions are specified along the y-axis, then initial values in the first two rows of the matrix are known and u_{xx} is the only derivative in the differential equation of Eq. 7.2.1 that does not appear in the first two rows of Eq. 7.2.6. Equation 7.2.1 must be used to obtain the initial value of u_{xx}, because u_{xx} is the only one of all the derivatives that appear in Eq. 7.2.1 whose initial value is not obtainable from the initial conditions alone. Thus, the assumption that Eq. 7.2.1 can be expressed as in Eq. 7.2.4 (i.e., solved algebraically for u_{xx} in terms of the other variables that appear in Eq. 7.2.1) is directly related to the form of the initial conditions. Similarly, if the initial curve is the x-axis, it should be possible to solve the differential equation algebraically for u_{yy}. Otherwise, the x-axis is a characteristic curve.

Suppose that initial conditions are given along some curve Q(x, y) = 0 whose tangent is not horizontal. Furthermore, suppose ν is not a unit vector in the y-direction; i.e., $\nu \neq [\ 0, 1\ ]^T$. To determine whether this curve is a characteristic curve, introduce new independent variables ξ and η, according to the transformation $\xi = Q(x, y)$ and $\eta = y$. In the new coordinate system, the initial curve will be the line $\xi = 0$; i.e., Q(x, y) = 0. Thus, curves Q(x, y) = c, for any real number c, are characteristic curves if the differential equation cannot be solved algebraically for $u_{\xi\xi}$.

Example 7.3.1

Consider the partial differential equation

$$u_{xx} = u_{xy} \tag{7.3.2}$$

Note that it is not possible to solve the equation algebraically for u_{yy}, so the x-axis is a characteristic curve. Since a translation of the type $\bar{x} = x$ and $\bar{y} = y - y_0$, with y_0

constant, would not alter the form of the differential equation, all lines y = constant are characteristic curves. However, these are not all the characteristics for this equation. To show that lines x + y = c are also characteristic curves, define ξ and η as $\xi = x + y$ and $\eta = y$. Then,

$$\begin{aligned} u_x &= u_\xi \xi_x + u_\eta \eta_x = u_\xi \\ u_{xx} &= u_{\xi\xi}\xi_x + u_{\xi\eta}\eta_x = u_{\xi\xi} \\ u_{xy} &= u_{\xi\xi}\xi_y + u_{\xi\eta}\eta_y = u_{\xi\xi} + u_{\xi\eta} \end{aligned} \tag{7.3.3}$$

In the new variables ξ and η, the partial differential equation of Eq. 7.3.2 becomes

$$u_{\xi\eta} = 0 \tag{7.3.4}$$

Since this equation cannot be solved for $u_{\xi\xi}$, lines ξ = constant are characteristic curves. In terms of the original variables x and y, this means that lines x + y = c are characteristic curves, which was to be shown. ■

Prototype Cauchy Problems

To show the significance of characteristic curves, three prototype Cauchy problems are studied. Even though they appear to be very similar, they have quite different properties regarding solutions. All three problems involve the differential equation of Example 7.3.1 and differ only in their initial conditions.

Example 7.3.2

Consider the following initial-value problems:

Problem A. Solve Eq. 7.3.2, subject to the initial conditions

$$\begin{aligned} u(0, y) &= \sin y \\ u_x(0, y) &= \sin y \end{aligned} \tag{7.3.5}$$

Problem B. Solve Eq. 7.3.2, subject to the initial conditions

$$\begin{aligned} u(x, 0) &= \sin x \\ u_y(x, 0) &= \sin x \end{aligned} \tag{7.3.6}$$

Problem C. Solve Eq. 7.3.2, subject to the initial conditions

$$\begin{aligned} u(x, 0) &= \sin x \\ u_y(x, 0) &= \cos x \end{aligned} \tag{7.3.7}$$

In problems B and C, the initial conditions are specified along the x-axis, which was shown in Example 7.3.1 to be a characteristic curve. This situation normally destroys either existence or uniqueness of solutions to an initial-value problem. In fact, it may be shown that Problem A has a solution that is unique, Problem B has no solution, and Problem C has infinitely many solutions. The solution of Problem A will be given here. Analyses of Problems B and C are left as exercises. To solve Problem A, the power series technique used previously could be applied. In this case, however, because of the simplicity of the differential equation, an easier method is available. First write the differential equation of Eq. 7.3.2 in the form

$$u_{xx} - u_{xy} = \frac{\partial}{\partial x}(u_x - u_y) = 0$$

A general solution is the first-order differential equation

$$u_x - u_y = f(y)$$

where f(y) is an arbitrary function. Writing this equation as

$$u_x - u_y - f(y) = 0$$

or

$$[\, u_x, u_y, -1 \,]\,[\, 1, -1, f(y) \,]^T = 0$$

indicates that at each point (x, y, u), the normal $[\, u_x, u_y, -1 \,]^T$ to the surface $u(x, y) - u = 0$ is perpendicular to the vector $[\, 1, -1, f(y) \,]^T$. The family of curves whose tangents at each point (x, y, u) are in the direction $[\, 1, -1, f(y) \,]^T$ must thus lie in a solution surface $u = u(x, y)$. The differential equations of these curves are

$$\frac{dx}{1} = \frac{dy}{-1} = \frac{du}{f(y)}$$

The first equality implies that

$$x + y = c_1$$

and the second equality implies that

$$u + F(y) = c_2$$

where $F'(y) = f(y)$. A solution surface may be generated from this two-parameter family of curves by establishing a relationship between the parameters; i.e.,

$$c_2 = G(c_1)$$

This leads to the **general solution**

$$u = G(x + y) - F(y) \tag{7.3.8}$$

of Eq. 7.3.2, where F and G are arbitrary functions of a single argument.

Applying the initial conditions of Problem A to Eq. 7.3.8,

$$u(0, y) = G(y) - F(y) = \sin y$$

$$u_x(0, y) = G'(y) = \sin y$$

for all y. From the second condition, which is an ordinary differential equation in the variable y,

$$G(y) = -\cos y + k \tag{7.3.9}$$

where k is a constant. Substituting this result into the first condition yields

$$-\cos y + k - F(y) = \sin y$$

or

$$F(y) = -\sin y - \cos y + k \tag{7.3.10}$$

Note that Eqs. 7.3.9 and 7.3.10 define the functions G and F as functions of a single variable, here denoted y. Substituting the results of Eqs. 7.3.9 and 7.3.10 into Eq. 7.3.8 for u, the solution of Problem A is

$$u = -\cos(x + y) + \sin y + \cos y$$

which may be verified by direct substitution. Since Problem A satisfies the hypotheses of the Cauchy–Kowalewski theorem, this solution is unique. As already noted, the situation is different in Problems B and C. ■

It must be emphasized that the technique used to solve Problem A in Example 7.3.2 is not typical of methods that are generally applied to solve partial differential equations. In practical situations, it will seldom be possible to find a general solution of the partial differential equation and then to find the desired particular solution, by using the initial conditions to determine arbitrary functions that appear in the general solution.

EXERCISES 7.3

1. Use the general solution of Eq. 7.3.8 and the initial conditions of Eq. 7.3.6 to show that Problem B of Example 7.3.2 has no solution.

2. Use the general solution of Eq. 7.3.8 and the initial conditions of Eq. 7.3.7 to show that Problem C of Example 7.3.2 has infinitely many solutions.

3. Determine which of the following curves are characteristic curves of the differential equation $u_{tt} - a^2 u_{xx} = 0$:

   (a) $t = 0$

   (b) $x = 0$

   (c) $x + at = 0$

   (d) $x - at = 0$

4. Determine which of the following curves are characteristic curves of the differential equation $4u_{xx} + 10u_{xy} + 4u_{yy} = 0$:

   (a) $y - 2x = 0$

   (b) $y + 4x = 0$

   (c) $2y - x = 0$

7.4 STABILITY OF SOLUTIONS AND WELL-POSED PROBLEMS

Having introduced the subject of existence and uniqueness of solutions with Problems A, B, and C of Example 7.3.2, the matter of **stability**, or **continuous dependence on data** of the problem, can be considered. This is of concern in physical applications, since data are often the result of measurements, which are never exact. It is normally expected in applications that a small change in data will cause only a small change in the solution. If this is not the case, trouble is encountered. For initial-value problems of the type considered thus far, stability of the solution with respect to the initial conditions can be defined as follows.

Definition 7.4.1. Let $\{ f_m(y) \}$ and $\{ g_n(y) \}$ be sequences of analytic functions such that $\lim_{m\to\infty} f_m(y) = f(y)$ and $\lim_{n\to\infty} g_n(y) = g(y)$, pointwise. Then the solution $u(x, y)$ of the initial-value problem

$$\begin{aligned} u_{xx} &= G(x, y, u, u_x, u_y, u_{xy}, u_{yy}) \\ u(0, y) &= f(y), \quad u_x(0, y) = g(y) \end{aligned} \tag{7.4.1}$$

is **stable with respect to the initial conditions** if the solution $u_{mn}(x, y)$ of the initial-value problem

$$\begin{aligned} u_{xx} &= G(x, y, u, u_x, u_y, u_{xy}, u_{yy}) \\ u(0, y) &= f_m(y), \quad u_x(0, y) = g_n(y) \end{aligned} \tag{7.4.2}$$

approaches $u(x, y)$, pointwise, as m and n approach ∞. ■

Example 7.4.1

A well-known example, due to Hadamard, shows that initial-value problems with unstable solutions do exist. His example is

$$\begin{aligned} u_{xx} + u_{yy} &= 0 \\ u(0, y) = 0, \quad u_x(0, y) &= 0 \end{aligned} \tag{7.4.3}$$

An obvious solution of this problem is $u(x, y) = 0$. The Cauchy–Kowalewski theorem assures that this solution is unique. It is not a stable solution, however, as can be seen by considering the sequence of initial-value problems

$$u_{xx} + u_{yy} = 0$$
$$u(0, y) = 0, \quad u_x(0, y) = \frac{\sin ny}{n}$$

For each $n = 1, 2, 3, \ldots$, the solution of this problem is

$$u_n(x, y) = \frac{\sinh nx \sin ny}{n^2}$$

Also,

$$\lim_{m \to \infty} f_m(y) = 0$$

$$\lim_{n \to \infty} g_n(y) = \lim_{n \to \infty} \frac{\sin ny}{n} = 0$$

In fact, convergence is uniform. However,

$$\lim_{n \to \infty} u_n(x, y) = \lim_{n \to \infty} \frac{\sinh nx \sin ny}{n^2} \neq 0$$

for $x \neq 0$. This shows that the solution of the original problem does not depend continuously on the initial conditions. Thus, the initial-value problem of Eq. 7.4.3 is unstable. ■

Definition 7.4.2. A partial differential equation and its initial or boundary conditions is said to be a **well-posed problem** if (1) it has a solution, (2) the solution is unique, and (3) the solution is stable. ■

A great deal of effort is often devoted to determining whether a problem is well posed, before an attempt is made to solve the problem. The reason for this should be clear. Without assurance that a unique, stable solution exists, a great deal of time and effort could be expanded on a computer program, only to produce a result that is essentially worthless, or to obtain no result at all.

7.5 CLASSIFICATION OF LINEAR EQUATIONS IN TWO INDEPENDENT VARIABLES

Consideration is restricted in this and the following sections to second-order linear partial differential equations, various forms of which appear in applications. The general **second-order linear partial differential equation** in two independent variables may be written in the form

$$A(x, y)\, u_{xx} + 2B(x, y)\, u_{xy} + C(x, y)\, u_{yy} + D(x, y)\, u_x + E(x, y)\, u_y + F(x, y)\, u = G(x, y) \qquad (7.5.1)$$

The factor 2 that appears before the coefficient B is introduced for convenience, the reason for which will become clear later. If $G(x, y) = 0$, Eq. 7.5.1 is said to be a **homogeneous equation**.

Direct substitution verifies that if $u^1(x, y)$ and $u^2(x, y)$ are solutions of Eq. 7.5.1 with $G = 0$, then so is $a_1u^1 + a_2u^2$, where a_1 and a_2 are arbitrary constants. The same is true for any finite linear combination of solutions,

$$u(x, y) = \sum_{k=1}^{n} a_k u^k(x, y) \qquad (7.5.2)$$

This result holds for $n = \infty$, provided the series converges uniformly to a sufficiently differentiable function. This is often called the **principle of superposition** for homogeneous linear equations. Note that it is not valid for nonhomogeneous equations.

In contrast to homogeneous equations, if $v(x, y)$ is a solution of the nonhomogeneous equation of Eq. 7.5.1; i.e., if $G \neq 0$, then so is

$$v(x, y) + \sum_{k=1}^{n} a_k u^k(x, y) \qquad (7.5.3)$$

provided every $u^k(x, y)$ is a solution of the corresponding homogeneous equation; i.e., Eq. 7.5.1 with G set to zero. Once again, the finite sum may be extended to an infinite series, provided the series converges uniformly to a sufficiently differentiable function.

Canonical Forms of Linear Equations

The next objective is to classify equations of the form of Eq. 7.5.1, according to the nature of the coefficients $A(x, y)$, $B(x, y)$, and $C(x, y)$. A striking analogy exists between Eq. 7.5.1 and a quadratic algebraic equation in two variables. The most general such algebraic equation may be written in the form

$$Ax^2 + 2Bxy + Cy^2 + Dx + Ey + F = 0 \qquad (7.5.4)$$

It is shown in analytic geometry that this equation represents a hyperbola, parabola, or ellipse, according to whether the **discriminant** $B^2 - AC$ is greater than, equal to, or less than zero, respectively. By introduction of new variables $\xi = \xi(x, y)$ and $\eta = \eta(x, y)$, the quadratic form of Eq. 7.5.4 may be reduced to one of the following so-called **canonical forms**:

(1) If it is a **hyperbola** ($B^2 - AC > 0$),

$$\xi^2 - \eta^2 + \text{lower-degree terms} = 0 \tag{7.5.5}$$

or

$$\xi\eta + \text{lower-degree terms} = 0 \tag{7.5.6}$$

(2) If it is a **parabola** ($B^2 - AC = 0$),

$$\eta^2 + \text{lower-degree terms} = 0 \tag{7.5.7}$$

(3) If it is an **ellipse** ($B^2 - AC < 0$),

$$\xi^2 + \eta^2 + \text{lower-degree terms} = 0 \tag{7.5.8}$$

In order to study the **canonical form** of a second-order linear differential equation, Eq. 7.5.1 is written in the form

$$A(x, y)\, u_{xx} + 2B(x, y)\, u_{xy} + C(x, y)\, u_{yy} + H = 0 \tag{7.5.9}$$

where H is an expression that involves only u(x, y) and its first derivatives. Introducing new independent variables

$$\begin{aligned} \xi &= \xi(x, y) \\ \eta &= \eta(x, y) \end{aligned} \tag{7.5.10}$$

where $\xi(x, y)$ and $\eta(x, y)$ are functions that are to be determined, the derivatives of u(x, y) with respect to ξ and η are

$$\begin{aligned} u_x &= u_\xi \xi_x + u_\eta \eta_x \\ u_y &= u_\xi \xi_y + u_\eta \eta_y \\ u_{xx} &= u_{\xi\xi}\xi_x^2 + 2u_{\xi\eta}\xi_x\eta_x + u_{\eta\eta}\eta_x^2 + u_\xi \xi_{xx} + u_\eta \eta_{xx} \\ u_{xy} &= u_{\xi\xi}\xi_x\xi_y + u_{\xi\eta}(\xi_x\eta_y + \xi_y\eta_x) + u_{\eta\eta}\eta_x\eta_y \\ &\qquad + u_\xi \xi_{xy} + u_\eta \eta_{xy} \\ u_{yy} &= u_{\xi\xi}\xi_y^2 + 2u_{\xi\eta}\xi_y\eta_y + u_{\eta\eta}\eta_y^2 + u_\xi \xi_{yy} + u_\eta \eta_{yy} \end{aligned} \tag{7.5.11}$$

Substituting these expressions into Eq. 7.5.9 and collecting terms yields

$$\hat{A}(\xi, \eta)\, u_{\xi\xi} + 2\hat{B}(\xi, \eta)\, u_{\xi\eta} + \hat{C}(\xi, \eta)\, u_{\eta\eta} + \hat{H} = 0 \tag{7.5.12}$$

where

$$\hat{A}(\xi, \eta) = A\xi_x^2 + 2B\xi_x\xi_y + C\xi_y^2 \tag{7.5.13}$$

$$\hat{B}(\xi, \eta) = A\xi_x\eta_x + B(\xi_x\eta_y + \xi_y\eta_x) + C\xi_y\eta_y \tag{7.5.14}$$

$$\hat{C}(\xi, \eta) = A\eta_x^2 + 2B\eta_x\eta_y + C\eta_y^2 \tag{7.5.15}$$

and $\hat{H}$ is a new expression in u and its first derivatives with respect to ξ and η.

If a function $\xi(x, y)$ can be found so that the coefficient $\hat{A}$ of $u_{\xi\xi}$ in Eq. 7.5.12 is zero, then Eq. 7.5.12 fails to determine $u_{\xi\xi}$ and the curve $\xi(x, y) = \text{constant}$ is a **characteristic curve** for Eq. 7.5.12. The condition $\hat{A} = 0$ yields the first-order partial differential equation

$$A\xi_x^{\ 2} + 2B\xi_x\xi_y + C\xi_y^{\ 2} = 0 \tag{7.5.16}$$

Equation 7.5.16 is not a linear equation, but its left side may be reduced to the product of two linear functions, as follows. Consider the quadratic algebraic equation

$$Ar^2 + 2Br + C = 0 \tag{7.5.17}$$

If $A \neq 0$, the roots of Eq. 7.5.17 are

$$\begin{aligned} r_1 &= \frac{-B + \sqrt{B^2 - AC}}{A} \\ r_2 &= \frac{-B - \sqrt{B^2 - AC}}{A} \end{aligned} \tag{7.5.18}$$

Using these roots, Eq. 7.5.17 can be rewritten as

$$(r - r_1)(r - r_2) = 0$$

With $r = \xi_x / \xi_y$, this implies that Eq. 7.5.16 may be rewritten in the form

$$(\xi_x - r_1\xi_y)(\xi_x - r_2\xi_y) = 0 \tag{7.5.19}$$

which is satisfied if either factor is zero. Thus, the objective is to solve

$$\xi_x - r_k\xi_y = 0, \quad k = 1, 2 \tag{7.5.20}$$

where r_k are the functions of x and y given in Eq. 7.5.18.

Since Eq. 7.5.20 is a first-order equation of the type treated in Section 7.1, a two-parameter family of curves is sought that satisfies the ordinary differential equations

$$\frac{dx}{1} = \frac{dy}{-r_k(x, y)}$$
$$d\xi = 0 \tag{7.5.21}$$

By the theory of ordinary differential equations, the first equation has a solution

$$R_k(x, y) = c_1$$

and the solution of the second equation is

$$\xi = c_2$$

A solution surface of Eq. 7.5.20 may be established by an arbitrary relation between the parameters c_1 and c_2; i.e.,

$$c_2 = W(c_1)$$

where $W(\bullet)$ is an arbitrary function. This yields

$$\xi = W(R_k(x, y)), \quad k = 1, 2$$

In particular, selecting $W(R) = R$ yields

$$\xi = R_k(x, y), \quad k = 1, 2 \tag{7.5.22}$$

Hyperbolic Equations

If

$$B^2 - AC > 0 \tag{7.5.23}$$

then r_1 and r_2 of Eq. 7.5.18 are real and distinct and Eq. 7.5.9 is called a **hyperbolic partial differential equation**. In this case, not only is the coefficient of $u_{\xi\xi}$ in Eq. 7.5.12 equal to zero, but if $\eta(x, y)$ is the second solution of Eq. 7.5.22; i.e., if η satisfies

$$A\eta_x^{\,2} + 2B\eta_x\eta_y + C\eta_y^{\,2} = 0$$

then the coefficient $\hat{C}(\xi, \eta)$ in Eq. 7.5.12 is zero.

The new variables ξ and η are independent when r_1 and r_2 are real and distinct, by using Eq. 7.5.22 to define

$$\xi = R_1(x, y)$$
$$\eta = R_2(x, y) \tag{7.5.24}$$

The reason is that

$$\begin{vmatrix} \xi_x & \xi_y \\ \eta_x & \eta_y \end{vmatrix} = \xi_x\eta_y - \xi_y\eta_x = \xi_y\eta_y\left(\frac{\xi_x}{\xi_y} - \frac{\eta_x}{\eta_y}\right)$$

$$= \xi_y\eta_y(r_1 - r_2) = \xi_y\eta_y \frac{2\sqrt{B^2 - AC}}{A} \neq 0$$

since $\xi_y \neq 0$ and $\eta_y \neq 0$. Thus, there is no functional relationship between ξ and η. For this choice of ξ and η, $\hat{A} = \hat{C} = 0$.

To investigate the nature of $\hat{B}$, consider first the identity

$$(\hat{B}^2 - \hat{A}\hat{C}) = (B^2 - AC)(\xi_x\eta_y - \xi_y\eta_x)^2 \tag{7.5.25}$$

which is obtained from Eqs. 7.5.13, 7.5.14, and 7.5.15. This relationship shows that the sign of the discriminant is unaffected by the transformation of Eq. 7.5.10. In the hyperbolic case being considered, Eq. 7.5.25 shows that $\hat{B} \neq 0$. Hence, under the transformation of Eq. 7.5.24, Eq. 7.5.9 reduces to the **canonical form of a hyperbolic equation,**

$$u_{\xi\eta} + \text{lower-order terms} = 0 \tag{7.5.26}$$

It is left as an exercise to show that another canonical form for a hyperbolic equation,

$$u_{\alpha\alpha} - u_{\beta\beta} + \text{lower-order terms} = 0 \tag{7.5.27}$$

may be obtained by the further transformation

$$\alpha = \frac{1}{\sqrt{2}}(\xi + \eta)$$

$$\beta = \frac{1}{\sqrt{2}}(\xi - \eta)$$

The prototype hyperbolic equation is the **wave equation,** $a^2u_{xx} - u_{tt} = 0$.

Parabolic Equations

Consider next the **parabolic partial differential equation**; i.e., Eq. 7.5.9, with

$$B^2 - AC = 0 \tag{7.5.28}$$

Note that Eq. 7.5.28 requires that A and C have the same sign, which can be made positive by multiplying both sides of Eq. 7.5.1 by -1, if necessary. Here, the roots of Eq. 7.5.17 are real and equal; i.e.,

$$r_1 = r_2 = -\frac{B}{A} = -\sqrt{\frac{C}{A}}$$

and Eq. 7.5.19 may be written as

$$(\sqrt{A}\,\xi_x + \sqrt{C}\,\xi_y)^2 = 0 \tag{7.5.29}$$

The first-order partial differential equation

$$\sqrt{A}\,\xi_x + \sqrt{C}\,\xi_y = 0$$

which is equivalent to Eq. 7.5.29, may be solved to obtain $\xi = R(x, y)$. For this choice of ξ, $\hat{A} = 0$ and Eqs. 7.5.28 and 7.5.25 imply that $\hat{B} = 0$. The function $\eta(x, y)$ may be chosen arbitrarily, as long as $\xi_x\eta_y - \xi_y\eta_x \neq 0$, which guarantees that $\hat{C} = (\sqrt{A}\,\eta_x + \sqrt{C}\,\eta_y)^2 = (A/\xi_y^{\,2})\,(\xi_y\eta_x - \xi_x\eta_y)^2 \neq 0$. Thus, the partial differential equation of Eq. 7.5.9 reduces to

$$u_{\eta\eta} + \text{lower-order terms} = 0 \tag{7.5.30}$$

which is the **canonical form of a parabolic equation.**

The prototype parabolic equation is the **heat equation,** $k^2u_{xx} - u_t = 0$.

Elliptic Equations

Finally, consider the **elliptic partial differential equation**; i.e., Eq. 7.5.9 with

$$B^2 - AC < 0 \tag{7.5.31}$$

In this case, the roots r_1 and r_2 of Eq. 7.5.17 are **complex conjugates** of each other. The first-order differential equation of Eq. 7.5.21, for $k = 1$, becomes

$$\frac{dy}{dx} = -\,r_1(x, y) = \frac{B}{A} - i\,\frac{\sqrt{AC - B^2}}{A} \tag{7.5.32}$$

Until now, x and y have been regarded as real variables. It is clear that the derivative of one real variable with respect to another cannot be a complex number, so the easiest way to proceed is to allow x and y to take complex values and to assume that the coefficients A, B, and C are analytic functions. Then the right side of Eq. 7.5.32 is also an analytic function. The theory of ordinary differential equations [20] guarantees an analytic solution of the form

$$y = H(x, y_0) \tag{7.5.33}$$

where y_0 is an arbitrary value of y at an initial point x_0, which may be chosen as real. By solving Eq. 7.5.33 for y_0, an equation of the form

$$Q(x, y) = y_0 \tag{7.5.34}$$

is obtained, where Q is a complex-valued function of two complex arguments.

Differentiating Eq. 7.5.34 gives

$$Q_x \, dx + Q_y \, dy = 0$$

or

$$\frac{dy}{dx} = -\frac{Q_x}{Q_y} \tag{7.5.35}$$

which, by construction, must be identical to Eq. 7.5.32; i.e.,

$$-\frac{Q_x}{Q_y} = \frac{B}{A} - i\frac{\sqrt{AC - B^2}}{A}$$

Since the complex conjugate of a quotient equals the quotient of the conjugates,

$$-\frac{\bar{Q}_x}{\bar{Q}_y} = \frac{B}{A} + i\frac{\sqrt{AC - B^2}}{A}$$

where over bars denote complex conjugates. Thus, if $Q(x, y) = c_1$ is the general solution of

$$\frac{dy}{dx} = -r_1(x, y)$$

then $\bar{Q}(x, y) = c_2$ is the general solution of

$$\frac{dy}{dx} = -r_2(x, y)$$

Recall that in the hyperbolic case, the transformation

$$\xi = R_1(x, y)$$

$$\eta = R_2(x, y)$$

of Eq. 7.5.22 was introduced, where $R_k(x, y) = c_k$ is the general solution of

$$\frac{dy}{dx} = -r_k(x, y), \quad k = 1, 2 \tag{7.5.36}$$

This caused no problem, because in the hyperbolic case R_1 and R_2 are real and distinct. However, in the elliptic case, R_1 and R_2 are complex conjugates of each other. Therefore, new independent variables must be chosen differently, so that they will be real. Specifically, new variables α and β are introduced by the transformation

$$\alpha = \frac{Q(x, y) + \bar{Q}(x, y)}{2}$$
$$\beta = \frac{Q(x, y) - \bar{Q}(x, y)}{2i} \tag{7.5.37}$$

That is, α is the real and β is the imaginary part of Q. Since

$$\xi = Q(x, y) = \alpha(x, y) + i\beta(x, y) \tag{7.5.38}$$

is a solution of Eq. 7.5.16, derivatives of ξ may be expressed in terms of α and β and substituted into Eq. 7.5.16, to obtain

$$\begin{aligned} &A\xi_x^2 + 2B\xi_x\xi_y + C\xi_y^2 \\ &\quad = (A\alpha_x^2 + 2B\alpha_x\alpha_y + C\alpha_y^2) - (A\beta_x^2 + 2B\beta_x\beta_y + C\beta_y^2) \\ &\quad\quad + 2i [A\alpha_x\beta_x + B (\alpha_x\beta_y + \alpha_y\beta_x) + C\alpha_y\beta_y] = 0 \end{aligned}$$

This implies that

$$\begin{aligned} \hat{A}(\alpha, \beta) &\equiv A\alpha_x^2 + 2B\alpha_x\alpha_y + C\alpha_y^2 \\ &= A\beta_x^2 + 2B\beta_x\beta_y + C\beta_y^2 \equiv \hat{C}(\alpha, \beta) \\ \hat{B}(\alpha, \beta) &\equiv A\alpha_x\beta_x + B (\alpha_x\beta_y + \alpha_y\beta_x) + C\alpha_y\beta_y = 0 \end{aligned} \tag{7.5.39}$$

In the independent variables α and β, the coefficients $\hat{A}$ of $u_{\alpha\alpha}$ and $\hat{C}$ of $u_{\beta\beta}$ are equal and they cannot be zero, since, from Eq. 7.5.39,

$$(\hat{B}^2 - \hat{A}\hat{C}) = (B^2 - AC) (\alpha_x\beta_y - \alpha_y\beta_x)^2 \neq 0 \tag{7.5.40}$$

For the elliptic case being considered, it can be concluded that $\hat{A} = \hat{C} \neq 0$. Thus, dividing through Eq. 7.5.12 by $\hat{A}$, the desired **canonical form of an elliptic equation** is

$$u_{\alpha\alpha} + u_{\beta\beta} + \text{lower-order terms} = 0 \tag{7.5.41}$$

In discussion of the Cauchy–Kowalewski theorem, it was pointed out that the assumption of analyticity is, in general, too stringent for applications. Yet, in reducing Eq. 7.5.9 to canonical form in the elliptic case, it was assumed that the coefficients A(x, y), B(x, y), and C(x, y) are analytic. It is natural to ask whether such an assumption is always necessary in the elliptic case. The answer is no, but it would take a substantial development to show how this assumption can be relaxed. For a discussion of this point, see Ref. 19, pp. 66 – 69.

It should not be surprising that there is a relation between analytic functions and elliptic equations, since the prototype of elliptic equations is the **Laplace equation**

$$u_{xx} + u_{yy} = 0 \tag{7.5.42}$$

It is known from complex variable theory that this equation is satisfied by the real and imaginary parts of every analytic function of a complex variable. Conversely, every solution of Eq. 7.5.42 is the real part of some analytic function of a complex variable [19].

It should be noted that the mathematical classification developed in this section results in exactly the three standard equations that arose in the physical classification of the equations of mechanics in Section 6.6. This suggests that these three basic equations, and their theoretical properties, correspond to the three types of physical applications studied briefly in Chapter 6.

7.6 CHARACTERISTIC VARIABLES

There is a second reason why solutions of elliptic equations with regular coefficients tend to be regular functions. As has been seen in the theory of linear second-order partial differential equations (i.e., Eq. 7.5.1), the auxiliary equation of Eq. 7.5.16,

$$A(x, y)\,\xi_x^2 + 2B(x, y)\,\xi_x\xi_y + C(x, y)\,\xi_y^2 = 0 \tag{7.6.1}$$

plays a key role. If $\xi(x, y)$ is a solution of this first-order differential equation, then Eq. 7.5.21 shows that the one-parameter family of curves

$$\xi(x, y) = c \tag{7.6.2}$$

are **characteristic curves** for the partial differential equation of Eq. 7.5.1. Either existence or uniqueness of the solution may be lost if initial conditions are prescribed along such curves.

From Section 7.1, Eq. 7.6.2 is equivalent to

$$\xi_x\,dx + \xi_y\,dy = 0$$

or

$$\frac{\xi_x}{\xi_y} = -\frac{dy}{dx} \tag{7.6.3}$$

Equation 7.6.1 may be divided by $\xi_y{}^2$ and, using Eq. 7.6.3, written as

$$A\left(\frac{dy}{dx}\right)^2 - 2B\,\frac{dy}{dx} + C = 0$$

or

$$\frac{dy}{dx} = \frac{B \pm \sqrt{B^2 - AC}}{A} = -r_k \tag{7.6.4}$$

This is the ordinary differential equation for a characteristic curve of Eq. 7.5.21, which was found earlier by a different method. Thus, a hyperbolic equation has two distinct families of real characteristic curves, a parabolic equation has one such family, and an elliptic equation has none.

Example 7.6.1

Consider the equation

$$3u_{xx} + 10u_{xy} + 3u_{yy} = 0 \tag{7.6.5}$$

where $A = 3$, $B = 5$, and $C = 3$. Thus, $B^2 - AC = 25 - 9 = 16 > 0$, so the equation is hyperbolic. The transformation that reduces the equation to canonical form is found by solving the first-order equation

$$\hat{A} = 3\xi_x^2 + 10\xi_x\xi_y + 3\xi_y^2 = (3\xi_x + \xi_y)(\xi_x + 3\xi_y) = 0$$

whose solutions are

$$\xi = x - 3y$$

$$\eta = 3x - y$$

Note that the characteristic curves are families of straight lines,

$$x - 3y = c_1$$

$$3x - y = c_2$$

Calculating derivatives of u with respect to ξ and η,

$$\xi_x = 1, \quad \xi_y = -3, \quad \eta_x = 3, \quad \eta_y = -1$$

$$u_x = \xi_x u_\xi + \eta_x u_\eta = u_\xi + 3u_\eta$$

$$u_y = \xi_y u_\xi + \eta_y u_\eta = -3u_\xi - u_\eta$$

$$u_{xx} = \xi_x^2 u_{\xi\xi} + 2\xi_x\eta_x u_{\xi\eta} + \eta_x^2 u_{\eta\eta} = u_{\xi\xi} + 6u_{\xi\eta} + 9u_{\eta\eta}$$

$$u_{xy} = \xi_x\xi_y u_{\xi\xi} + (\xi_x\eta_y + \xi_y\eta_x)u_{\xi\eta} + \eta_x\eta_y u_{\eta\eta}$$

$$= -3u_{\xi\xi} - 10u_{\xi\eta} - 3u_{\eta\eta}$$

$$u_{yy} = \xi_y^2 u_{\xi\xi} + 2\xi_y\eta_y u_{\xi\eta} + \eta_y^2 u_{\eta\eta} = 9u_{\xi\xi} + 6u_{\xi\eta} + u_{\eta\eta}$$

Hence, Eq. 7.6.5 reduces to

$$3u_{xx} + 10u_{xy} + 3u_{yy} = -64u_{\xi\eta} = 0$$

or simply to

$$u_{\xi\eta} = 0$$

This equation may be solved, since it implies that u_ξ is a function of ξ alone; i.e.,

$$u_\xi = f'(\xi)$$

where $f'(\xi)$ is the derivative of some function $f(\xi)$. Integrating this,

$$u = f(\xi) + g(\eta)$$

where the functions $f(\xi)$ and $g(\eta)$ are arbitrary. Finally, in terms of the original variables,

$$u = f(x - 3y) + g(3x - y)$$

is the general solution of Eq. 7.6.5. ■

Note that characteristic curves are straight lines that are parallel to the ξ and η axes in Example 7.6.1. For this reason, when the hyperbolic equation is reduced to the canonical form

$$u_{\xi\eta} + \text{lower-order terms} = 0$$

the variables ξ and η are called **characteristic variables**. In this form, arbitrary initial conditions may not be prescribed along any horizontal or vertical line in the ξ-η plane, without destroying either existence or uniqueness of the solution.

Example 7.6.2

Consider the equation

$$yu_{xx} + u_{yy} = 0 \tag{7.6.6}$$

which is called the **Tricomi equation**. Note that, since $B^2 - AC = -y$, it is

(a) hyperbolic in the lower half-plane; i.e., $y < 0$
(b) elliptic in the upper half-plane; i.e., $y > 0$

To put Eq. 7.6.6 in canonical form in the upper half-plane, where $A = y$, $B = 0$, $C = 1$, solutions of the equation

$$\hat{A} = A\xi_x^2 + 2B\xi_x\xi_y + C\xi_y^2 = y\xi_x^2 + \xi_y^2 = 0$$

must be obtained by solving the ordinary differential equation of Eq. 7.5.32, which reduces here to

$$\frac{dy}{dx} = -i\frac{\sqrt{y}}{y} = \frac{-i}{\sqrt{y}}$$

or

$$\sqrt{y}\,dy = -i\,dx$$

Thus,

$$\frac{2}{3}y^{3/2} = -ix + c$$

or

$$c = (2/3)\,y^{3/2} + ix$$

In order to avoid complex quantities, new variables α and β are chosen as

$$\alpha = (2/3)\,y^{3/2}$$

$$\beta = x$$

Then

$$\alpha_x = 0, \quad \alpha_y = y^{1/2}$$

$$\beta_x = 1, \quad \beta_y = 0$$

$$u_x = \alpha_x u_\alpha + \beta_x u_\beta = u_\beta$$

$$u_y = \alpha_y u_\alpha + \beta_y u_\beta = y^{1/2}u_\alpha$$

$$u_{xx} = \alpha_x u_{\alpha\beta} + \beta_x u_{\beta\beta} = u_{\beta\beta}$$

$$u_{yy} = y^{1/2}(\alpha_y u_{\alpha\alpha} + \beta_y u_{\alpha\beta}) + (1/2)\,y^{-1/2}u_\alpha$$

$$= yu_{\alpha\alpha} + (1/2)\,y^{-1/2}u_\alpha$$

Thus,

$$yu_{xx} + u_{yy} = yu_{\beta\beta} + yu_{\alpha\alpha} + \frac{u_\alpha}{2y^{1/2}} = 0$$

$$u_{\alpha\alpha} + u_{\beta\beta} + \frac{u_\alpha}{2y^{3/2}} = 0$$

or

$$u_{\alpha\alpha} + u_{\beta\beta} + \frac{u_\alpha}{3\alpha} = 0$$

which is the canonical form of the Tricomi equation of Eq. 7.6.5, for $y > 0$. ■

EXERCISES 7.6

1. Reduce each of the following equations to canonical form:

   (a) $u_{xx} + 2u_{xy} + u_{yy} + u_x - u_y = 0$

   (b) $u_{xx} + 2u_{xy} + 5u_{yy} + 3u_x + u = 0$

   (c) $3u_{xx} + 14u_{xy} + 8u_{yy} = 0$

   (d) $3u_{xx} + 8u_{xy} - 3u_{yy} = 0$

   (e) $u_{xx} - x^2 u_{yy} = 0$

   (f) $u_{xx} + x^2 u_{yy} = 0$

2. Find the characteristic curves for Exercises 1(c) and 1(d) and construct general solutions of the differential equations.

3. The differential equation for steady isentropic irrotational flow, from Eq. 6.5.26, is

$$\left(1 - \frac{u^2}{c^2}\right)\phi_{xx} - \left(\frac{2uv}{c^2}\right)\phi_{xy} + \left(1 - \frac{v^2}{c^2}\right)\phi_{yy} = 0$$

   where $c^2 = \frac{dp}{d\rho} = A\gamma\rho^{\gamma-1}$. Regarding this as a second-order equation for the unknown ϕ, with variable coefficients,

   (a) Under what conditions can characteristic curves occur?

   (b) What physical meaning can be associated with the parameter c?

4. Determine the equation type and find the characteristic curves, if any exist, for the equation

$$u_{xx} + (y^2 - 1)\,u_{yy} = 0$$

5. Derive the solution of the initial-value problem for $3u_{xx} + 10u_{xt} + 3u_{tt} = 0$, on $-\infty < x < \infty$, $t \geq 0$, with

$$u(x, 0) = f(x)$$

$$u_t(x, 0) = g(x)$$

7.7 LINEAR EQUATIONS IN MORE THAN TWO INDEPENDENT VARIABLES

Until now, only second-order partial differential equations in two independent variables have been considered. This chapter is concluded with a brief introduction to the subject of second-order partial differential equations in more than two independent variables. A reexamination of the prototype physical equations derived in Chapter 6 is relied upon to extend the classification system to linear equations with more than two independent variables.

Canonical Form of Linear Equations

A general second-order partial differential equation in n independent variables $x_1, x_2, \ldots, x_n$ is

$$\sum_{i,j=1}^{n} \frac{\partial}{\partial x_i}\left(a_{ij} \frac{\partial u}{\partial x_j} \right) + \sum_{i=1}^{n} b_i \frac{\partial u}{\partial x_i} + cu = f \tag{7.7.1}$$

where each a_{ij}, b_i, c, and f is a function of $x_1, x_2, \ldots, x_n$ and $a_{ij} = a_{ji}$, for all i and j.

As in the procedure followed with two independent variables, new variables are introduced by the transformation

$$\begin{aligned} z_1 &= z_1(x_1, x_2, \ldots, x_n) \\ z_2 &= z_2(x_1, x_2, \ldots, x_n) \\ &\cdots\cdots\cdots\cdots\cdots \\ z_n &= z_n(x_1, x_2, \ldots, x_n) \end{aligned} \tag{7.7.2}$$

which is to be chosen in such a way that (1) no mixed partial derivatives $\frac{\partial^2 u}{\partial z_i \partial z_k}$ appear in the transformed differential equation and (2) the coefficient of every second partial derivative $\frac{\partial^2 u}{\partial z_i^2}$ in the transformed differential equation is either +1, −1, or 0.

Unfortunately, it is not always possible to do this for n > 2, because the number of conditions that the z_i must satisfy is too great. It can be done only if the a_{ij} in Eq. 7.7.1 are all constants. If they are not all constants, the best that can be done is to consider the nature of the equation in the neighborhood of a fixed point $\mathbf{x} = [\, x_1, x_2, \ldots, x_n \,]^T = [\, x_1^0, x_2^0, \ldots, x_n^0 \,]^T \equiv \mathbf{x}^0$, where each x_i^0 is a constant. This means that in each function $a_{ij}(x_1, x_2, \ldots, x_n)$, each x_m is replaced by the corresponding constant x_m^0. Thus, each a_{ij} is treated as a constant, for the purpose of determining to which canonical form the differential equation reduces in a neighborhood of the point $\mathbf{x}^0$.

If the a_{ij} are all constants, it is possible to find functions z_i in Eq. 7.7.2 so that the transformed differential equation satisfies conditions (1) and (2) above. Motivated by diagonalization of quadratic forms in Section 3.2 and the occurrence of the quadratic form of Eq. 7.5.4 in Section 7.5 by the **modal transformation**, let

$$\mathbf{x} = \mathbf{Q}\mathbf{y} \tag{7.7.3}$$

where $\mathbf{Q} = [\, \mathbf{x}^1, \mathbf{x}^2, \ldots, \mathbf{x}^n \,]$ is made up of orthonormalized eigenvectors of the symmetric matrix $\mathbf{A} = [\, a_{ij} \,]$,

$$\mathbf{D} = \begin{bmatrix} \lambda_1 & & & & \mathbf{0} \\ & \lambda_2 & & & \\ & & \cdot & & \\ & & & \cdot & \\ & & & & \cdot \\ \mathbf{0} & & & & \lambda_n \end{bmatrix} \tag{7.7.4}$$

and λ_i is the eigenvalue of $\mathbf{A}$ corresponding to the eigenvector $\mathbf{x}^i$.

If the eigenvalues λ_i of $\mathbf{A}$ are ordered so that $\lambda_1 < 0, \ldots, \lambda_p < 0, \lambda_{p+1} > 0, \ldots, \lambda_q > 0$, and $\lambda_{q+1} = 0, \ldots, \lambda_n = 0$, then the matrix

$$\mathbf{V} \equiv \begin{bmatrix} \sqrt{-\lambda_1} & & & & & & & \mathbf{0} \\ & \ddots & & & & & & \\ & & \sqrt{-\lambda_p} & & & & & \\ & & & \sqrt{\lambda_{p+1}} & & & & \\ & & & & \ddots & & & \\ & & & & & \sqrt{\lambda_q} & & \\ & & & & & & 1 & \\ & \mathbf{0} & & & & & & \ddots \\ & & & & & & & & 1 \end{bmatrix} \tag{7.7.5}$$

can be defined. Next, define a new transformation $\mathbf{y} = \mathbf{V}\mathbf{z}$, or

$$\mathbf{x} = \mathbf{Q}\mathbf{V}\mathbf{z} \tag{7.7.6}$$

By the chain rule of differentiation,

$$\frac{\partial}{\partial x_i} = \frac{\partial y_j}{\partial x_i}\frac{\partial}{\partial y_j}, \quad i = 1, \ldots, n$$

where summation notation is used. Since the matrix $\mathbf{Q}$ is orthogonal,

$$\mathbf{y} = \mathbf{Q}^T\mathbf{x}$$

and

$$\frac{\partial}{\partial x_i} = Q_{ji}^T \frac{\partial}{\partial y_j} = Q_{ij} \frac{\partial}{\partial y_j}$$

The second-order term in Eq. 7.7.1 may now be written as

$$\frac{\partial}{\partial y_j} Q_{ij} \left(a_{il} Q_{lk} \frac{\partial u}{\partial y_k} \right) = \frac{\partial}{\partial y_j} \left\{ (Q_{ij} a_{il} Q_{lk}) \frac{\partial u}{\partial y_k} \right\}$$

$$= \sum_{j=1}^{n} \frac{\partial}{\partial y_j} \left(\lambda_j \frac{\partial u}{\partial y_j} \right) = \lambda_j \frac{\partial^2 u}{\partial y_j^2} \qquad (7.7.7)$$

where the summation sign on the left is retained, because the index j appears in three places. Since $\lambda_j = 0$, $j = q+1, \ldots, n$; $y_i = \sqrt{-\lambda_i}\, z_i$, $i = 1, \ldots, p$; and $y_i = \sqrt{\lambda_i}\, z_i$, $i = p+1, \ldots, q$, it follows from Eq. 7.7.7 that Eq. 7.7.1 reduces to the **canonical form**

$$\sum_{i=1}^{n} \alpha_i \frac{\partial^2 u}{\partial z_i^2} \equiv -\sum_{i=1}^{p} \frac{\partial^2 u}{\partial z_i^2} + \sum_{i=p+1}^{q} \frac{\partial^2 u}{\partial z_i^2}$$

$$= f(\mathbf{QVz}) + \text{lower-order terms} \qquad (7.7.8)$$

Classification of Linear Equations

Consider now the possible results of the foregoing procedure. Equation 7.7.1, with a_{ij} constant, reduces to one of the following types:

(1) **Elliptic Equation.** If all the α_i in Eq. 7.7.8 are +1, the differential equation becomes

$$\frac{\partial^2 u}{\partial z_1^2} + \frac{\partial^2 u}{\partial z_2^2} + \ldots + \frac{\partial^2 u}{\partial z_n^2} + \text{lower-order terms} = f \qquad (7.7.9)$$

The classical example of such an equation is the **Laplace equation** $\nabla^2 u = 0$, which, in three independent variables, is

$$u_{xx} + u_{yy} + u_{zz} = 0 \qquad (7.7.10)$$

Note that Eq. 7.7.1 is elliptic if and only if its matrix $\mathbf{A} = [\, a_{ij} \,]$ of coefficients is positive definite.

(2) **Hyperbolic Equation.** If all $\alpha_i \neq 0$, but at least one is -1 and at least one is $+1$, the equation is hyperbolic. Equations of hyperbolic type are further subclassified as follows:

(a) **Normal Hyperbolic.** If exactly one α_i, say α_1, is -1 and all the rest are $+1$, the differential equation becomes

$$-\frac{\partial^2 u}{\partial z_1^2} + \frac{\partial^2 u}{\partial z_2^2} + \ldots + \frac{\partial^2 u}{\partial z_n^2} + \text{lower-order terms} = f \qquad (7.7.11)$$

An example would be the equation of a vibrating membrane of Chapter 6, which may be reduced to

$$u_{xx} + u_{yy} - u_{tt} = 0$$

(b) **Ultrahyperbolic.** If at least two α_i are -1 and at least two are $+1$, the equation is called ultrahyperbolic. This type does not arise in the applications studied in Chapter 6.

(3) **Parabolic Equation.** If at least one α_i is 0, say $\alpha_i = 0$ for $r < i \leq n$, then the differential equation becomes

$$\sum_{i=1}^{r} \alpha_i \frac{\partial^2 u}{\partial z_i^2} + \text{lower-order terms} = f \qquad (7.7.12)$$

where each α_i, $1 \leq i \leq r$, is either $+1$ or -1. An example of this type is the equation for conduction of heat in solids encountered in Chapter 6,

$$u_{xx} + u_{yy} + u_{zz} - ku_t = 0$$

Referring to Table 6.6.1 of Chapter 6, note that the physical situations listed in the first column are modeled by normal hyperbolic equations, those in the second column by parabolic equations, and those in the third column by elliptic equations.

EXERCISES 7.7

1. Show that the transformation $\mathbf{x} = \mathbf{QVz}$ in Eq. 7.7.6 is nonsingular.

2. For a_{ij} constant in Eq. 7.7.1, show that the change in variable of Eq. 7.7.2 transforms Eq. 7.7.1 to

$$\sum_{k,l=1}^{n} b_{kl} \frac{\partial^2 u}{\partial z_k \partial z_l} + \text{lower-order terms} = f$$

where

$$b_{kl} = \sum_{i,j=1}^{n} a_{ij} \frac{\partial z_k}{\partial x_i} \frac{\partial z_l}{\partial x_j}$$

3. A **characteristic surface** in R^n for the differential equation of Eq. 7.7.1 is defined by the equation $z(\mathbf{x}) = 0$, where $z(\mathbf{x})$ satisfies the first-order equation

$$\sum_{i,j=1}^{n} a_{ij} \frac{\partial z}{\partial x_i} \frac{\partial z}{\partial x_j} = 0$$

(a) Show that if the equation is elliptic, then there are no characteristic surfaces. [Hint: Use the result of Exercise 2.]

(b) Show that if a differential equation is in the canonical form of Eq. 7.7.8, then on a characteristic surface, the differential equation fails to determine one of the second derivatives of u. [Hint: Use the result of Exercise 2.]

8

METHODS OF SOLVING SECOND-ORDER PARTIAL DIFFERENTIAL EQUATIONS

It was found in Sections 6.5 and 7.5, from both physical and mathematical points of view, that linear second-order partial differential equations reduce naturally to one of three distinct types of equations; wave, heat, and Laplace. Before delving into the theory of these three standard classes of problems in Sections 8.4 through 8.6, it is instructive to briefly study prototype problems in these three classes. This is done in Sections 8.1 through 8.3.

8.1 THE WAVE EQUATION

In Section 6.3, it was found that the equation that governs transverse displacement of a vibrating string, with no applied load, is the **homogeneous wave equation**,

$$u_{tt} - a^2 u_{xx} = 0 \tag{8.1.1}$$

where a is constant. This equation is hyperbolic and the two families of characteristic curves are $x + at = c_1$ and $x - at = c_2$.

General Solution of the Wave Equation

Introducing the characteristic variables

$$\begin{aligned} \xi &= x + at \\ \eta &= x - at \end{aligned} \tag{8.1.2}$$

Eq. 8.1.1 reduces to the canonical form

$$u_{\xi\eta} = \frac{\partial}{\partial\eta}(u_\xi) = 0$$

This equation says that u_ξ is a function of ξ alone; i.e.,

$$u_\xi = \phi'(\xi)$$

where $\phi'(\bullet)$ is the derivative of a function $\phi(\bullet)$. Integrating with respect to ξ,

$$u = \phi(\xi) + \psi(\eta) \tag{8.1.3}$$

where ϕ and ψ are arbitrary functions in $D^1 \cap C^0$. Substituting for ξ and η in terms of x and t, from Eq. 8.1.2, yields

$$u(x, t) = \phi(x + at) + \psi(x - at) \tag{8.1.4}$$

which is the **general solution of the wave equation** of Eq. 8.1.1.

To obtain a physically meaningful solution for a string of length 2ℓ, initial and boundary conditions must be specified. Initial conditions have the form

$$\begin{aligned} u(x, 0) &= f(x) \\ u_t(x, 0) &= g(x) \end{aligned} \tag{8.1.5}$$

for $|x| < \ell$, and boundary conditions for a string with fixed ends are

$$\begin{aligned} u(-\ell, t) &= 0 \\ u(\ell, t) &= 0 \end{aligned} \tag{8.1.6}$$

for $t \geq 0$.

To be specific, consider the **plucked string problem** in which the string is initially displaced at its midpoint and is released with zero initial velocity. The initial conditions are thus

$$\begin{aligned} u(x, 0) &= m(\ell - |x|) \\ u_t(x, 0) &= 0 \end{aligned} \tag{8.1.7}$$

for $|x| < \ell$, where m is a positive number that is much less than 1. The initial shape of the string is shown in Fig. 8.1.1. Notice that there is a discontinuity in u_x at the point (0, 0), so u_{xx} is not defined there and u_{tt} must therefore be undefined in Eq. 8.1.1. A solution of Eq. 8.1.1 is now sought that satisfies the boundary conditions of Eq. 8.1.6 and the initial conditions of Eq. 8.1.7.

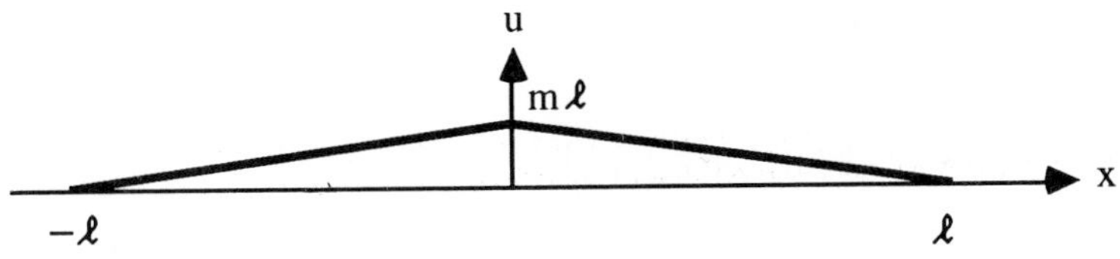

Figure 8.1.1 Initial Condition of String

Applying the first initial condition of Eq. 8.1.7 to Eq. 8.1.4 yields

$$\phi(x) + \psi(x) = m(\ell - |x|) \tag{8.1.8}$$

which shows that either ϕ' or ψ', or both, must be discontinuous at $x = 0$. Applying the second initial condition of Eq. 8.1.7,

$$a\phi'(x) - a\psi'(x) = 0 \tag{8.1.9}$$

which shows that ϕ and ψ differ by a constant. However, because of discontinuities in the derivatives at $x = 0$, their difference may have one constant value for $x < 0$ and a different constant value for $x > 0$; i.e.,

$$\begin{aligned} \psi(x) &= \phi(x) + c_1, \quad \text{for } x < 0 \\ \psi(x) &= \phi(x) + c_2, \quad \text{for } x > 0 \end{aligned} \tag{8.1.10}$$

The two cases defined by Eq. 8.1.10 may be studied separately. For $x < 0$, $|x| = -x$. Substituting this expression and the first of Eq. 8.1.10 into Eq. 8.1.8,

$$2\phi(x) + c_1 = m(\ell + x) \tag{8.1.11}$$

for $x < 0$. For $x > 0$, $|x| = x$ and

$$2\phi(x) + c_2 = m(\ell - x) \tag{8.1.12}$$

Since u is continuous, ϕ and ψ must also be continuous. Hence, letting x approach 0 in Eqs. 8.1.11 and 8.1.12,

$$2\phi(0-) + c_1 = 2\phi(0) + c_1 = m\ell$$

$$2\phi(0+) + c_2 = 2\phi(0) + c_2 = m\ell$$

which shows that $c_1 = c_2 = m\ell - 2\phi(0)$. Substituting from Eq. 8.1.10 into Eq. 8.1.8,

$$\phi(x) = \phi(0) - \frac{m|x|}{2} \tag{8.1.13}$$

Finally, substituting from Eq. 8.1.13 into Eq. 8.1.10,

$$\psi(x) = m\ell - \frac{m|x|}{2} - \phi(0) \tag{8.1.14}$$

Propagation of Waves

Note that Eqs. 8.1.13 and 8.1.14 hold for $-\ell < x < \ell$, so they define the functions ϕ and ψ in this range. Substituting the arguments x + at and x − at into Eqs. 8.1.13 and 8.1.14, the solution of Eq. 8.1.4 is obtained as

$$u(x, t) = m\ell - \frac{m|x + at|}{2} - \frac{m|x - at|}{2} \tag{8.1.15}$$

This shows that u_x and u_t are discontinuous along the characteristic curves $x + at = 0$ and $x - at = 0$ and are continuous everywhere else. In other words, the original discontinuity at (0, 0) is propagated along characteristic curves that pass through that point. This may be interpreted as **propagation of a wave** along the string, hence the name "wave equation."

It should be noted that the solution given by Eq. 8.1.15 is valid only in the rectangle $-\ell \le x \le \ell$, $0 \le t \le \ell/a$ of the x-t plane. The limits on x are obvious, since there is no physical meaning for $|x| > \ell$. The upper limit on t follows from the boundary conditions $u(\pm\ell, t) = 0$, since these conditions are not satisfied by the function u(x, t) of Eq. 8.1.15 if $t > \ell/a$.

Knowing the lines along which derivatives of the function in Eq. 8.1.15 are not continuous, the rectangle $|x| \le \ell$, $0 \le t \le \ell/a$ may be divided into regions within which u is smooth, and distinct expressions for u may be written in each such region. There are three such regions, as shown in Fig. 8.1.2.

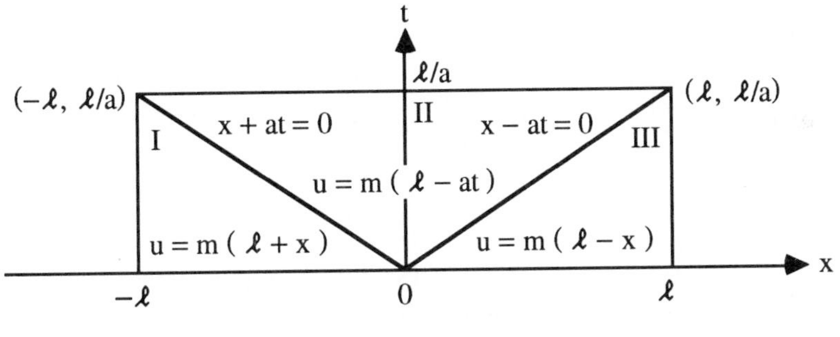

Figure 8.1.2 Wave Propagation in a String

The three regions are defined as follows: (I) $x + at < 0$, $x - at < 0$, (II) $x + at > 0$, $x - at < 0$, and (III) $x + at > 0$, $x - at > 0$. Equation 8.1.15 reduces to the following analytic expressions in the respective regions: (I) $u(x, t) = m(\ell + x)$, which does not depend on t; (II) $u(x, t) = m(\ell - at)$, which does not depend on x; and (III) $u(x, t) = m(\ell - x)$, which does not depend on t. Each of these expressions satisfies the differential equation of Eq. 8.1.1 and the appropriate boundary conditions. They also show the nature of discontinuities in first derivatives of u along characteristic curves. For example, u_t equals zero in regions I and III, but jumps to the value $-ma$, which is a nonzero constant, in region II.

Referring to Fig. 8.1.2, the physical significance of the constant a in the wave equation of Eq. 8.1.1 becomes clear. Consider the characteristic curve $x - at = 0$. At any time t, the **wave front**; i.e., points at which derivatives of u are discontinuous, is a distance $x = at$ from the center of the string. Hence, the velocity at which this wave front travels to the right along the string is $dx/dt = a$. Similarly, the velocity at which the wave front travels to the left on the characteristic curve $x + at = 0$ is $-a$. The parameter a in Eq. 8.1.1 is thus the **wave speed**.

To determine the behavior of the vibrating string for $t > \ell/a$, note that when $t = \ell/a$, $u = 0$ and $u_t = -ma$ for $-\ell < x < \ell$. Thus, for $t > \ell/a$, the wave equation is again solved, subject to the new initial conditions

$$\begin{aligned} u(x, \ell/a) &= 0 \\ u_t(x, \ell/a) &= -ma \end{aligned} \qquad (8.1.16)$$

for $|x| < \ell$, and boundary conditions

$$\begin{aligned} u(\ell, t) &= 0 \\ u(-\ell, t) &= 0 \end{aligned} \qquad (8.1.17)$$

for $t \geq \ell/a$.

Note that at $x = \ell$ and $t = \ell/a$, the second of Eq. 8.1.16 and the first of Eq. 8.1.17 yield different values of $u_t(\ell, \ell/a)$, the first giving the value $-ma$ and the second giving 0. Thus, derivatives of the solution must have different values for $x + at > 2\ell$ and $x + at < 2\ell$, so the line $x + at - 2\ell = 0$ plays a special role in the solution for $t > \ell/a$. Similarly, a discrepancy in values of u_t occurs from the second of Eq. 8.1.16 and the second of Eq. 8.1.17 at $x = -\ell$, $t = \ell/a$. This leads to the related conclusion that the line $x - at + 2\ell = 0$ plays a special role in the solution for $t > \ell/a$. This is not too surprising, since these are both characteristic curves that pass through points in the x-t plane where u_t has a discontinuity.

The form of solution in Eq. 8.1.15, as a sum of absolute values of expressions that define characteristic curves through the point of discontinuity of u_t, suggests considering a function of the form

$$u(x, t) = c + \frac{m}{2}\left\{ |x + at - 2\ell| + |x - at + 2\ell| \right\} \qquad (8.1.18)$$

To see if there is a constant c such that this is a solution, note first that except for points on characteristic curves, u is linear in x and t, so Eq. 8.1.1 is trivially satisfied.

Next consider the first initial condition of Eq. 8.1.16, with u from Eq. 8.1.18; i.e.,

$$u(x, \ell/a) = c + \frac{m}{2}\left\{ |x - \ell| + |x + \ell| \right\}$$

For $-\ell < x < \ell$, $|x + \ell| + |x - \ell| = -(x - \ell) + (x + \ell) = 2\ell$ and the first initial condition is satisfied if $c = -m\ell$. As long as $t - \ell/a > 0$ is small, the above sign arguments hold, so

$$u_t(x, \ell/a) = \frac{m}{2}\{-a - a\} = -ma$$

which satisfies the second initial condition of Eq. 8.1.16.

Finally, consider the first boundary condition of Eq. 8.1.17, with u from Eq. 8.1.18; i.e.,

$$u(\ell, t) = -m\ell + \frac{m}{2}\Big\{ |at - \ell| + |3\ell - at| \Big\}$$

As long as $\ell/a \le t \le 3\ell/a$, $at - \ell \ge 0$ and $3\ell - at \ge 0$. Thus,

$$u(\ell, t) = -m\ell + \frac{m}{2}\{at - \ell + 3\ell - at\} = 0$$

Similarly, the second boundary condition of Eq. 8.1.17 is satisfied in the same rectangle of the x-t plane.

The function suggested in Eq. 8.1.18, with $c = -m\ell$, is thus the desired extension of the solution of Eq. 8.1.15 over the time interval $\ell/a \le t \le 3\ell/a$. Whereas solutions of initial-boundary-value problems of partial differential equations are almost never found this easily, this example does bring out some general properties of solutions of hyperbolic equations.

The solution in this example shows that discontinuities in initial data are propagated along characteristic curves. They are waves that are **reflected** by the boundary and are propagated back along characteristic curves through the point $(\pm\ell, \ell/a)$. Since there is no damping, the motion continues indefinitely in this manner. A sequence of deflected shapes of the string is shown in Fig. 8.1.3.

D'Alembert's Formula

Consider again the vibrating string described by Eq. 8.1.1 and boundary conditions of Eq. 8.1.6, but with the more general initial conditions

$$\begin{aligned} u(x, 0) &= f(x) \\ u_t(x, 0) &= g(x) \end{aligned} \tag{8.1.19}$$

for $|x| < \ell$. Assume that the function f(x) is twice continuously differentiable and g(x) is continuously differentiable, which was not the case in the plucked string problem. Equation 8.1.4 is still valid, so applying the initial conditions of Eq. 8.1.19 yields

$$u(x, 0) = \phi(x) + \psi(x) = f(x) \tag{8.1.20}$$

$$u_t(x, 0) = a\phi'(x) - a\psi'(x) = g(x) \tag{8.1.21}$$

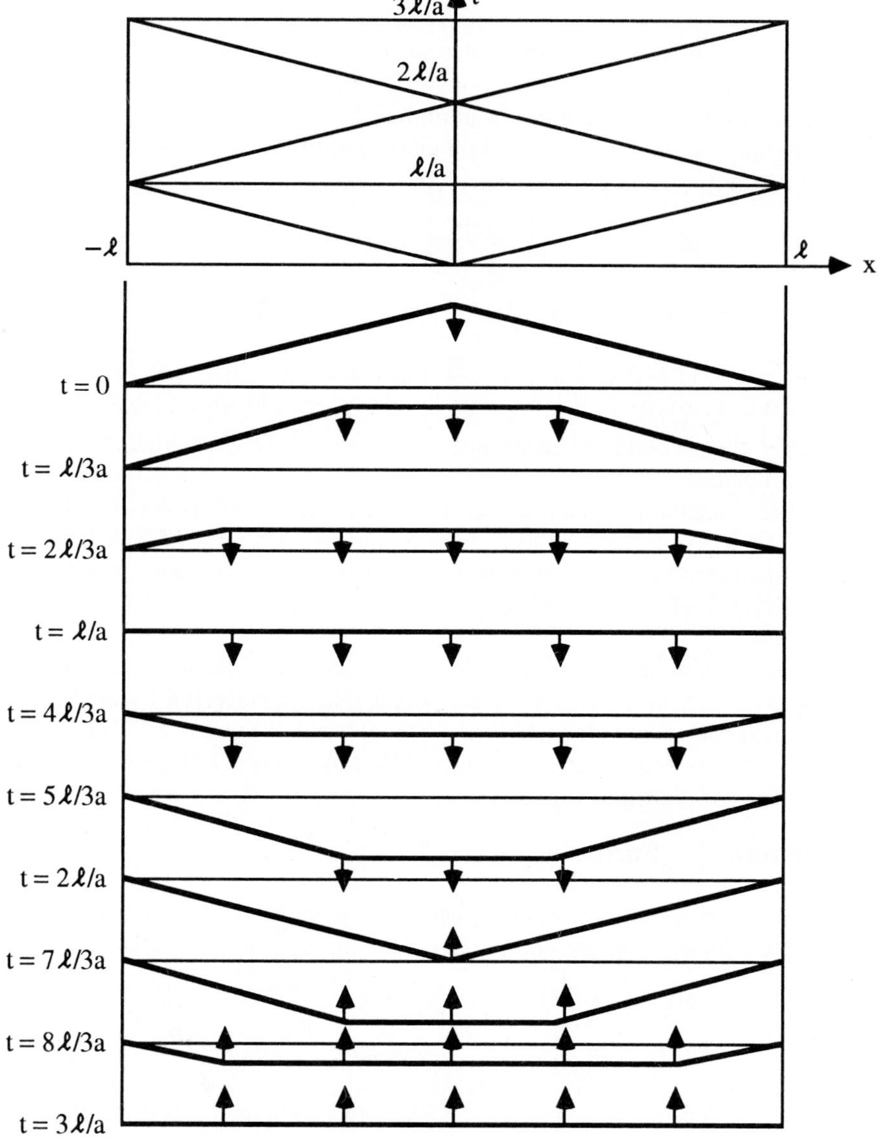

Figure 8.1.3 Vibration of a Plucked String

for $-\ell < x < \ell$. Differentiating Eq. 8.1.20 and multiplying by the constant a gives

$$a\phi'(x) + a\psi'(x) = af'(x) \tag{8.1.22}$$

Equations 8.1.21 and 8.1.22 may be solved for ϕ' and ψ' and integrated from 0 to x. Assuming f(x) of Eq. 8.1.20 is consistent with the boundary conditions; i.e., $f(\pm\ell) = 0$, this process leads to

$$\phi(x) = \frac{f(x)}{2} + \frac{1}{2a}\int_0^x g(\xi)\,d\xi + k_1$$

$$\psi(x) = \frac{f(x)}{2} - \frac{1}{2a}\int_0^x g(\xi)\,d\xi + k_2$$

Substituting these expressions into Eq. 8.1.20 shows that $k_1 + k_2 = 0$.

Substituting the arguments $x + at$ and $x - at$ into the above expressions for ϕ and ψ, Eq. 8.1.4 yields **D'Alembert's formula**

$$u(x, t) = \frac{f(x + at) + f(x - at)}{2} + \frac{1}{2a}\int_{x-at}^{x+at} g(\xi)\,d\xi \tag{8.1.23}$$

as the desired solution.

Uniqueness of Solution

To show that the solution of the initial-boundary-value problem treated here is unique, suppose that there are two solutions $u_1(x, t)$ and $u_2(x, t)$ and let

$$v(x, t) = u_1(x, t) - u_2(x, t)$$

Thus, v(x, t) is a solution of the problem

$$\begin{aligned}
v_{tt} &= a^2 v_{xx}, && -\ell < x < \ell, \quad t > 0 \\
v(-\ell, t) &= 0, && t \geq 0 \\
v(\ell, t) &= 0, && t \geq 0 \\
v(x, 0) &= 0, && -\ell \leq x \leq \ell \\
v_t(x, 0) &= 0, && -\ell \leq x \leq \ell
\end{aligned} \tag{8.1.24}$$

If it can be shown that the function v(x, t) is identically zero, then the solution is unique. To do so, define the function

$$J(t) = \frac{1}{2}\int_{-\ell}^{\ell} (a^2 v_x^2 + v_t^2)\,dx \tag{8.1.25}$$

which physically represents the **total energy** of the vibrating string at time t.

Since the function v(x, t) is twice continuously differentiable, differentiate J(t) with respect to t, to obtain

$$J'(t) = \int_{-\ell}^{\ell} (a^2 v_x v_{xt} + v_t v_{tt}) \, dx \tag{8.1.26}$$

Integrating by parts,

$$\int_{-\ell}^{\ell} a^2 v_x v_{xt} \, dx = a^2 v_x v_t \Big|_{-\ell}^{\ell} - \int_{-\ell}^{\ell} a^2 v_t v_{xx} \, dx$$

But from the conditions $v(-\ell, t) = 0$ and $v(\ell, t) = 0$, $v_t(-\ell, t) = 0$ and $v_t(\ell, t) = 0$. Hence, the boundary term vanishes and Eq. 8.1.26 becomes

$$J'(t) = \int_{-\ell}^{\ell} v_t (v_{tt} - a^2 v_{xx}) \, dx \tag{8.1.27}$$

Since $v_{tt} - a^2 v_{xx} = 0$, Eq. 8.1.27 reduces to

$$J'(t) = 0$$

which means that

$$J(t) = \text{constant} \equiv C$$

Since $v(x, 0) = 0$, $v_x(x, 0) = 0$. Taking into account the condition $v_t(x, 0) = 0$, evaluation of J(0) yields

$$J(0) = C = \frac{1}{2} \int_{-\ell}^{\ell} [a^2 v_x^2 + v_t^2] \Big|_{t=0} dx = 0$$

This implies that $J(t) = 0$, for all $t \geq 0$, which can happen only when $v_x(x, t) = v_t(x, t) = 0$, for $-\ell < x < \ell$ and $t > 0$. To satisfy both of these conditions, v(x, t) = constant must hold. Employing the condition $v(x, 0) = 0$, $v(x, t) \equiv 0$. Therefore $u_1(x, t) \equiv u_2(x, t)$ and the solution u(x, t) is unique.

Note that, even though continuous differentiability of the functions f and g was an assumption in deriving D'Alembert's formula, it may still yield the solution of a Cauchy problem for the wave equation when f and g are not everywhere differentiable. To see this, formally apply D'Alembert's formula to the initial conditions of Eq. 8.1.7 for the plucked string problem. The fact that the solution of Eq. 8.1.15 is obtained, however, does not rigorously justify the use of the formula.

More generally, note that in deriving the general solution of the wave equation in the form $u_{\xi\eta} = 0$, the functions ϕ and ψ need only be in $C^0 \cap D^1$ for the solution to make physical sense. Further, it is suggested that D'Alembert's formula is meaningful for functions f and g that are less regular than C^2 and C^1, respectively, which would be required to rigorously satisfy Eq. 8.1.1. It thus appears that in deriving the wave equation for the vibrating string in Section 6.3, the assumption that the solution is in C^2 is unnecessarily restrictive. Less regular solutions arose for a string with a point load in Section 6.3, when it was found that the derivative of displacement had a discontinuity at the point of applied load.

It is interesting to determine whether the momentum balance laws of Eq. 6.3.6, with the applied load set equal to zero, are satisfied by D'Alembert's formula with g in D^0 and f in $C^0 \cap D^1$.

From Eqs. 6.3.4 and 6.3.5,

$$\begin{aligned} A &= \int_{x_1}^{x_2} \rho \, [\, u_t(\xi, t_2) - u_t(\xi, t_1) \,] \, d\xi \\ B &= T \int_{t_1}^{t_2} [\, u_x(x_2, \tau) - u_x(x_1, \tau) \,] \, d\tau \end{aligned} \tag{8.1.28}$$

Substituting from D'Alembert's formula and noting that $a^2 = T/\rho$, after performing a number of integrations,

$$\begin{aligned} A = \frac{\sqrt{\tau\rho}}{2} &[\, f(x_2 + at_2) - f(x_1 + at_2) - f(x_2 - at_2) + f(x_1 - at_2) \\ &- f(x_2 + at_1) + f(x_1 + at_1) + f(x_2 - at_1) - f(x_1 - at_1) \,] \\ &+ \frac{\rho}{2} \left\{ \int_{x_1+at_2}^{x_2+at_2} g(\tau)\, d\tau - \int_{x_1-at_2}^{x_2-at_2} g(\tau)\, d\tau - \int_{x_1+at_1}^{x_2+at_1} g(\tau)\, d\tau \right. \\ &\left. + \int_{x_1-at_1}^{x_2-at_1} g(\tau)\, d\tau \right\} \end{aligned} \tag{8.1.29}$$

$$\begin{aligned} B = \frac{\sqrt{\tau\rho}}{2} &[\, f(x_2 + at_2) - f(x_2 + at_1) - f(x_1 + at_2) + f(x_1 + at_1) \\ &- f(x_2 - at_2) + f(x_2 - at_1) + f(x_1 - at_2) - f(x_1 - at_1) \,] \\ &+ \frac{\rho}{2} \left\{ \int_{x_2+at_1}^{x_2+at_2} g(\tau)\, d\tau - \int_{x_1+at_1}^{x_1+at_2} g(\tau)\, d\tau - \int_{x_2-at_1}^{x_2-at_2} g(\tau)\, d\tau \right. \\ &\left. + \int_{x_1-at_1}^{x_1-at_2} g(\tau)\, d\tau \right\} \end{aligned} \tag{8.1.30}$$

Note that the terms in A and B that involve f cancel, due to continuity of f. Further, the following integrals of $g(\tau)$ in B may be transformed:

$$E = \int_{x_2+at_1}^{x_2+at_2} g(\tau)\, d\tau + \int_{x_1+at_2}^{x_1+at_1} g(\tau)\, d\tau + \int_{x_2-at_1}^{x_2-at_2} g(\tau)\, d\tau + \int_{x_1-at_1}^{x_1-at_2} g(\tau)\, d\tau$$

Writing the first pair of integrals from $x_1 + at_2$ to $x_2 + at_2$, subtracting the integral from $x_1 + at_1$ to $x_2 + at_1$, writing the second pair of integrals from $x_1 - at_2$ to $x_2 - at_2$, and subtracting the missing integral from $x_1 - at_1$ to $x_2 - at_2$, yield

$$E = \int_{x_1+at_2}^{x_2+at_2} g(\tau)\, d\tau - \int_{x_1+at_1}^{x_2+at_1} g(\tau)\, d\tau - \int_{x_1-at_2}^{x_2-at_2} g(\tau)\, d\tau + \int_{x_1-at_1}^{x_2-at_1} g(\tau)\, d\tau$$

This is precisely the set of terms that involve g on the right side of Eq. 8.1.29 for the term A. Thus, A = B and the momentum balance law that governs motion of the string is satisfied for f in $C^0 \cap D^1$ and g in D^0.

This result justifies the solution of the plucked string problem, by substituting $f(x) = m(\ell - |x|)$ from Eq. 8.1.7 into D'Alembert's formula of Eq. 8.1.23. The situation illustrated here, wherein the regularity properties that are assumed in the derivation of differential equations are far too restrictive, is common in practice. It is important, therefore, to carefully assess the expected regularity properties of the solution, before using the equations.

EXERCISES 8.1

1. Solve the initial-boundary-value problem for the wave equation

$$u_{tt} - a^2 u_{xx} = 0$$

with the initial conditions

$$u(x, 3\ell/a) = 0$$

$$u_t(x, 3\ell/a) = ma$$

for $|x| \leq \ell$, and the boundary conditions

$$u(\pm\ell, t) = 0$$

for $t \geq 3\ell/a$. In what region is the solution valid?

2. Expand the plot of characteristics for the plucked string problem in Fig. 8.1.2 in the t direction to $t = 5\ell/a$ and indicate the direction of the velocity discontinuity across the characteristics.

3. Apply D'Alembert's formula to the initial conditions of Eq. 8.1.7 and verify that the result is consistent with Eq. 8.1.15.

8.2 THE HEAT EQUATION

In Section 6.4, the heat equation was derived. In one space variable, it is

$$u_t = k^2 u_{xx} \tag{8.2.1}$$

This equation is parabolic and the one-parameter family of lines

$$t = c \tag{8.2.2}$$

are characteristic curves.

In order to use the Cauchy–Kowalewski theorem to assure existence of a unique solution of a second-order equation, two initial conditions must be specified; the values of u and its normal derivative along a noncharacteristic curve. Since the line $t = 0$ (the x-axis) is a characteristic for Eq. 8.2.1 and since u_{xx} is the only second derivative that appears, the Cauchy–Kowalewski theorem applies only if initial conditions of the form $u(0, t) = f(t)$ and $u_x(0, t) = g(t)$ are specified. Initial conditions of this form, however, are not meaningful for physical situations that are modeled by the heat equation. Rather, the physically motivated initial condition encountered in Chapter 6 prescribes only the value of temperature u along the characteristic line $t = 0$. This means that the Cauchy–Kowalewski theorem fails to determine whether a solution exists and is unique. It is clear, however, that it would be too much to prescribe both u and u_t along $t = 0$, because $t = 0$ is a characteristic curve.

If, by some constructive method, a solution to the heat equation can be found that satisfies initial conditions along the characteristic $t = 0$, the problem will only be half solved. Existence will have been proved by actually finding a solution, but some other method will be required to prove that the solution is unique. This is in fact the procedure that is normally followed with the heat equation. A solution is found and it is then shown to be unique. In going through this procedure, a broadly applicable technique for solving linear partial differential equations will be illustrated.

Consider the physical problem of determining the temperature distribution in a thin homogeneous rod of length ℓ. Assume that the surface of the rod is insulated, so that there is no heat loss through it, and that the ends of the rod are both kept at a constant temperature; e.g., $u(0) = u(\ell) = 0$. Finally, assume that the initial temperature distribution in the rod is known. The physical problem may thus be formulated as follows: Solve

$$u_t = k^2 u_{xx} \tag{8.2.3}$$

for $0 < x < \ell$, $t > 0$, subject to the initial condition

$$u(x, 0) = f(x) \tag{8.2.4}$$

for $0 \le x \le \ell$, and the boundary conditions

$$\begin{aligned} u(0, t) &= 0 \\ u(\ell, t) &= 0 \end{aligned} \tag{8.2.5}$$

for $t \geq 0$. The question of what restrictions should be placed on the function f(x) is left open for the moment, except to note that it should be a function that describes a temperature distribution that could actually be realized in a bar.

Separation of Variables

The method of **separation of variables** consists of seeking solutions of a partial differential equation in variables x and t that may be expressed in the form

$$u(x, t) = X(x)\,T(t) \tag{8.2.6}$$

i.e., as the product of a function of x alone and a function of t alone, such that the initial and boundary conditions of Eqs. 8.2.4 and 8.2.5 are satisfied. Substituting the assumed expression for u of Eq. 8.2.6 into Eq. 8.2.3,

$$X(x)\,T'(t) - k^2 X''(x)\,T(t) = 0$$

or

$$\frac{T'(t)}{k^2 T(t)} = \frac{X''(x)}{X(x)} \tag{8.2.7}$$

for $0 < x < \ell$, $t > 0$. Equation 8.2.7 is said to be a **separated equation** because the left side is a function of t alone and the right side is a function of x alone. Since x and t are independent, Eq. 8.2.7 can hold only if both sides are equal to a constant. For reasons that will become apparent shortly, it is assumed that the constant is negative; i.e.,

$$\frac{T'}{k^2 T} = \frac{X''}{X} = -\alpha^2 \tag{8.2.8}$$

from which, two separate equations are obtained; i.e.,

$$\frac{T'}{k^2 T} = -\alpha^2 \tag{8.2.9}$$

$$\frac{X''}{X} = -\alpha^2 \tag{8.2.10}$$

The solution of Eq. 8.2.9 is

$$T(t) = T(0)\,e^{-\alpha^2 k^2 t} \tag{8.2.11}$$

If $T(0) \neq 0$, then T(t) is never zero and if $T(0) = 0$, then T(t) is identically zero. But if T(t) is identically zero, then so is u(x, t). If $f(x) \neq 0$ in Eq. 8.2.4, then it is required that $T(t) \neq 0$, for all t.

Since u(x, t) of Eq. 8.2.6 must satisfy the boundary conditions of Eq. 8.2.5,

$$u(0, t) = X(0)\,T(t) = 0$$

$$u(\ell, t) = X(\ell)\,T(t) = 0$$

Since $T(t) \neq 0$, this yields

$$X(0) = 0 \qquad X(\ell) = 0 \tag{8.2.12}$$

The general solution of Eq. 8.2.10 (i.e., $X'' + \alpha^2 X = 0$) is

$$X(x) = A \sin \alpha x + B \cos \alpha x$$

The first of Eq. 8.2.12 requires that $B = 0$, so

$$X(x) = A \sin \alpha x \tag{8.2.13}$$

It is required that $A \neq 0$, since otherwise u(x, t) would be identically zero. Applying the second boundary condition of Eq. 8.2.12 to the function X(x) in Eq. 8.2.13, nonzero values of α are sought such that

$$\sin \alpha\ell = 0 \tag{8.2.14}$$

At this point, it should be clear why the separation constant was chosen negative in Eq. 8.2.8. It is left as an exercise to show that for any other choice, the boundary conditions of Eq. 8.2.5 could not be satisfied. Furthermore, if the separation constant were positive, Eq. 8.2.11 would have a positive exponential coefficient and the temperature would grow exponentially, which is not physically meaningful.

Values of α that satisfy Eq. 8.2.14 are

$$\alpha = \frac{\pi}{\ell}, \frac{2\pi}{\ell}, \frac{3\pi}{\ell}, \ldots, \frac{n\pi}{\ell}, \ldots$$

Thus, an infinite number of solutions of Eq. 8.2.3 that satisfy the homogeneous boundary conditions of Eq. 8.2.5 exist; i.e.,

$$u_n(x, t) = e^{-\left(\frac{n\pi}{\ell}\right)^2 k^2 t} \sin\left(\frac{n\pi}{\ell}\right) x, \quad n = 1, 2, 3, \ldots \tag{8.2.15}$$

A countably infinite set of solutions of Eq. 8.2.3 has been found that satisfy the boundary conditions of Eq. 8.2.5. It remains to see if the solutions of Eq. 8.2.15 can be combined so that the initial condition of Eq. 8.2.4 is also satisfied. In other words, can constants a_i, $i = 1, 2, 3, \ldots$, be found so that the sequence

$$S_m(x, t) = \sum_{i=1}^{m} a_i u_i(x, t) \tag{8.2.16}$$

converges to a function u(x, t) that satisfies all the conditions of the problem? Summation notation is not used here, because the number m of terms in the partial sum in Eq. 8.2.16 is to be varied.

It is clear that, if the sequence of functions S_m in Eq. 8.2.16 converges, the limit function will satisfy the homogeneous boundary conditions, since they are satisfied by each function $u_i(x, t)$. What is not clear is the following:

(1) Can the coefficients a_i be chosen so that the initial condition of Eq. 8.2.4 is satisfied? If so, how?
(2) Having chosen the a_i, will the sequence of Eq. 8.2.16 converge to a differentiable function?
(3) If a limit function exists, will it satisfy the differential equation of Eq. 8.2.3?

Existence of Solution

Fortunately, it is not too difficult to show, from the theory of Fourier series, that the answers to all three questions are favorable. Assume, for the moment, that the series of Eq. 8.2.16 converges to a limit function u(x, t). Applying Eq. 8.2.4 to this limit function at $t = 0$ gives

$$\sum_{i=1}^{\infty} a_i \sin\left(\frac{i\pi}{\ell}\right) x = f(x) \tag{8.2.17}$$

which simply says that the a_i are coefficients in the Fourier sine series for f(x), in the interval $(0, \ell)$. Thus, from Section 5.4,

$$a_i = \frac{2}{\ell} \int_0^{\ell} f(x) \sin\left(\frac{i\pi}{\ell}\right) x \, dx \tag{8.2.18}$$

It is assumed that the function f(x) is at least piecewise continuous; i.e., $f(x) \in D^0(0, \ell)$, so the integral in Eq. 8.2.18 exists and the theory of Fourier series of Section 5.5 assures that the series of Eq. 8.2.17 converges.

It is now to be shown that the series of Eq. 8.2.16 converges uniformly as m approaches infinity and can be differentiated term by term twice; i.e., both

$$u(x, t) \approx \sum_{n=1}^{\infty} a_n e^{-\left(\frac{n\pi}{\ell}\right)^2 k^2 t} \sin\left(\frac{n\pi}{\ell}\right) x \tag{8.2.19}$$

and

$$u_{xx} \approx \sum_{n=1}^{\infty} -a_n \left(\frac{n^2\pi^2}{\ell^2}\right) e^{-\left(\frac{n\pi}{\ell}\right)^2 k^2 t} \sin\left(\frac{n\pi}{\ell}\right) x \tag{8.2.20}$$

converge uniformly. Note that, from Eqs. 8.2.19 and 8.2.20, $u_t \approx k^2 u_{xx}$. For any $\tau > 0$, if $\tau \le t < \infty$ and $0 \le x \le \ell$,

$$\left| a_n \frac{n^2\pi^2}{\ell^2} e^{-\left(\frac{n\pi}{\ell}\right)^2 k^2 t} \sin\left(\frac{n\pi}{\ell}\right) x \right| \le |a_n| \left(\frac{n^2\pi^2}{\ell^2} e^{-\left(\frac{n\pi}{\ell}\right)^2 k^2 \tau} \right) \le \frac{|a_n|}{n^2}$$

if $n > N$, where N is an integer that is large enough so that $e^{-\left(\frac{n\pi}{\ell}\right)^2 k^2 \tau} < \dfrac{\ell^2}{n^4\pi^2}$. Such an N exists, since

$$\lim_{n\to\infty} \frac{e^{-\left(\frac{n\pi}{\ell}\right)^2 k^2 t}}{\ell^2/(n^4\pi^2)} = \lim_{n\to\infty} \frac{\pi^2}{\ell^2} \frac{n^4}{e^{\left(\frac{n\pi}{\ell}\right)^2 k^2 t}} = \lim_{n\to\infty} \frac{4n^3}{2nk^2 t\, e^{\left(\frac{n\pi}{\ell}\right)^2 k^2 t}}$$

$$= \lim_{n\to\infty} \frac{\ell^2}{\pi^2} \frac{2}{(k^2 t)^2 e^{\left(\frac{n\pi}{\ell}\right)^2 k^2 t}} = 0$$

Since $\lim_{n\to\infty} |a_n| = 0$, there exists an integer M such that $|a_n| < 1$ for all $n > M$ and the series $\sum_{n=1}^{\infty} \frac{|a_n|}{n^2}$ converges (see Example 4.1.6). By the Weierstrass test of Section 4.2, the series for u_{xx} and u_t converge uniformly. Similarly, Theorem 4.2.3 assures that all derivatives of u(x, t) in Eq. 8.2.19 can be calculated by termwise differentiation.

A C^2 function u(x, t) has thus been found in Eq. 8.2.19. For $t > 0$, u(x, t) satisfies the heat equation of Eq. 8.2.3; at $t = 0$, it satisfies the initial condition of Eq. 8.2.4; and for all t, it satisfies the boundary conditions of Eq. 8.2.5. This proves existence of a solution of the initial-boundary-value problem, if f(x) is in $D^0(0, \ell)$, by actually finding a solution. To obtain continuity of the solution at $t = 0$, however, it is required that f(x) is in $D^1(0, \ell) \cap C^0(0, \ell)$, since the uniform convergence of the series in Eq. 8.2.17 is required, as shown in Section 3, Chapter 2 of Ref. 18.

Note that a discontinuity in u, u_x, or u_t at $t = 0$ immediately disappears, rather than propagating with finite speed. This attribute of solutions of the heat equation is very different from those of the wave equation.

Uniqueness of Solution

To show that the solution of the initial-boundary-value problem found here is unique, suppose there are two solutions of the problem, $u_1(x, t)$ and $u_2(x, t)$. Let

$$v(x, t) = u_1(x, t) - u_2(x, t)$$

Thus, v(x, t) verifies

$$\begin{aligned} v_t &= k^2 v_{xx}, \quad 0 < x < \ell, \quad t > 0 \\ v(0, t) &= 0, \quad t \geq 0 \\ v(\ell, t) &= 0, \quad t \geq 0 \\ v(x, 0) &= 0, \quad 0 \leq x \leq \ell \end{aligned} \tag{8.2.21}$$

Define the function

$$J(t) = \frac{1}{2k^2} \int_0^\ell v^2(x, t)\, dx$$

Differentiating both sides with respect to t and employing the relation $v_t = k^2 v_{xx}$,

$$J'(t) = \frac{1}{k^2} \int_0^\ell v v_t\, dx = \int_0^\ell v v_{xx}\, dx$$

Integrating by parts,

$$\int_0^\ell v v_{xx}\, dx = v v_x \Big|_0^\ell - \int_0^\ell v_x^2\, dx$$

Since

$$v(0, t) = v(\ell, t) = 0$$

$$J'(t) = -\int_0^\ell v_x^2\, dx \leq 0$$

for all $t \geq 0$. From the condition $v(x, 0) = 0$, $J(0) = 0$. This condition and $J'(t) \leq 0$ imply that

$$J(t) \leq 0$$

But by the definition of J(t),

$$J(t) \geq 0$$

Hence,

$$J(t) = 0$$

for all $t \geq 0$. Since v(x, t) is continuous for $t > 0$, $J(t) = 0$ implies

$$v(x, t) = 0$$

for $0 \leq x \leq \ell$, $t \geq 0$. Therefore, $u_1(x, t) \equiv u_2(x, t)$ and the solution is unique.

EXERCISES 8.2

1. Show that, if the separation constant $-\alpha^2$ in Eq. 8.2.8 were chosen to be positive or zero, instead of negative, it is not possible to find any functions u(x, t) of the form u(x, t) = X(x) T(t) that satisfy Eq. 8.2.3 and the boundary conditions of Eq. 8.2.5.

2. Solve the following initial-boundary-value problem by the method of separation of variables:

$$u_t - k^2 u_{xx} = 0, \quad 0 < x < 1, \quad t > 0$$

$$u(x, 0) = u_0 (1 - 2x - \cos \pi x), \quad 0 \le x \le 1$$

$$u(0, t) = u(1, t) = 0, \quad t > 0$$

8.3 LAPLACE'S EQUATION

It was shown in Section 7.4 that an initial-value problem for Laplace's equation $u_{xx} + u_{yy} = 0$ is unstable, thus it is not well posed. This does not imply that all problems that involve Laplace's equation fail to be well posed. It does indicate that an initial-value problem is not a formulation that makes sense with Laplace's equation in physical situations. Rather, physical situations that are described by Laplace's equation involve boundary conditions, rather than initial conditions. This might be expected, since initial conditions generally relate to the time variable t, whereas the independent variables in Laplace's equation are normally space variables. These observations are based on the following boundary-value-problems in Chapter 6, in which Laplace's equation arises:

(1) Static deflection of a membrane; Eqs. 6.3.40 and 6.3.41
(2) Steady-state temperature distribution; Eqs. 6.4.9 – 6.4.10
(3) Steady incompressible, irrotational fluid flow; Eq. 6.5.17

In each of these cases, which involve two space dimensions, there was a closed curve Γ that bounded some region Ω of interest and values of the unknown function u or its normal derivative that were prescribed along Γ. Problems of this type will later be shown to be well posed. For now, an example of a boundary-value problem that involves Laplace's equation is given to illustrate a technique for finding a solution by the method of separation of variables.

A steady state temperature distribution is sought in the thin rectangular plate shown in Fig. 8.3.1, in which two edges are insulated. One edge is kept at zero temperature and the remaining edge has a prescribed temperature distribution. The mathematical formulation is as follows: Solve the differential equation

$$u_{xx} + u_{yy} = 0 \tag{8.3.1}$$

for $0 < x < a$, $0 < y < b$, subject to boundary conditions

$$
\begin{aligned}
u(x, 0) &= 0 \\
u(x, b) &= f(x) \\
u_x(0, y) &= 0 \\
u_x(a, y) &= 0
\end{aligned}
\tag{8.3.2}
$$

The insulated edges are taken as x = 0 and x = a. Insulation implies that $u_x = 0$ along these edges. The edge that is kept at zero temperature is taken as y = 0. The function f(x) is the prescribed temperature distribution along the remaining edge, y = b.

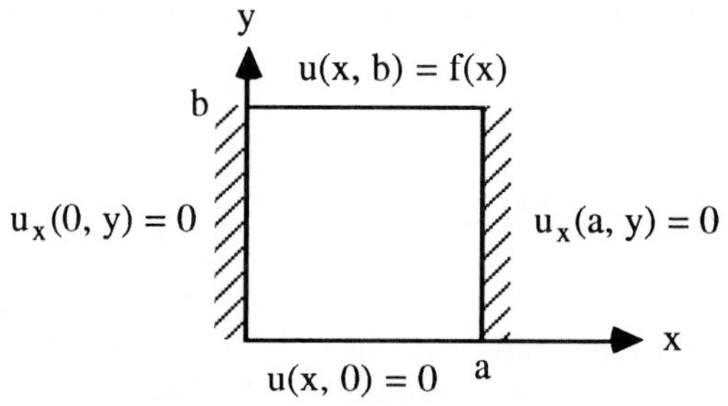

Figure 8.3.1 Partially Insulated Plate

Separation of Variables

To solve this problem by separation of variables, first assume a solution of the form

$$u(x, y) = X(x)\,Y(y) \tag{8.3.3}$$

Substituting this expression into the differential equation of Eq. 8.3.1,

$$X''Y + XY'' = 0$$

or

$$\frac{X''}{X} = -\frac{Y''}{Y} = -\alpha^2 \tag{8.3.4}$$

where both ratios must be constant, since one depends only on x and the other depends only on y. The sign of the separation constant has been selected as negative, so that X(x) will be a trigonometric function. This is because the boundary conditions indicate that u_x must vanish for two different values of x, which is not possible if the separation constant is positive.

The problem of finding X(x) is equivalent to the eigenvalue problem of the differential equation

$$X'' + \alpha^2 X = 0 \tag{8.3.5}$$

which arises from Eq. 8.3.4, and the boundary conditions

$$X'(0) = X'(a) = 0 \tag{8.3.6}$$

that follow from the last two of Eq. 8.3.2. The eigenfunctions of Eqs. 8.3.5 and 8.3.6 are

$$1, \cos(\pi x/a), \cos(2\pi x/a), \ldots, \cos(n\pi x/a), \ldots \tag{8.3.7}$$

and the corresponding eigenvalues are

$$\alpha^2 = 0, (\pi/a)^2, (2\pi/a)^2, \ldots, (n\pi/a)^2, \ldots \tag{8.3.8}$$

Consider next the problem of finding Y(y), by solving the differential equation

$$Y'' - (n\pi/a)^2 Y = 0, \quad n = 0, 1, 2, \ldots$$

subject to the condition

$$Y(0) = 0$$

which follows from the first of Eq. 8.3.2. This leads to

$$\begin{aligned} Y_0 &= b_0 y \\ Y_n &= b_n \sinh\left(\frac{n\pi y}{a}\right), \quad n = 1, 2, 3, \ldots \end{aligned} \tag{8.3.9}$$

where the b_n are to be determined. From these results, a solution of the original problem is constructed in the form

$$u(x, y) = b_0 y + \sum_{n=1}^{\infty} b_n \cos\frac{n\pi x}{a} \sinh\frac{n\pi y}{a} \tag{8.3.10}$$

If convergence is uniform, if u(x, y) has the required differentiability properties, and if the second of Eq. 8.3.2 is satisfied; i.e.,

$$u(x, b) = b_0 b + \sum_{n=1}^{\infty} \left(b_n \sinh\frac{n\pi b}{a} \right) \cos\frac{n\pi x}{a} = f(x) \tag{8.3.11}$$

then Eq. 8.3.10 provides the desired solution. Since the Fourier cosine series for f(x) in the interval (0, a) is

$$f(x) = \frac{a_0}{2} + \sum_{n=1}^{\infty} a_n \cos\frac{n\pi x}{a} \tag{8.3.12}$$

where

$$
\begin{aligned}
a_n &= \frac{2}{a}\int_0^a f(x)\cos\frac{n\pi x}{a}\,dx, \quad n = 0, 1, 2, \ldots \\
b_0 &= \frac{1}{ab}\int_0^a f(x)\,dx \\
b_n &= \frac{2}{a\sinh\dfrac{n\pi b}{a}}\int_0^a f(x)\cos\frac{n\pi x}{a}\,dx \\
&= \frac{a_n}{\sinh\dfrac{n\pi b}{a}}, \quad n = 1, 2, \ldots
\end{aligned}
\tag{8.3.13}
$$

the formal solution of the original problem is given by Eq. 8.3.10, where the constants b_n, $n = 0, 1, 2, \ldots, m$, are determined by Eq. 8.3.13.

Existence of Solution

To prove convergence of the series in Eq. 8.3.10, substitute from Eq. 8.3.13 into Eq. 8.3.10, to obtain

$$
u(x, y) \approx b_0 y + \sum_{n=1}^{\infty} a_n \frac{\sinh\dfrac{n\pi y}{a}}{\sinh\dfrac{n\pi b}{a}} \cos\frac{n\pi x}{a}
$$

Since $\sinh\theta = (1/2)(e^{\theta} - e^{-\theta})$, as θ approaches ∞, $\sinh\theta \approx e^{\theta}/2$ and, for m large,

$$
u(x, y) \approx b_0 y + \sum_{n=m+1}^{\infty} a_n e^{\frac{n\pi}{a}(y-b)} \cos\frac{n\pi x}{a} + \sum_{n=1}^{m} a_n \frac{\sinh\dfrac{n\pi y}{a}}{\sinh\dfrac{n\pi b}{a}} \cos\frac{n\pi x}{a}
$$

For $0 \le y \le \bar{y} < b$, the convergence argument of Eqs. 8.2.19 and 8.2.20 leads to the conclusion that the series converges uniformly and can be differentiated term by term as many times as desired. Thus, Eq. 8.3.10, with coefficients of Eq. 8.3.13, provides a solution of the boundary-value problem for $0 \le y \le \bar{y} < b$, if $f(x) \in D^0(0, a)$. As in the heat equation, however, to obtain continuity at $y = b$, $f(x)$ must be in $D^1(0, a) \cap C^0(0, a)$, since uniform convergence of the series in Eq. 8.3.12 is required. The fact that this solution is unique will follow from results presented in Section 8.4.

8.4 ELLIPTIC DIFFERENTIAL EQUATIONS

Elliptic Boundary-value Problems

As noted in Chapter 6 and earlier in this chapter, important classes of differential equations that arise in membrane deflection, steady state heat transfer, and potential flow fall under the general classification of **elliptic differential equations.** As noted in Section 7.7, the most general second-order equations of this type are of the form

$$-\sum_{j,k=1}^{m} \frac{\partial}{\partial x_j}\left[a_{jk}(\mathbf{x}) \frac{\partial u}{\partial x_k} \right] + C(\mathbf{x})\, u = f(\mathbf{x}) \tag{8.4.1}$$

on a bounded domain Ω in R^m, where $a_{jk} = a_{kj}$, subject to one of the following boundary conditions:

$$u \,\Big|_{\Gamma} = 0 \tag{8.4.2}$$

$$\left(N(u) + \sigma(\mathbf{x})\, u \right) \Big|_{\Gamma} = 0 \tag{8.4.3}$$

$$N(u) \,\Big|_{\Gamma} = 0 \tag{8.4.4}$$

where

$$N(u) = \sum_{j,k=1}^{m} a_{jk}(\mathbf{x}) \frac{\partial u}{\partial x_k} n_j \tag{8.4.5}$$

Γ is the boundary of Ω, n_j is the jth direction cosine of the normal to Γ, and $a_{jk}(\mathbf{x})$ satisfy the inequality

$$\sum_{j,k=1}^{m} a_{jk}(\mathbf{x})\, t_j t_k \geq \mu_0 \sum_{k=1}^{m} t_k^2$$

for some $\mu_0 > 0$, for all real numbers $t_1, t_2, \ldots, t_m$, and for all $\mathbf{x} \in \Omega$.

A comprehensive study of this class of equations requires variational and function space concepts that are presented in Chapter 9. In this section, some classical results are developed and applied to the following simple problems of elliptic type: Determine a function u(x, y, z) that satisfies **Poisson's equation (Laplace's equation** if f = 0):

$$\nabla^2 u = f \tag{8.4.6}$$

and one of the following boundary conditions on the boundary Γ of Ω:

(1) **first boundary-value problem**, or **Dirichlet problem**,

$$u = f_1, \text{ on } \Gamma \tag{8.4.7}$$

(2) **second boundary-value problem**, or **Neumann problem**,

$$\frac{\partial u}{\partial n} = f_2, \text{ on } \Gamma \tag{8.4.8}$$

(3) **third boundary-value problem**, or **mixed problem**,

$$\frac{\partial u}{\partial n} + h(u - f_3) = 0, \text{ on } \Gamma \tag{8.4.9}$$

The following coordinate systems are often used in applications:

(a) **Spherical Coordinates**,

$$x = r \sin\theta \cos\phi, \; y = r \sin\theta \sin\phi, \; z = r\cos\theta \tag{8.4.10}$$

for which Laplace's equation $\nabla^2 u = 0$ takes the form

$$\frac{1}{r^2}\frac{\partial}{\partial r}\left(r^2 \frac{\partial u}{\partial r}\right) + \frac{1}{r^2 \sin\theta}\frac{\partial}{\partial \theta}\left(\sin\theta \frac{\partial u}{\partial \theta}\right) + \frac{1}{r^2 \sin^2\theta}\frac{\partial^2 u}{\partial \phi^2} = 0 \tag{8.4.11}$$

(b) **Cylindrical Coordinates**,

$$x = \rho\cos\phi, \; y = \rho \sin\phi, \; z = z \tag{8.4.12}$$

for which Laplace's equation takes the form

$$\frac{1}{\rho}\frac{\partial}{\partial \rho}\left(\rho \frac{\partial u}{\partial \rho}\right) + \frac{1}{\rho^2}\frac{\partial^2 u}{\partial \phi^2} + \frac{\partial^2 u}{\partial z^2} = 0 \tag{8.4.13}$$

Example 8.4.1

To see how Eq. 8.4.13 can be obtained, consider the coordinate transformation of Eq. 8.4.12, from (x, y) to (ρ, ϕ). Once the differential equation is obtained in terms of ρ and ϕ, u_{zz} can simply be added to obtain the Laplace equation.

Since $x = \rho\cos\phi$ and $y = \rho\sin\phi$,

$$\begin{aligned} u_\rho &= u_x \cos\phi + u_y \sin\phi \\ u_\phi &= -u_x\rho \sin\phi + u_y\rho\cos\phi \end{aligned} \tag{8.4.14}$$

where subscripts denote partial derivatives. Solving Eq. 8.4.14 for u_x and u_y,

$$u_x = u_\rho \cos\phi - u_\phi \frac{\sin\phi}{\rho}$$
$$u_y = u_\rho \sin\phi + u_\phi \frac{\cos\phi}{\rho} \tag{8.4.15}$$

Differentiating Eq. 8.4.15,

$$u_{xx} = (u_x)_x = \left(u_\rho \cos\phi - u_\phi \frac{\sin\phi}{\rho} \right)_\rho \cos\phi - \left(u_\rho \cos\phi - u_\phi \frac{\sin\phi}{\rho} \right)_\phi \frac{\sin\phi}{\rho}$$

$$= \left(u_{\rho\rho} \cos\phi + u_\phi \frac{\sin\phi}{\rho^2} - u_{\rho\phi} \frac{\sin\phi}{\rho} \right) \cos\phi - \left(-u_\rho \sin\phi + u_{\phi\rho} \cos\phi - u_\phi \frac{\cos\phi}{\rho} - u_{\phi\phi} \frac{\sin\phi}{\rho} \right) \frac{\sin\phi}{\rho}$$

$$u_{yy} = \left(u_{\rho\rho} \sin\phi - u_\phi \frac{\cos\phi}{\rho^2} + u_{\rho\phi} \frac{\cos\phi}{\rho} \right) \sin\phi + \left(u_\rho \cos\phi + u_{\phi\rho} \sin\phi - u_\phi \frac{\sin\phi}{\rho} + u_{\phi\phi} \frac{\cos\phi}{\rho} \right) \frac{\cos\phi}{\rho}$$

Substituting these results into Laplace's equation,

$$u_{xx} + u_{yy} = u_{\rho\rho} + \frac{1}{\rho^2} u_{\phi\phi} + \frac{1}{\rho} u_\rho$$
$$= \frac{1}{\rho} (\rho u_\rho)_\rho + \frac{1}{\rho^2} u_{\phi\phi} = 0 \tag{8.4.16}$$

and the result of Eq. 8.4.13 can be obtained by adding u_{zz} to both sides of Eq. 8.4.16. ■

In the special case of a **spherically symmetric solution** u(r) of Laplace's equation, Eq. 8.4.11 reduces to

$$\frac{d}{dr}\left(r^2 \frac{du}{dr} \right) = 0 \tag{8.4.17}$$

whose **general integral** is given by

$$u = \frac{c_1}{r} + c_2 \tag{8.4.18}$$

Setting $c_1 = 1$ and $c_2 = 0$, the solution $u_0 = 1/r$ is obtained, called the **fundamental solution of Laplace's equation in three dimensions.**

By setting $u = u(\rho)$, the general form of a **cylindrically symmetric solution** of Eq. 8.4.13 is

$$u(\rho) = c_1 \ln \rho + c_2$$

or, for $c_1 = -1$ and $c_2 = 0$,

$$u_0 = \ln \frac{1}{\rho} \tag{8.4.19}$$

which is called the **fundamental solution of Laplace's equation in two dimensions.**

The function $u_0 = 1/r$ coincides, within a constant factor, with the **potential** due to a charge e that is placed at the origin. The potential of this field is actually $u = e/r$. The function $\ln(1/\rho)$ coincides, within a constant factor, with the potential of a uniform charge distributed along an infinitely long wire. The potential of this field is given by $u(\rho) = 2e_1 \ln(1/\rho)$, where e_1 is the charge density per unit of length. Hence, the body of theory associated with Laplace's equation is often called **potential theory**.

General Properties of Harmonic Functions

Let u(x) be a **harmonic function**; i.e., $\nabla^2 u = 0$, that has continuous first and second derivatives on an open connected set Ω in R^3 and on its boundary Γ. For $\mathbf{x}_0$ and $\mathbf{x}$ in Ω, let r be the distance between $\mathbf{x}_0$ and $\mathbf{x}$. Further, let K_ε be a sphere of radius ε, with center at $\mathbf{x}_0$ and boundary Γ_ε, as shown in Fig. 8.4.1.

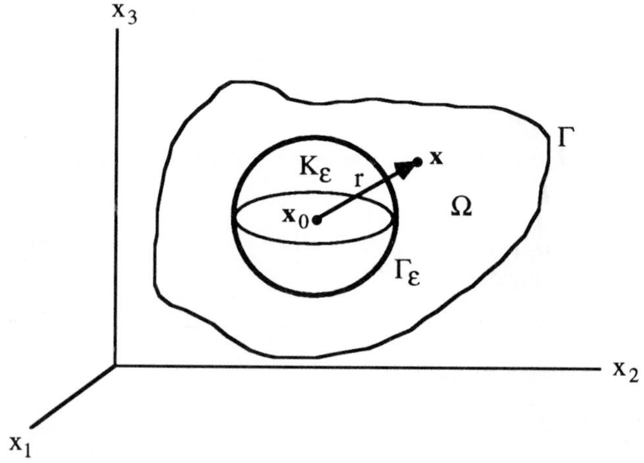

Figure 8.4.1 Neighborhood of x_0

Employing the divergence theorem twice yields the identity

$$\iiint_{\Omega-K_\varepsilon}\left[u\,\nabla^2\left(\frac{1}{r}\right)-\frac{1}{r}\nabla^2 u\right]d\Omega$$

$$=\iint_{\Gamma}\left[u\frac{\partial}{\partial n}\left(\frac{1}{r}\right)-\frac{1}{r}\frac{\partial u}{\partial n}\right]d\sigma$$

$$+\iint_{\Gamma_\varepsilon}u\frac{\partial}{\partial n}\left(\frac{1}{r}\right)d\sigma-\iint_{\Gamma_\varepsilon}\frac{1}{r}\frac{\partial u}{\partial n}d\sigma \qquad (8.4.20)$$

Since the outward normal of the boundary of the set $\Omega - K_\varepsilon$ points to the interior of K_ε and the radius vector **r** of Γ_ε points out of K_ε,

$$\frac{\partial}{\partial n}\left(\frac{1}{r}\right)\Big|_{\Gamma_\varepsilon}=-\frac{\partial}{\partial r}\left(\frac{1}{r}\right)\Big|_{r=\varepsilon}=\frac{1}{\varepsilon^2} \qquad (8.4.21)$$

It follows that

$$\iint_{\Gamma_\varepsilon}u\frac{\partial}{\partial n}\left(\frac{1}{r}\right)d\sigma=\frac{1}{\varepsilon^2}\iint_{\Gamma_\varepsilon}u\,d\sigma=4\pi u^*$$

where u^* is the average value of u on Γ_ε, defined by

$$u^*=\frac{1}{4\pi\varepsilon^2}\iint_{\Gamma_\varepsilon}u\,d\sigma$$

Similarly, the third integral on the right of Eq. 8.4.20 can be written as

$$\iint_{\Gamma_\varepsilon}\frac{1}{r}\frac{\partial u}{\partial n}d\sigma=\frac{1}{\varepsilon}\iint_{\Gamma_\varepsilon}\frac{\partial u}{\partial n}d\sigma=4\pi\varepsilon\left(\frac{\partial u}{\partial n}\right)^*$$

where $(\partial u/\partial n)^*$ is the average value of $(\partial u/\partial n)$ on Γ_ε. It follows, by substitution into Eq. 8.4.20 and by the fact that $\nabla^2(1/r) = 0$ on $\Omega - K_\varepsilon$, that

$$\iiint_{\Omega-K_\varepsilon}\left(-\frac{1}{r}\right)\nabla^2 u\,d\Omega=\iint_{\Gamma}\left[u\frac{\partial}{\partial n}\left(\frac{1}{r}\right)-\frac{1}{r}\frac{\partial u}{\partial n}\right]d\sigma$$

$$+\,4\pi u^*-4\pi\varepsilon\left(\frac{\partial u}{\partial n}\right)^* \qquad (8.4.22)$$

The following facts (for proof see Chapter 1 of Ref. 18) are cited for future use:

(a) $\lim_{\varepsilon\to 0} u^* = u(\mathbf{x}_0)$

(b) Since $\left(\frac{\partial u}{\partial n}\right)^*$ is finite, $\lim_{\varepsilon\to 0} 4\pi\varepsilon\left(\frac{\partial u}{\partial n}\right)^* = 0$

(c) $\lim_{\varepsilon\to 0}\iiint_{\Omega-K_\varepsilon}\left(-\frac{1}{r}\nabla^2 u\right)d\Omega = \iiint_{\Omega}\left(-\frac{1}{r}\nabla^2 u\right)d\Omega$

The last result follows from the definition of **improper integral.**

Let ε approach zero in Eq. 8.4.22, to obtain

$$4\pi\, u(\mathbf{x}_0) = -\iint_\Gamma\left[u\frac{\partial}{\partial n}\left(\frac{1}{r}\right) - \frac{1}{r}\frac{\partial u}{\partial n}\right]d\sigma - \iiint_\Omega \frac{\nabla^2 u}{r}\,d\Omega \quad (8.4.23)$$

If u is harmonic in Ω, Eq. 8.4.23 reduces to

$$u(\mathbf{x}_0) = \frac{1}{4\pi}\iint_\Gamma\left[\frac{1}{r}\frac{\partial u}{\partial n} - u\frac{\partial}{\partial n}\left(\frac{1}{r}\right)\right]d\sigma \quad (8.4.24)$$

Therefore, the value of a harmonic function at an arbitrary interior point $\mathbf{x}_0$ of Ω can be expressed in terms of its value and the value of its normal derivative on the boundary Γ of Ω.

Theorem 8.4.1. The value of a harmonic function u at any point $\mathbf{x}_0$ is equal to the average value of the function on an arbitrary spherical surface Γ_a with center at $\mathbf{x}_0$, provided Γ_a lies entirely in the region in which u is harmonic. ■

This theorem follows by recalling that $1/r = 1/a$ on Γ_a, noting that $\partial u/\partial n = \nabla u^T n$, and using the divergence theorem to obtain

$$0 = \iiint_{K_a}\nabla^2 u\, d\Omega = \iiint_{K_a}\nabla^T(\nabla u)\, d\Omega$$

$$= \iint_{\Gamma_a}\nabla u^T n\, d\sigma = \iint_{\Gamma_a}\frac{\partial u}{\partial n}\, d\sigma$$

and

$$\frac{\partial}{\partial n}\left(\frac{1}{r}\right)\Big|_{r=a} = \frac{\partial}{\partial r}\left(\frac{1}{r}\right)\Big|_{r=a} = -\frac{1}{a^2}$$

on Γ_a. Thus, from Eq. 8.4.24,

$$u(\mathbf{x}_0) = \frac{1}{4\pi a^2}\iint_{\Gamma_a} u\, d\sigma \quad (8.4.25)$$

The formula corresponding to Eq. 8.4.24 for functions on a domain A in R^2 is similarly derived as

$$2\pi u(\mathbf{x}_0) = -\int_C \left[u \frac{\partial}{\partial n}\left(\ln \frac{1}{r} \right) - \ln \frac{1}{r} \frac{\partial u}{\partial n} \right] d\sigma - \iint_A \left(\ln \frac{1}{r} \right) \nabla^2 u \, dA$$

which, for a harmonic function, gives

$$u(\mathbf{x}_0) = \frac{1}{2\pi} \int_C \left[\ln\left(\frac{1}{r}\right) \frac{\partial u}{\partial n} - u \frac{\partial}{\partial n} \ln\left(\frac{1}{r}\right) \right] d\sigma \tag{8.4.26}$$

For two independent variables, the analogous formula to Eq. 8.4.25 is

$$u(\mathbf{x}_0) = \frac{1}{2\pi a} \int_{C_a} u \, ds \tag{8.4.27}$$

■

Theorem 8.4.2 (Maximum Principle). If a function is continuous on a closed region $\Omega \cup \Gamma$ in R^3 and is **harmonic** in its interior Ω, then it assumes its maximum and minimum values on the boundary Γ. ■

For purposes of proof, assume that u achieves its maximum value at an interior point $\mathbf{x}_0$. Then $u(\mathbf{x}_0) \ge u(\mathbf{x})$, for all $\mathbf{x} \in \Omega$. Constructing a sphere of radius ρ around $\mathbf{x}_0$, with boundary Γ_ρ that is contained in Ω, Theorem 8.4.1 implies that

$$u(\mathbf{x}_0) = \frac{1}{4\pi\rho^2} \iint_{\Gamma_\rho} u \, d\sigma \le \frac{1}{4\pi\rho^2} \iint_{\Gamma_\rho} u(\mathbf{x}_0) \, d\sigma = u(\mathbf{x}_0)$$

If $u(\mathbf{x}) < u(\mathbf{x}_0)$ at some point $\mathbf{x}$ on Γ_ρ, then since u is continuous on Γ_ρ, the contradiction $u(\mathbf{x}_0) < u(\mathbf{x}_0)$ is obtained. Hence, $u(\mathbf{x}) = u(\mathbf{x}_0)$ everywhere on Γ_ρ.

Letting $\bar{\rho}$ be the minimum distance from $\mathbf{x}_0$ to Γ, $u(\mathbf{x}) = u(\mathbf{x}_0)$ for all points on Γ_ρ. By continuity, it follows that $u(\mathbf{x}^*) = u(\mathbf{x}_0)$ for those points $\mathbf{x}^*$ that belong to the intersection of $\Gamma_{\bar{\rho}}$ and Γ. This proves the maximum principle and shows that the maximum value $u(\mathbf{x}_0)$ is assumed at a point on the boundary. For minimum values on the boundary Γ, the argument is repeated with the negative of $u(\mathbf{x})$. ■

Corollary 8.4.1. Two conclusions follow directly from the maximum principle;

(1) If u and v are continuous in $\Omega \cup \Gamma$ and harmonic in Ω, with $u \le v$ on Γ, then $u \le v$ in Ω.

(2) If u and v are continuous in the region $\Omega \cup \Gamma$ and harmonic in Ω, with $|u| \le v$ on Γ, then $|u| \le v$ on Ω. ■

Uniqueness of the solution for the first boundary-value problem (see Eq. 8.4.7) follows from the maximum principle. That is, if u_1 and u_2 are two solutions, then $\nabla^2(u_1 - u_2) = 0$ in Ω and $u_1 - u_2 = 0$ on Γ. Thus, $u_1 - u_2 = 0$ in Ω.

The function u is said to be a **stable solution** of a boundary-value problem if it depends continuously on the boundary conditions; i.e., if a small change in the boundary conditions implies a small change in the solution. Let u_1 and u_2 be two functions that are continuous in $\Omega \cup \Gamma$, harmonic in Ω, and satisfy

$$\max_{\mathbf{x} \in \Gamma} |u_1(\mathbf{x}) - u_2(\mathbf{x})| \leq \varepsilon$$

Then, by the maximum principle,

$$\max_{\mathbf{x} \in \Omega \cup \Gamma} |u_1(\mathbf{x}) - u_2(\mathbf{x})| \leq \varepsilon$$

which proves stability of the first boundary-value problem.

Separation of Variables for the Laplace Equation on a Circle

Consider the problem of finding a function u that satisfies Laplace's equation $\nabla^2 u = 0$ inside a circle of radius a and the condition $u(a, \phi) = f(\phi)$ on its circumference. Introducing cylindrical coordinates, from Eq. 8.4.13, Laplace's equation takes the form

$$\nabla^2 u = \frac{1}{\rho}\frac{\partial}{\partial \rho}\left(\rho \frac{\partial u}{\partial \rho}\right) + \frac{1}{\rho^2}\frac{\partial^2 u}{\partial \phi^2} = 0 \tag{8.4.28}$$

Writing the solution in the form $u(\rho, \phi) = R(\rho)\,\Phi(\phi)$, Eq. 8.4.28 yields

$$\frac{\dfrac{d}{d\rho}\left(\rho \dfrac{dR}{d\rho}\right)}{R/\rho} = \frac{-\Phi''}{\Phi} = \lambda$$

where λ is a constant. This leads to two ordinary differential equations,

$$\Phi'' + \lambda \Phi = 0$$

$$\rho \frac{d}{d\rho}\left(\rho \frac{dR}{d\rho}\right) - \lambda R = 0$$

It follows that

$$\Phi(\phi) = A \cos \sqrt{\lambda}\,\phi + B \sin \sqrt{\lambda}\,\phi$$

and since Φ is continuous, it must be periodic in ϕ with period 2π. Then $\sqrt{\lambda} = n$, where n is an integer, and

$$\Phi_n(\phi) = A_n \cos n\phi + B_n \sin n\phi$$

Taking $R(\rho)$ to be of the form $R(\rho) = \rho^{\mu}$, the differential equation for $R = R(\rho)$ yields $\mu^2\rho^{\mu} - \lambda\rho^{\mu} = 0$. Thus, $n^2 = \lambda = \mu^2$, or $n = \pm\mu$. Thus, for $n \neq 0, R(\rho) = C\rho^{n} + D\rho^{-n}$. For the solution inside a circle, $R(\rho) = C\rho^{n}$, since if $D \neq 0$, then the solution is unbounded as $\rho \to 0$. For $n = 0$, $R = E \ln \rho + F$. Since $\lim_{\rho\to 0} \ln \rho = -\infty$, $E = 0$ is required.

A particular solution of Laplace's equation is, therefore,

$$u_n(\rho, \phi) = \rho^{n} (A_n \cos n\phi + B_n \sin n\phi), \quad n = 0, 1, \ldots$$

for $\rho \leq a$, and $n = 0$ yields the constant solution. The sum of these solutions,

$$u(\rho, \phi) = \sum_{n=0}^{\infty} \rho^{n} (A_n \cos n\phi + B_n \sin n\phi) \tag{8.4.29}$$

is harmonic, provided it converges to a twice differentiable function. The coefficients A_n and B_n are determined from the boundary condition

$$u(a, \phi) = \sum_{n=0}^{\infty} a^{n} (A_n \cos n\phi + B_n \sin n\phi) = f(\phi)$$

Let the Fourier series for $f(\phi)$ be $f(\phi) = \frac{\alpha_0}{2} + \sum_{n=1}^{\infty} (\alpha_n \cos n\phi + \beta_n \sin n\phi)$, where $\alpha_0 = \frac{1}{\pi}\int_{-\pi}^{\pi} f(\psi)\, d\psi$, $\alpha_n = \frac{1}{\pi}\int_{-\pi}^{\pi} f(\psi) \cos n\psi\, d\psi$, and $\beta_n = \frac{1}{\pi}\int_{-\pi}^{\pi} f(\psi) \sin n\psi\, d\psi$. Then, $A_0 = \alpha_0/2$, $A_n = \alpha_n/a^n$, and $B_n = \beta_n/a^n$. The solution can therefore be written as

$$u(\rho, \phi) = \frac{\alpha_0}{2} + \sum_{n=1}^{\infty} \left(\frac{\rho}{a}\right)^{n} (\alpha_n \cos n\phi + \beta_n \sin n\phi) \tag{8.4.30}$$

In order to prove that this function represents the desired solution, it is necessary to show convergence of the series, termwise differentiability, and continuity of the limit function on the circumference of the circle. Writing the series expansion of Eq. 8.4.30 as

$$u(\rho, \phi) = \frac{\alpha_0}{2} + \sum_{n=1}^{\infty} t^{n} (\alpha_n \cos n\phi + \beta_n \sin n\phi) \tag{8.4.31}$$

where $t = \rho/a \leq 1$, and letting $u_n = t^n (\alpha_n \cos n\phi + \beta_n \sin n\phi)$, the kth derivative of u_n with respect to ϕ may be computed and transformed, with the aid of trigonometric identities, to

$$\frac{\partial^k u_n}{\partial \phi^k} = t^n n^k \left[\alpha_n \cos\left(n\phi + k\frac{\pi}{2} \right) + \beta_n \sin\left(n\phi + k\frac{\pi}{2} \right) \right]$$

Hence,

$$\left| \frac{\partial^k u_n}{\partial \phi^k} \right| \le t^n n^k 2m$$

where m is the maximum of the absolute values of the Fourier coefficients α_n and β_n; i.e.,

$$|\alpha_n| < m, \quad |\beta_n| < m, \quad n = 1, 2, \ldots$$

Choosing a fixed value $\rho_0 < a$, where $t_0 = \frac{\rho_0}{a} < 1$, note that

$$\sum_{n=1}^{\infty} t^n n^k (|\alpha_n| + |\beta_n|) \le 2m \sum_{n=1}^{\infty} t_0^n n^k$$

for $t \le t_0$. Application of the ratio test shows that the series converges, implying uniform convergence of the series

$$\sum_{n=1}^{\infty} \frac{\partial^k u_n}{\partial \phi^k} = \sum_{n=1}^{\infty} t^n n^k \left[\alpha_n \cos\left(n\phi + \frac{k\pi}{2} \right) + \beta_n \sin\left(n\phi + \frac{k\pi}{2} \right) \right]$$

for $t \le t_0 \le 1$. Therefore, the series representation for u can be differentiated arbitrarily many times with respect to ϕ. A similar argument shows that the series expansion for u can be differentiated arbitrarily many times with respect to ρ and the equation $\nabla^2 u = 0$ is satisfied. Since this result holds for any series representation given by Eq. 8.4.31, with $t < 1$ and with α_n and β_n bounded, it follows that Eq. 8.4.31 defines functions u in C^∞ that satisfy $\nabla^2 u = 0$ for $t < 1$, for any square integrable function f that is defined on the interval $[-\pi, \pi]$.

In order to prove continuity of the solution in the closed region $t \le 1$, it is necessary to impose additional regularity assumptions on the function f. In particular, it is required that f be periodic, continuous, and piecewise differentiable; i.e., $f \in D^1 \cap C^0$. Defining A_n and B_n as the Fourier coefficients of $f'(x)$, it was shown in the proof of Theorem 5.5.2 that $\alpha_n = -B_n/n$ and $\beta_n = A_n/n$. Note that $(|A_n| - 1/n)^2 = A_n^2 - 2|A_n|/n + 1/n^2 \ge 0$ and similarly for B_n. Adding equations yields $|A_n|/n + |B_n|/n \le (1/2)(A_n^2 + B_n^2) + 1/n^2$. Since the series of both terms on the right converges,

$$\sum_{n=1}^{\infty} (|\alpha_n| + |\beta_n|) = \sum_{n=1}^{\infty} \left(\frac{|A_n|}{n} + \frac{|B_n|}{n} \right)$$

converges. Since $| t^n \alpha_n \cos n\phi | \le | \alpha_n |$ and $| t^n \beta_n \sin n\phi | \le | \beta_n |$, the series of Eqs. 8.4.30 and 8.4.31 converge uniformly for $t \le 1$. It follows that the functions represented by their sums are continuous on the circumference of the circle.

Well-posed Boundary-value Problem

With existence of a solution of the first boundary-value problem established, and since uniqueness and continuity with respect to boundary data were established as a consequence of the maximum principle, the result that follows is established.

Theorem 8.4.3. The boundary-value problem consisting of the Laplace equation of Eq. 8.4.6 and the boundary condition of Eq. 8.4.7 is well posed. ■

Green's Function

In this subsection, the definition and basic properties of Green's function for Laplace's equation are treated. Let $u(\mathbf{x})$ be a C^2 function in a bounded domain $\Omega \subset R^3$ with boundary Γ. Assume further that $u(\mathbf{x})$ is C^1 on the union of Ω and its boundary Γ, which is a closed set. By Eq. 8.4.23,

$$u(\mathbf{x}_0) = \frac{1}{4\pi} \iint_\Gamma \left[\frac{1}{r} \frac{\partial u}{\partial n} - u \frac{\partial}{\partial n}\left(\frac{1}{r}\right) \right] d\sigma - \frac{1}{4\pi} \iiint_\Omega \frac{\nabla^2 u}{r} \, d\Omega \tag{8.4.32}$$

If $u(\mathbf{x})$ is harmonic, then the volume integral is zero and, if $u(\mathbf{x})$ satisfies Poisson's equation, the volume integral represents a known function.

Let $v(\mathbf{x})$ be any harmonic function that has no singularities. Two integrations by parts yield the relation

$$-\iiint_\Omega u \nabla^2 v \, d\Omega = 0 = \iint_\Gamma \left(v \frac{\partial u}{\partial n} - u \frac{\partial v}{\partial n} \right) d\sigma - \iiint_\Omega v \nabla^2 u \, d\Omega$$

Adding Eq. 8.4.32 and this equation, it follows that

$$u(\mathbf{x}_0) = \iint_\Gamma \left(G \frac{\partial u}{\partial n} - u \frac{\partial G}{\partial n} \right) d\sigma - \iiint_\Omega \nabla^2 u \, G \, d\Omega \tag{8.4.33}$$

where $G(\mathbf{x}, \mathbf{x}_0) = (1/4\pi r) + v$, r is the distance from $\mathbf{x}$ to $\mathbf{x}_0$, and v is any harmonic function that has no singularities.

In the open domain Ω, the function $G(\mathbf{x}, \mathbf{x}_0)$ satisfies the equation $\nabla^2 G = 0$, except at the point $\mathbf{x} = \mathbf{x}_0$, where it has a singularity of the form $1/4\pi r$. Choosing $v(\mathbf{x})$ to satisfy $G = 0$ on Γ; i.e., $v = -1/4\pi r$ on Γ, the function G is called **Green's function** of the first boundary-value problem for Laplace's equation $\nabla^2 u = 0$. Using this function and the fact that $G = 0$ on Γ, Eq. 8.4.33 can be used to represent the solution of $\nabla^2 u = 0$, with $u = f$ on Γ, as

$$u(\mathbf{x}_0) = -\iint_{\Gamma} f \frac{\partial G}{\partial n}\, d\sigma \tag{8.4.34}$$

It should be noted that two assumptions were made in the derivation of Eq. 8.4.34; (1) $\partial G/\partial n$ exists on Γ, and (2) there exists a harmonic function $u(\mathbf{x})$ in Ω that takes on the value f on Γ. The use of Green's function reduces the solution of the first boundary-value problem with arbitrary boundary conditions to the solution of Laplace's equation with the boundary condition $v = -1/4\pi r$ on Γ.

In **electrostatics**, the Green function

$$G(\mathbf{x}, \mathbf{x}_0) = \frac{1}{4\pi r} + v \tag{8.4.35}$$

represents the **potential** at a point $\mathbf{x}$ of a point charge at $\mathbf{x}_0$ inside a grounded conducting surface Γ. In this expression for $G(\mathbf{x}, \mathbf{x}_0)$, $1/4\pi r$ is the potential of the point charge in R^3 and the term v represents the potential of the field that arises from a charge induced on the conducting surface Γ.

In the following discussion of properties of Green's function, it is assumed that the region considered is such that Green's function exists and possesses a continuous normal derivative $\partial G/\partial n$. It is first to be shown that Green's function is a **symmetric function**; i.e., $G(\mathbf{x}, \mathbf{x}_0) = G(\mathbf{x}_0, \mathbf{x})$. To prove this, consider two spheres Γ_1 and Γ_2 of radius ε, with $\mathbf{x}'_0$ and $\mathbf{x}''_0$ as their centers, where $\mathbf{x}'_0$ and $\mathbf{x}''_0$ are arbitrary fixed points in Ω. Let Ω_ε represent the region Ω, with the regions enclosed by Γ_1 and Γ_2 removed, as shown in Fig. 8.4.2. Letting $u = G(\mathbf{x}, \mathbf{x}'_0)$ and $v = G(\mathbf{x}, \mathbf{x}''_0)$ and performing two integrations by parts yields

$$\iiint_{\Omega_\varepsilon} [\, u\nabla^2 v - v\nabla^2 u \,]\, d\tau = \iint_{\Gamma_1 \cup \Gamma_2 \cup \Gamma} \left[u \frac{\partial v}{\partial n} - v \frac{\partial u}{\partial n} \right] d\sigma$$

Since $\nabla^2 G = 0$ in Ω_ε and $G = 0$ on Γ, this equation reduces to

$$\iint_{\Gamma_1} \left[G(\mathbf{x}, \mathbf{x}'_0) \frac{\partial G(\mathbf{x}, \mathbf{x}''_0)}{\partial n} - G(\mathbf{x}, \mathbf{x}''_0) \frac{\partial G(\mathbf{x}, \mathbf{x}'_0)}{\partial n} \right] d\sigma$$

$$+ \iint_{\Gamma_2} \left[G(\mathbf{x}, \mathbf{x}'_0) \frac{\partial G(\mathbf{x}, \mathbf{x}''_0)}{\partial n} - G(\mathbf{x}, \mathbf{x}''_0) \frac{\partial G(\mathbf{x}, \mathbf{x}'_0)}{\partial n} \right] d\sigma = 0$$

Applying Eq. 8.4.34, noting that $\dfrac{\partial G(\mathbf{x}, \mathbf{x}''_0)}{\partial n}$ is smooth on Γ_1 and $\dfrac{\partial G(\mathbf{x}, \mathbf{x}'_0)}{\partial n}$ is smooth on Γ_2, and taking the limit as ε approaches zero yields the desired symmetry relation,

$$G(\mathbf{x}'_0, \mathbf{x}''_0) = G(\mathbf{x}''_0, \mathbf{x}'_0)$$

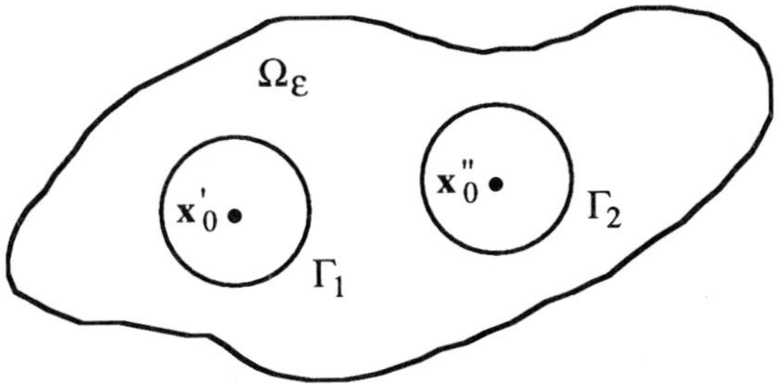

Figure 8.4.2 Domain With Spheres Removed

In a two dimensional domain A, with boundary C, Green's function is characterized by the following conditions:

(1) $\nabla^2 G(\mathbf{x}, \mathbf{x}_0) = 0$, on A, except at the point $\mathbf{x} = \mathbf{x}_0$

(2) At $\mathbf{x} = \mathbf{x}_0$, $G(\mathbf{x}, \mathbf{x}_0)$ has a singularity of the form $\frac{1}{2\pi} \ln \frac{1}{r}$

(3) $G(\mathbf{x}, \mathbf{x}_0) = 0$, on C

In this case, Green's function has the form

$$G(\mathbf{x}, \mathbf{x}_0) = \frac{1}{2\pi} \ln \frac{1}{r} + v(\mathbf{x}) \tag{8.4.36}$$

where $v(\mathbf{x})$ is any harmonic function that is continuous everywhere and satisfies the boundary condition

$$v = -\frac{1}{2\pi} \ln \frac{1}{r}, \quad \text{on C}$$

The solution of the first boundary-value problem for $\nabla^2 u = 0$ is then given by

$$u(\mathbf{x}_0) = -\iint_C f \frac{\partial G}{\partial n} d\sigma \tag{8.4.37}$$

The value of Green's function for Laplace's equation is primarily as a theoretical tool. Green's function is used extensively in Refs. 11 and 18 to prove existence theorems and to investigate properties of solutions of elliptic problems. Exploitation of Green's function in the study of specific mechanics problems tends to be ad hoc in nature, so it will not be discussed here.

EXERCISES 8.4

1. Transform Laplace's equation $\nabla^2 u = 0$ and the boundary condition $u = f(x, y, z)$ on the boundary of a sphere of radius 1; i.e., $x^2 + y^2 + z^2 = 1$, into spherical coordinates using $x = r \sin\theta \cos\phi$, $y = r \sin\theta \sin\phi$, and $z = r \cos\theta$.

2. Solve the boundary-value problem

$$\nabla^2 u = \frac{1}{\rho}\frac{\partial}{\partial \rho}\left(\rho \frac{\partial u}{\partial \rho}\right) + \frac{1}{\rho^2}\frac{\partial^2 u}{\partial \phi^2} = 0$$

on a circle with radius $\rho = 1$, and the boundary condition

$$u(1, \phi) = \sin 3\phi$$

3. Let u be a harmonic function in a domain Ω that is bounded by the surface Γ. Show that $\iint_\Gamma \frac{\partial u}{\partial n}\, d\sigma = 0$ and hence that the second boundary-value problem ($\nabla^2 u = 0$ in Ω and $\frac{\partial u}{\partial n} = f$ on Γ) has a solution only if $\iint_\Gamma f\, d\sigma = 0$.

4. Construct a formal series solution of Laplace's equation inside a circle of radius $\rho = a$, with the boundary condition

$$\frac{\partial u}{\partial \rho}(a, \phi) = f(\phi)$$

where $\int_0^{2\pi} f(\phi)\, d\phi = 0$.

5. Construct solutions of the following boundary-value problems on a circle $r \le 1$, $0 \le \phi \le 2\pi$, if a solution exists:

(a) $\nabla^2 u = 0, \quad u(1, \phi) = c_1$

(b) $\nabla^2 u = 0, \quad u(1, \phi) = c_2 \cos \phi$

(c) $\nabla^2 u = 0, \quad \frac{\partial u}{\partial n}(1, \phi) = c_3 \neq 0$

(d) $\nabla^2 u = 0, \quad \frac{\partial u}{\partial n}(1, \phi) = c_4 \cos \phi$

6. Derive Eq. 8.4.26.

7. Derive Eq. 8.4.27.

8. Find the steady state temperature distribution in a tube, for which the inner temperature is $u(a, \phi) = u_0$ and the outer temperature is $u(b, \phi) = u_1$, where $0 < a < b$.

9. Construct a formal series solution of Laplace's equation outside a circle of radius $\rho = a$ (**exterior problem**), with the boundary condition

$$u(a, \phi) = f(\phi)$$

subject to the condition that u is finite as ρ approaches ∞.

10. Let $\nabla^2 u = 0$ in the circle $r < 1$ and satisfy the boundary condition

$$\frac{\partial u}{\partial n}(1, \theta) = a \cos \theta$$

Derive Green's function for this Neumann problem. Apply this Green's function to obtain the solution.

11. Solve the exterior problem of Exercise 10, using Green's functions.

8.5 PARABOLIC DIFFERENTIAL EQUATIONS

Parabolic Initial-boundary-value Problems

As shown in Section 7.5, the most general linear homogeneous **parabolic differential equation** of second order in two independent variables may be written as

$$A(x, t)\, u_{xx} + B(x, t)\, u_x + C(x, t)\, u = u_t \tag{8.5.1}$$

Defining Lu as the left side of Eq. 8.5.1 yields the more condensed form

$$Lu = u_t \tag{8.5.2}$$

for a general parabolic equation. The left side of Eq. 8.5.2 involves spatial derivatives only. This form carries over into higher space dimensions; i.e., to situations in which u is a function of the space variables x, y, and z and time t.

The most important example of Eq. 8.5.2 is the case in which L is proportional to the **Laplace operator** ∇^2. In this case, Eq. 8.5.2 reduces to the **heat equation**

$$k^2 \nabla^2 u = u_t \tag{8.5.3}$$

As seen in Chapter 6, this equation is the mathematical model for both conduction of heat in a solid and diffusion of gas in a porous medium. Characteristics of Eq. 8.5.3 are the curves, surfaces, or volumes $t = \text{constant}$, since Eq. 8.5.3 fails to determine u_{tt}. As seen in the one space dimension example of Section 8.2, the typical problem that involves this equation is a **characteristic initial-value problem**; i.e., the initial values of u are prescribed along the characteristic $t = 0$, with certain boundary conditions also prescribed. Specifically, in Section 8.2, the initial condition was

$$u(x, 0) = f(x)$$

$0 \leq x \leq \ell$, and the boundary conditions were

$$u(0, t) = 0$$

$$u(\ell, t) = 0$$

for $t \geq 0$.

In higher dimensions (say, three space dimensions) consider a homogeneous body Ω with boundary Γ. The initial condition then takes the form

$$u(x, y, z, 0) = f(x, y, z) \tag{8.5.4}$$

in Ω, which prescribes the initial temperature distribution in Ω. The boundary condition can be expressed in the more general form

$$\alpha \frac{\partial u}{\partial n} + \beta u = 0 \tag{8.5.5}$$

on Γ, where α and β are constants. For example, if $\alpha = 1$ and $\beta \neq 0$, Eq. 8.5.5 states that the rate of change of temperature on Γ in the direction of the inner normal is proportional to the jump in temperature from the exterior to the interior of Ω. For simplicity, however, consider first the case in which $\alpha = 0$ and $\beta = 1$, so Eq. 8.5.5 reduces to

$$u = 0 \tag{8.5.6}$$

on Γ.

Separation of Variables

The method of **separation of variables** that was used in Section 8.2 may also be applied to the initial-boundary-value problem of Eqs. 8.5.3, 8.5.4, and 8.5.6. Substituting

$$u = V(x, y, z)\, T(t) \tag{8.5.7}$$

into Eq. 8.5.3 and manipulating yields

$$\frac{\nabla^2 V}{V} = \frac{T'}{k^2 T} = -\lambda$$

where λ is a constant, since the first and second ratios are functions of only space and time, respectively. This leads to the eigenvalue problem

$$-\nabla^2 V = \lambda V \tag{8.5.8}$$

in Ω and

$$V = 0 \tag{8.5.9}$$

on Γ. Integrating the product of V with both sides of Eq. 8.5.8 over Ω, using the divergence theorem and Eq. 8.5.9,

$$-\iiint_\Omega V\nabla^2 V \, d\Omega = \iiint_\Omega \nabla V^T \nabla V \, d\Omega = \lambda \iiint_\Omega V^2 \, d\Omega$$

which implies $\lambda \geq 0$ and $\lambda = 0$ only if $V = 0$. Thus, $\lambda > 0$.

For a given **eigenvalue** λ_n of Eqs. 8.5.8 and 8.5.9 and its associated **eigenfunction** V_n, a solution of Eq. 8.5.3 has the form

$$u_n = c_n V_n e^{-\lambda_n k^2 t} \tag{8.5.10}$$

As was shown above, the eigenvalues λ_n are all positive. Further, the normalized eigenfunctions V_n form a complete system [18], which may be taken to be orthonormal. This means that f(x, y, z) may be expanded in terms of the V_n as

$$f(x, y, z) = \sum_{n=1}^{\infty} c_n V_n(x, y, z) \tag{8.5.11}$$

where

$$c_n = \iiint_\Omega f V_n \, d\Omega \tag{8.5.12}$$

The formal solution of the initial-boundary-value problem of Eqs. 8.5.3, 8.5.4, and 8.5.6 is then

$$u(x, y, z, t) = \sum_{n=1}^{\infty} c_n V_n(x, y, z) \, e^{-\lambda_n k^2 t} \tag{8.5.13}$$

Note that the positive character of the eigenvalues λ_n implies that u approaches zero asymptotically as t increases, as would be expected from the physical meaning of the problem. Just as in Section 8.4, the factor $e^{-\lambda_n k^2 t}$ leads to the conclusion that, for $t \geq t_0 > 0$, the series of Eq. 8.5.13 converges and can be differentiated term by term.

Example 8.5.1

A copper rod of length 20 units has insulated lateral surfaces. Its left end is maintained at a temperature u(0) = 0°C, and its right end is maintained at a temperature

$u(20) = 100°C$. Determine the temperature of the rod, as a function of x and t, if the initial temperature is

$$u(x, 0) = f(x) = \begin{cases} 10x, & 0 < x < 10 \\ 10, & 10 < x < 20 \end{cases}$$

and, for copper, $k^2 = 1.15 \times 10^{-4}\ m^2/s$.

Since the boundary condition at $x = 20$ is nonhomogeneous, seek first a time-independent solution $\bar{u} = \bar{u}(x)$ of Eq. 8.5.3 that satisfies the boundary condition; i.e., $\bar{u}_{xx} = 0$, which has the general solution

$$\bar{u}(x) = Ax + B$$

To satisfy the boundary conditions, $\bar{u}(0) = 0$ and $\bar{u}(20) = 100$, it is necessary to require that $B = 0$ and $A = 5$. Then the particular solution

$$\bar{u}(x) = 5x$$

satisfies the differential equation and nonhomogeneous boundary conditions.

Writing $u(x, t) = 5x + v(x, t)$, $v(x, t)$ satisfies the differential equation of Eq. 8.5.3 and homogeneous boundary conditions. Separating variables for $v(x, t)$, as in Eq. 8.5.7, and letting $\lambda = \beta^2$,

$$v_\beta(x, t) = e^{-\beta^2 k^2 t}\,[\,C \sin \beta x + D \cos \beta x\,]$$

The boundary condition $v(0, t) = 0$ requires that $D = 0$ and

$$v(20, t) = 0 = Ce^{-\beta^2 k^2 t} \sin 20\beta$$

requires that $\sin 20\beta = 0$, which occurs when

$$\beta = n\pi/20, \quad n = 1, 2, 3, \ldots$$

The solution is obtained by summing these solutions for $v(x, t)$ and adding the particular solution $\bar{u}(x) = 5x$; i.e.,

$$u(x, t) = 5x + \sum_{n=1}^{\infty} C_n e^{-n^2 \pi^2 k^2 t/400} \sin \frac{n\pi x}{20}$$

Finally, the initial condition must be satisfied; i.e.,

$$f(x) = 5x + \sum_{n=1}^{\infty} C_n \sin \frac{n\pi x}{20}$$

This is just the condition that f(x) – 5x can be expanded in a Fourier sine series; i.e.,

$$C_n = \frac{2}{20}\int_0^{20} [\, f(x) - 5x \,] \sin\frac{n\pi x}{20}\, dx$$

$$= \frac{1}{10}\int_0^{10} (\, 10x - 5x \,) \sin\frac{n\pi x}{20}\, dx$$

$$+ \frac{1}{10}\int_{10}^{20} (\, 10 - 5x \,) \sin\frac{n\pi x}{20}\, dx$$

$$= \frac{1}{2}\left[-\frac{20}{n\pi}\, x \cos\frac{n\pi x}{20} + \frac{400}{n^2\pi^2}\sin\frac{n\pi x}{20} \right]_0^{10}$$

$$- \frac{1}{2}\left[\frac{20}{n\pi} \cos\frac{n\pi x}{20} \right]_{10}^{20}$$

$$- \frac{1}{2}\left[-\frac{20}{n\pi}\, x \cos\frac{n\pi x}{20} + \frac{400}{n^2\pi^2}\sin\frac{n\pi x}{20} \right]_{10}^{20}$$

$$= \frac{400}{n^2\pi^2}\sin\frac{n\pi}{2}$$

The solution is thus

$$u(x, t) = 5x + \sum_{n=1}^{\infty} \frac{40.5}{n^2}\sin\frac{n\pi}{2}\, e^{-2.8375\times 10^{-4} n^2 t}\sin\frac{n\pi x}{20}$$ ■

In general, it is necessary to expand the initial condition in a Fourier series in order to find the solution. This is also true for two- or three-dimensional cases. For problems that can be formulated in terms of polar or cylindrical coordinates, Bessel functions are found more suitable. If spherical coordinates are used, Legendre polynomials would appear.

Maximum Principle

Much as in the case of elliptic boundary-value problems, there is a **maximum principle for parabolic equations** that bounds the value of the solution by initial and boundary data. Consider the problem

$$u_t = k^2 u_{xx} \tag{8.5.14}$$

for $0 < t < T$ and $0 < x < \ell$.

Theorem 8.5.1 (Maximum Principle). A continuous function u(x, t) that satisfies the heat equation of Eq. 8.5.14 in the closed region $0 \le t \le T$, $0 \le x \le \ell$ assumes its maximum and minimum values at the initial time t = 0, or at boundary points x = 0 or $x = \ell$ for some time $0 \le t \le T$. ■

Before proving this theorem, note that the function u(x, t) = constant satisfies the heat equation and assumes its maximum and minimum values at each point. However, this does not contradict the theorem, because it means that when a maximum or minimum is assumed in the interior of the region, it is also assumed for t = 0 and for x = 0 or $x = \ell$. The physical significance of the maximum principle is that if the temperature on the boundary and at the initial time does not exceed a value M, then no temperature higher than M can ever be attained in the interior of the body.

To prove Theorem 8.5.1, define M as the maximum value of u(x, t) for t = 0 over $0 \le x \le \ell$ and for x = 0 or $x = \ell$ over $0 \le t \le T$. Let the function u(x, t) take on a relative maximum at an interior point (x_0, t_0); i.e., $0 < x_0 < \ell$ and $0 < t_0 \le T$. For purposes of proof by contradiction, assume that

$$u(x_0, t_0) = M + \varepsilon$$

where $\varepsilon > 0$.

First compare the algebraic signs of terms in Eq. 8.5.14 at point (x_0, t_0). Since the function $u(x, t_0)$ has a relative maximum at (x_0, t_0) in the interior of $(0, \ell)$, it is necessary [8] that

$$u_x(x_0, t_0) = 0$$

and

$$u_{xx}(x_0, t_0) \le 0$$

Since $u(x_0, t)$ has a relative maximum at $t = t_0$, where t_0 could equal T, all that can be concluded is that

$$u_t(x_0, t_0) \ge 0$$

Since the signs on the left and right sides of Eq. 8.5.14 must be the same, it follows that $u_t = u_{xx} = 0$ at (x_0, t_0).

Consider now the auxiliary function

$$v(x, t) = u(x, t) + a(t_0 - t)$$

where $a > 0$ is a constant. Then

$$v(x_0, t_0) = u(x_0, t_0) = M + \varepsilon$$

and

$$a (t_0 - t) \leq aT$$

Selecting $a > 0$ such that $aT < \varepsilon/2$, the maximum of $v(x, t)$, for $t = 0$ and for $x = 0$ or $x = \ell$, does not exceed the value $M + \varepsilon/2$; i.e.,

$$v(x, t) \leq M + \varepsilon/2, \text{ for } t = 0 \text{ or } x = 0 \text{ or } x = \ell \tag{8.5.15}$$

Since $v(x, t)$ is a continuous function, a point (x_1, t_1) exists at which it takes on its maximum. Then

$$v(x_1, t_1) \geq v(x_0, t_0) = M + \varepsilon$$

Therefore, $t_1 > 0$ and $0 < x_1 < \ell$, since for $t = 0$ and for $x = 0$ or $x = \ell$ the inequality of Eq. 8.5.15 is valid. It follows that

$$v_{xx}(x_1, t_1) = u_{xx}(x_1, t_1) \leq 0$$

and

$$v_t(x_1, t_1) = u_t(x_1, t_1) - a \geq 0$$

or

$$u_t(x_1, t_1) \geq a > 0$$

By comparison of the signs on the right and the left sides of Eq. 8.5.14 at the point (x_1, t_1), a contradiction is obtained. Therefore, the first part of the theorem is proved.

The statement for the minimum value can be proved by simply applying the above result to $u_1 = -u$. ■

Consider now a series of consequences of the maximum principle.

Theorem 8.5.2. The solution of the heat equation

$$u_t = k^2 u_{xx} \tag{8.5.16}$$

for $0 < x < \ell$ and $t > 0$, with the initial and boundary conditions

$$u(x, 0) = f(x), \quad 0 \leq x \leq \ell$$

$$u(0, t) = g_1(t), \quad t \geq 0$$

$$u(\ell, t) = g_2(t), \quad t \geq 0$$

is unique. ■

For proof of this theorem, suppose there are two solutions $u_1(x, t)$ and $u_2(x, t)$ of the problem and let

$$v(x, t) = u_2(x, t) - u_1(x, t)$$

Since $u_1(x, t)$ and $u_2(x, t)$ are continuous, their difference is continuous and $v(x, t)$ is also a solution of the heat equation. Consequently, the maximum principle can be applied to this function and the maximum and the minimum of $v(x, t)$ occur for $t = 0$ or for $x = 0$ or $x = \ell$. According to the hypotheses,

$$v(x, 0) = 0, \quad v(0, t) = v(\ell, t) = 0$$

Therefore, $v(x, t) \equiv 0$; i.e.,

$$u_1(x, t) \equiv u_2(x, t)$$

from which uniqueness of the solution of the initial-boundary-value problem follows. ■

Well-posed Initial-boundary-value Problem

Theorem 8.5.3. (1) If two solutions $u_1(x, t)$ and $u_2(x, t)$ of the heat equation satisfy the conditions

$$u_1(x, 0) \le u_2(x, 0), \quad \text{for } 0 \le x \le \ell$$

$$u_1(0, t) \le u_2(0, t), \quad u_1(\ell, t) \le u_2(\ell, t), \quad \text{for } 0 \le t \le T$$

then

$$u_1(x, t) \le u_2(x, t)$$

for all $0 \le x \le \ell$, $0 \le t \le T$.

(2) If three solutions $u(x, t)$, $\underline{u}(x, t)$, and $\bar{u}(x, t)$ of the heat equation satisfy the conditions

$$\underline{u}(x, t) \le u(x, t) \le \bar{u}(x, t)$$

for $t = 0$ and $0 \le x \le \ell$, as well as for $x = 0$ and ℓ and $0 \le t \le T$, then this inequality holds for all x in $0 \le x \le \ell$ and for all t in $0 \le t \le T$.

(3) If, for two solutions $u_1(x, t)$ and $u_2(x, t)$ of the heat equation, the inequality

$$| u_1(x, t) - u_2(x, t) | \le \varepsilon,$$

holds for $t = 0$ and $0 \le x \le \ell$, as well as for $x = 0$ and $x = \ell$ and $0 \le t \le T$, then

$$| u_1(x, t) - u_2(x, t) | \le \varepsilon$$

for all x and t in $0 \le x \le \ell$, $0 \le t \le T$. ■

To see that conclusion (1) is true, note that the difference $v(x, t) = u_2(x, t) - u_1(x, t)$ satisfies the conditions on which the maximum principle is based and

$$v(x, 0) \geq 0, \quad \text{for } 0 \leq x \leq \ell$$

$$v(0, t) \geq 0, \quad v(\ell, t) \geq 0, \quad \text{for } 0 \leq t \leq T$$

Therefore,

$$v(x, t) \geq 0$$

for $0 < x < \ell$ and $0 < t \leq T$, since $v(x, t)$ in the region $0 < x < \ell$, $0 < t \leq T$ would otherwise have a negative value.

Conclusion (2) follows from an application of conclusion (1) to the pairs of functions

$$u(x, t), \; \bar{u}(x, t) \quad \text{and} \quad u(x, t), \; \underline{u}(x, t)$$

Conclusion (3) follows from conclusion (2), when it is applied to solutions of the heat equation,

$$\underline{u}(x, t) = -\varepsilon$$

$$u(x, t) = u_1(x, t) - u_2(x, t)$$

$$\bar{u}(x, t) = \varepsilon$$

■

The question of continuous dependence of the solution of the heat equation on initial and boundary conditions is answered completely by conclusion (3) of Theorem 8.5.3. To see this, consider the solution $u(x, t)$ of the heat equation that corresponds to the initial and boundary conditions

$$u(x, 0) = f(x), \quad u(0, t) = g_1(t), \quad u(\ell, t) = g_2(t)$$

Let $u_1(x, t)$ be a solution that satisfies other initial and boundary conditions given by functions $f^*(x)$, $g_1^*(t)$, and $g_2^*(t)$, which differ by less than ε from the functions $f(x)$, $g_1(t)$, and $g_2(t)$; i.e.,

$$|f(x) - f^*(x)| \leq \varepsilon, \quad |g_1(t) - g_1^*(t)| \leq \varepsilon, \quad |g_2(t) - g_2^*(t)| \leq \varepsilon$$

According to conclusion (3), the function $u_1(x, t)$ differs by less than ε from the function $u(x, t)$; i.e.,

$$|u(x, t) - u_1(x, t)| \leq \varepsilon$$

for all $t \geq 0$ and $0 \leq x \leq \ell$.

The question of uniqueness and continuous dependence of the solution on data for the initial-boundary-value problem with the heat equation of Eq. 8.5.14 has now been investigated in detail. The uniqueness theorem for the first boundary-value problem, for a two- or three-dimensional bounded region, can be proved by repetition of these arguments.

In Section 8.2, it was shown that if f is in $D^1(0, \ell)$ and $g_1 = g_2 = 0$, there is a solution. These conclusions may be summarized by the following theorem.

Theorem 8.5.4. The initial-boundary-value problem for the heat equation is well posed. ■

EXERCISES 8.5

1. The initial temperature of a laterally insulated rod of length ℓ is constant; i.e., $u(x, 0) = u_0$, $0 < x < \ell$, and the ends of the rod are held at constant temperatures $u(0, t) = u_1$ and $u(\ell, t) = u_2$. Find the temperature distribution u(x, t) and the steady state temperature distribution

$$\bar{u}(x) = \lim_{t \to \infty} u(x, t)$$

2. Find the temperature u(x, t) of a rod that is laterally thermally insulated and thermally insulated at $x = 0$. The initial temperature of the rod is zero and a constant heat flow is fed into the rod through the end $x = \ell$.

3. Find the temperature distribution in a laterally insulated rod with end temperatures $u(0, t) = 0$ and $u(\ell, t) = c_1 t$. The initial temperature distribution is $u(x, 0) = 0$. [Hint: Make a change of variable from u(x, t) to v(x, t) such that v(x, t) satisfies homogeneous boundary conditions.]

4. The pressure and temperature of the air in a cylinder of length ℓ are atmospheric. One end of the cylinder is sealed at all times and the other end is opened at time $t = 0$. The concentration of a foreign gas in the surrounding air is u_0 and is initially zero in the cylinder. Find the amount of gas that diffuses into the tube through the open end, as a function of time.

8.6 HYPERBOLIC DIFFERENTIAL EQUATIONS

Fundamental Solutions of the Cauchy Problem

Results derived in Sections 7.5 and 7.7 support the conclusions that follow.

Theorem 8.6.1. (1) When a linear second-order equation in two variables is hyperbolic, it may be reduced to either of the following canonical forms:

$$u_{\xi\xi} - u_{\eta\eta} + \text{lower-order terms} = G_1(\xi, \eta) \tag{8.6.1}$$

or

$$u_{\xi\eta} + \text{lower-order terms} = G_2(\xi, \eta) \tag{8.6.2}$$

(2) The **initial-value problem**, or **Cauchy problem**, consists of a hyperbolic differential equation plus specified values of u and one of its first derivatives along an initial curve Γ, provided the first derivative is not taken in a direction tangent to Γ. Some common ways of satisfying the last requirement are as follows:

(a) Specify u and u_x along a vertical line, such as the y-axis, or specify u and u_y along a horizontal line, such as the x-axis.
(b) Specify u and $\partial u/\partial \nu$ along an arbitrary smooth curve, where ν is a unit vector that is not tangent to the curve.
(c) Specify u and u_x along any smooth curve whose tangent is never horizontal, or specify u and u_y along any curve whose tangent is never vertical.

(3) The Cauchy problem is appropriate for a hyperbolic equation, in the following sense:

(a) Under mild restrictions, the Cauchy problem is well posed; i.e., it has a solution that is both unique and stable.
(b) Cauchy problems that involve hyperbolic equations arise naturally in applications.

(4) Characteristic curves of a hyperbolic differential equation of the form of Eq. 7.5.1 are solutions of the ordinary differential equation

$$A(x, y)\left(\frac{dy}{dx}\right)^2 - 2B(x, y)\frac{dy}{dx} + C(x, y) = 0 \qquad (8.6.3)$$

Such curves are exceptional initial curves for Cauchy problems, in the sense that the differential equation and Cauchy conditions along these curves do not uniquely determine all second derivatives of u. Hence, the Cauchy problem will either not have a solution or it will have many solutions; i.e., either existence or uniqueness is lost.

(5) For a hyperbolic equation, there are two distinct families of characteristics. If the equation is in the first canonical form of Eq. 8.6.1, with leading terms $u_{xx} - u_{yy}$, the characteristics are

$$\begin{aligned} x + y &= \text{constant} \\ x - y &= \text{constant} \end{aligned} \qquad (8.6.4)$$

If the equation is in the second canonical form of Eq. 8.6.2, where u_{xy} is the only second partial derivative in the equation, the characteristics are

$$\begin{aligned} x &= \text{constant} \\ y &= \text{constant} \end{aligned} \qquad (8.6.5)$$

In the latter case, the independent variables x and y are called **characteristic variables**.

(6) For the prototype Cauchy problem, namely, the homogeneous wave equation

$$u_{tt} - a^2 u_{xx} = 0 \tag{8.6.6}$$

and the initial conditions

$$u(x, 0) = f(x) \tag{8.6.7}$$

$$u_t(x, 0) = g(x) \tag{8.6.8}$$

the solution is given by **D'Alembert's formula**

$$u(x, t) = \frac{f(x + at) + f(x - at)}{2} + \frac{1}{2a} \int_{x-at}^{x+at} g(\xi)\, d\xi \tag{8.6.9}$$

This formula provides a classical solution if g has one continuous derivative and f has two continuous derivatives. It yields a physically meaningful solution even if f is in $D^1 \cap C^0$ and g is in D^0. ■

To prove well-posedness, first the fact that Eq. 8.6.9 provides a solution of the Cauchy problem with the wave equation demonstrates existence of a solution. Further, it was shown in Section 8.1 that this solution is unique. Finally, if the initial data f(x) and g(x) are changed slightly, only a small change in the solution occurs. Thus, the Cauchy problem with the wave equation of Eq. 8.6.6 and initial conditions of Eqs. 8.6.7 and 8.6.8 is well posed. This result follows directly from the closed form solution of Eq. 8.6.9. For more general problems, it will not be possible to construct a closed form solution and the proof that the problem is well posed is more difficult. ■

Of the two canonical forms of Eqs. 8.6.1 and 8.6.2, the first arises more naturally in applications. On the other hand, the canonical form of Eq. 8.6.2 is frequently more useful for theoretical purposes. Since it is a simple matter to convert from one form to the other, the most convenient form is used in a particular investigation.

Domains of Dependence and Influence

Besides solving the Cauchy problem for the homogeneous wave equation, D'Alembert's formula of Eq. 8.6.9 yields a property of hyperbolic equations that is not shared by equations of parabolic or elliptic type. This property pertains to the manner in which the solution at any point depends on the initial conditions.

For any fixed point (x_0, t_0), $u(x_0, t_0)$ is given by Eq. 8.6.9 as

$$u(x_0, t_0) = \frac{f(x_0 + at_0) + f(x_0 - at_0)}{2} + \frac{1}{2a} \int_{x_0-at_0}^{x_0+at_0} g(\xi)\, d\xi \tag{8.6.10}$$

This says that the solution of the homogeneous wave equation at (x_0, t_0) is determined solely by the values of f and g in the interval $[\, x_0 - at_0, x_0 + at_0 \,]$ of the x-axis. For this reason, the interval $[\, x_0 - at_0, x_0 + at_0 \,]$ is called the **domain of dependence** of the solution u at (x_0, t_0). It is significant that this domain is the portion of the initial line $(t = 0)$ that is bounded by the two characteristic curves that pass through (x_0, t_0), as shown in Fig. 8.6.1.

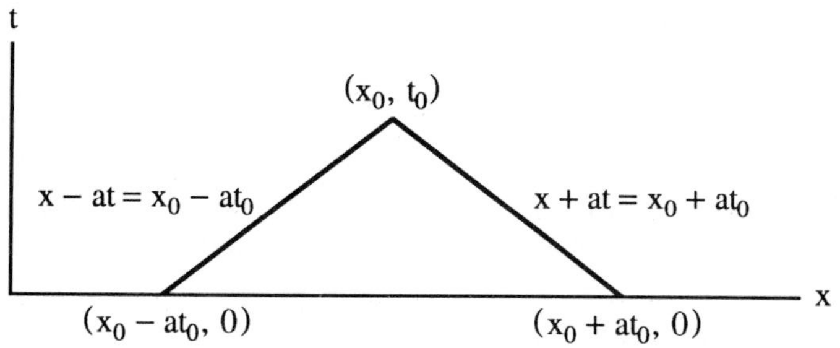

Figure 8.6.1 Domain of Dependence

Consider next a small interval; say, a neighborhood of a point x_0, on the x-axis. At what points in the x-t plane will the value of u be influenced by the values of f and g on this interval? Clearly, all points whose domain of dependence includes the point x_0 will be affected. From the prior observation about the characteristic curves, this must include all points in the infinite sector of the half-plane $t \geq 0$ that lie on or between the characteristic curves that pass through $(x_0, 0)$. If a point (x, t) lies outside this sector, the value of u at this point will not depend on the values of f and g at $(x_0, 0)$.

Instead of considering the influence of values of u and u_t along the x-axis on values of u elsewhere, consider the following more general question: At what points (x, t) do values of u and u_t near a point (x_0, t_0) influence the value $u(x, t)$? To answer this question, consider a Cauchy problem with the Cauchy data prescribed along $t = t_0$, instead of $t = 0$. Then the **domain of influence** of the point (x_0, t_0) on u is the finite sector above (x_0, t_0) between the characteristic lines through (x_0, t_0), as shown in Fig. 8.6.2. The analytic expression for the domain of influence of (x_0, t_0) is $\{\, (x, t) : t \geq t_0, x_0 - a\,(t - t_0) \leq x \leq x_0 + a\,(t - t_0)\, \}$.

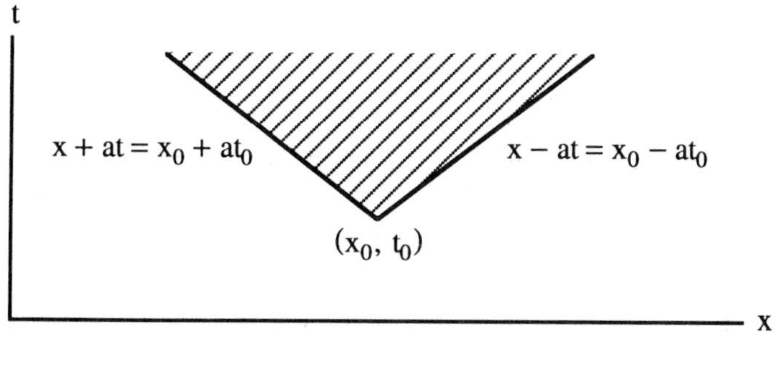

Figure 8.6.2 Domain of Influence

Understanding the domain of dependence and the ability to solve a homogeneous linear equation help in solving the corresponding nonhomogeneous linear equation. For this, consider the nonhomogeneous wave equation

$$u_{tt} - a^2 u_{xx} = h(x, t) \tag{8.6.11}$$

As before, Cauchy conditions are

$$u(x, 0) = f(x), \quad u_t(x, 0) = g(x) \tag{8.6.12}$$

Consider a point $P_0 = (x_0, t_0)$, with $t_0 > 0$, and draw the characteristic curves through P_0. They intersect the x-axis at $P_1 = (x_0 - at_0, 0)$ and $P_2 = (x_0 + at_0, 0)$, as shown in Fig. 8.6.3. Let R be the closed region in Fig. 8.6.3 that is bounded by the triangle $P_0P_1P_2$. Both sides of Eq. 8.6.11 may be integrated over R to obtain

$$\iint_R (u_{tt} - a^2 u_{xx})\, dx\, dt = \iint_R h(x, t)\, dx\, dt \tag{8.6.13}$$

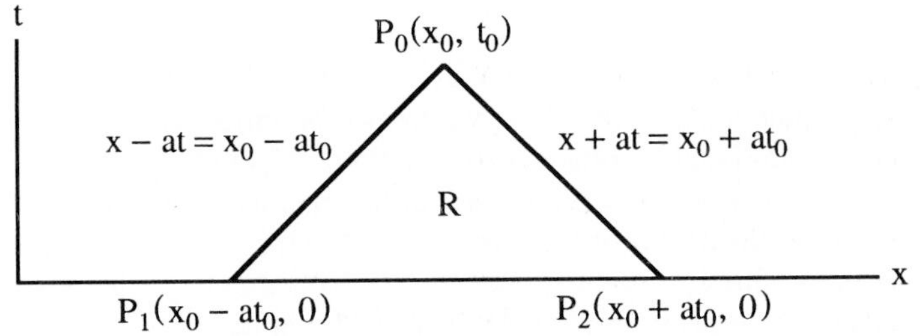

Figure 8.6.3 Domain of Dependence

Applying Green's theorem to the left side of Eq. 8.6.13,

$$\iint_R (u_{tt} - a^2 u_{xx})\, dx\, dt = -\int_C (u_t\, dx + a^2 u_x\, dt) \tag{8.6.14}$$

where C is the boundary of R, traversed counterclockwise; i.e., $C = P_1P_2 + P_2P_0 + P_0P_1$. On P_1P_2, $dt = 0$, so

$$\int_{P_1}^{P_2} (u_t\, dx + a^2 u_x\, dt) = \int_{x_0 - at_0}^{x_0 + at_0} u_t(x, 0)\, dx \tag{8.6.15}$$

On P_2P_0, $dx = -a\, dt$, so

$$\int_{P_2}^{P_0} (u_t\, dx + a^2 u_x\, dt) = \int_{P_2}^{P_0} (-au_t\, dt - au_x\, dx)$$

$$= -\int_{P_2}^{P_0} a\, du = a[u(P_2) - u(P_0)] \tag{8.6.16}$$

On P_0P_1, dx = a dt, so

$$\int_{P_0}^{P_1} (u_t\, dx + a^2 u_x\, dt) = \int_{P_0}^{P_1} (a u_t\, dt + a u_x\, dx)$$

$$= \int_{P_0}^{P_1} a\, du = a\,[\, u(P_1) - u(P_0)\,] \tag{8.6.17}$$

Combining Eqs. 8.6.15 through 8.6.17,

$$\int_C (u_t\, dx + a^2 u_x\, dt)$$

$$= \int_{x_0 - at_0}^{x_0 + at_0} u_t\, dx + a\,u(P_1) + a\,u(P_2) - 2a\,u(P_0) \tag{8.6.18}$$

Substituting Eq. 8.6.18 into Eq. 8.6.14 and using Eq. 8.6.13,

$$u(P_0) = u(x_0, t_0) = \frac{u(P_1) + u(P_2)}{2}$$

$$+ \frac{1}{2a} \int_{x_0 - at_0}^{x_0 + at_0} u_t(x, 0)\, dx + \frac{1}{2a} \iint_R h(x, t)\, dx\, dt \tag{8.6.19}$$

Since (x_0, t_0) is an arbitrary point, the subscripts may be deleted and the coordinates of p_0 written simply as (x, t). The integral over R in Eq. 8.6.19 may also be expressed as a double integral. Finally, substituting the initial values from Eq. 8.6.12 into Eq. 8.6.19 yields

$$u(x, t) = \frac{f(x + at) + f(x - at)}{2} + \frac{1}{2a} \int_{x-at}^{x+at} g(\xi)\, d\xi$$

$$+ \frac{1}{2a} \int_0^t \int_{a\eta + x - at}^{-a\eta + x + at} h(\xi, \eta)\, d\xi\, d\eta \tag{8.6.20}$$

This is the solution of the Cauchy problem for the nonhomogeneous wave equation. Note that it reduces to D'Alembert's formula if h = 0.

Initial-boundary-value Problem

Initial-boundary-value problems, as in Section 6.3, arise more often than pure Cauchy initial-value problems for vibrating strings and membranes. Propagation of waves along a semi-infinite string and along a string of finite length is considered in this section.

A solution of the wave equation

$$a^2 u_{xx} = u_{tt} \tag{8.6.21}$$

for $0 < x < \infty$, $t > 0$, is sought that satisfies the boundary condition

$$u(0, t) = \mu(t) \tag{8.6.22}$$

and the initial conditions

$$\begin{aligned} u(x, 0) &= f(x) \\ u_t(x, 0) &= g(x) \end{aligned} \tag{8.6.23}$$

Consider first the homogeneous boundary condition $u(0, t) = 0$; i.e., propagation of an initial displacement of a string that is fixed at its end $x = 0$. It is to be shown that solutions of the wave equation over an infinite straight line have the following properties:

(1) If the functions $f(x)$ and $g(x)$ in the initial conditions of Eq. 8.6.23 are odd with respect to any point x_0, the solution at this point is equal to zero.
(2) If the functions $f(x)$ and $g(x)$ in the initial conditions of Eq. 8.6.23 are even with respect to any point x_0, then the derivative of the solution with respect to x at this point is equal to zero.

To prove conclusion (1), select x_0 as the origin; i.e., $x_0 = 0$, so $f(x) = -f(-x)$ and $g(x) = -g(-x)$. For $x = 0$, the function $u(x, t)$ defined by Eq. 8.6.9 is

$$u(0, t) = \frac{f(at) + f(-at)}{2} + \frac{1}{2a}\int_{-at}^{at} g(\alpha)\, d\alpha = 0$$

The first term vanishes, since $f(x)$ is odd and the integral of the odd function $g(x)$ is equal to zero. This proves conclusion (1).

For conclusion (2), the functions f and g are even, so $f(x) = f(-x)$ and $g(x) = g(-x)$. Since the derivative of an even function is odd; i.e., $f'(x) = -f'(-x)$, it follows from Eq. 8.6.9 that

$$u_x(0, t) = \frac{f'(at) + f'(-at)}{2} + \frac{1}{2a}\left[\, g(at) - g(-at)\,\right] = 0$$

The first term of the sum vanishes because $f'(x)$ is odd and the second vanishes because $g(x)$ is even. This proves conclusion (2).

With the help of these two observations, the following problem can be solved: Find a solution of Eq. 8.6.21 that satisfies the initial conditions of Eq. 8.6.23 and the boundary condition $u(0, t) = 0$.

The functions $\phi(x)$ and $\psi(x)$ defined by the relations

$$\phi(x) = \begin{cases} f(x), & x > 0 \\ -f(-x), & x < 0 \end{cases}$$

$$\psi(x) = \begin{cases} g(x), & x > 0 \\ -g(-x), & x < 0 \end{cases}$$

are odd extensions of f(x) and g(x). The function

$$u(x, t) = \frac{\phi(x + at) + \phi(x - at)}{2} + \frac{1}{2a}\int_{x-at}^{x+at} \psi(\alpha)\, d\alpha \qquad (8.6.24)$$

is defined for all x and $t > 0$ and satisfies the wave equation. According to conclusion (1) above, $u(0, t) = 0$. Moreover, u(x, t) satisfies the initial conditions

$$u(x, 0) = \phi(x) = f(x)$$

$$u_t(x, 0) = \psi(x) = g(x)$$

for $x > 0$. Thus, the function u(x, t), for $x > 0$ and $t > 0$, satisfies all the requirements of the given problem.

With regard to the original functions f and g,

$$u(x, t) = \begin{cases} \dfrac{f(x + at) + f(x - at)}{2} + \dfrac{1}{2a}\displaystyle\int_{x-at}^{x+at} g(\alpha)\, d\alpha, & t < \dfrac{x}{a},\ x > 0 \\ \dfrac{f(x + at) - f(at - x)}{2} + \dfrac{1}{2a}\displaystyle\int_{at-x}^{x+at} g(\alpha)\, d\alpha, & t > \dfrac{x}{a},\ x > 0 \end{cases}$$

(8.6.25)

As expected, in the region $t < x/a$, the boundary condition has no influence on the solution.

A corresponding development is possible when a free end exists at the point $x = 0$; i.e., $u_x(0, t) = 0$. Form even extensions of f(x) and g(x),

$$\phi(x) = \begin{cases} f(x), & x > 0 \\ f(-x), & x < 0 \end{cases}$$

$$\psi(x) = \begin{cases} g(x), & x > 0 \\ g(-x), & x < 0 \end{cases}$$

As a solution of the wave equation,

$$u(x, t) = \frac{\phi(x + at) + \phi(x - at)}{2} + \frac{1}{2a}\int_{x-at}^{x+at} \psi(\alpha)\, d\alpha$$

or

$$u(x, t) = \begin{cases} \dfrac{f(x + at) + f(x - at)}{2} + \dfrac{1}{2a}\displaystyle\int_{x-at}^{x+at} g(\alpha)\, d\alpha, & t < \dfrac{x}{a} \\[2ex] \dfrac{f(x + at) + f(at - x)}{2} \\ \quad + \dfrac{1}{2a}\left\{ \displaystyle\int_{0}^{x+at} g(\alpha)\, d\alpha + \int_{0}^{at-x} g(\alpha)\, d\alpha \right\}, & t > \dfrac{x}{a} \end{cases}$$

(8.6.26)

In the following, the above application of the **method of even and odd extensions** is of value, even if the initial conditions are defined only for a finite subregion. Results obtained thus far are summarized as the following rules:

(1) For solution of a problem of a semi-infinite string with the boundary condition $u(0, t) = 0$, the initial conditions are continued as odd functions along the entire x-axis.

(2) For solution of a problem of a semi-infinite string with the boundary condition $u_x(0, t) = 0$, the initial conditions are continued as even functions along the entire x-axis.

Example 8.6.1

Let initial conditions be defined on the semi-infinite line $x \geq 0$, with an initial displacement given by the function $f(x)$ that is represented by the triangle shown in Fig. 8.6.4 at $t = 0$ and with $g(x) = 0$. The solution of this problem is obtained by continuing the initial conditions as odd functions along the entire x-axis.

Figure 8.6.4 shows the odd extension at $t = 0$ and the time history of wave propagation. Equation 8.6.25 shows that the initial displacement is distributed as two waves of half the original displacement that progress at a constant velocity toward the origin and two others that propagate away from the origin. This continues as long as the half-waves have not reached the point $x = 0$; i.e., for $t < t_1$ in Fig. 8.6.4. When the half-waves reach the point $x = 0$, a wave with opposite phase develops, as shown by the bold deflected curve. Accordingly, a reflection of the half-wave occurs at the fixed end. Figure 8.6.4 shows the reflection process in several stages. The deflected segment of the string is shortened between t_1 and t_2, it vanishes at t_3, a negative displacement develops between t_3 and t_4, and finally the reflected half-wave moves to the right after t_5. Consequently, the wave due to reflection changes sign at the fixed boundary.

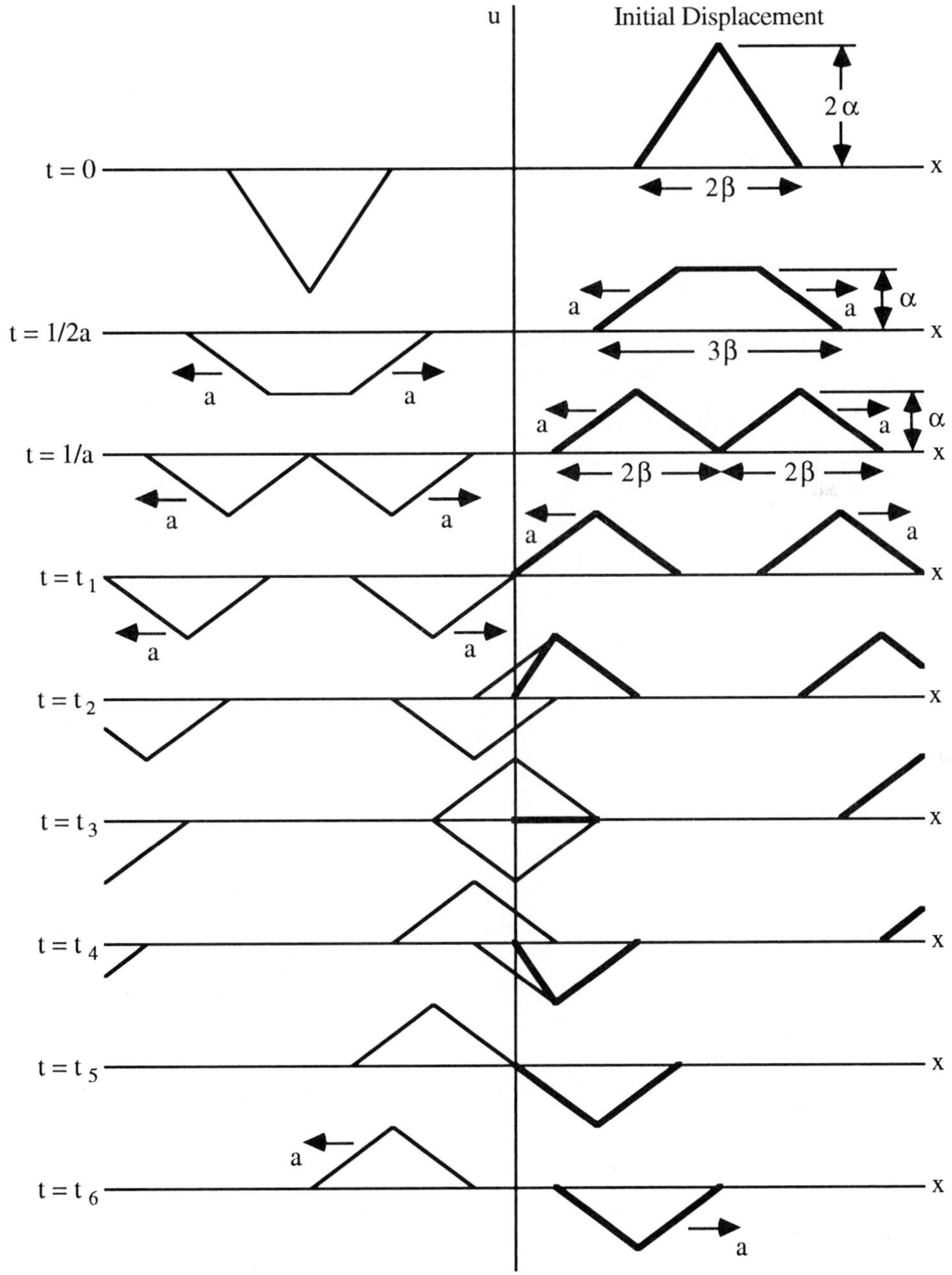

Figure 8.6.4 Reflection of a Wave With Boundary Condition u(0, t) = 0 ■

Consider next the homogeneous wave equation with the homogeneous initial conditions

$$u(x, 0) = 0 = u_t(x, 0)$$

and the nonhomogeneous boundary condition

$$u(0, t) = \mu(t)$$

Such a boundary condition produces a wave that moves toward the right with velocity a. From Eq. 8.1.4, this wave has the analytical form

$$u(x, t) = r(x - at)$$

where r is an arbitrary function, since $x - at = \text{constant}$ is a characteristic curve. The function r may be found using the boundary condition; i.e.,

$$u(0, t) = r(-at) = \mu(t)$$

Thus,

$$r(z) = \mu(-z/a)$$

where $z = -at$, and

$$u(x, t) = \mu[-(x - at)/a] = \mu(t - x/a)$$

This function is defined only for the region $x - at \le 0$, since $\mu(t)$ is defined only for $t \ge 0$. In order to determine u(x, t) for all the values of the arguments, define $\mu(t)$ for negative values of t, setting $\mu(t) = 0$ for $t < 0$. Then

$$u(x, t) = \mu(t - x/a)$$

is defined for all values of the argument and satisfies homogeneous initial conditions.

Recall that Eq. 8.6.25 gives the solution of the homogeneous wave equation, with the boundary condition $u(0, t) = 0$ and general initial conditions. Adding the above solution to the solution of Eq. 8.6.25 yields the solution of the initial-boundary-value problem for the homogeneous wave equation with the boundary condition of Eq. 8.6.22 and initial conditions of Eq. 8.6.23; i.e.,

$$u(x, t) = \begin{cases} \dfrac{f(x + at) + f(x - at)}{2} + \dfrac{1}{2a}\displaystyle\int_{x-at}^{x+at} g(\alpha)\, d\alpha, & t < \dfrac{x}{a} \\[2ex] \mu\left(t - \dfrac{x}{a}\right) + \dfrac{f(x + at) - f(at - x)}{2} \\ \qquad + \dfrac{1}{2a}\displaystyle\int_{at-x}^{x+at} g(\alpha)\, d\alpha, & t > \dfrac{x}{a} \end{cases} \tag{8.6.27}$$

Finally, consider a bounded interval $(0, \ell)$ and seek a solution of

$$u_{tt} = a^2 u_{xx}$$

for $0 < x < \ell$ and $t > 0$ that satisfies the boundary conditions

$$u(0, t) = \mu_1(t)$$

$$u(\ell, t) = \mu_2(t)$$

for $t \geq 0$, and the initial conditions

$$u(x, 0) = f(x)$$

$$u_t(x, 0) = g(x)$$

for $0 \leq x \leq \ell$.

The solution may be found by extension of functions. First, construct the expression

$$u(x, t) = \frac{\phi(x + at) + \phi(x - at)}{2} + \frac{1}{2a}\int_{x-at}^{x+at} \psi(\alpha)\, d\alpha \qquad (8.6.28)$$

where the functions ϕ and ψ are to be defined appropriately. The functions ϕ and ψ are determined on the interval $(0, \ell)$ by the initial conditions

$$u(x, 0) = \phi(x) = f(x)$$

$$u_t(x, 0) = \psi(x) = g(x)$$

In order that $\phi(x)$ and $\psi(x)$ satisfy homogeneous boundary conditions, which are considered first, let $\phi(x)$ and $\psi(x)$ be odd with respect to the points $x = 0$ and $x = \ell$; i.e.,

$$\phi(x) = -\phi(-x), \quad \phi(x) = -\phi(2\ell - x)$$

$$\psi(x) = -\psi(-x), \quad \psi(x) = -\psi(2\ell - x)$$

From these equations, it follows that

$$\phi(x') = \phi(x' + 2\ell)$$

where $x' = -x$. A corresponding relation holds for $\psi(x)$; i.e., ϕ and ψ are periodic functions, with period 2ℓ.

The extensions of $\phi(x)$ and $\psi(x)$ are such that these functions are odd, periodic functions with respect to the origin of coordinates and are defined on the entire line $-\infty < x < \infty$. By introducing these extensions into Eq. 8.6.28, the solution of the problem is obtained.

Finally, consider the propagation of boundary effects. For this purpose the solution of the wave equation is sought with the homogeneous initial conditions

$$u(x, 0) = f(x) = 0, \quad u_t(x, 0) = g(x) = 0$$

and boundary conditions

$$u(0, t) = \mu(t)$$

$$u(\ell, t) = 0$$

It was shown that for $t < \ell/a$, the function $u(x, t) = \hat{\mu}(t - x/a)$, with

$$\hat{\mu}(t) \equiv \begin{cases} \mu(t), & t > 0 \\ 0, & t < 0 \end{cases}$$

is a solution of the wave equation that satisfies the first boundary condition. This function, however, does not satisfy the second boundary condition.

If the value $\hat{\mu}(t - \ell/a)$ is taken as an assigned boundary value at $x = \ell$, for $t \geq \ell/a$, a solution of the form $\tilde{u}(x, t) \equiv s(x + at)$ could be constructed, where s is an arbitrary function, since $x + at =$ constant is a characteristic curve. This requires that $s(\ell + at) = \hat{\mu}(t - \ell/a)$, for $\ell/a \leq t \leq 2\ell/a$. For clarity, put $\ell + at = \theta$. Then $s(\theta) = \hat{\mu}[(\ell + at)/a - 2\ell/a] = \hat{\mu}(\theta/a - 2\ell/a)$. Thus,

$$\tilde{u}(x, t) = s(x + at) = \hat{\mu}\left(\frac{x + at}{a} - \frac{2\ell}{a}\right) = \hat{\mu}\left(t - \frac{2\ell}{a} + \frac{x}{a}\right)$$

The difference between these solutions now satisfies $u(\ell, t) = 0$. Thus, for $0 \leq t \leq 2\ell/a$,

$$u(x, t) = \hat{\mu}\left(t - \frac{x}{a}\right) - \hat{\mu}\left(t - \frac{2\ell}{a} + \frac{x}{a}\right)$$

By repeating this process, a solution is obtained in the form of the series

$$u(x, t) = \sum_{n=0}^{\infty} \hat{\mu}\left(t - \frac{2n\ell}{a} - \frac{x}{a}\right) - \sum_{n=1}^{\infty} \hat{\mu}\left(t - \frac{2n\ell}{a} + \frac{x}{a}\right) \tag{8.6.29}$$

This sum contains only a finite number of nonzero terms, since the argument with each new reflection about $2n\ell/a$ is decreased and $\hat{\mu}(t) = 0$ for $t < 0$. To show that the boundary conditions are satisfied, set $x = 0$. The term for $n = 0$ in the first sum is equal to $\mu(t)$, whereas the remaining terms of the first and second sums cancel pairwise, for equal values of n. Thus, $u(0, t) = \mu(t)$. Replacing n by $n - 1$ in the first sum and varying the limits of summation, the first sum may be written as

$$\sum_{n=1}^{\infty} \hat{\mu}\left(t - \frac{2n\ell}{a} + \frac{2\ell - x}{a}\right)$$

Setting $x = \ell$, it is seen that the terms of the first and second sums cancel each other and $u(\ell, t) = 0$.

Equation 8.6.29 has a simple physical significance. First, $\hat{\mu}(t - x/a)$ represents a wave that is due to the effect of the boundary at $x = 0$, independent of the effect at the point $x = \ell$, as though it were an infinitely long string. The remaining terms represent successive reflections at the point $x = \ell$ (the second sum) and the point $x = 0$ (the first sum).

Similarly,

$$u(x, t) = \sum_{n=0}^{\infty} \hat{\mu}\left(t - \frac{(2n+1)\ell}{a} + \frac{x}{a}\right) - \sum_{n=1}^{\infty} \hat{\mu}\left(t - \frac{(2n+1)\ell}{a} - \frac{x}{a}\right) \tag{8.6.30}$$

is the solution of the homogeneous wave equation with homogeneous initial conditions

$$u(x, 0) = u_t(x, 0) = 0$$

and boundary conditions

$$u(0, t) = 0$$

$$u(\ell, t) = \mu(t)$$

Wave Propagation

Having a general solution of the wave equation permits study of the progression of motion of a string due to the effects of both initial and boundary conditions. It was noted that superposition of terms in D'Alembert's formula and its generalizations could be interpreted as **traveling waves**, with characteristic curves playing a unique role as curves or surfaces across which discontinuities of derivatives of the solution may occur. These concepts can now be developed for a more definitive treatment of wave propagation.

The function $u(x, t)$ that appears in Eq. 8.1.23 as a solution of the wave equation, with initial conditions of Eq. 8.1.19, is the sum of the functions

$$u_1(x, t) = \frac{f(x + at) + f(x - at)}{2}$$

$$u_2(x, t) = \frac{1}{2a} \int_{x-at}^{x+at} g(\xi)\, d\xi$$

The first function $u_1(x, t)$ represents propagation of the initial displacement, with zero initial velocity. The second function $u_2(x, t)$ represents the effect of initial velocity, with zero initial displacement. The function $u(x, t) = u_1(x, t) + u_2(x, t)$ can be interpreted geometrically as a surface in u-x-t space, as shown in Fig. 8.6.5(a). The intersection of this surface with the plane $t = \bar{t}$ is analytically given as $u = u(x, \bar{t})$ and yields the profile of the string at time $\bar{t}$. On the other hand, the intersection of the surface u(x, t) with the plane $x = \bar{x}$ gives $u = u(\bar{x}, t)$, which is the path of motion of the points $\bar{x}$.

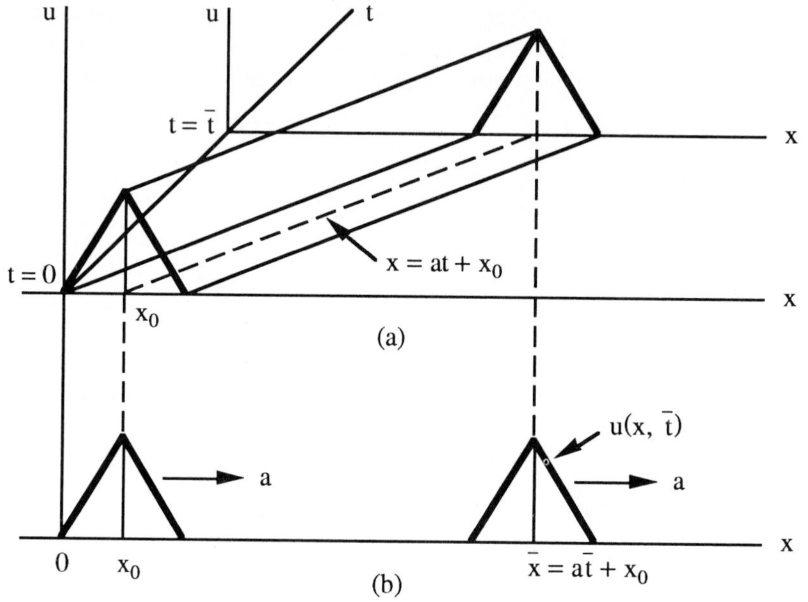

Figure 8.6.5 Propagation of a Disturbance

The function u(x, t) given by $u(x, t) = h(x - at + x_0)$ is described in mechanics as a propagating wave. The displaced profile defined by this function at time t can be illustrated by assuming that an observer moves parallel to the x-axis with velocity a, as shown in Fig. 8.6.5(b). If the observer is at the position x_0 at time t = 0, then at time t he has moved along the path toward the right. If a new coordinate system $x' = x - at + x_0$, $t' = t$ is introduced, which moves with the observer, then $u(x, t) = h(x - at + x_0)$ is defined in the new coordinate system by

$$u(x', t') = h(x')$$

i.e., the observer always sees one profile f(x'), which coincides with the profile h(x) at the initial time t = 0. Therefore, $h(x - at + x_0)$ represents a fixed profile h(x') that moves to the right with velocity a.

Considering the x-t plane, the function $u = h(x - at)$ remains constant on straight lines $x - at = \text{constant}$. The surface $u = h(x - at)$ is a cylindrical surface whose generators are parallel to the straight line x = at. Thus, the form of the cylindrical surface is determined by the profile of the initial displacement.

From an analytical point of view, it has been found that the differential equation fails to determine all second derivatives along characteristic curves, so discontinuities in second derivatives of the solution could occur along characteristic curves, which are called

waves. It is next shown that this interpretation is correct and that something can be said about the magnitude of the jump discontinuity.

Let C be a curve defined by $\phi(x, y) = 0$ that separates the regions $\phi > 0$ and $\phi < 0$ and let u(x, y) be a solution of a differential equation

$$au_{xx} + 2bu_{xy} + cu_{yy} + du_x + eu_y + fu = 0 \tag{8.6.31}$$

in each of these regions, such that u and its first derivatives are continuous across C, but second derivatives of u may suffer jumps across C. Tangential derivatives, however, remain continuous along $\phi = 0$, in the following sense. If $\lambda = \psi$ and $\eta = \phi$ are coordinates on the integral surface J in a neighborhood of C, where λ is arclength on C, then derivatives of u, u_x, and u_y with respect to λ remain continuous across C.

Denote by [h] the jump of a function h across C, in the direction of increasing values of ϕ; i.e., for a point (x, y) on C, $[\ h(x, y)\ ] = h(x, y)_{\phi>0} - h(x, y)_{\phi<0}$. By assumption, u_x is continuous and so is its tangential derivative, given by $u_{xx}\phi_y - u_{xy}\phi_x$, as well as the tangential derivatives of u_y, given by $u_{xy}\phi_y - u_{yy}\phi_x$. Thus, any jumps in second derivatives must satisfy the relations

$$[\ u_{xx}\ ]\,\phi_y - [\ u_{xy}\ ]\,\phi_x = 0$$

$$[\ u_{xy}\ ]\,\phi_y - [\ u_{yy}\ ]\,\phi_x = 0$$

which have many solutions, one of which is

$$[\ u_{xx}\ ] = \kappa\phi_x^{\,2}, \quad [\ u_{xy}\ ] = \kappa\phi_x\phi_y, \quad [\ u_{yy}\ ] = \kappa\phi_y^{\,2} \tag{8.6.32}$$

for any factor of proportionality κ.

Consider now the differential equation of Eq. 8.6.31 at two points P_1 and P_2 on different sides of the curve C. Subtracting the two resulting equations from each other and allowing P_1 and P_2 to converge to a point P on C, the continuous lower-order terms cancel and

$$a\,[\ u_{xx}\ ] + b\,[\ u_{xy}\ ] + c\,[\ u_{yy}\ ] = 0$$

Substituting for the jump terms from Eq. 8.6.32 and canceling the factor κ,

$$a\phi_x^{\,2} + 2b\phi_x\phi_y + c\phi_y^{\,2} = 0$$

which is simply Eq. 7.5.16, which defines characteristic curves for Eq. 8.6.31. Thus, it is confirmed that discontinuities in the solution of the type indicated can occur only along **characteristic curves.**

For physical interpretation, consider y = t as time and think of the solution u(x, t). If this wave has a discontinuity along the characteristic curve $\phi(x, t) = 0$, solve $\phi = 0$ for x = x(t), with

$$\frac{dx}{dt} = -\frac{\phi_t}{\phi_x}$$

The curves $\phi = 0$ of discontinuity in the x-t plane can then be interpreted as paths of points of discontinuity that move along the x-axis with velocity dx/dt.

Separation of Variables

The method of separation of variables, which has been used to solve parabolic and selected elliptic equations, is also applicable in solving wave equations. The purpose of this subsection is threefold: (1) It provides practice and insight in using the separation of variables technique, (2) it provides motivation for the theoretical development of the general eigenfunction expansion method, and (3) it provides insights into problems of physical interest.

Example 8.6.2 (The Vibrating String)

Paralleling the idea of the separation of variables technique for the heat equation in Section 8.2, consider the vibrating string problem studied in Section 6.3. A homogeneous differential equation will be considered here, so Eqs. 6.3.8, 6.3.9, and 6.3.10 are

$$u_{tt} - a^2 u_{xx} = 0 \tag{8.6.33}$$

$$\begin{aligned} u(x, 0) &= f(x) \\ u_t(x, 0) &= g(x) \end{aligned} \tag{8.6.34}$$

for $0 \le x \le \ell$, and

$$\begin{aligned} u(0, t) &= 0 \\ u(\ell, t) &= 0 \end{aligned} \tag{8.6.35}$$

for $t \ge 0$, where f(x) and g(x) are the initial displacement and initial velocity, respectively.

For separation of variables, a solution is sought of the form

$$u(x, t) = X(x)\, T(t) \tag{8.6.36}$$

Substituting this into Eq. 8.6.33,

$$XT'' = a^2 X'' T$$

This equation can be written as

$$\frac{X''}{X} = \frac{1}{a^2}\frac{T''}{T} \tag{8.6.37}$$

providing $XT \neq 0$. Since the left side of this equation is independent of t and the right side is independent of x,

$$\frac{X''}{X} = \frac{1}{a^2}\frac{T''}{T} = -\alpha^2$$

where the separation constant $-\alpha^2$ is shown below to be negative. Thus,

$$X'' + \alpha^2 X = 0 \tag{8.6.38}$$

and

$$T'' + \alpha^2 a^2 T = 0 \tag{8.6.39}$$

The first boundary condition of Eq. 8.6.35 requires that

$$u(0, t) = X(0)\, T(t) = 0$$

Since T(t) must not be identically zero,

$$X(0) = 0 \tag{8.6.40}$$

In a similar manner, the second boundary condition of Eq. 8.6.35 implies

$$X(\ell) = 0 \tag{8.6.41}$$

To determine X(x), first solve the eigenvalue problem

$$\begin{aligned} -X'' &= \alpha^2 X \\ X(0) &= X(\ell) = 0 \end{aligned} \tag{8.6.42}$$

Note that this eigenvalue problem is identical to Eqs. 8.2.10 and 8.2.12, which arose in solving the heat equation. The resulting infinite sequence of positive eigenvalues is

$$\alpha_n^2 = \left(\frac{n\pi}{\ell}\right)^2, \quad n = 1, 2, 3, \ldots \tag{8.6.43}$$

Corresponding to these eigenvalues, the eigenfunctions are

$$\sin\left(\frac{n\pi}{\ell}\right)x, \quad n = 1, 2, 3, \ldots$$

It is not necessary to consider negative values of n, since

$$\sin(-n)\frac{\pi x}{\ell} = -\sin\frac{n\pi x}{\ell}$$

so no new solution is obtained with negative n.

The solutions of Eq. 8.6.42 are thus

$$X_n(x) = B_n \sin \frac{n\pi x}{\ell} \tag{8.6.44}$$

where the B_n are arbitrary constants.

For $\alpha = \alpha_n$, the general solution of Eq. 8.6.39 may be written in the form

$$T_n(t) = C_n \cos \frac{n\pi a}{\ell} t + D_n \sin \frac{n\pi a}{\ell} t \tag{8.6.45}$$

where C_n and D_n are arbitrary constants.

Thus, the functions

$$u_n(x, t) = X_n(x)\, T_n(t) = \left(a_n \cos \frac{n\pi a}{\ell} t + b_n \sin \frac{n\pi a}{\ell} t \right) \sin \frac{n\pi x}{\ell}$$

satisfy Eq. 8.6.33 and the boundary conditions of Eq. 8.6.35, for arbitrary values of $a_n \equiv B_n C_n$ and $b_n \equiv B_n D_n$.

Since Eq. 8.6.33 is linear and homogeneous, the series

$$u(x, t) = \sum_{n=1}^{\infty} \left(a_n \cos \frac{n\pi a}{\ell} t + b_n \sin \frac{n\pi a}{\ell} t \right) \sin \frac{n\pi x}{\ell} \tag{8.6.46}$$

is also a solution, provided it converges and its limit is twice differentiable with respect to x and t. Since each term of the series satisfies Eq. 8.6.35, the series satisfies these conditions. Two initial conditions remain to be satisfied, from which the constants a_n and b_n will be determined.

First differentiate the series in Eq. 8.6.46 with respect to t, to obtain

$$u_t = \sum_{n=1}^{\infty} \frac{n\pi a}{\ell} \left(-a_n \sin \frac{n\pi a}{\ell} t + b_n \cos \frac{n\pi a}{\ell} t \right) \sin \frac{n\pi x}{\ell} \tag{8.6.47}$$

Applying the initial conditions of Eq. 8.6.34,

$$u(x, 0) = f(x) = \sum_{n=1}^{\infty} a_n \sin \frac{n\pi x}{\ell}$$

$$u_t(x, 0) = g(x) = \sum_{n=1}^{\infty} b_n \left(\frac{n\pi a}{\ell} \right) \sin \frac{n\pi x}{\ell}$$

These equations will be satisfied if f(x) and g(x) are expandable in Fourier sine series. The coefficients are given by

$$a_n = \frac{2}{\ell}\int_0^{\ell} f(x)\sin\frac{n\pi x}{\ell}\,dx$$

$$b_n = \frac{2}{n\pi a}\int_0^{\ell} g(x)\sin\frac{n\pi x}{\ell}\,dx \tag{8.6.48}$$

The formal solution of the vibrating string problem is therefore given by Eq. 8.6.46, where the coefficients a_n and b_n are determined by Eq. 8.6.48. ■

While the separation of variables technique has been applied thus far only to problems with no external forcing function, it is applicable to problems with nonhomogeneous differential equations. The treatment presented here will serve as the model for a more general treatment of forced vibration problems, known as the **eigenfunction expansion technique**. In the next example, a formal solution is developed for the general problem of forced motion of the vibrating string of Example 8.6.2.

Example 8.6.3

The governing equations for the forced motion of a string are the differential equation

$$u_{tt} - a^2u_{xx} = h(x, t) \tag{8.6.49}$$

the initial conditions

$$u(x, 0) = f(x)$$
$$u_t(x, 0) = g(x) \tag{8.6.50}$$

for $0 \le x \le \ell$, and the boundary conditions

$$u(0, t) = 0$$
$$u(\ell, t) = 0 \tag{8.6.51}$$

for all $t \ge 0$.

From Eq. 8.6.42, the associated eigenvalue problem is

$$-X'' = \alpha^2 X$$

$$X(0) = X(\ell) = 0$$

The eigenvalues for this problem, according to the solution of the homogeneous problems are

$$\alpha_n^2 = \left(\frac{n\pi}{\ell}\right)^2, \quad n = 1, 2, 3, \ldots$$

and the corresponding eigenfunctions are

$$\sin\left(\frac{n\pi}{\ell}\right)x, \quad n = 1, 2, 3, \ldots$$

A solution to the nonhomogeneous equation is sought in the form

$$u(x, t) = \sum_{n=1}^{\infty} T_n(t) \sin \frac{n\pi x}{\ell} \tag{8.6.52}$$

and the forcing function f(x, t) is expanded as

$$h(x, t) = \sum_{n=1}^{\infty} h_n(t) \sin \frac{n\pi x}{\ell} \tag{8.6.53}$$

where $h_n(t)$ are given by

$$h_n(t) = \frac{2}{\ell}\int_0^{\ell} h(x, t) \sin \frac{n\pi x}{\ell}\, dx \tag{8.6.54}$$

Note that u(x, t) of Eq. 8.6.52 automatically satisfies the homogeneous boundary conditions.

Inserting u(x, t) and h(x, t) in the wave equation, coefficients of $\sin(n\pi x/\ell)$ are obtained as

$$T_n'' + \alpha_n^2 a^2 T_n - h_n = 0$$

where it is assumed that u is twice continuously differentiable with respect to t. Thus, the solution of this ordinary differential equation takes the form

$$T_n(t) = A_n \cos \alpha_n at + B_n \sin \alpha_n at$$

$$+ \frac{1}{\alpha_n a}\int_0^t h_n(\tau) \sin \alpha_n a\,(t - \tau)\, d\tau \tag{8.6.55}$$

The initial conditions give

$$u(x, 0) = f(x) = \sum_{n=1}^{\infty} A_n \sin \frac{n\pi x}{\ell}$$

Assuming f(x) is continuous in x, the coefficient A_n of the Fourier series is given by

$$A_n = \frac{2}{\ell}\int_0^{\ell} f(x)\sin\frac{n\pi x}{\ell}\,dx$$

Similarly, from the remaining initial condition,

$$u_t(x, 0) = g(x) = \sum_{n=1}^{\infty} B_n\alpha_n a \sin\frac{n\pi x}{\ell}$$

Hence, for continuous g(x),

$$B_n = \frac{2}{\ell\alpha_n a}\int_0^{\ell} g(x)\sin\frac{n\pi x}{\ell}\,dx$$

The solution of the initial-boundary-value problem is, therefore,

$$u(x, t) = \sum_{n=1}^{\infty} T_n(t)\sin\frac{n\pi x}{\ell} \tag{8.6.56}$$

provided the series for u(x, t) and its first and second derivatives converge uniformly.

As a specific example, let $h(x, t) = e^x \cos\omega t$ and $f(x) = g(x) = 0$. Then

$$\begin{aligned} h_n(t) &= \frac{2n\pi^2}{(n^2\pi^2 + \ell^2)}[1 + (-1)^{n+1}e^{\ell}]\cos\omega t \\ &\equiv c_n \cos\omega t \end{aligned}$$

Hence,

$$\begin{aligned} T_n(t) &= \frac{1}{\alpha_n a}\int_0^t c_n \cos\omega\tau \,\sin\alpha_n a(t-\tau)\,d\tau \\ &= \frac{c_n}{(\alpha_n^2 a^2 - \omega^2)}(\cos\omega t - \cos\alpha_n at) \end{aligned}$$

provided $\omega \neq \alpha_n a$. Thus, the formal solution may be written in the form

$$u(x, t) = \sum_{n=1}^{\infty} \frac{c_n}{(\alpha_n^2 a^2 - \omega^2)}(\cos\omega t - \cos\alpha_n at)\sin\frac{n\pi x}{\ell}$$

■

EXERCISES 8.6

1. Let Γ be a smooth curve that is defined parametrically by

$$x = \phi(t), \quad y = \psi(t)$$

with $\psi'(t) \neq 0$ for any t. Let u and u_x be specified along Γ by

$$u = f(t), \quad u_x = g(t)$$

where f is differentiable. Let $v = [\cos\theta, \sin\theta]^T$ be a unit vector in an arbitrary direction. Show that $\partial u/\partial v$ can be found as a function of t along Γ.

2. Verify by direct substitution that Eq. 8.6.20 satisfies the partial differential equation of Eq. 8.6.11 and the initial conditions of Eq. 8.6.12.

3. Verify that if f is piecewise continuously differentiable and if g and h are just piecewise continuous, then u from Eq. 8.6.20 satisfies the momentum balance law of Eq. 6.3.6 for the vibrating string.

4. Let the ends $x = 0$ and $x = \ell$ of a string be fixed, let the initial deflection be

$$u(x, 0) = C_1 \sin \frac{\pi x}{\ell}$$

for $0 \leq x \leq \ell$, and let the initial velocity be zero. Find the deflection for all $t > 0$, by making an appropriate extension of the initial condition to the whole line.

5. Solve the initial-boundary-value problem of Eqs. 8.6.33 – 8.6.35, with $f(x) = \sin x$, $g(x) = 0$, and $\ell = \pi$.

6. Solve the initial-boundary-value problem of Eqs. 8.6.49 – 8.6.51, with $h(x, t) = kx$, $f(x) = g(x) = 0$, and $\ell = 1$.

9

LINEAR OPERATOR THEORY IN MECHANICS

One of the most useful concepts in the study of mechanics is the linear operator. Finite-dimensional linear operators (namely, matrices) have been studied in Chapters 1 to 3. Many of the techniques and results developed for matrices also apply in the study of more general linear operators. In infinite-dimensional function spaces, however, technical complexities arise. Fortunately, most work can be done effectively in L_2, which is a Hilbert space. In this setting, there is a strong mathematical relation between force and displacement functions, which also provides physical insight into problems of mechanics. Finally, boundary-value problems that arise naturally in applications are studied and completeness of eigenfunctions is established.

9.1 FUNCTIONALS

Supremum and Infimum of Real Valued Functions

A real valued function $f(\mathbf{x})$, $\mathbf{x} \in \Omega \subset R^n$, is said to be **bounded from above** if there exists a real number a such that $f(\mathbf{x}) \leq a$, for all $\mathbf{x} \in \Omega \subset R^n$. The real number a is called an **upper bound** of the function. Similarly, the function is said to be **bounded from below** if there exists a real number b such that $b \leq f(\mathbf{x})$, for all $\mathbf{x} \in \Omega \subset R^n$. The real number b is called a **lower bound** of the function. If a function is bounded from above and from below, then the function is said to be bounded. An upper (or lower) bound α (or β) of the function is said to be the **maximum** (or **minimum**) of $f(\mathbf{x})$ on W if there exists an $\mathbf{x}_0 \in \Omega \subset R^n$ such that $f(\mathbf{x}_0) = \alpha$ [or $f(\mathbf{x}_0) = \beta$]. A bounded function may or may not have a maximum or minimum. If a real valued function is continuous on a closed and bounded set W, then it has a maximum and minimum [8, 14]. This is the reason that the maximum was used for stability of solutions in Section 8.4 (below Corollary 8.4.1).

Every real valued function that is bounded from above has a least upper bound, and every real valued function that is bounded from below has a greatest lower bound. Let $f(\mathbf{x})$, $\mathbf{x} \in \Omega \subset R^n$, be a function that is bounded above, and let A be the set of all real numbers that are upper bounds of $f(\mathbf{x})$, $\mathbf{x} \in \Omega \subset R^n$. The **least upper bound** of $f(\mathbf{x})$, $\mathbf{x} \in \Omega \subset R^n$, is the smallest number in A, called the **supremum** of $f(\mathbf{x})$, $\mathbf{x} \in \Omega \subset R^n$, and is denoted by $\sup_{\mathbf{x} \in \Omega} f(\mathbf{x})$. Similarly, for a function that is bounded below, let B denote the set of all real numbers that are lower bounds of $f(\mathbf{x})$, $\mathbf{x} \in \Omega \subset R^n$. The largest number in B is called the **infimum** of $f(\mathbf{x})$, $\mathbf{x} \in \Omega \subset R^n$, and is denoted by $\inf_{\mathbf{x} \in \Omega} f(\mathbf{x})$.

Example 9.1.1

Let f(x) be the function

$$f(x) = x, \quad x \in (0, 1]$$

The set A of all upper bounds of f(x) and the set B of all lower bounds of f(x) are

$$A = \{ a: a \in R^1, 1 \le a < \infty \}, \quad B = \{ b: b \in R^1, -\infty < b \le 0 \}$$

The maximum of f(x) is 1, which is achieved at $x_0 = 1 \in (0, 1]$. However, f(x) has no minimum, since $f(\varepsilon) = \varepsilon > 0$ is achieved for arbitrary small ε, but for $f(x_0) = 0$, $x_0 = 0 \notin (0, 1]$. Notice that, even though f(x) is continuous, the set (0, 1] is not closed. The least upper bound of f(x) is 1; i.e., $\sup_{x \in (0, 1]} f(x) = 1$, and the greatest lower bound of f(x) is 0; i.e., $\inf_{x \in (0, 1]} f(x) = 0$.

Example 9.1.2

Let f(x) be the function

$$f(x) = \begin{cases} x, & 0 \le x < 1 \\ x/3, & 1 \le x \le 2 \end{cases}$$

The set A of all upper bounds of f(x) and the set B of all lower bounds of f(x) are

$$A = \{ a: a \in R^1, 1 \le a < \infty \}, \quad B = \{ b: b \in R^1, -\infty < b \le 0 \}$$

The minimum of f(x) is 0, but f(x) has no maximum. Notice that, even though the domain $0 \le x \le 2$ is closed and bounded, the function f(x) is not continuous. The least upper bound of f(x) is 1; i.e., $\sup_{x \in [0, 2]} f(x) = 1$, and the greatest lower bound of f(x) is 0; i.e., $\inf_{x \in [0, 2]} f(x) = 0$.

Bounded Functionals

Let H be a Hilbert space and S be a subset of H. If to each u in S there corresponds a real number T(u), then T is called a **functional** on S. The subset S is called the **domain** of the functional T and is denoted D_T. A functional T is said to be a **bounded functional** if there exists a finite constant c such that

$$|T(u)| \le c\,\|u\| \tag{9.1.1}$$

for all u in D_T. The smallest c for which this inequality holds is called the **norm of the functional** T, denoted as $\|T\|$, and is given by

$$\|T\| = \sup_{u \in D_T,\, u \neq 0} \frac{|T(u)|}{\|u\|} \tag{9.1.2}$$

Thus, $|T(u)| \le \|T\|\,\|u\|$, for all u in D_T. A functional T is a **continuous functional** at u^0 if, for any $\varepsilon > 0$, there exists a $\delta > 0$ such that if $\|u^0 - u\| < \delta$, then $|T(u^0) - T(u)| < \varepsilon$. Equivalently, T is continuous at u^0 if $\lim_{n \to \infty} u^n = u^0$ implies that $\lim_{n \to \infty} T(u^n) = T(u^0)$ [15, 18].

Example 9.1.3

On a Hilbert space H, the scalar product is real valued and defines the functional

$$T(u) = (u, u) = \|u\|^2 \tag{9.1.3}$$

To see that this functional is not bounded, note that for any $v \neq 0$ in H, $u = \alpha v$ is in H for any scalar $\alpha \ge 0$. For any finite constant c, choosing $\alpha > c/\|v\|$,

$$T(u) = \|u\|\,\|u\| = \alpha\,\|v\|\,\|u\| > c\,\|u\|$$

so T is not bounded.

Even though T is not bounded, it is continuous at all u^0 in H. To see this, let $u^0 \neq 0$ be in H. Then, for any u in H,

$$\begin{aligned} |T(u^0) - T(u)| &= |(u^0, u^0) - (u, u)| \\ &= |(u^0, u^0) - (u, u^0) + (u, u^0) - (u, u)| \\ &= |(u^0 - u, u^0) - (u, u - u^0)| \\ &\le |(u^0 - u, u^0)| + |(u, u - u^0)| \\ &\le \|u^0 - u\|\,\|u^0\| + \|u\|\,\|u - u^0\| \end{aligned} \tag{9.1.4}$$

For any $\varepsilon > 0$, choose $\delta < (\varepsilon/4)\,\|u^0\|$ small enough so that for $\|u - u^0\| < \delta$, $\|u\| < 2\,\|u^0\|$. This is possible since

$$\|u\| - \|u^0\| \le \|u - u^0\| \le \delta$$

implies that

$$\| u \| \le \| u^0 \| + \delta < 2 \| u^0 \|$$

Now, Eq. 9.1.4 yields

$$| T(u^0) - T(u) | \le \frac{\varepsilon}{4} + \frac{\varepsilon}{2} = \frac{3}{4}\varepsilon < \varepsilon$$

if $\| u - u^0 \| < \delta$. For the remaining possibility, $u^0 = 0$,

$$| T(u^0) - T(u) | = | T(u) | = \| u \|^2$$

For any $\varepsilon > 0$, let $\delta < \sqrt{\varepsilon}$. Then, $\| u - u^0 \| = \| u \| < \delta$ implies that $| T(u^0) - T(u) | < \varepsilon$. Thus, even though T is not bounded, it is continuous at all u^0 in H. ■

Linear Functionals

Definition 9.1.1. A functional T is said to be a **linear functional** if (1) its domain D_T is a linear space and (2)

$$\begin{aligned} T(u^1 + u^2) &= T(u^1) + T(u^2) \\ T(\alpha u) &= \alpha T(u) \end{aligned} \tag{9.1.5}$$

for all u^1 and u^2 in D_T and all real α. ■

Note that Eq. 9.1.5 is equivalent to $T(\alpha u^1 + \beta u^2) = \alpha T(u^1) + \beta T(u^2)$ and, if T is linear, then $T(0) = 0$.

Example 9.1.4

For any f in a Hilbert space H, the functional

$$T(u) = (f, u) \tag{9.1.6}$$

with domain $D_T = H$ is a linear functional, since its domain is a linear space and it satisfies Eq. 9.1.5. By the Schwartz inequality,

$$| T(u) | = | (f, u) | \le \| f \| \| u \|$$

Since $\| f \|$ is finite, T is bounded and $\| T \| \le \| f \|$. In fact, $\| T \| = \| f \|$. To show this, for purposes of contradiction, assume that $\| T \| < \| f \|$. Evaluating Eq. 9.1.6 at f,

$$T(f) = (f, f) = \| f \| \| f \| > \| T \| \| f \|$$

which is a contradiction. Thus, indeed $\| T \| = \| f \|$.

For any u^0 in H,

$$| T(u^0) - T(u) | = | (f, u^0 - u) |$$

$$\leq \| f \| \| u^0 - u \|$$

If $\| f \| \neq 0$, choose $\delta < \varepsilon / \| f \|$. Thus, $\| u^0 - u \| < \delta$ implies $| T(u^0) - T(u) | < \varepsilon$, so T is continuous. If $\| f \| = 0$, $T(u) = 0$ is trivially continuous. ■

Example 9.1.5

Let T be a linear functional on R^n and $\mathbf{e}^1, \mathbf{e}^2, \ldots, \mathbf{e}^n$ be the usual basis vectors for R^n; i.e., $\mathbf{e}^i = [\, \delta_{ij} \,]_{n \times 1}$. Then, any $\mathbf{u}$ in R^n can be written as $\mathbf{u} = u_i \mathbf{e}^i$ and $T(\mathbf{u}) = u_i T(\mathbf{e}^i)$. Let $\mathbf{f} = [\, T(\mathbf{e}^1), \ldots, T(\mathbf{e}^n) \,]^T$. Since any vector $\mathbf{u}$ can be represented as $\mathbf{u} = u_i \mathbf{e}^i$ and T is linear, $T(\mathbf{u}) = u_i T(\mathbf{e}^i) = \mathbf{f}^T \mathbf{u} = (\mathbf{f}, \mathbf{u})$. That is, for any linear functional T on R^n, there exists a vector $\mathbf{f}$ in R^n such that $T(\mathbf{u}) = (\mathbf{f}, \mathbf{u})$. As shown in Example 9.1.4, this is a bounded linear functional. ■

The theorems that follow demonstrate some of the remarkable regularity properties of linear functionals.

Theorem 9.1.1. If a linear functional T is continuous at $u = 0$, it is continuous on its entire domain D_T. ■

To prove Theorem 9.1.1, let u^n be a sequence in D_T with $\lim_{n \to \infty} u^n = u \neq 0$ in D_T. It is to be shown that $\lim_{n \to \infty} T(u^n) = T(u)$. By linearity of T, $T(u) - T(u^n) = T(u - u^n)$. Since D_T is a linear space, the sequence $v^n = u - u^n$ is in D_T and v^n approaches 0. Since T is continuous at 0, $T(v^n) = T(u - u^n) = T(u) - T(u^n)$ approaches $T(0) = 0$, as required. ■

As a result of Theorem 9.1.1, it is not necessary to speak of a linear functional as continuous at a particular u^0 in its domain. A linear functional is either continuous or not continuous everywhere in its domain.

Theorem 9.1.2. A linear functional T is continuous if and only if it is bounded. ■

To prove Theorem 9.1.2, suppose first that T is bounded. Then

$$| T(u^n) - T(u) | = | T(u^n - u) | \leq \| T \| \| u^n - u \|$$

Thus, $\lim_{n \to \infty} T(u^n) = T(u)$, which implies that T is continuous.

Suppose now that T is continuous. Assume, for purposes of proof by contradiction, that T is not bounded. Then for each n there must exist a vector u^n such that $| T(u^n) | \geq n \| u^n \|$. The sequence $v^n = \dfrac{u^n}{n \| u^n \|}$ approaches zero and has the property

$$T(v^n) = \frac{1}{n \| u^n \|} T(u^n) \geq 1$$

Thus, $\lim_{n \to \infty} T(v^n) \geq 1 \neq 0$, which violates the continuity assumption. Thus, T is bounded. ■

It is interesting to note that the functional of Example 9.1.3 was shown to be continuous, but not bounded. This does not contradict Theorem 9.1.2, since the functional of Example 9.1.3 is not linear. The reader is cautioned that linearity is a very special property of a functional and that characteristics of linear functionals are not shared by nonlinear functionals. As will be seen later, some properties of functionals on finite-dimensional spaces do not carry over to infinite-dimensional spaces. The very important property of R^n shown in Example 9.1.5 is, however, valid on all Hilbert spaces, even if they are infinite dimensional.

Riesz Representation Theorem

Theorem 9.1.3 (Riesz Representation Theorem). Every continuous linear functional T on a Hilbert space H can be expressed in the form $T(u) = (u, f)$, where f is in H. Furthermore, f is unique. ■

To prove this, let N be the null space of T; i.e., $N = \{ u: T(u) = 0 \}$. It is easily shown that N is a closed subspace of H. If N = H, then select f = 0. If $N \neq H$, then write $H = N + N^{\perp}$, where $N^{\perp}$ is the set of all vectors that are orthogonal to each vector in N. Since $N \neq H$, $N^{\perp}$ contains a nonzero element; say, f_0. By normalization, set $\| f_0 \| = 1$. Let $v = T(u) f_0 - u T(f_0)$, where u is arbitrary. Clearly v is in N, since T(v) = 0. Thus, with f_0 in $N^{\perp}$,

$$(v, f_0) = (T(u) f_0 - u T(f_0), f_0) = 0$$

or

$$T(u) - T(f_0) (u, f_0) = 0$$

and

$$T(u) = T(f_0) (u, f_0) = (u, T(f_0) f_0)$$

Thus, $f = T(f_0) f_0$. To prove uniqueness, suppose that $T(u) = (u, f) = (u, g)$, for all u in H. Then $(u, f - g) = 0$ for all u, in particular for $u = f - g$, so $\| f - g \|^2 = 0$, which implies f = g. ■

Consider n **generalized coordinates** $q_1, \ldots, q_n$ of a mechanical system. The linear space R^n is the space of generalized coordinates. The **generalized force Q**, for any **q** in R^n, is a vector, such that the **virtual work** $\delta W = (\mathbf{Q}, \delta\mathbf{q})$ due to a differential $\delta\mathbf{q}$, called a **virtual displacement**, is a continuous linear functional. The component Q_i of **Q** in the scalar product $(\mathbf{Q}, \delta\mathbf{q})$ is called the component of generalized force corresponding to the generalized coordinate q_i. If q_i has the physical dimension of length, then Q_i will have the physical dimension of force. If q_i is dimensionless, then Q_i will have the dimension of energy.

Example 9.1.6

The configuration of the double pendulum in Fig. 9.1.1 is uniquely determined by the angles q_1 and q_2. The moments M_1 and M_2 are the corresponding generalized forces, since the work performed by the moments as they act through differential rotations δq_1 and δq_2 is

$$\begin{aligned} \delta W &= M_i\, \delta q_i \\ &= (\mathbf{Q}^M, \delta\mathbf{q}) \end{aligned}$$

where $\mathbf{Q}^M = [\, M_1, M_2 \,]^T$.

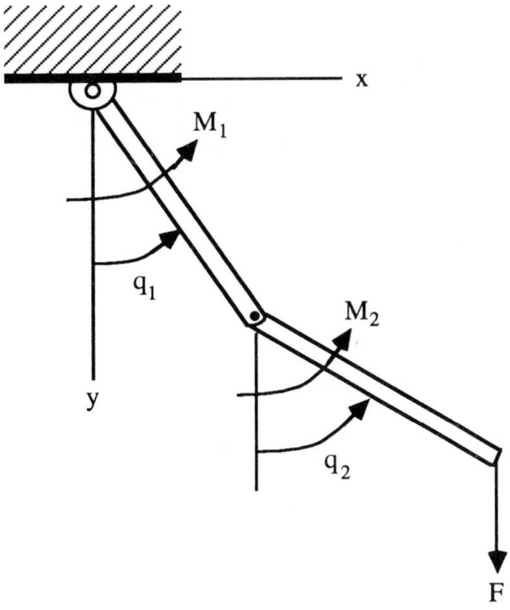

Figure 9.1.1 Double Pendulum

For the force F that acts at the end of the second bar, the work done is

$$\delta W = \delta y \, F$$

where δy is a small vertical displacement of the end of the second bar. If the bars have unit length,

$$y = \cos q_1 + \cos q_2$$

Taking the total differential of y,

$$\delta y = -\delta q_1 \sin q_1 - \delta q_2 \sin q_2$$

Thus,

$$\delta W = (-F \sin q_1)\, \delta q_1 + (-F \sin q_2)\, \delta q_2$$
$$= (\mathbf{Q}^F, \delta \mathbf{q})$$

and the generalized force corresponding to F is $\mathbf{Q}^F = [\, -F \sin q_1, -F \sin q_2 \,]^T$. ■

Example 9.1.7

Consider the beam shown in Fig. 9.1.2, under the action of a distributed force f(x). Let $\delta u(x)$ be a differential displacement of the beam, in the interval $0 \le x \le \ell$. This generalized displacement is in the Hilbert space $L_2(0, \ell)$, so the virtual work, which is a bounded linear functional, must be of the form

$$\delta W = \int_0^{\ell} \delta u(x)\, f(x)\, dx = (\, f, \delta u \,)$$

where, by the Riesz representation theorem, f is also in $L_2(0, \ell)$. Thus, the generalized force for a flexible beam is the distributed load f(x).

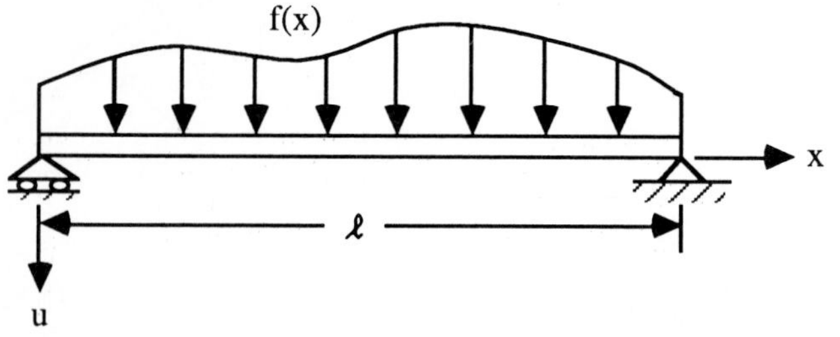

Figure 9.1.2 Beam ■

EXERCISES 9.1

1. Prove that every linear functional on R^n is continuous.

2. Show that if f is in $L_2(0, \ell)$, then the linear functional

$$T(u) = (u, f)$$

for u in $L_2(0, \ell)$, is bounded.

3. Show that a linear functional

$$T(u) = (u, f)$$

where u is in $L_2(0, \ell)$, need not be bounded if f is not in $L_2(0, \ell)$. [Hint: Let $f(x) = 1/x$.]

9.2 LINEAR OPERATOR EQUATIONS

Linear algebra in R^n and linearity properties of matrix operations played a key role in the theory and solution methods for matrix equations in Chapters 1 to 3. Linear operator concepts are introduced in this section and form the foundation for treating elliptic boundary-value problems as extensions of matrix equations. The reader is encouraged to note both the similarities and differences between matrix and differential operators. A goal is to use knowledge gained in matrix theory to exploit similarities between the two. A second goal is to identify fundamental differences, so that pitfalls due to inappropriate use of matrix ideas in manipulating differential operators can be avoided.

Linear Operators

Definition 9.2.1. Let V be a function space and A an **operator** that assigns to each element u in a linear subspace D_A of V a vector $g = Au$ in V, such that for any u and v in D_A and for any real α and β,

$$A(\alpha u + \beta v) = \alpha Au + \beta Av \tag{9.2.1}$$

The operator is called a **linear operator** with **domain** D_A and **range** R_A, where

$$R_A = \{ g: Au = g, \text{ for some u in } D_A \} \tag{9.2.2}$$

■

Example 9.2.1

Let the subspace $D_A = C^2(0, \ell)$ of the function space $L_2(0, \ell)$ be the domain of the **second-order differential operator**

$$Au = -\frac{d^2u}{dx^2} \tag{9.2.3}$$

for $0 < x < \ell$. Application of the second derivative operator to a function u in $C^2(0, \ell)$ yields a continuous function. Thus,

$$R_A = C^0(0, \ell) \tag{9.2.4}$$

Linearity properties of differentiation imply that A satisfies Eq. 9.2.1, so it is a linear operator. ■

Example 9.2.2

Let the space $D_A = L_2(0, \ell)$ be the domain of the **integral operator**

$$Au = \int_0^x u(\xi)\, d\xi \tag{9.2.5}$$

for $0 < x < \ell$. Note that Au is a function of x. Since integration yields a continuous function,

$$R_A = \{\, g \in C^0(0, \ell) : Au = g,\ u \in L_2(0, \ell) \,\} \tag{9.2.6}$$

Linearity properties of integration imply that A satisfies Eq. 9.2.1, so it is a linear operator. ■

Bounded Linear Operators

A linear operator A with domain D_A is said to be a **bounded linear operator** if there exists a real number c such that

$$\| Au \| \le c \| u \| \tag{9.2.7}$$

for all u in D_A. The smallest number c that satisfies the above inequality is called the **norm of the linear operator** A, denoted $\| A \|$, and given by

$$\| A \| = \sup_{u \in D_A,\, u \neq 0} \frac{\| Au \|}{\| u \|} \tag{9.2.8}$$

Thus, $\| Au \| \le \| A \| \, \| u \|$, for all u in D_A. The linear operator A is said to be a **continuous linear operator** if $\lim_{n\to\infty} \| u - u^n \| = 0$ implies that $\lim_{n\to\infty} \| Au - Au^n \| = 0$.

Theorem 9.2.1. A linear operator is continuous if and only if it is bounded. Furthermore, if it is continuous at 0, then it is continuous on its entire domain. ■

The proof of Theorem 9.2.1 follows the same arguments used in the proofs of Theorems 9.1.1 and 9.1.2. ■

Example 9.2.3

Consider the linear operator $Au = -d^2u/dx^2$ defined on $D_A = C^2(0, 2\pi)$, which is a subspace of $L_2(0, 2\pi)$. To see that A is unbounded, as an operator from D_A to $L_2(0, 2\pi)$, consider the sequence of functions

$$\phi^n = \frac{\cos nx}{n\sqrt{\pi}}, \quad n = 1, 2, \ldots$$

Since

$$A\phi^n = n\frac{\cos nx}{\sqrt{\pi}}$$

the norm of $A\phi^n$ is

$$\| A\phi^n \| = \left[\int_0^{2\pi} \left(n \frac{\cos nx}{\sqrt{\pi}} \right)^2 dx \right]^{1/2} = n$$

whereas $\| \phi^n \| \to 0$ as $n \to \infty$, so

$$\sup_n \frac{\| A\phi^n \|}{\| \phi^n \|} = +\infty$$

which means that there is no real constant c such that $\| A\phi^n \| \leq c \| \phi^n \|$ for all n. Thus, the operator A is unbounded. Moreover, since $\| \phi^n \| \to 0$ and $\| A\phi^n \| \to \infty$ as $n \to \infty$, A is not a continuous operator. This result is consistent with Theorem 9.2.1. ■

Example 9.2.4

Consider next the integral operator of Example 9.2.2, whose domain D_A is all of $L_2(0, \ell)$. Note that

$$\begin{aligned} \| Au \|^2 &= \int_0^\ell \left(\int_0^x u(\xi)\, d\xi \right)^2 dx \\ &= \int_0^\ell \left(\left| \int_0^x 1 \times u(\xi)\, d\xi \right| \right)^2 dx \\ &\leq \int_0^\ell \left(\int_0^x 1^2\, d\xi \right) \left(\int_0^x u^2(\xi)\, d\xi \right) dx \\ &\leq \int_0^\ell \left(\int_0^\ell 1^2\, d\xi \right) \left(\int_0^\ell u^2(\xi)\, d\xi \right) dx \\ &= \ell^2 \| u \|^2 \end{aligned}$$

Thus,

$$\| Au \| \leq \ell \| u \|$$

so A is a bounded linear operator and $\| A \| \leq \ell$. ■

Symmetric Operators

Definition 9.2.2. A linear operator A, with domain D_A and range R_A that are subspaces of a Hilbert space H, is said to be a **symmetric linear operator** if

$$(Au, v) = (u, Av) \tag{9.2.9}$$

for all u and v in D_A. ■

Example 9.2.5

Since D_A and R_A for the second-order differential operator A of Example 9.2.1 are subspaces of the Hilbert space $L_2(0, \ell)$, the operator may be checked for symmetry. For u and v in $D_A = C^2(0, \ell)$, using integration by parts,

$$(Au, v) = \int_0^\ell - \frac{d^2u}{dx^2} v \, dx$$

$$= \left[- \frac{du}{dx} v \right]_0^\ell + \int_0^\ell \frac{du}{dx} \frac{dv}{dx} dx \tag{9.2.10}$$

Whereas the second term on the right of Eq. 9.2.10 is symmetric in u and v, the first term is not. Thus, the linear differential operator of Example 9.2.1, whose domain is all of $C^2(0, \ell)$, is not symmetric. ■

Example 9.2.6

Consider the linear differential operator A of Example 9.2.1, but with its domain a proper subset of $C^2(0, \ell)$ that satisfies homogeneous boundary conditions; i.e.,

$$Au = - \frac{d^2u}{dx^2}, \quad 0 < x < \ell$$

$$D_A = \{ u \text{ in } C^2(0, \ell): u(0) = u(\ell) = 0 \} \tag{9.2.11}$$

Even though the differential formula in the definition of the operators here and in Example 9.2.1 is the same, the operators have different domains. Thus, they are different operators.

To see that the operator of Eq. 9.2.11 is symmetric, note that Eq. 9.2.10 is valid, but for v in D_A, $v(0) = v(\ell) = 0$. Thus,

$$(Au, v) = \int_0^{\ell} \frac{du}{dx}\frac{dv}{dx}\, dx \tag{9.2.12}$$

for all u and v in D_A. Since the right side of Eq. 9.2.12 is symmetric in u and v, $(Au, v) = (Av, u) = (u, Av)$. Thus, the operator of Eq. 9.2.11 is symmetric. ■

Positive Operators

Definition 9.2.3. Let A be a symmetric linear operator that is defined on a dense subspace D_A of a Hilbert space H, with range in H. Then A is said to be a **positive semidefinite linear operator** if

$$(Au, u) \geq 0 \tag{9.2.13}$$

for all u and if $(Au, u) = 0$ for some $u \neq 0$ in D_A. It is said to be a **positive definite linear operator** if

$$(Au, u) > 0 \tag{9.2.14}$$

for all $u \neq 0$ in D_A. In other words, if A is positive definite, $(Au, u) = 0$ implies $u = 0$. Finally, A is said to be a **positive bounded below linear operator** if there exists a constant $c > 0$ such that

$$(Au, u) \geq c (u, u) = c \| u \|^2 \tag{9.2.15}$$

for all u in D_A. Clearly, if A is positive bounded below, it is positive definite. ■

In many mechanics problems, u is a deflection and Au is a force, so (Au, u) is proportional to the energy that is required to produce the deflection u. The property $(Au, u) > 0$, for all admissible $u \neq 0$, is a statement that a positive amount of energy is required to produce a nonzero deflection. The property $(Au, u) \geq c \| u \|^2$, for $c > 0$, is a statement that a lower bound exists for the amount of energy that must be expended in achieving a displacement of fixed norm, which implies **stability** of the system. That is, large deflections can be produced only by large expenditures of energy. If A is positive definite, but not positive bounded below, then for any integer n there is a u_n in D_A such that $(Au_n, u_n) \leq \| u_n \|^2/n$. Putting $\bar{u}_n = u_n/\| u_n \|$, $\| \bar{u}_n \| = 1$ and $(A\bar{u}_n, \bar{u}_n) \leq (1/n) \to 0$ as $n \to \infty$. This means that a nonzero deflection $\bar{u}_n$ can be produced with a very small expenditure of energy $(A\bar{u}_n, \bar{u}_n)$.

Example 9.2.7

For the operator A of Example 9.2.6, Eq. 9.2.12 yields

$$(Au, u) = \int_0^{\ell} \left(\frac{du}{dx} \right)^2 dx \geq 0$$

In fact, (Au, u) = 0 implies that du/dx = 0, or u = c. Due to the boundary conditions of Eq. 9.2.11, c = 0, so u(x) = 0. Thus, A is positive definite. The fact that it is positive bounded below follows as a special case of the following example. ■

Example 9.2.8

Consider the Laplace operator

$$Au = -\nabla^2 u \tag{9.2.16}$$

for functions in the domain

$$D_A = \{ u \text{ in } C^2(\Omega): u(x) = 0, \text{ on } \Gamma \} \tag{9.2.17}$$

where Γ is the boundary of Ω, which is a bounded subset of R^n. The second Green's formula in Exercise 3 (b) of Section 6.1 shows that the operator is symmetric and that

$$(Au, v) = \iint_\Omega (-\nabla^2 u)\, v\, d\Omega = \iint_\Omega \nabla u^T \nabla v\, d\Omega \tag{9.2.18}$$

for all u and v in D_A. Setting v = u in this relation,

$$(Au, u) = \iint_\Omega \nabla u^T \nabla u\, d\Omega \geq 0 \tag{9.2.19}$$

for all u in D_A. It remains to show that if (Au, u) = 0, then u = 0. This is true, since if (Au, u) = 0, then by Eq. 9.2.19, $\nabla u = 0$, which implies that u is constant on Ω. Since u = 0 on Γ, the constant is zero and u = 0. Thus, the Laplace operator, with domain D_A of Eq. 9.2.17, is positive definite.

In fact, the Laplace operator of Eq. 9.2.16 with domain of Eq. 9.2.17 is positive bounded below. To simplify the proof of this result, consider Ω in R^2. Then, by Eq. 9.2.19,

$$(Au, u) = \iint_\Omega \left[\left(\frac{\partial u}{\partial x} \right)^2 + \left(\frac{\partial u}{\partial y} \right)^2 \right] d\Omega \tag{9.2.20}$$

Since Ω is assumed to be bounded, it can be enclosed in some rectangle Ω_1 with the coordinate axes along two of its sides, as shown in Fig. 9.2.1.

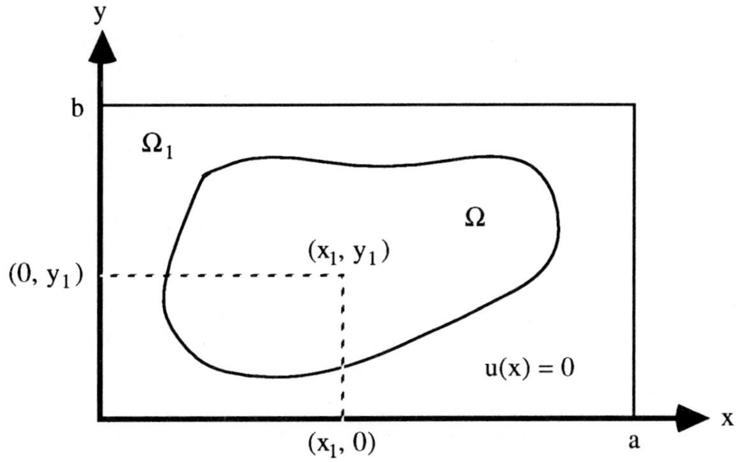

Figure 9.2.1 Rectangular Enclosure of Ω

Extending the physical domain from Ω to Ω_1 and setting admissible functions equal to zero on $\Omega_1 - \Omega$,

$$\int_0^{x_1} \frac{\partial u(x, y_1)}{\partial x} dx = u(x_1, y_1) - u(0, y_1) = u(x_1, y_1)$$

for any point (x_1, y_1) in Ω. Applying the Schwartz inequality,

$$\begin{aligned} u^2(x_1, y_1) &= \left(\int_0^{x_1} 1 \times \frac{\partial u(x, y_1)}{\partial x} dx \right)^2 \\ &\leq \int_0^{x_1} (1)^2 dx \int_0^{x_1} \left[\frac{\partial u(x, y_1)}{\partial x} \right]^2 dx \\ &= x_1 \int_0^{x_1} \left[\frac{\partial u(x, y_1)}{\partial x} \right]^2 dx \\ &\leq a \int_0^{a} \left[\frac{\partial u(x, y_1)}{\partial x} \right]^2 dx \end{aligned}$$

Integrating this inequality over the rectangle $0 \leq x_1 \leq a$, $0 \leq y_1 \leq b$,

$$\iint_{\Omega_1} u^2(x_1, y_1)\, dx_1\, dy_1$$

$$\leq \int_0^b \int_0^a \left\{ a \int_0^a \left[\frac{\partial u(x, y_1)}{\partial x} \right]^2 dx \right\} dx_1\, dy_1$$

$$= a^2 \int_0^b \int_0^a \left[\frac{\partial u(x, y_1)}{\partial x} \right]^2 dx\, dy_1$$

$$= a^2 \int_0^b \int_0^a \left[\frac{\partial u(x, y)}{\partial x} \right]^2 dx\, dy$$

$$\leq a^2 \iint_{\Omega_1} \left[\left(\frac{\partial u}{\partial x} \right)^2 + \left(\frac{\partial u}{\partial y} \right)^2 \right] dx\, dy$$

Letting $a^2 = 1/\alpha$ and noting that the integrands are zero outside Ω, this inequality becomes,

$$\iint_{\Omega} \left[\left(\frac{\partial u}{\partial x} \right)^2 + \left(\frac{\partial u}{\partial y} \right)^2 \right] d\Omega \geq \alpha \iint_{\Omega} u^2\, d\Omega \qquad (9.2.21)$$

which is known as **Friedrichs' inequality**. From Eq. 9.2.20, this inequality is

$$(- \nabla^2 u, u) \geq \alpha \| u \|^2 \qquad (9.2.22)$$

which shows that the Laplace operator of Eq. 9.2.16, with the domain of Eq. 9.2.17, is positive bounded below. ■

Matrices as Linear Operators

While the power of linear operator theory is best exploited in dealing with boundary-value problems, it is instructive to see that the concepts introduced thus far in this section are related to properties of matrices. Let $H = R^n$ and $V = R^m$, with $m \leq n$, be Hilbert spaces with the norm of Eq. 2.4.4. Consider the $m \times n$ matrix $\mathbf{A}$, written both in terms of its elements and its columns as

$$\mathbf{A} = [\, a_{ij} \,]_{m\times n} = [\, \mathbf{a}_1, \ldots, \mathbf{a}_n \,] \qquad (9.2.23)$$

where $\mathbf{a}_j$ are $m \times 1$ column matrices. With the domain $D_{\mathbf{A}} = H$, for any $\mathbf{u}$ in $D_{\mathbf{A}}$, the matrix $\mathbf{A}$ defines the operator

$$\mathbf{Au} = [\, a_{ij} u_j \,]_{m\times 1} = u_j \mathbf{a}_j \qquad (9.2.24)$$

The range $R_{\mathbf{A}}$ of this operator is thus a subspace of V.

If the matrix $\mathbf{A}$ has full row rank, then there are m linearly independent columns $\mathbf{a}_j$ in Eq. 9.2.23 that form a basis for V. Thus, for any $\mathbf{v}$ in V, there exist constants u_j^* corresponding to the $\mathbf{a}_j$ that form a basis of V, such that

$$\mathbf{v} = u_j^* \mathbf{a}_j = \mathbf{Au}^* \qquad (9.2.25)$$

where $\mathbf{u}^*$ is defined as an n-vector that contains the parameter u_j^* in row j. However, $\mathbf{u}^*$ may not be unique if m < n. Thus, if the matrix **A** has full row rank, $R_{\mathbf{A}} = V$.

Since $D_{\mathbf{A}} = H$ is a vector space and

$$\mathbf{A}\,(\,\alpha\mathbf{u} + \beta\mathbf{w}\,) = \alpha\mathbf{A}\mathbf{u} + \beta\mathbf{A}\mathbf{w} \tag{9.2.26}$$

follows from properties of matrix multiplication and addition, for all real α and β and vectors **u** and **w** in $D_{\mathbf{A}}$, **A** is a linear operator.

To see that any matrix operator **A** is bounded, using the Schwartz inequality, note that

$$\begin{aligned}
\|\,\mathbf{A}\mathbf{u}\,\| &= \left[\sum_{i=1}^{m}\left(\sum_{j=1}^{n} a_{ij}u_j\right)^2\right]^{1/2} \\
&= \left[\sum_{i=1}^{m}\left(\,[\,a_{i1}, a_{i2}, \ldots, a_{in}\,]\,[\,u_1, u_2, \ldots, u_n\,]^T\,\right)^2\right]^{1/2} \\
&\le \left[\sum_{i=1}^{m}\|\,[\,a_{i1}, a_{i2}, \ldots, a_{in}\,]^T\,\|^2\,\|\,\mathbf{u}\,\|^2\right]^{1/2} \\
&= \left[\sum_{i=1}^{m}\left\{\sum_{j=1}^{n} a_{ij}^2\right\}\left\{\sum_{j=1}^{n} u_j^2\right\}\right]^{1/2} \\
&= F(\mathbf{A})\,\|\,\mathbf{u}\,\|
\end{aligned} \tag{9.2.27}$$

where

$$F(\mathbf{A}) = \left[\sum_{i=1}^{m}\sum_{j=1}^{n} a_{ij}^2\right]^{1/2} \tag{9.2.28}$$

is called the **Frobenius norm** of **A**. Thus, **A** is a bounded linear operator from H to V, with its **operator norm** $\|\,\mathbf{A}\,\|$ bounded by

$$\|\,\mathbf{A}\,\| \le F(\mathbf{A}) \tag{9.2.29}$$

To determine the operator norm of **A**, the theorem that follows can be used.

Theorem 9.2.2. The operator norm of **A** is

$$\| \mathbf{A} \| = \max_{\lambda \in \sigma(\mathbf{A}^T\mathbf{A})} | \lambda | \tag{9.2.30}$$

where $\sigma(\mathbf{A}^T\mathbf{A})$ is the **spectrum** of $\mathbf{A}^T\mathbf{A}$, which is the set of all eigenvalues of $\mathbf{A}^T\mathbf{A}$. ■

For proof, note that the eigenvalues of $\mathbf{A}^T\mathbf{A}$ are nonnegative. To see this, let

$$\mathbf{A}^T\mathbf{A}\mathbf{u} = \lambda \mathbf{u}$$

where $\mathbf{u} \neq \mathbf{0}$. Then,

$$(\mathbf{u}, \mathbf{A}^T\mathbf{A}\mathbf{u}) = (\mathbf{u}, \lambda\mathbf{u}) = \lambda \| \mathbf{u} \|^2$$

and

$$(\mathbf{u}, \mathbf{A}^T\mathbf{A}\mathbf{u}) = \mathbf{u}^T\mathbf{A}^T\mathbf{A}\mathbf{u} = (\mathbf{A}\mathbf{u})^T\mathbf{A}\mathbf{u} = (\mathbf{A}\mathbf{u}, \mathbf{A}\mathbf{u}) = \| \mathbf{A}\mathbf{u} \|^2 \geq 0$$

Thus $\lambda \geq 0$. Let the eigenvalues of $\mathbf{A}^T\mathbf{A}$ be arranged in ascending order,

$$0 \leq \lambda_1 \leq \lambda_2 \leq \ldots \leq \lambda_n$$

and let $\phi^1, \phi^2, \ldots, \phi^n$, be the corresponding orthonormalized eigenvectors.

For any $\mathbf{u} \in R^n$,

$$\| \mathbf{A}\mathbf{u} \|^2 = (\mathbf{A}\mathbf{u}, \mathbf{A}\mathbf{u}) = (\mathbf{u}, \mathbf{A}^T\mathbf{A}\mathbf{u})$$

Since $\{ \phi^i \}$ forms a basis for R^n, **u** can be expressed as

$$\mathbf{u} = \sum_{i=1}^{n} a_i \phi^i$$

and

$$a_i = (\mathbf{u}, \phi^i)$$

Then

$$\mathbf{A}^T\mathbf{A}\mathbf{u} = \sum_{i=1}^{n} a_i \mathbf{A}^T\mathbf{A}\phi^i = \sum_{i=1}^{n} a_i \lambda_i \phi^i$$

and

$$\| \mathbf{Au} \|^2 = \left(\sum_{i=1}^{n} a_i \phi^i, \sum_{j=1}^{n} a_j \lambda_j \phi^j \right) = \sum_{j=1}^{n} \lambda_j a_j^2$$

$$\leq \lambda_n \sum_{j=1}^{n} a_j^2 = \lambda_n \sum_{i=1}^{n} \sum_{j=1}^{n} a_i a_j \delta_{ij}$$

$$= \lambda_n \sum_{i=1}^{n} \sum_{j=1}^{n} a_i \phi^{iT} a_j \phi^j = \lambda_n \| \mathbf{u} \|^2$$

Thus,

$$\| \mathbf{A} \| \leq \lambda_n$$

However, if $\mathbf{u} = \phi^n$, then $\| \mathbf{u} \| = 1$ and

$$\| \mathbf{Au} \|^2 = (\mathbf{u}, \mathbf{A}^T\mathbf{Au}) = (\phi^n, \mathbf{A}^T\mathbf{A}\phi^n) = \lambda_n$$

Thus,

$$\| \mathbf{A} \| = \lambda_n = \max_{\lambda \in \sigma(\mathbf{A}^T\mathbf{A})} | \lambda | \qquad \blacksquare$$

Consider next the case in which m = n; i.e., **A** is a square matrix. Using properties of matrix transpose and the scalar product on R^n,

$$(\mathbf{Au}, \mathbf{v}) = (\mathbf{Au})^T\mathbf{v} = \mathbf{u}^T\mathbf{A}^T\mathbf{v} \tag{9.2.31}$$

If the operator **A** is symmetric, then

$$(\mathbf{Au}, \mathbf{v}) \equiv \mathbf{u}^T\mathbf{A}^T\mathbf{v} = \mathbf{u}^T\mathbf{Av} \equiv (\mathbf{u}, \mathbf{Av}) \tag{9.2.32}$$

for all **u** and **v** in R^n. In particular, for $\mathbf{u} = \mathbf{e}_k \equiv [\delta_{ki}]$ and $\mathbf{v} = \mathbf{e}_j \equiv [\delta_{ji}]$, Eq. 9.2.32 becomes

$$a_{jk} = a_{kj} \tag{9.2.33}$$

Thus, the matrix **A** associated with a symmetric operator is a symmetric matrix.

Conversely, if **A** is a symmetric matrix, then for all **u** and **v** in R^n,

$$(\mathbf{Au}, \mathbf{v}) = \mathbf{u}^T\mathbf{A}^T\mathbf{v} = \mathbf{u}^T\mathbf{Av} = (\mathbf{u}, \mathbf{Av}) \tag{9.2.34}$$

So the operator associated with a symmetric matrix **A** is a symmetric operator.

For m = n and **A** a symmetric matrix, if the operator **A** is positive semidefinite,

$$(\mathbf{u}, \mathbf{Au}) = \mathbf{u}^T\mathbf{Au} \geq 0 \tag{9.2.35}$$

for all **u** in R^n and there is a $\mathbf{u}^* \neq \mathbf{0}$ such that $(\mathbf{u}^*, \mathbf{Au}^*) = 0$. Thus, the matrix **A** is positive semidefinite.

If the symmetric operator **A** is positive definite,

$$(\mathbf{u}, \mathbf{Au}) = \mathbf{u}^T\mathbf{Au} > 0 \tag{9.2.36}$$

for all $\mathbf{u} \neq \mathbf{0}$ in R^n. Thus, the matrix **A** that defines the operator is positive definite.

Let ϕ^i be orthonormal eigenvectors of the $n \times n$ positive definite matrix **A**, with corresponding eigenvalues $\lambda_i > 0$, $i = 1, \ldots, n$. For any **u** in R^n,

$$\mathbf{u} = b_i\phi^i \tag{9.2.37}$$

Thus,

$$\begin{aligned} \| \mathbf{u} \|^2 &= b_i\phi^{iT}b_j\phi^j \\ &= b_ib_j\delta_{ij} = \sum_{i=1}^{n} b_i^2 \end{aligned} \tag{9.2.38}$$

Finally,

$$\begin{aligned} (\mathbf{u}, \mathbf{Au}) &= b_i\phi^{iT}b_j\mathbf{A}\phi^j \\ &= \sum_{i=1}^{n}\sum_{j=1}^{n} b_ib_j\lambda_j\phi^{iT}\phi^j = \sum_{i=1}^{n} b_i^2\lambda_i \\ &\geq \min_j \lambda_j \sum_{i=1}^{n} b_i^2 = \min_j \lambda_j \| \mathbf{u} \|^2 \end{aligned} \tag{9.2.39}$$

This shows that the linear operator associated with a positive definite matrix is also positive bounded below. This result is valid only for finite-dimensional vector spaces and operators that can be defined by matrices. It is not in general true for infinite-dimensional, differential operators on $L_2(\Omega)$.

Operator Eigenvalue Problems

As in use of separation of variables in Section 8.5, eigenvalue problems associated with operator equations are often encountered. For symmetric linear operators A and B whose common domain is a subspace D of a Hilbert space H and whose ranges are in H, define the eigenvalue operator equation

$$Au = \lambda Bu \tag{9.2.40}$$

where $u \neq 0$ is in D. The function u is called an **eigenfunction** for the operators and λ is the associated **eigenvalue**.

Let u^1 and u^2 be eigenfunctions that correspond to different eigenvalues, $\lambda_1 \neq \lambda_2$; i.e.,

$$\begin{aligned} Au^1 &= \lambda_1 Bu^1 \\ Au^2 &= \lambda_2 Bu^2 \end{aligned} \tag{9.2.41}$$

Taking the scalar product of both sides of Eq. 9.2.41 with u^2 and u^1,

$$\begin{aligned} (u^2, Au^1) &= \lambda_1 (u^2, Bu^1) \\ (u^1, Au^2) &= \lambda_2 (u^1, Bu^2) \end{aligned} \tag{9.2.42}$$

Since u^1 and u^2 are both in D and A and B are symmetric,

$$(u^1, Au^2) = (Au^1, u^2) = (u^2, Au^1)$$

$$(u^1, Bu^2) = (Bu^1, u^2) = (u^2, Bu^1)$$

These results can be substituted into the second of Eq. 9.2.42 and the result subtracted from the first, to obtain

$$0 = (\lambda_1 - \lambda_2)(u^1, Bu^2) \tag{9.2.43}$$

Here, (u^1, Bu^2) is said to be the **scalar product with respect to the operator B**. This proves the theorem that follows.

Theorem 9.2.3. Eigenfunctions of the operator equation of Eq. 9.2.40, with symmetric operators A and B, that correspond to different eigenvalues are orthogonal with respect to the operator B. Further, if the operator B is positive definite, the eigenvalues are real. ■

The proof of Theorem 3.4.2 is applicable here, with a modification of the definition of the scalar product. ■

EXERCISES 9.2

1. Find the null space of the operator

$$Au = -\frac{d^2u}{dx^2}$$

from $D_A = C^2(0, 2\pi)$ to $C^0(0, 2\pi)$; i.e., $N_A = \{ u \text{ in } D_A: u'' = 0 \}$, and determine its dimension.

2. Let A be the operator in Exercise 1, with domain restricted to those functions that satisfy $u(0) = 0$. Determine the null space of this operator and its dimension. Answer the same question for the operator A, with domain restricted to functions that satisfy $u(0) = u(2\pi) = 0$.

3. Let A be the operator defined on $L_2(0, 1)$ into $L_2(0, 1)$ by

$$Au = \int_0^1 K(s, t)\, u(t)\, dt$$

where

$$\int_0^1 \int_0^1 K(s, t)^2\, dt\, ds < \infty$$

Is this operator bounded? If so, what is its norm?

4. Show that for a symmetric, positive definite, linear operator A, if the equation $Au = f$ has a solution, it is unique.

5. Let A be a symmetric, positive definite operator that is defined on a dense subspace D_A of a Hilbert space H. Verify that

$$[\, u, v\,]_A \equiv (\, Au, v\,)$$

is a scalar product on D_A. This scalar product is referred to as the **energy scalar product** with respect to A. Also show that $\| u \|_A = [\, u, u\,]_A^{1/2}$ is a norm.

6. Let Ω be a region in R^3, with boundary Γ, and let A denote the Laplace operator $-\nabla^2$, defined on the domain

$$D_A = \left\{ u \text{ in } C^2(\Omega): \frac{\partial u}{\partial n} = 0 \text{ on } \Gamma \right\}$$

Determine whether the operator is symmetric.

9.3 STURM–LIOUVILLE PROBLEMS

In this section, results concerning sets of orthogonal functions that are generated as solutions of certain types of boundary-value problems are summarized.

Eigenfunctions and Eigenvalues

A problem that arises often in mechanics consists of a homogeneous linear differential equation of the form

$$-\frac{d}{dx}\left[p(x) \frac{du}{dx} \right] + q(x)\, u = \lambda\, r(x)\, u \tag{9.3.1}$$

together with homogeneous boundary conditions that are prescribed at the end points of an interval [a, b]. This problem has a nontrivial solution only if the parameter λ has a certain value; say, $\lambda = \lambda_k$. The permissible values of λ are known as **eigenvalues** and the corresponding functions $u = \phi^k(x)$, which satisfy Eq. 9.3.1 with $\lambda = \lambda_k$, are known as **eigenfunctions.**

In most cases that occur in practice, the functions p(x) and r(x) are positive in the interval [a, b], except possibly at one or both of the end points. Defining the linear differential operator

$$Au = -\frac{d}{dx}\left[p \frac{du}{dx} \right] + qu \tag{9.3.2}$$

and the scalar operator

$$Bu = ru \tag{9.3.3}$$

the differential equation of Eq. 9.3.1 takes the operator form

$$Au = \lambda Bu \tag{9.3.4}$$

Note that

$$\begin{aligned}(Au, v) &= \int_a^b [-(pu')' + qu]\, v\, dx \\ &= \int_a^b u[-(pv')' + qv]\, dx - [p(u'v - uv')]_a^b \end{aligned} \tag{9.3.5}$$

so the operator A is symmetric if its domain is limited to functions that satisfy

$$[p(u'v - uv')]_a^b = 0 \tag{9.3.6}$$

Note also that

$$(Bu, v) = \int_a^b ruv\, dx = (u, Bv) \tag{9.3.7}$$

so B is symmetric. Furthermore, if $r(x) \geq \alpha > 0$, for $a \leq x \leq b$, then

$$(Bu, u) = \int_a^b ru^2\, dx \geq \alpha \int_a^b u^2\, dx \geq \alpha \| u \|^2$$

so B is positive bounded below, hence positive definite.

Assuming that $\lambda_j \neq \lambda_i$ are eigenvalues of Eq. 9.3.4, Theorem 9.2.2 shows that if the specified boundary conditions imply Eq. 9.3.6, then the corresponding eigenfunctions ϕ^i and ϕ^j are orthogonal relative to the weighting function r(x); i.e.,

$$[\phi^i, \phi^j]_B \equiv \int_a^b r\phi^i\phi^j\, dx \equiv (\phi^i, \phi^j)_r = 0 \tag{9.3.8}$$

Boundary conditions that give rise to this situation include the following:

(1) At each end point of the interval, u, du/dx, or $\alpha u + \beta(du/dx)$ vanish.
(2) If it happens that p(x) vanishes at x = a or at x = b, u and du/dx remain finite at that point and impose one of the conditions in (1) at the other end point.
(3) If it happens that p(b) = p(a), u(b) = u(a) and u'(b) = u'(a).

A collection of functions u in $C^2(a, b)$ that satisfy boundary conditions (1), (2), or (3) is defined to be the domain D of the operators A and B of Eqs. 9.3.2 and 9.3.3.

In most practical cases; in particular if p, q, and r are in $C^1(a, b)$ and both p and r are positive throughout the interval (a, b) that is of finite length, it will be shown in Section 9.7 that there exists an infinite set of distinct eigenvalues $\lambda_1, \lambda_2, \ldots$. If $q(x) \geq 0$ in (a, b) and if

$$[\, p\phi^i (\phi^i)' \,]_a^b \geq 0 \tag{9.3.9}$$

then the eigenvalues are all nonnegative (see Theorems 9.2.2 and 9.7.1). Further, except in the case of the periodicity condition of (3) above, to each eigenvalue there corresponds one and only one eigenfunction. In case (3), two linearly independent eigenfunctions generally correspond to each eigenvalue (see Example 9.3.1). Such pairs of functions can be orthogonalized by the Gram–Schmidt procedure.

The boundary-value problem considered here is known as a **Sturm–Liouville problem**. The importance of such problems stems from the fact that the sets of orthogonal eigenfunctions that are generated by these problems are complete in $L_2(a, b)$. Further, a positive statement can be made about pointwise convergence of the series representation of a sufficiently well behaved function f(x) in terms of the eigenfunctions.

In practice, it is often inconvenient to normalize the eigenfunctions. In such cases, the coefficients in a series representation

$$f(x) \approx \sum_{i=1}^{\infty} c_i \phi^i(x) \tag{9.3.10}$$

are given by the formula

$$c_i \int_a^b r\,(\phi^i)^2\,dx = \int_a^b r f \phi^i\,dx \tag{9.3.11}$$

or

$$c_i \,\|\phi^i\|_r^2 = (\, f, \phi^i \,)_r \tag{9.3.12}$$

This result is obtained formally by multiplying both sides of Eq. 9.3.10 by the product $r\phi^i$, integrating the result term by term over (a, b), and taking into account the orthogonality of the eigenfunctions relative to the weighting function r(x).

Convergence theorems for series expansion of functions in the eigenfunctions of the Sturm–Liouville problem play a central role in engineering mathematics. Two theorems on convergence and completeness are stated here, one of which is proved in Section 9.7. The first theorem concerns completeness, in the L_2 sense, of the eigenfunctions of the Sturm–Liouville problem.

Completeness of Eigenfunctions in L_2

Theorem 9.3.1. Let the functions p(x), q(x), and r(x) be in $C^1(a, b)$ with $p(x) > 0$ and $r(x) > 0$ in $a \le x \le b$. Then the eigenfunctions of Eq. 9.3.4, with boundary conditions of type (1), are complete in $L_2(a, b)$. That is, for any f in $L_2(a, b)$,

$$\left\| \sum_{i=1}^{n} c_i \phi^i - f \right\| \to 0$$

as $n \to \infty$, where the c_i are given by Eq. 9.3.12. ■

The proof of this theorem requires mathematical tools of Section 9.7, so it will be postponed. The value of this theorem is that it establishes the L_2 convergence of the generalized Fourier series of Eq. 9.3.10, for sequences of eigenfunctions $\{ \phi^i(x) \}$ of Sturm–Liouville problems that arise in solution of specific problems of mechanics. These eigenfunctions are called **special functions** and were introduced in Section 4.4.

Example 9.3.1

Consider the differential equation

$$-\frac{d^2u}{dx^2} = \lambda u \tag{9.3.13}$$

which is a special case of Eq. 9.3.1, in which $p(x) = r(x) = 1$ and $q(x) = 0$. Consider the interval $(0, \ell)$ and impose the boundary conditions

$$u(0) = 0, \quad u(\ell) = 0 \tag{9.3.14}$$

Note that the left side of Eq. 9.3.13, with the boundary conditions of Eq. 9.3.14, is just the operator of Eq. 9.2.11, which was shown in Examples 9.2.6 and 9.2.7 to be symmetric and positive definite. The eigenvalues are of the form $\lambda_k = k^2\pi^2/\ell^2$, where k is any positive integer. The corresponding eigenfunctions are

$$\phi^k = \sin \frac{k\pi x}{\ell}, \quad k = 1, 2, \ldots$$

Thus, Eq. 9.3.10 is just the Fourier sine series representation

$$f(x) = \sum_{k=1}^{\infty} c_k \sin \frac{k\pi x}{\ell}$$

for $0 < x < \ell$, where, with $r(x) = 1$, Eq. 9.3.11 yields

$$c_k = \frac{2}{\ell} \int_0^{\ell} f(x) \sin \frac{k\pi x}{\ell} \, dx$$

In a similar way, the conditions $u'(0) = u'(\ell) = 0$ give rise to a Fourier cosine series representation, while the periodicity conditions $u(-\ell) = u(\ell)$ and $u'(-\ell) = u'(\ell)$ on the interval $(-\ell, \ell)$ lead to the general Fourier series representation over that interval, involving both sines and cosines of period 2ℓ. This establishes the L_2 completeness of the Fourier sine-cosine series that was stated in Theorem 5.3.4. ■

In addition to the important L_2 completeness result for eigenfunctions of Sturm–Liouville problems, a theorem is stated here that guarantees pointwise convergence results similar to those proved for sine-cosine series in Section 5.5. For proof see Ref. 20.

Theorem 9.3.2. Let the Sturm–Liouville problem of Eq. 9.3.4 satisfy the hypotheses of Theorem 9.3.1. Then,

(1) If f(x) is in $D^1(a, b)$, the generalized Fourier series of Eq. 9.3.10 converges to

$$\frac{1}{2} [\, f(x+0) + f(x-0) \,] = \sum_{k=1}^{\infty} c_k \phi^k(x)$$

for $a < x < b$.

(2) If f(x) is in $C^2(a, b)$ and satisfies the boundary condition of type (1), then the generalized Fourier series of Eq. 9.3.10 converges uniformly to f(x) in $a \le x \le b$. ■

EXERCISES 9.3

1. Find the eigenvalues and eigenfunctions of the following Sturm–Liouville problems:

(a) $u'' + \lambda u = 0$

$u(0) = u'(\pi) = 0$

(b) $u'' + \lambda u = 0$

$u(0) = u(2\pi), \quad u'(0) = u'(2\pi)$

2. Expand the function $f(x) = \sin x$, $0 \le x \le \pi$, in the eigenfunctions of Exercise 1 (a).

9.4 SEPARATION OF VARIABLES AND EIGENFUNCTION EXPANSIONS

Thus far, the study of dynamics problems has been limited to linear second-order equations in two independent variables. While the study of this special class of problems is illuminating, extensions to equations with more independent variables and higher-order equations that arise in applications should be considered. In this section, the separation of variables method is extended to higher-order and higher-dimensional problems. For these problems, the separation method and the formal theory of eigenfunction expansions are found to be quite broadly applicable.

Example 9.4.1 (Vibration of a Cantilever Beam)

Natural vibration of a beam is governed by the differential equation

$$\frac{1}{a^2}\frac{\partial^2 u}{\partial t^2} + \frac{\partial^4 u}{\partial x^4} \equiv \frac{1}{a^2} u_{tt} + Au = 0 \tag{9.4.1}$$

where $a^2 = EI/\rho S$, E is Young's modulus, I is the second moment of the cross-sectional area S, ρ is mass density, and A is a fourth-order differential operator. For the cantilever beam of Fig. 9.4.1, the boundary conditions at the fixed end $x = 0$ are

$$u(0, t) = u_x(0, t) = 0 \tag{9.4.2}$$

At the free end, the bending moment and the shear force must be equal to zero, so

$$u_{xx}(\ell, t) = u_{xxx}(\ell, t) = 0 \tag{9.4.3}$$

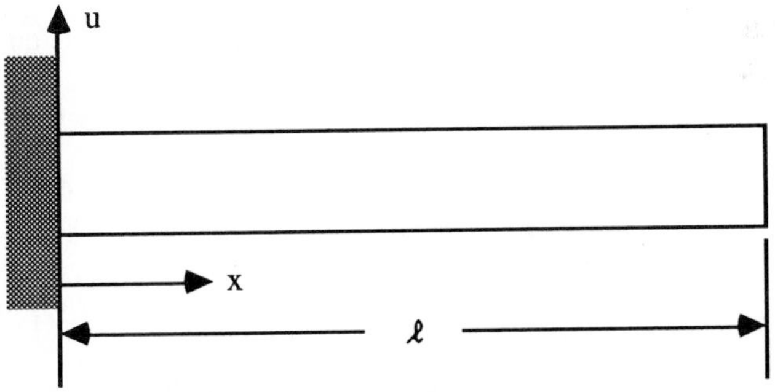

Figure 9.4.1 Cantilever Beam

In order to completely define the motion of the beam, initial conditions that prescribe the displacement and velocity of the beam at $t = 0$ must be given, in the form

$$\begin{aligned} u(x, 0) &= f(x) \\ u_t(x, 0) &= g(x) \end{aligned} \tag{9.4.4}$$

The problem considered therefore leads to the solution of Eq. 9.4.1, with boundary conditions of Eqs. 9.4.2 and 9.4.3 and initial conditions of Eq. 9.4.4.

Following the idea of separation of variables introduced in Section 8.2, a solution of the following form is sought:

$$u(x, t) = X(x)\,T(t) \tag{9.4.5}$$

Substituting from Eq. 9.4.5 into Eq. 9.4.1,

$$-\frac{\ddot{T}(t)}{a^2 T(t)} = \frac{AX(x)}{X(x)} = \alpha^4 \tag{9.4.6}$$

where $AX = d^4X/dx^4$ and the positive separation constant is selected to avoid exponential growth of the solution in time. The second of Eq. 9.4.6 yields the eigenvalue problem

$$AX = \alpha^4 X \tag{9.4.7}$$

$$X(0) = 0,\; X'(0) = 0,\; X''(\ell) = 0,\; X'''(\ell) = 0 \tag{9.4.8}$$

for X(x). To see that the constant in Eq. 9.4.6 must be positive, from another point of view, observe that A is a positive definite operator. By taking the scalar product on both sides of Eq. 9.4.7 with X, $(AX, X) = \alpha^4 (X, X) > 0$, since $X \neq 0$. Thus, $\alpha^4 > 0$ is justified.

The general solution of Eq. 9.4.7 is

$$X(x) = A\cosh\alpha x + B\sinh\alpha x + C\cos\alpha x + D\sin\alpha x$$

From the conditions $X(0) = 0$ and $X'(0) = 0$ of Eq. 9.4.8, $C = -A$ and $D = -B$. It follows that

$$X(x) = A[\cosh\alpha x - \cos\alpha x] + B[\sinh\alpha x - \sin\alpha x] \tag{9.4.9}$$

From the conditions $X''(\ell) = 0$ and $X'''(\ell) = 0$ of Eq. 9.4.8,

$$A[\cosh\alpha\ell + \cos\alpha\ell] + B[\sinh\alpha\ell + \sin\alpha\ell] = 0$$

$$A[\sinh\alpha\ell - \sin\alpha\ell] + B[\cosh\alpha\ell + \cos\alpha\ell] = 0$$

This system of homogeneous linear equations has a nontrivial solution for A and B only if the determinant of the coefficient matrix is zero. Setting the determinant equal to zero, eigenvalues are obtained from the transcendental equation

$$\sinh^2\alpha\ell - \sin^2\alpha\ell = \cosh^2\alpha\ell + 2\cosh\alpha\ell\cos\alpha\ell + \cos^2\alpha\ell$$

Because $\cosh^2 \alpha \ell - \sinh^2 \alpha \ell = 1$ and $\sin^2 \alpha \ell + \cos^2 \alpha \ell = 1$, this equation can be reduced to

$$(\cosh \alpha \ell)(\cos \alpha \ell) = -1 \tag{9.4.10}$$

The roots of Eq. 9.4.10 can be calculated numerically as [21]

$$\begin{aligned} \alpha_1 \ell &= 1.875 \\ \alpha_2 \ell &= 4.694 \\ \alpha_3 \ell &= 7.854 \\ &\dots \\ \alpha_n \ell &\approx \frac{\pi}{2}(2n - 1), \quad \text{for } n > 3 \end{aligned} \tag{9.4.11}$$

The last formula gives a value of α_n for n = 3 that is accurate to the third decimal place and for n = 7 that is accurate to the sixth place.

The time dependence in this problem is governed by the first equality of Eq. 9.4.6; i.e.,

$$\ddot{T} + a^2 \alpha_n^4 T = 0 \tag{9.4.12}$$

which is satisfied by the trigonometric functions

$$T_n(t) = a_n \cos 2\pi \nu_n t + b_n \sin 2\pi \nu_n t \tag{9.4.13}$$

with the **natural frequencies** of vibration

$$\nu_n = \frac{a\alpha_n^2}{2\pi} = \frac{\alpha_n^2}{2\pi}\sqrt{\frac{EI}{\rho S}} \tag{9.4.14}$$

i.e., the natural frequencies ν_n of vibration behave as the square of α_n. Because

$$\frac{\alpha_2^2}{\alpha_1^2} = 6.267 > 2^{2.5}, \quad \frac{\alpha_3^2}{\alpha_1^2} = 17.548 > 2^4$$

the second natural frequency is more than two and one-half octaves higher than the base tone, while the third frequency is more than four octaves higher than the base tone.

If the cantilever beam is a tuning fork with a basic frequency of 440 cycles per second, then the second frequency of the tuning fork is 2757.5 cycles per second, while the third frequency of 7721.1 cycles per second is already higher than frequencies that are used in music. If the tuning fork is set in vibration by a blow, higher frequencies appear in addition to the first, which explains the initial metallic sound. The higher harmonics are quickly damped, so the tuning fork soon rings out with the pure basic tone.

Substituting the α_i of Eq. 9.4.11 into Eq. 9.4.9, with the associated normalized eigenvectors $[A_i, B_i]^T$, eigenfunctions $X_i(x)$ are obtained that satisfy the homogeneous boundary conditions of the problem. Solutions of the vibration problem may thus be written in the form

$$\begin{aligned} u(x, t) &= \sum_{i=1}^{\infty} X_i(x)\, T_i(t) \\ &= \sum_{i=1}^{\infty} X_i(x)\,[\, a_i \cos 2\pi\nu_i t + b_i \sin 2\pi\nu_i t\,] \end{aligned} \tag{9.4.15}$$

where the constants a_i and b_i must be chosen to satisfy the initial conditions of Eq. 9.4.4. More specifically,

$$\begin{aligned} u(x, 0) &= f(x) = \sum_{i=1}^{\infty} a_i X_i(x) \\ u_t(x, 0) &= g(x) = \sum_{i=1}^{\infty} b_i 2\pi\nu_i X_i(x) \end{aligned} \tag{9.4.16}$$

so a_i and $2\pi\nu_i b_i$ are Fourier coefficients of functions f(x) and g(x), respectively, with respect to coordinate functions $X_i(x)$.

If there had been a lateral force $\rho S\, h(x, t)$ applied to the beam of Fig. 9.4.1, the differential equation of motion would have been

$$u_{tt} + a^2 u_{xxxx} = h(x, t) \tag{9.4.17}$$

To solve this **forced vibration** problem, with homogeneous initial conditions, assume that the eigenfunctions $X_n(x)$ are complete in $L_2(0, \ell)$, so

$$h(x, t) = \sum_{n=1}^{\infty} h_n(t)\, X_n(x) \tag{9.4.18}$$

where

$$h_n(t) = \int_0^{\ell} h(x, t)\, X_n(x)\, dx \tag{9.4.19}$$

Assuming a particular solution of the form

$$u_p(x, t) = \sum_{n=1}^{\infty} T_{p_n}(t)\, X_n(x) \tag{9.4.20}$$

and substituting Eqs. 9.4.18 and 9.4.20 into Eq. 9.4.17,

$$\ddot{T}_{p_n} + a^2\alpha_n^4 T_{p_n} = h_n(t) \tag{9.4.21}$$

A particular solution T_{p_n} of Eq. 9.4.21 is

$$T_{p_n}(t) = \frac{1}{\alpha_n^2 a} \int_0^t h_n(\tau) \sin \{ \alpha_n^2 a (t - \tau) \} \, d\tau \tag{9.4.22}$$

and, since

$$T_{p_n}(0) = T_{p_n}(0) = 0$$

it follows that

$$u_p(x, t) = \sum_{n=1}^{\infty} T_{p_n}(t)\, X_n(x)$$

is a particular solution of Eq. 9.4.17 that satisfies homogeneous initial and boundary conditions. The solution for forced vibration of a cantilever beam is obtained by superposing the solutions of Eqs. 9.4.15 and 9.4.22. That is, the forced solution $\bar{u}(x, t)$ is given by

$$\bar{u}(x, t) = u(x, t) + u_p(x, t) \tag{9.4.23}$$

The completeness assumption for the eigenfunctions $X_n(x)$ in the foregoing analysis is motivated by the fact that the eigenfunctions of the Sturm–Liouville problem are complete. It will be shown later in this chapter that these eigenfunctions are indeed complete in $L_2(0, \ell)$. ■

Example 9.4.2 (Vibrating Membrane)

From Section 6.3, the equation of a vibrating membrane on a rectangular domain $\Omega = \{ \mathbf{x} = (x_1, x_2): 0 \le x_1 \le a, 0 \le x_2 \le b \}$, with boundary Γ, is

$$Au = -\nabla^2 u = -\frac{1}{m^2} u_{tt} \tag{9.4.24}$$

for $\mathbf{x}$ in Ω,

$$u(\mathbf{x}, t) = 0 \tag{9.4.25}$$

for $\mathbf{x}$ on Γ, and

$$u(\mathbf{x}, 0) = f(\mathbf{x}) \tag{9.4.26}$$

$$u_t(\mathbf{x}, 0) = g(\mathbf{x}) \tag{9.4.27}$$

for $\mathbf{x}$ in Ω, where $m^2 = T/\rho$. This is similar to the form of problem treated in Example 8.6.2. The problem separates with $u(\mathbf{x}, t) = T(t)\, X(\mathbf{x})$, and the solution is

$$u(\mathbf{x}, t) = \sum_{i=1}^{\infty} (\, a_i \cos m\alpha_i t \; + \; b_i \sin m\alpha_i t\,)\, X_i(\mathbf{x}) \tag{9.4.28}$$

where $X_i(\mathbf{x})$ is the normalized solution of the eigenvalue problem

$$\begin{aligned} AX &= -\nabla^2 X = \alpha^2 X, && \mathbf{x} \text{ in } \Omega \\ X &= 0, && \mathbf{x} \text{ on } \Gamma \end{aligned} \tag{9.4.29}$$

On the rectangular domain Ω, the solution $X(\mathbf{x})$ of Eq. 9.4.29 may be attempted by separation of variables; i.e.,

$$X(\mathbf{x}) = U(x_1)\, V(x_2) \tag{9.4.30}$$

The boundary condition of Eq. 9.4.29 implies that

$$U(0) = U(a) = V(0) = V(b) = 0$$

Substitution from Eq. 9.4.30 into the first of Eq. 9.4.29 yields

$$U''(x_1)\, V(x_2) + U(x_1)\, V''(x_2) + \alpha^2 U(x_1)\, V(x_2) = 0 \tag{9.4.31}$$

Dividing by $-U(x_1)\, V(x_2)$ yields

$$-\frac{U''(x_1)}{U(x_1)} = \frac{V''(x_2)}{V(x_2)} + \alpha^2 \tag{9.4.32}$$

In order for this equation to hold for all x_1 and x_2, the left and right sides must be equal to a constant. Since $-U''$ is a positive definite operator with the boundary conditions treated here, the constant must be positive, say β^2. Thus,

$$-U''(x_1) = \beta^2 U(x_1) \tag{9.4.33}$$

$$-V''(x_2) = (\, \alpha^2 - \beta^2\,)\, V(x_2) \tag{9.4.34}$$

Since the operator $-V''$ with these boundary conditions is also positive definite, $\alpha^2 - \beta^2 \equiv \gamma^2 > 0$.

Equations 9.4.33 and 9.4.34 have solutions

$$U = C \cos \beta x_1 + D \sin \beta x_1 \tag{9.4.35}$$

$$V = E \cos \gamma x_2 + F \sin \gamma x_2 \tag{9.4.36}$$

Since $U(0) = V(0) = 0$, $C = E = 0$. Further, since $U(a) = V(b) = 0$, $\sin \beta a = 0$ and $\sin \gamma b = 0$. Thus,

$$\begin{aligned} \beta_i &= \frac{i\pi}{a}, \quad i = 1, 2, \ldots \\ \gamma_j &= \frac{j\pi}{b}, \quad j = 1, 2, \ldots \end{aligned} \tag{9.4.37}$$

but since $\gamma^2 = \alpha^2 - \beta^2$, a double sequence of eigenvalues α_{ij} is obtained,

$$\alpha_{ij} = (\gamma_j^2 + \beta_i^2)^{1/2} = \pi\left[\left(\frac{i}{a}\right)^2 + \left(\frac{j}{b}\right)^2\right]^{1/2} \tag{9.4.38}$$

with eigenfunctions

$$X_{ij}(\mathbf{x}) = \sin \frac{i\pi x_1}{a} \sin \frac{j\pi x_2}{b} \tag{9.4.39}$$

Thus, the general solution is

$$u(\mathbf{x}, t) = \sum_{i=1}^{\infty}\sum_{j=1}^{\infty} [\, a_{ij} \cos m\alpha_{ij}t + b_{ij} \sin m\alpha_{ij}t \,] \sin \frac{i\pi x_1}{a} \sin \frac{j\pi x_2}{b}$$

The initial conditions may now be written as

$$u(\mathbf{x}, 0) = f(\mathbf{x}) = \sum_{i=1}^{\infty}\sum_{j=1}^{\infty} a_{ij} \sin \frac{i\pi x_1}{a} \sin \frac{j\pi x_2}{b}$$

$$u_t(\mathbf{x}, 0) = g(\mathbf{x}) = \sum_{i=1}^{\infty}\sum_{j=1}^{\infty} m\alpha_{ij} b_{ij} \sin \frac{i\pi x_1}{a} \sin \frac{j\pi x_2}{b}$$

Thus, coefficients a_{ij} and b_{ij} are Fourier coefficients of the functions $f(\mathbf{x})$ and $g(\mathbf{x})$ in double sine series expansions; i.e.,

$$a_{ij} = \frac{4}{ab}\int_0^a\int_0^b f(x_1, x_2) \sin \frac{i\pi x_1}{a} \sin \frac{j\pi x_2}{b}\, dx_1\, dx_2$$

$$b_{ij} = \frac{4}{m\alpha_{ij}ab}\int_0^a\int_0^b g(x_1, x_2) \sin \frac{i\pi x_1}{a} \sin \frac{j\pi x_2}{b}\, dx_1\, dx_2$$

■

EXERCISES 9.4

1. Obtain equations for the frequency of vibration of a clamped-clamped beam; i.e., for boundary conditions $u(0, t) = u_x(0, t) = u(\ell, t) = u_x(\ell, t) = 0$.

2. Show that the selection of a nonnegative constant on the right of Eq. 9.4.6 is justified by the fact that the operator

$$AX = X^{(4)}$$

with domain

$$D_A = \{ X \in C^4(0, \ell): X(0) = X'(0) = X''(\ell) = X'''(\ell) = 0 \}$$

is positive definite.

3. Derive formulas for forced vibration of the membrane of Example 9.4.2; i.e., replace Eq. 9.4.24 by

$$\frac{1}{m^2} u_{tt} - \nabla^2 u = h(\mathbf{x}, t)$$

9.5 A FORMAL TREATMENT OF THE EIGENFUNCTION EXPANSION METHOD

As seen in Chapter 8 and Section 9.4, the Fourier expansion of a function f(x) in terms of the eigenfunctions $u_i(x)$ of a symmetric, positive definite operator is often possible. Applications studied in Section 9.4 show that when this happens, the separation of variables technique can be employed to construct solutions of linear dynamics problems that arise in mechanics. In light of this apparent success, it is desirable to formalize and carefully justify the eigenfunction expansion method, which is the purpose of this section. It is noted here, and further developed later, that existence of a Green function for an operator is a key to proving completeness of the eigenfunctions.

Let Ω be the domain of the independent variables $\mathbf{x}$ (i.e., an interval on the x-axis, an area in the x_1-x_2-plane, or a volume in x_1-x_2-x_3-space) that has a piecewise smooth boundary Γ. Suppose the state of a continuum that occupies Ω is characterized by a function $u(\mathbf{x}, t)$ that vanishes identically if the system is in stable equilibrium under zero load. Let A be a differential operator in the independent variables $\mathbf{x}$, defined on Ω, let $\rho(\mathbf{x})$ represent the mass density of the material at any point in Ω, and let $Q(\mathbf{x}, t)$ represent a given external distributed force. A solution of the differential equation

$$\rho u_{tt} + Au = Q \tag{9.5.1}$$

is sought that satisfies homogeneous time-independent boundary conditions on the boundary Γ of Ω and the initial conditions

$$\begin{aligned} u(\mathbf{x}, 0) &= f(\mathbf{x}) \\ u_t(\mathbf{x}, 0) &= g(\mathbf{x}) \end{aligned} \tag{9.5.2}$$

All derivatives that occur are assumed to be continuous. For applications with homogeneous boundary conditions considered here, the operator A has been shown to be positive definite.

Separation of Variables

Consider first free vibration that is characterized by solutions of the homogeneous differential equation

$$\rho u_{tt} + Au = 0 \tag{9.5.3}$$

that satisfy prescribed homogeneous boundary conditions. Specifically, consider solutions of the form $u(\mathbf{x}, t) = X(\mathbf{x})\,T(t)$. Each such solution is associated with an eigenvalue α^2, with the properties

$$\ddot{T} = -\alpha^2 T \tag{9.5.4}$$

$$AX = \alpha^2 \rho X \tag{9.5.5}$$

where $X(\mathbf{x})$ must satisfy the boundary conditions that are imposed on u, so the eigenvalue α^2 of the positive definite operator A is positive. Thus,

$$T(t) = a \cos \alpha t + b \sin \alpha t$$

The problem of Eq. 9.5.5 is to determine values of the eigenvalues α^2 for which the homogeneous differential equation of Eq. 9.5.5 has eigenfunctions that satisfy the prescribed boundary conditions. The state of vibration that satisfies Eq. 9.5.3 is then represented by

$$u(\mathbf{x}, t) = T(t)\,X(\mathbf{x}) = (\,a \cos \alpha t + b \sin \alpha t\,)\,X(\mathbf{x}) \tag{9.5.6}$$

Completeness of Eigenfunctions

In the case of a bounded domain Ω, the following statements are true, but remain to be proved:

(1) The eigenvalues α^2 form a countably infinite ascending sequence $\alpha_1^2 \le \alpha_2^2 \le \alpha_3^2 \le \ldots$

(2) There exists a sequence of associated eigenfunctions $X_1(\mathbf{x}), X_2(\mathbf{x}), \ldots$ that is complete in the L_2 sense and satisfy the orthonormality relations

$$\int_\Omega \rho X_i X_k \, d\mathbf{x} = 0, \quad i \neq k$$
$$\int_\Omega \rho X_i^2 \, d\mathbf{x} = 1, \quad i = 1, 2, \ldots \tag{9.5.7}$$

(3) Every function $w(\mathbf{x})$ that satisfies the prescribed homogeneous boundary conditions, and for which Aw is continuous, may be expanded in an absolutely and uniformly convergent series in the eigenfunctions $X_i(\mathbf{x})$; i.e.,

$$w(\mathbf{x}) = \sum_{i=1}^{\infty} c_i X_i(\mathbf{x})$$
$$c_i = \int_\Omega \rho w X_i \, d\mathbf{x} \tag{9.5.8}$$

This pseudo-theorem will be justified in subsequent sections.

On the basis of these properties, which must be proved for each problem, an infinite sequence of functions $(a_i \cos \alpha_i t + b_i \sin \alpha_i t) X_i(\mathbf{x})$ is obtained. From these functions, solutions of the initial-boundary-value problem for the differential equation of Eq. 9.5.3 are obtained by superposition,

$$u(\mathbf{x}, t) = \sum_{i=1}^{\infty} T_i(t) X_i(\mathbf{x}) \tag{9.5.9}$$

if the constants a_i and b_i are chosen as

$$a_i = \int_\Omega \rho f X_i \, d\mathbf{x}$$
$$b_i = \frac{1}{\alpha_i} \int_\Omega \rho g X_i \, d\mathbf{x}$$

For the nonhomogeneous equation of Eq. 9.5.1, with homogeneous boundary conditions, $u(\mathbf{x}, t)$ is found by determining its time-dependent expansion coefficients $T_i(t)$. To this end, use

$$u_p(\mathbf{x}, t) = \sum_{i=1}^{\infty} T_{p_i}(t) X_i(\mathbf{x})$$

and

$$Q(\mathbf{x}, t) = \rho \sum_{i=1}^{\infty} Q_i(t) X_i(\mathbf{x})$$

in Eq. 9.5.1 to obtain

$$\sum_{j=1}^{\infty} \rho \ddot{T}_{p_j} + \sum_{j=1}^{\infty} T_{p_j} A X_j = \rho \sum_{j=1}^{\infty} Q_j X_j$$

Multiplying both sides of this equation by X_i, integrating over Ω, and using Eqs. 9.5.5 and 9.5.7,

$$\ddot{T}_{p_i} + \alpha_i^2 T_{p_i} = Q_i(t) \tag{9.5.10}$$

where $Q_i(t)$ is the expansion coefficient of $Q(\mathbf{x}, t)\,\rho^{-1}$ with respect to the orthonormalized eigenfunction $X_i(\mathbf{x})$. A particular solution $T_{p_i}(t)$ of this equation is given by

$$T_{p_i}(t) = \frac{1}{\alpha_i} \int_0^t Q_i(\tau) \sin \alpha_i \,(t - \tau)\, d\tau$$

The function formed using these expansion coefficients is a particular solution of Eq. 9.5.1. Other solutions are obtained by adding solutions of Eq. 9.5.3. Thus, the initial value problem under consideration is reduced to the problem of solving the homogeneous equation of Eq. 9.5.3.

Application of the separation of variables method to the heat conduction problem also leads to eigenvalue problems. If the units of time and length are suitably chosen, the differential equation of heat conduction takes the form

$$u_t + Au = 0 \tag{9.5.11}$$

where u denotes temperature, which is a function of position **x** and time t. The convection of heat from a homogeneous body Ω with surface Γ into an infinite medium of zero temperature is characterized at the surface Γ by a boundary condition of the form

$$\frac{\partial u}{\partial n} + \sigma u = 0 \tag{9.5.12}$$

where σ is a positive constant. This condition states that the rate of change of temperature in the direction of the inner normal is proportional to the difference in temperature between the exterior and the interior of the body. A solution of the equation of heat conduction is sought that satisfies this boundary condition and a prescribed initial temperature condition at time $t = 0$.

Writing $u(\mathbf{x}, t)$ in the form $u(\mathbf{x}, t) = X(\mathbf{x})\,T(t)$ yields

$$\frac{AX}{X} = -\frac{\dot{T}}{T} = \alpha^2 \tag{9.5.13}$$

since A is positive definite. This yields the following eigenvalue problem for $X(\mathbf{x})$:

$$\begin{aligned} AX &= \alpha^2 X, && \text{in } \Omega \\ \frac{\partial X}{\partial n} + \sigma X &= 0, && \text{on } \Gamma \end{aligned} \tag{9.5.14}$$

For a given eigenvalue α^2 and its eigenfunction $X(\mathbf{x})$, the corresponding solution of the differential equation has the form

$$u(\mathbf{x}, t) = c\, X(\mathbf{x})\, e^{-\alpha^2 t} \tag{9.5.15}$$

By the eigenfunction expansion method, the solution can be made to satisfy a given initial condition $u(\mathbf{x}, 0) = f(\mathbf{x})$, where $f(\mathbf{x})$ is continuous in Ω, together with its derivatives of first and second order, and satisfies the boundary conditions. If the normalized eigenfunctions $X_1(\mathbf{x}), X_2(\mathbf{x}), \ldots,$ form a complete sequence in $L_2(\Omega)$, the desired solution is given by

$$\begin{aligned} u(\mathbf{x}, t) &= \sum_{i=1}^{\infty} c_i\, X_i(\mathbf{x})\, e^{-\alpha^2 t} \\ c_i &= \int_\Omega f(\mathbf{x})\, X_i(\mathbf{x})\, d\mathbf{x} \end{aligned} \tag{9.5.16}$$

Note that the positive character of the eigenvalues α^2 implies that the solution $u(\mathbf{x}, t)$ approaches zero asymptotically as t increases, as is expected from the physical meaning of the problem.

Green's Function

Eigenfunctions can also be used to solve the equilibrium problem; i.e., the boundary-value problem for the differential equation $Au = Q(\mathbf{x})$. Let

$$Q(\mathbf{x}) = \sum_{i=1}^{\infty} X_i(\mathbf{x}) \int_\Omega Q(\xi)\, X_i(\xi)\, d\xi$$

$$u(\mathbf{x}) = \sum_{i=1}^{\infty} c_i\, X_i(\mathbf{x})$$

To find $c_i(t)$, substitute Q and u into $Au = Q$, to obtain

$$\sum_{i=1}^{\infty} c_i\, AX_i(\mathbf{x}) = \sum_{i=1}^{\infty} c_i \alpha_i^2\, X_i(\mathbf{x}) = \sum_{i=1}^{\infty} X_i(\mathbf{x}) \int_\Omega Q(\xi)\, X_i(\xi)\, d\xi$$

Equating coefficients of $X_i(\mathbf{x})$,

$$c_i = \frac{1}{\alpha_i^2} \int_\Omega Q(\xi)\, X_i(\xi)\, d\xi \tag{9.5.17}$$

Hence, the solution is given formally as

$$u(\mathbf{x}) = \sum_{i=1}^{\infty} \frac{X_i(\mathbf{x})}{\alpha_i^2} \int_\Omega Q(\xi)\, X_i(\xi)\, d\xi \tag{9.5.18}$$

Formally interchanging summation and integration in this expression, a function

$$K(\mathbf{x}, \xi) = \sum_{i=1}^{\infty} \frac{X_i(\mathbf{x})\, X_i(\xi)}{\alpha_i^2} \tag{9.5.19}$$

is defined, so that the solution of the boundary-value problem can formally be written in the form

$$u(\mathbf{x}) = \int_\Omega Q(\xi)\, K(\mathbf{x}, \xi)\, d\xi \tag{9.5.20}$$

The function $K(\mathbf{x}, \xi)$ is called **Green's function** for the operator. It was characterized in Section 8.4 in a quite different manner. It forms the basis for a more detailed investigation, which reaches beyond the formal structure of the present treatment.

9.6 GREEN'S FUNCTION FOR ORDINARY BOUNDARY-VALUE PROBLEMS

Consider the linear ordinary differential operator A of second order,

$$Au = -(pu')' + qu \tag{9.6.1}$$

where $u(x)$ is defined in the interval $a \le x \le b$, $p(x)$ is in $C^1(a, b)$, $q(x)$ is in $C^0(a, b)$, and $p(x) > 0$. The associated nonhomogeneous differential equation is of the form

$$Au = h(x) \tag{9.6.2}$$

where $h(x)$ is in $D^0(a, b)$. The goal in solving the boundary-value problem is to find a solution of Eq. 9.6.2 that satisfies homogeneous boundary conditions at a and b; i.e.,

$$u(a) = u(b) = 0 \tag{9.6.3}$$

Influence Function

It is natural to start with physical considerations. Consider Eq. 9.6.2 as the condition for static equilibrium of a string, under the influence of a time-independent force that is distributed with density h(x) over the string. A limiting process may be used to transform the continuously distributed force to a **point force**; i.e., to a force that acts at a single point $x = \xi$ with a given intensity. Let $K(x, \xi)$ denote the deflection of the string at point x, as a result of the action of a point force of unit intensity that is applied at point ξ. The effect at x of the distributed force h(x) can then be considered as superposition of the effects of differential forces $h(\xi)\, d\xi$. On physical grounds, it is expected that the desired solution is of the form

$$u(x) = \int_a^b K(x, \xi)\, h(\xi)\, d\xi \tag{9.6.4}$$

Note that if $h(\xi)$ is a point load at ξ, then $h(\xi) = \delta(\xi)$ and $u(x) = K(x, \xi)$. The function $K(x, \xi)$, which satisfies the prescribed boundary conditions at a and b, is called the **influence function** or **Green's function** for the differential operator A. It follows that the function u(x) that is represented in Eq. 9.6.4 by an integral in terms of the **kernel** $K(x, \xi)$ and the load density h(x) also satisfies these boundary conditions.

The Green function $K(x, \xi)$, for ξ fixed, satisfies the differential equation

$$AK = 0 \tag{9.6.5}$$

everywhere except at the point $x = \xi$, since K corresponds to a zero force when $x \neq \xi$. At the point $x = \xi$, the function $K(x, \xi)$ must have a singularity, which can be found in the following way. Consider the point force as the limiting case of a force $h_\varepsilon(x)$ that vanishes in (a, b) if $|x - \xi| > \varepsilon$, but for which the total intensity is

$$\int_{\xi-\delta}^{\xi+\delta} h_\varepsilon(x)\, dx = 1 \tag{9.6.6}$$

Denote the associated deflection of the string as $K_\varepsilon(x, \xi)$. Thus,

$$AK_\varepsilon = -(pK_\varepsilon')' + qK_\varepsilon = h_\varepsilon(x) \tag{9.6.7}$$

Integrating this equation between the limits $\xi - \delta$ and $\xi + \delta$, where $\delta \geq \varepsilon$ may be chosen arbitrarily provided the interval of integration remains in (a, b),

$$\int_{\xi-\delta}^{\xi+\delta} \left[-\frac{d}{dx}\left(p\,\frac{dK_\varepsilon}{dx} \right) + qK_\varepsilon \right] dx = 1$$

Taking the limit as $\varepsilon \to 0$, assuming that K_ε converges to a continuous function $K(x, \xi)$ that is continuously differentiable except at $x = \xi$,

$$\int_{\xi-\delta}^{\xi+\delta}\left[-\frac{d}{dx}\left(p\frac{dK}{dx}\right)-qK\right]dx = 1$$

Now, as $\delta \to 0$, integration of the second term qK becomes zero and

$$\lim_{\delta\to 0}\frac{\partial K(x,\xi)}{\partial x}\Bigg|_{x=\xi-\delta}^{x=\xi+\delta} = -\frac{1}{p(\xi)} \tag{9.6.8}$$

which characterizes the singularity of the Green function at $x = \xi$.

Properties of Green's Functions

Summarizing the foregoing heuristic discussion, a rigorous mathematical theory can be obtained.

Theorem 9.6.1. Let $K(x, \xi)$ be a Green function of the differential operator A and homogeneous boundary conditions, with the following properties:

(1) For fixed ξ, $K(x, \xi)$ is a continuous function of x that satisfies the prescribed boundary conditions.
(2) Except at the point $x = \xi$, the first and second derivatives of K with respect to x are continuous. At the point $x = \xi$, the first derivative has a jump discontinuity that is given by

$$\frac{\partial K(x,\xi)}{\partial x}\Bigg|_{x=\xi-0}^{x=\xi+0} = -\frac{1}{p(\xi)} \tag{9.6.9}$$

(3) Considered as a function of x, $K(x, \xi)$ satisfies the differential equation $AK = 0$, except at the point $x = \xi$.

If h(x) is a continuous or piecewise continuous function of x, then the function u(x) in Eq. 9.6.4 is a solution of the differential equation of Eq. 9.6.2 that satisfies the boundary conditions. Conversely, if the function u(x) satisfies Eq. 9.6.2 and the boundary conditions, it can be represented by Eq. 9.6.4. ■

To prove the first conclusion in Theorem 9.6.1, Leibniz's rule for differentiation of an integral with respect to a parameter of Eq. 6.1.8 is needed. Differentiating Eq. 9.6.4,

$$u'(x) = \int_a^b K'(x,\xi)\,h(\xi)\,d\xi$$

$$= \int_a^x K'(x,\xi)\,h(\xi)\,d\xi + \int_x^b K'(x,\xi)\,h(\xi)\,d\xi \tag{9.6.10}$$

Multiplying both sides of Eq. 9.6.10 by p(x) and differentiating again,

$$(pu')'(x) = \int_a^x (pK')'(x, \xi)\, h(x, \xi)\, d\xi + \int_x^b (pK')'(x, \xi)\, h(x, \xi)\, d\xi$$

$$+ p\, K'(x, x - 0)\, h(x) - p\, K'(x, x + 0)\, h(x)$$

$$= \int_a^b (pK')'(x, \xi)\, h(\xi)\, d\xi$$

$$+ p\, [\, K'(x + 0, x) - K'(x - 0, x)\,]\, h(x)$$

$$= \int_a^b (pK')(x, \xi)\, h(\xi)\, d\xi - h(x)$$

where $K'(x, x - 0) = K'(x + 0, x)$, $K'(x, x + 0) = K'(x - 0, x)$, and Eq. 9.6.9 have been used. Therefore,

$$Au = - (pK')' + qu$$

$$= \int_a^b (- (pK')' + qK)\, h(\xi)\, d\xi + h(x)$$

$$= \int_a^b (AK)\, h(\xi)\, d\xi + h(x)$$

This proves that u is a solution, since $AK = 0$, except at $\xi = x$.

To prove the converse, form the integral identity

$$\int_{x_1}^{x_2} (vAu - uAv)\, dx = - p\, (u'v - v'u) \Big|_{x_1}^{x_2} \tag{9.6.11}$$

With $v = K(x, \xi)$ in the intervals $a \le x \le \xi$ and $\xi \le x \le b$,

$$\int_a^b K(x, \xi)\, Au(x)\, dx = - p(\xi)\, [\, u'(\xi)\, (K(\xi - 0, \xi) - K(\xi + 0, \xi))$$

$$- u(\xi)\, (K'(\xi - 0, \xi) - K'(\xi + 0, \xi))\,]$$

Since $K(x, \xi)$ is continuous, the first term on the right vanishes. By Eq. 9.6.9, the second term on the right becomes $u(\xi)$. By interchanging ξ and x, this is just

$$u(x) = \int_a^b K(\xi, x)\, Au(\xi)\, d\xi$$

Substituting for Au from Eq. 9.6.2, the desired formula of Eq. 9.6.4 is obtained for the solution. ∎

Green's function for a symmetric differential operator is a symmetric function of the arguments ξ and x; i.e.,

$$K(x, \xi) = K(\xi, x) \tag{9.6.12}$$

This follows almost immediately from Eq. 9.6.11 by substituting $v = K(x,\eta)$ and $u = K(x, \xi)$, dividing the domain of integration into the intervals $a \le x \le \xi$, $\xi \le x \le \eta$, and $\eta \le x \le b$, and treating each interval separately. The proof is completed by taking into account both the discontinuity relation of Eq. 9.6.10 at the points $x = \xi$ and $x = \eta$ and the boundary conditions. Symmetry of Green's function expresses a **reciprocity** that frequently occurs in mechanics; i.e., a unit force that is applied at point ξ produces the displacement $K(x, \xi)$ at point x and a unit force that acts at x produces the same result at ξ.

To construct Green's function for Au, with boundary conditions $u(a) = u(b) = 0$, consider any solution $u_0(x)$ of the differential equation $Au = 0$ that satisfies the given boundary condition at $x = a$. Then, $c_0u_0(x)$ is the most general such solution. Similarly, let $c_1u_1(x)$ be the family of solutions of $Au = 0$ that satisfy the boundary condition at $x = b$. There are two possible cases. Either the two families of solutions are distinct, which is the general case, or they are identical.

In the first case, the functions $u_0(x)$ and $u_1(x)$ are linearly independent. In this case, the constants c_0 and c_1 can be chosen such that the point of intersection of the solutions is $x = \xi$ and such that the discontinuity of the derivative at this point has precisely the value $-1/p(\xi)$. In this way, Green's function $K(x, \xi)$ is obtained explicitly as

$$K(x, \xi) = \begin{cases} \dfrac{u_1(\xi)\, u_0(x)}{p(\xi)\, [\, u_0'(\xi)\, u_1(\xi) - u_0(\xi)\, u_1'(\xi)\,]}, & a \le x \le \xi \\[2ex] \dfrac{u_0(\xi)\, u_1(x)}{p(\xi)\, [\, u_0'(\xi)\, u_1(\xi) - u_0(\xi)\, u_1'(\xi)\,]}, & \xi \le x \le b \end{cases} \tag{9.6.13}$$

In the second case, $u_0(x)$ and $u_1(x)$ differ only by a constant factor. Every solution that belongs to one family also belongs to the other. The function $u_0(x)$ satisfies not only the boundary condition at $x = a$, but also the boundary condition at $x = b$. Thus, the equation $Au = 0$ has a nontrivial solution $u_0(x)$ that satisfies the boundary conditions. This can also be expressed by stating that $\lambda = 0$ is an eigenvalue of $Au = \lambda u$. Hence, the above construction fails and no Green's function exists.

The existence of Green's function is equivalent to the existence of a unique solution of the homogeneous boundary-value problem for the differential equation $Au = h(x)$. Therefore, the following alternative exists: Under given homogeneous boundary conditions, either the equation $Au = h(x)$ has a unique solution $u(x)$ for any given $h(x)$, or the homogeneous equation $Au = 0$ has a nontrivial solution.

Ordinary differential equations of higher order are not essentially different. The discussion here is restricted to a typical example associated with the differential equation $u'''' = f$, the uniform beam of Example 9.4.1. As before, the influence function, or

Green's function $\hat{K}(x, \xi)$ is introduced as the displacement of the beam under the influence of a unit point force that acts at the point $x = \xi$ and satisfies the prescribed homogeneous boundary conditions. In the same manner as above, the following typical conditions for the Green function are obtained:

(1) For every value of the parameter ξ, the function $\hat{K}(x, \xi)$ and its first and second derivatives are continuous and $\hat{K}(x, \xi)$ satisfies the prescribed homogeneous boundary conditions.

(2) At any point $x \neq \xi$, the third and fourth derivatives with respect to x are continuous. However, at $x = \xi$ the following discontinuity condition holds:

$$\lim_{\varepsilon \to 0} [\, \hat{K}'''(\xi + \varepsilon, \xi) - \hat{K}'''(\xi - \varepsilon, \xi)\,] = 1 \tag{9.6.14}$$

(3) Except at the point $x = \xi$, the differential equation

$$\hat{K}''''(x, \xi) = 0 \tag{9.6.15}$$

is satisfied.

The fundamental property of Green's function for the fourth-order operator $\hat{A}u = u''''$ can be stated as follows: Let u(x) be a continuous function that satisfies the boundary conditions and has three piecewise continuous derivatives. Let h(x) be a piecewise continuous function. If u(x) and h(x) are connected by the relation

$$\hat{A}u = u'''' = h(x) \tag{9.6.16}$$

then

$$u(x) = \int_{x_0}^{x_1} \hat{K}(x, \xi)\, h(\xi)\, d\xi \tag{9.6.17}$$

and conversely.

EXERCISES 9.6

1. Carry out the calculations outlined in the text to obtain Eq. 9.6.8.

2. Derive Eq. 9.6.12.

3. Carry out the calculations outlined in the text, using properties (1), (2), and (3) of $K(x, \xi)$ in Theorem 9.6.1, to derive Eq. 9.6.13.

4. Show that u(x) of Eq. 9.6.17 is a solution of Eq. 9.6.16.

9.7 COMPLETENESS OF EIGENFUNCTIONS

With the aid of the Green function, completeness relations for eigenfunctions of positive bounded below ordinary differential operators are proved for the Sturm–Liouville operator. Literature is cited for more general results and an outline of the proof for general ordinary differential operators is given.

Green's Function for the Sturm–Liouville Operator

Consider first the Sturm–Liouville operator

$$\begin{aligned} Au &\equiv -(p(x)\,u')' + q(x)\,u \\ u(0) &= u(1) = 0 \end{aligned} \tag{9.7.1}$$

where $q(x)$ is in $C^0(0, 1)$, $p(x)$ is in $C^1(0, 1)$, and $h(x)$ is in $L_2(0, 1)$. Suppose that $p(x)$ is positive and that $q(x)$ is nonnegative for $0 \le x \le 1$. Since the operator $Au = -(pu')' + qu$ is positive bounded below on $D_A = \{ u \in C^2(0, 1): u(0) = u(1) = 0 \}$, it follows that the boundary-value problem

$$\begin{aligned} -(pu')' + qu &= h(x) \\ u(0) &= u(1) = 0 \end{aligned} \tag{9.7.2}$$

has a unique solution and, as shown in Section 9.6, the Green function $G(x, \xi)$ exists. That is, the solution of this problem is

$$u(x) = \int_0^1 G(x, \xi)\, h(\xi)\, d\xi$$

Setting $h(\xi) = \lambda\rho\, u(\xi)$ in Eq. 9.7.2, the relation

$$u(x) = \lambda \int_0^1 G(x, \xi)\, \rho(\xi)\, u(\xi)\, d\xi \tag{9.7.3}$$

is obtained for the eigenfunctions of $Au = \lambda\rho(x)\,u$, where $\rho(x) > 0$ is in $C^0(0, 1)$. Conversely, if $u(x)$ satisfies Eq. 9.7.3, it is an eigenfunction and λ is an eigenvalue of $Au = \lambda\rho u$.

Given the eigenfunctions $u_n(x)$ of A, the equation

$$\frac{u_n(x)}{\lambda_n K_n} = \int_0^1 G(x, \xi)\, \sqrt{\rho(\xi)} \left[\frac{\sqrt{\rho(\xi)}\; u_n(\xi)}{K_n} \right] d\xi \tag{9.7.4}$$

can be thought of as providing Fourier coefficients of $G(x, \xi)\,\sqrt{\rho(\xi)}$, in terms of normalized functions $\dfrac{\sqrt{\rho(\xi)}\, u_n(\xi)}{K_n}$, where $K_n = \left[\int_0^1 \rho(\xi)\, u_n^2(\xi)\, d\xi \right]^{1/2}$. The Bessel inequality of Eq. 5.4.14, applied to the Fourier coefficients of Eq. 9.7.4, yields

$$\sum_{n=1}^{m} \frac{u_n^2(x)}{\lambda_n^2 \int_0^1 \rho(\xi)\, u_n^2(\xi)\, d\xi} \leq \int_0^1 G^2(x, \xi)\, \rho(\xi)\, d\xi$$

Multiplying both sides by $\rho(x)$ and integrating from 0 to 1 yields the relation

$$\sum_{n=1}^{m} \frac{1}{\lambda_n^2} \leq \int_0^1 \int_0^1 G^2(x, \xi)\, \rho(x)\, \rho(\xi)\, d\xi\, dx \tag{9.7.5}$$

Since $G(x, \xi)$ is continuous in x and ξ, the right side of Eq. 9.7.5 is finite. Since $\lambda_1, \ldots, \lambda_m$ are eigenvalues, it follows that if there are an infinite number of eigenvalues, the series $\sum_{n=1}^{\infty} \frac{1}{\lambda_n^2}$ converges. In particular,

$$\lim_{k \to \infty} \lambda_k = \infty \tag{9.7.6}$$

It is shown in Ref. 20 that there are always an infinite number of eigenvalues for the Sturm–Liouville operator.

The following special case of a more general minimization principle for eigenvalues of positive bounded below differential operators is required for theoretical arguments. This result is a special case of the theory developed in Chapter 10.

Theorem 9.7.1. The smallest eigenvalue λ_1 of the Sturm–Liouville operator A is given by

$$\lambda_1 = \inf_{\phi \text{ in } D_A,\ \phi \neq 0} \frac{\int_0^1 \{ p\phi'^2 + q\phi^2 \}\, dx}{\int_0^1 \rho\phi^2\, dx} \tag{9.7.7}$$

where minimization is over functions in D_A that are continuous and have piecewise continuous derivatives. The normalized function u(x) for which the minimum is achieved is the corresponding eigenfunction. Successive eigenvalues λ_k are given by the relation

$$\lambda_k = \inf_{\substack{\phi \text{ in } D_A \\ \int_0^1 \rho\phi u_1 \, dx = 0 \\ \vdots \\ \int_0^1 \rho\phi u_{k-1} \, dx = 0}} \left\{ \frac{\int_0^1 \{ p\phi'^2 + q\phi^2 \} \, dx}{\int_0^1 \rho\phi^2 \, dx} \right\} \tag{9.7.8}$$

■

The principle theorem of this section is now stated and proved.

Completeness of Sturm–Liouville Equations

Theorem 9.7.2. The eigenfunctions $u_n(x)$ of $Au = \lambda\rho u$ for the Sturm–Liouville operator A are complete in the space of functions $L_2^\rho(0, 1)$; i.e., functions that satisfy

$$\| f \|_\rho^2 \equiv \int_0^1 \rho f^2 \, dx < \infty$$

■

To prove Theorem 9.7.2, let f be any continuous, piecewise continuously differentiable function with $f(0) = f(1) = 0$. The Fourier series for $f(x)$ in terms of the $u_n(x)$ is

$$f(x) \approx \sum_{n=1}^{\infty} c_n u_n(x)$$

where the coefficients are

$$c_n = \frac{\int_0^1 \rho f u_n \, dx}{\int_0^1 \rho u_n^2 \, dx}$$

By definition of c_n,

$$\int_0^1 \rho \left(f - \sum_{n=1}^{k-1} c_n u_n \right) u_i \, dx = 0, \quad i = 1, \ldots, k-1$$

Putting $\phi = \left(f - \sum_{n=1}^{k-1} c_n u_n \right)$ in Eq. 9.7.8,

$$\int_0^1 \rho \left(f - \sum_{n=1}^{k-1} c_n u_n \right)^2 dx$$

$$\leq \frac{1}{\lambda_k} \int_0^1 \left\{ p \left(f' - \sum_{n=1}^{k-1} c_n u_n' \right)^2 + q \left(f - \sum_{n=1}^{k-1} c_n u_n \right)^2 \right\} dx$$

$$= \frac{1}{\lambda_k} \left\{ \int_0^1 (pf'^2 + qf^2)\, dx - 2 \sum_{n=1}^{k-1} c_n \int_0^1 (pf'u_n' + qfu_n)\, dx \right.$$

$$\left. + \sum_{m=1}^{k-1} \sum_{n=1}^{k-1} c_n c_m \int_0^1 (pu_n'u_m' + qu_n'u_m')\, dx \right\}$$

Integration by parts yields

$$\int_0^1 (pf'u_n' + qfu_n)\, dx = \int_0^1 f \{ -(pu_n')' + qu_n \}\, dx$$

$$= \lambda_n \int_0^1 \rho f u_n\, dx$$

$$= \lambda_n c_n \int_0^1 \rho u_n^2\, dx$$

Also,

$$\int_0^1 (pu_n'u_m' + qu_n'u_m')\, dx = \lambda_n \int_0^1 \rho u_n u_m\, dx$$

$$= \begin{cases} \lambda_n \int_0^1 \rho u_n^2\, dx, & m = n \\ 0, & m \neq n \end{cases}$$

Thus,

$$0 \leq \int_0^1 \rho \left(f - \sum_{n=1}^{k-1} c_n u_n \right)^2 dx$$

$$\leq \frac{1}{\lambda_k} \left\{ \int_0^1 (pf'^2 + qf^2)\, dx - \sum_{n=1}^{k-1} \lambda_n c_n \int_0^1 \rho u_n\, dx \right\}$$

$$\leq \frac{1}{\lambda_k} \left\{ \int_0^1 (pf'^2 + qf^2)\, dx \right\}$$

Since $\lambda_k \to \infty$, it follows that any function f(x) that is continuously differentiable and satisfies f(0) = f(1) = 0 can be approximated in the mean by a finite linear combination of the $u_i(x)$. Since any function for which $\int_0^1 \rho f^2\,dx$ is finite can be approximated in the mean by a continuously differentiable function [14], it follows that the eigenfunctions $\{ u_i(x) \}$ are complete in L_2^ρ. This completes the proof of the theorem. ■

Completeness of Eigenfunctions of General Operators

Consider now the general linear boundary-value problem with the nth order ordinary differential equation

$$Au = p_0(x)\, u^{(n)} + p_1(x)\, u^{(n-1)} + \ldots + p_n(x)\, u = h(x) \tag{9.7.9}$$

in $a \le x \le b$, where $p_i(x)$ is in $C^{n-i}(a, b)$ and $p_0(x) \ne 0$ on $a \le x \le b$, with the boundary conditions

$$U_j u = \sum_{k=1}^{n} \left(M_{jk}\, u^{(k-1)}(a) + N_{jk}\, u^{(k-1)}(b) \right) = 0, \quad j = 1, \ldots, n \tag{9.7.10}$$

where M_{jk} and N_{jk} are constants.

Three theorems that are proved in Ref. 20 (see Section 7.4) are now stated. The first theorem provides existence of an infinite set of distinct eigenvalues. The second theorem is an extension of Theorem 9.7.2, which is proved in the same way. The third theorem provides results on pointwise convergence.

Theorem 9.7.3. If the operator A of the boundary-value problem of Eq. 9.7.9, with the boundary conditions of Eq. 9.7.10, is symmetric, then it has an infinite set of distinct eigenvalues with no finite cluster point. ■

Theorem 9.7.4. Let the boundary-value problem be as in Theorem 9.7.3. Then its normalized eigenfunctions $\{ u_n(x) \}$ are orthogonal and complete in $L_2(a, b)$. ■

Theorem 9.7.5. Let the boundary-value problem be as in Theorem 9.7.3. If f(x) is in $C^n(a, b)$ and satisfies the boundary conditions of Eq. 9.7.10, then

$$f(x) = \sum_{k=0}^{\infty} (f, u_k)\, u_k(x), \quad a \le x \le b \tag{9.7.11}$$

and convergence is uniform in $a \le x \le b$. ■

The foregoing results are next generalized to higher dimensions. Let A be a positive bounded below differential operator of order 2n, defined on a dense subspace D_A of

$L_2(\Omega)$. If there exists a Green function $G(\mathbf{x}, \xi)$ for $Au = h(\mathbf{x})$; i.e., a function $G(\mathbf{x}, \xi)$ in $L_2^\rho(\Omega)$ such that $u(\mathbf{x}) = \int_\Omega G(\mathbf{x}, \xi)\, h(\xi)\, d\xi$, and if there are an infinite number of distinct eigenvalues λ_n of A, then $\lambda_n \to \infty$ as $n \to \infty$.

To show that this is true, let $u_n(x)$ be eigenfunctions of $Au = \lambda\rho u$, with the corresponding eigenvalues λ_n. Then

$$\frac{u_n(\mathbf{x})}{\lambda_n\left[\iint_\Omega \rho\, u_n^2(\xi)\, d\xi\right]^{1/2}} = \frac{\iint_\Omega G(\mathbf{x}, \xi)\, \rho(\xi)\, u_n(\xi)\, d\xi}{\left[\iint_\Omega \rho\, u_n^2(\xi)\, d\xi\right]^{1/2}} \tag{9.7.12}$$

This equation may be thought of as providing the Fourier coefficients of $G(\mathbf{x}, \xi)\sqrt{\rho(\xi)}$, in terms of functions $\sqrt{\rho(\xi)}\, u_n(\xi)$ that are L_2-orthogonal. Bessel's inequality yields

$$\sum_{n=1}^{m} \frac{u_n^2(\mathbf{x})}{\lambda_n \iint_\Omega \rho(\xi)\, u_n^2(\xi)\, d\xi} \le \iint_\Omega G^2(\mathbf{x}, \xi)\, \rho(\xi)\, d\xi$$

Multiplying both sides of this inequality by $\rho(\mathbf{x})$ and integrating,

$$\sum_{n=1}^{m} \frac{1}{\lambda_n^2} \le \iint_\Omega \iint_\Omega G^2(\mathbf{x}, \xi)\, \rho(\mathbf{x})\, \rho(\xi)\, d\xi\, d\mathbf{x} \tag{9.7.13}$$

from which it follows that $\lambda_n \to \infty$ as $n \to \infty$.

The foregoing and results proved in Chapter 10 yield the result that follows.

Theorem 9.7.6. Let A be an operator with the above properties. Then the smallest eigenvalue of A is given by

$$\lambda_1 = \min_{\phi \text{ in } D_A,\ \phi \neq 0} \frac{(A\phi, \phi)}{\|\phi\|_\rho^2} \tag{9.7.14}$$

where $\|\phi\|_\rho^2 \equiv \iint_\Omega \rho\, \phi^2(\xi)\, d\xi$. The successive eigenvalues are given by

$$\lambda_k = \min_{\substack{\phi \text{ in } D_A,\ \phi \neq 0 \\ (\phi, u_i)_\rho = 0 \\ i = 1, \ldots, k-1}} \left\{ \frac{(A\phi, \phi)}{\|\phi\|_\rho^2} \right\} \tag{9.7.15}$$

■

Theorem 9.7.7. Let the operator A be as in Theorem 9.7.6. Then the eigenfunctions of A are complete in $L_2^\rho(\Omega)$. ■

To prove Theorem 9.7.7, let $f(\mathbf{x})$ be in D_A, which is a dense subspace of $L_2^\rho(\Omega)$. Expanding $f(\mathbf{x})$ in a Fourier series $f(\mathbf{x}) \sim \sum_{n=1}^{\infty} c_n u_n(\mathbf{x})$, where the $u_n(\mathbf{x})$ are normalized by $\| u_n \|_\rho = 1$, the coefficients are

$$c_n = (f, u_n)_\rho$$

where $(u, v)_\rho \equiv \iint_\Omega \rho uv \, d\Omega$. By definition of c_n,

$$\left(\left[f - \sum_{n=1}^{k-1} c_n u_n \right], u_i \right)_\rho = 0, \quad i = 1, \ldots, k-1$$

By Eq. 9.7.15,

$$\begin{aligned} \left\| f - \sum_{n=1}^{k-1} c_n u_n \right\|_\rho^2 &\leq \frac{1}{\lambda_k} \left(A\left(f - \sum_{n=1}^{k-1} c_n u_n \right), f - \sum_{n=1}^{k-1} c_n u_n \right) \\ &= \frac{1}{\lambda_k} \left\{ (Af, f) - \sum_{n=1}^{k-1} \lambda_n c_n (u_n, f)_\rho \right. \\ &\qquad \left. - \left(Af, \sum_{n=1}^{k-1} c_n u_n \right) + \sum_{n=1}^{k-1} \lambda_n c_n^2 \right\} \\ &= \frac{1}{\lambda_k} \left\{ (Af, f) - \sum_{n=1}^{k-1} \lambda_n c_n^2 \right\} \end{aligned} \tag{9.7.16}$$

Further, Af is in L_2^ρ, so its Fourier coefficients are

$$b_i = (u_i, Af)_\rho = \lambda_i c_i$$

Thus, both sequences $\{ c_i \}$ and $\{ \lambda_i c_i \}$ are in the space ℓ_2 defined in Section 5.3, so $\sum_{n=1}^{\infty} \lambda_n c_n^2 < \infty$. Therefore, the series $\sum_{n=1}^{\infty} \lambda_n c_n^2$ converges and the right side of Eq. 9.7.16

can be made arbitrarily small for k sufficiently large, which proves that any $f(\mathbf{x})$ in D_A can be approximated by a finite linear combination of the $u_i(\mathbf{x})$. Since D_A is dense in $L_2^\rho(\Omega)$, it follows that the sequence $\{ u_i(\mathbf{x}) \}$ is complete in $L_2^\rho(\Omega)$. ■

The proof in Section 9.2 that the Laplace operator with reasonable boundary conditions is positive bounded below, the construction of Green's function for this operator for simple sets Ω in Section 8.4, and Theorem 9.7.7 shows that its eigenfunctions are complete. While the hypotheses of Theorem 9.7.7 have not been verified for all conceivable linear operators, it has been shown to apply to a broad variety of problems.

It is worthy of note that physical intuition and reasoning indicates that Theorem 9.7.7 is generally applicable to linear conservative mechanics problems. Such problems are generally governed by variational principles, so the resulting operators should be symmetric [18] and those associated with elastic deformation (strain energy) are generally positive definite. The positive bounded below property was seen in Section 9.2 to be associated with stable equilibrium; hence it is also expected to be valid.

Finally, while Green functions have been constructed only for a few problems, the Green function has been shown to be the response at point ξ due to a unit input at point $\mathbf{x}$, which certainly exists for wide classes of mechanics problems. Indeed, there is an extensive literature that presents Green's functions for a variety of problems. The physical motivation for validity of the hypotheses of Theorem 9.7.7 and the extensive literature that rigorously verifies these hypotheses in specific cases provide confidence that the theorem is widely applicable. This theoretical support for the formal eigenfunction method presented in Section 9.5 provides one of the most useful mathematical tools for the study of dynamics of linear continua.

10

VARIATIONAL METHODS FOR BOUNDARY-VALUE PROBLEMS

The fundamental properties of symmetry and positive definiteness of linear operators of mechanics have been seen in Chapter 9 to yield desirable properties of eigenfunctions. These properties make eigenfunction expansion methods broadly applicable. In this chapter, even more broadly applicable methods of treating elliptic boundary-value problems are introduced, called variational methods. These methods are shown to yield theoretical and computational results that form the foundation for most modern computational methods in solid mechanics.

10.1 ENERGY CONVERGENCE

In this section, a new measure of closeness of two functions is introduced. Convergence associated with this measure of closeness plays an important role in the variational theory developed in this chapter.

Consider a positive definite linear operator A, with domain of definition D_A. If A is a differential operator, then D_A will be a dense set of functions in L_2 that are continuous and have continuous derivatives in a closed set $\bar{\Omega} = \Omega + \Gamma$. If A is a differential operator of order k, then D_A will consist of functions that, together with their derivatives of order $k - 1$, are continuous in Ω and whose derivatives of order k exist, except at a finite number of points, curves, or surfaces, and have finite norm in the open domain Ω. It follows that the functions Au for u in D_A also have finite norm. In addition to the above continuity assumptions, the functions in D_A must satisfy certain boundary conditions, which for the present will be assumed homogeneous.

Energy Scalar Product and Norm

Definition 10.1.1. If A is a positive definite linear operator, the quantity (Au, v) defines a scalar product on D_A, called the **energy scalar product**, denoted by

$$[\, u, v \,]_A \equiv (\, Au, v \,) = \iint_\Omega vAu \, d\Omega \tag{10.1.1}$$

■

To show that Eq. 10.1.1 defines a scalar product, note first that for any real α, $[\alpha u, v]_A = (A\alpha u, v) = (\alpha Au, v) = \alpha [u, v]_A$. Next, since A is symmetric, $[u, v]_A = (Au, v) = (u, Av) = (Av, u) = [v, u]_A$. Finally, since A is positive definite, $[u, u]_A = (Au, u) > 0$ if $u \neq 0$. ■

If D_A is interpreted as a space of admissible displacements, then for u and v in D_A, Au may be interpreted as the force that is required to produce the displacement u and the scalar product (Au, v) becomes the work done by the force Au acting through the displacement v. The quantity (Au, u) is proportional to the strain energy associated with the displacement u.

Definition 10.1.2. The quantity $\| u \|_A \equiv (Au, u)^{1/2}$ is called the **energy norm** of the function u. ■

Let u and v be two functions in D_A. A measure of their closeness is the square root of the energy of their difference; i.e.,

$$\| u - v \|_A \equiv \sqrt{(A(u-v), u-v)} \tag{10.1.2}$$

Energy Convergence

The formulation of the energy measure of Eq. 10.1.2 of the distance between two functions in D_A leads to a new concept of convergence of functions in D_A.

Definition 10.1.3. Let u and u_n be in D_A, $n = 1, 2, \ldots$, for a positive definite operator A. The sequence u_n is said to **converge in energy** to u if

$$\lim_{n\to\infty} \| u_n - u \|_A = 0$$

Symbolically, convergence in energy will be denoted by $u_n(x) \xrightarrow{E} u(x)$. ■

Note that the concept of energy convergence can be defined only after the corresponding positive definite operator has been defined.

Theorem 10.1.1. If the operator A is positive bounded below and $u_n \xrightarrow{E} u$, then $\lim_{n\to\infty} \| u_n - u \| = 0$, denoted $u_n \xrightarrow{\text{mean}} u$. ■

To prove this theorem, note that since A is positive bounded below, there exists a positive constant γ such that $(Au, u) \geq \gamma \| u \|^2$, for all u in D_A. It follows therefore that $\| u \|^2 \leq (1/\gamma) \| u \|_A^2$ and $\| u_n - u \|^2 \leq (1/\gamma) \| u_n - u \|_A^2$. Thus, if $\| u_n - u \|_A \to 0$, then $\| u_n - u \| \to 0$. ■

Example 10.1.1

Consider the operator defined by

$$Au = -\frac{d^2u}{dx^2} \tag{10.1.3}$$

$$D_A = \{ u \text{ in } C^2(0, 1): u(0) = u(1) = 0 \}$$

Then

$$\| u \|_A^2 = (Au, u) = -\int_0^1 \left(\frac{d^2u}{dx^2} \right) u \, dx = \int_0^1 \left(\frac{du}{dx} \right)^2 dx \tag{10.1.4}$$

The statement $u_n \overset{E}{\to} u$ is thus equivalent to

$$\lim_{n\to\infty} \int_0^1 [u_n'(x) - u'(x)]^2 \, dx = 0$$

It was shown in Example 9.2.8 that A is positive bounded below, so it follows that, for all u and u_n in D_A,

$$\lim_{n\to\infty} \int_0^1 [u_n'(x) - u'(x)]^2 \, dx = 0$$

implies that

$$\lim_{n\to\infty} \int_0^1 [u_n(x) - u(x)]^2 \, dx = 0$$

■

Example 10.1.2

Consider the Laplace operator $Au = -\nabla^2 u$ with domain $D_A = \{ u \in C^2(\Omega): u = 0$ on $\Gamma \}$. It was shown in Example 9.2.8 that this operator is positive bounded below. The energy of a function u in D_A is thus

$$\| u \|_A^2 = (-\nabla^2 u, u) = \iint_\Omega \sum_{i=1}^{m} \left(\frac{\partial u}{\partial x_i} \right)^2 d\Omega$$

If $u_n \overset{E}{\to} u$, for this operator,

$$\lim_{n\to\infty} \iint_\Omega \sum_{i=1}^{m} \left(\frac{\partial u_n}{\partial x_i} - \frac{\partial u}{\partial x_i} \right)^2 d\Omega = 0$$

Hence, it follows that

$$\lim_{n\to\infty} \iint_\Omega \left(\frac{\partial u_n}{\partial x_i} - \frac{\partial u}{\partial x_i} \right)^2 d\Omega = 0, \quad i = 1, 2, \ldots, m$$

Thus, energy convergence implies that each of the first derivatives of the functions u_n converge in the L_2 norm to the first derivative of the function $u(\mathbf{x})$. It also implies that u_n converges to u in the mean. ■

Definition 10.1.4. A sequence of functions $\{ \phi_i \}$ in D_A is said to be **energy orthogonal** if $[\phi_i, \phi_j]_A = 0$, $i \neq j$, and energy orthonormal if $[\phi_i, \phi_j]_A = \delta_{ij}$. An energy orthonormal sequence of functions $\{ \phi_i \}$ is said to be **complete in energy** if the **Parseval relation**

$$\| u \|_A^2 = \sum_{i=1}^{\infty} [\phi_i, u]_A^2 \tag{10.1.5}$$

holds for every function u in D_A. This is equivalent to the property that given any $\varepsilon > 0$, there exists an integer N and constants $\alpha_1, \alpha_2, \ldots, \alpha_N$ such that

$$\left\| u - \sum_{i=1}^{N} \alpha_i \phi_i \right\|_A < \varepsilon \tag{10.1.6}$$

■

Example 10.1.3

Let A be the operator of Example 10.1.1 on $D_A = \{ u \text{ in } C^2(0, \pi): u(0) = u(\pi) = 0 \}$. Then, for u and v in D_A,

$$[u, v]_A = -\int_0^\pi v \frac{d^2u}{dx^2} dx = \int_0^\pi \frac{du}{dx} \frac{dv}{dx} dx \tag{10.1.7}$$

and

$$\| u \|_A^2 = \int_0^\pi \left(\frac{du}{dx} \right)^2 dx \tag{10.1.8}$$

Energy orthogonality of the functions in this example is thus equivalent to L_2 orthogonality of their first derivatives.

It can be shown that the sequence of functions $\phi_n(x) = \frac{1}{n}\sqrt{\frac{2}{\pi}} \sin nx$, $n = 1, 2, \ldots$, is energy orthonormal and complete in D_A. Orthonormality follows from

$$[\phi_n, \phi_m]_A = \frac{2}{\pi}\int_0^{\pi} \cos nx \cos mx \, dx = \delta_{mn}$$

For energy completeness, since the sequence 1, cos x, cos 2x, . . . is complete in the sense of convergence in $L_2(0, \pi)$, any function $\psi(x)$ with finite norm can be expressed as

$$\psi(x) = b_0 + \sum_{i=1}^{\infty} b_n \cos nx$$

which converges in the L_2 norm. Setting $\psi(x) = u'(x)$, for u in D_A,

$$u(0) = u(\pi) = 0$$

Thus,

$$b_0 = \frac{1}{\pi}\int_0^{\pi} \psi(x)\, dx = \frac{1}{\pi}\int_0^{\pi} u'(x)\, dx = \frac{1}{\pi}[\, u(\pi) - u(0)\,] = 0$$

Hence,

$$u'(x) = \sum_{n=1}^{\infty} b_n \cos nx = \sum_{n=1}^{\infty} b_n^* \phi_n'(x)$$

where $b_n^* = \sqrt{\frac{\pi}{2}}\, b_n$ and $\phi_n'(x) = \sqrt{\frac{2}{\pi}} \cos nx$. This shows that the derivative of any function in D_A can be approximated in the mean with any degree of accuracy by linear combinations of the derivatives of the $\phi_n(x)$. This result and Example 10.1.1 show that the set $\phi_n(x) = \frac{1}{n}\sqrt{\frac{2}{\pi}} \sin nx$ is complete in energy. ■

EXERCISES 10.1

1. Construct the energy norm associated with the following positive definite operators, using integration by parts to minimize the highest-order derivatives that appear:

(a) $A_1 u = -\frac{d}{dx}\left(a(x) \frac{du}{dx} \right), \quad a(x) > 0 \text{ on } (0, \ell)$

$$D_{A_1} = \{\, u \text{ in } C^2(0, \ell)\text{: } u(0) = u(\ell) = 0 \,\}$$

(b) $$A_2u = \frac{d^2}{dx^2}\left[EI(x) \frac{d^2u}{dx^2} \right], \quad EI(x) > 0 \text{ on } (0, \ell)$$

$$D_{A_2} = \{ u \text{ in } C^4(0, \ell): u(0) = u'(0) = u(\ell) = u'(\ell) = 0 \}$$

(c) $$A_3u = \nabla^2(b(\mathbf{x}) \nabla^2 u), \quad b(\mathbf{x}) > 0 \text{ on } \Omega$$

$$D_{A_3} = \left\{ u \text{ in } C^4(\Omega): u = \frac{\partial u}{\partial n} = 0 \text{ on } \Gamma \right\}$$

2. The fourth-order operator A_2 of Exercise 1 (b) can be reformulated as the following system of two second-order equations:

$$\mathbf{Lz} = \begin{bmatrix} -z_2'' \\ -z_1'' - \dfrac{z_2}{EI} \end{bmatrix} = \begin{bmatrix} q(x) \\ 0 \end{bmatrix}$$

where $z_1 \equiv u$ and $z_2 = -EIz_1''$. For a simply supported beam, the domain of this operator is

$$D_L = \{ z_1, z_2 \text{ in } C^2(0, \ell): z_1(0) = z_2(0) = z_1(\ell) = z_2(\ell) = 0 \}$$

Show that this operator is symmetric, but not positive definite.

10.2 THE MINIMUM FUNCTIONAL THEOREM

The concepts of positive definite operators and energy norms were used in the 1940s, primarily by Soviet mathematicians, to create the modern foundation for variational methods in mechanics and the mathematical theory of functional analysis. The penetrating and inspiring treatment of this subject in Ref. 21 remains as a landmark in the field.

Let A be a positive definite linear operator and consider the problem of finding a solution u of

$$Au = f \tag{10.2.1}$$

By this is meant, find a function u in D_A that satisfies Eq. 10.2.1. The fact that u is in D_A implies that it satisfies the boundary conditions of the problem.

Minimum Principles for Operator Equations

Theorem 10.2.1 (Minimum Functional Theorem). Let A be a positive definite linear operator with domain D_A that is dense in $L_2(\Omega)$ and let Eq. 10.2.1 have a solution. Then the solution is unique and minimizes the **energy functional**

$$F(u) = (Au, u) - 2 (u, f) = \int_\Omega (uAu - 2uf) \, d\Omega \tag{10.2.2}$$

over all u in D_A. Conversely, if there exists a function u in D_A that minimizes the functional of Eq. 10.2.2, then it is the unique solution of Eq. 10.2.1. ■

To show that a solution of Eq. 10.2.1 is unique, assume there are two solutions u and v in D_A. Then, u − v satisfies A(u − v) = 0. Thus,

$$(A(u - v), (u - v)) = 0$$

Since A is positive definite, u − v = 0. Thus, if a solution of Eq. 10.2.1 exists, it is unique.

Next let u_0 in D_A be the solution to Eq. 10.2.1, so that $Au_0 = f$. Making this substitution for f in Eq. 10.2.2 yields

$$\begin{aligned} F(u) &= (Au, u) - 2 (Au_0, u) \\ &= [u, u]_A - 2 [u_0, u]_A \\ &= [u - u_0, u - u_0]_A - [u_0, u_0]_A \\ &= \| u - u_0 \|_A^2 - \| u_0 \|_A^2 \end{aligned}$$

It is clear that F(u) assumes its minimum value if and only if $u = u_0$, proving the first part of Theorem 10.2.1.

Now suppose that there exists a function u_0 in D_A that minimizes the functional F of Eq. 10.2.2. Let v(x) be an arbitrary function from D_A and let α be an arbitrary real number. Then $F(u_0 + \alpha v) - F(u_0) \geq 0$. Using symmetry of the operator A, the function

$$F(u_0 + \alpha v) - F(u_0) = 2\alpha (Au_0 - f, v) + \alpha^2 (Av, v) \geq 0$$

of α takes on its minimum value of zero at $\alpha = 0$. Thus, its derivative with respect to α at $\alpha = 0$ must be zero; i.e., $2 (Au_0 - f, v) = 0$, for all v in D_A. Since D_A is dense in $L_2(\Omega)$, it follows that $Au_0 - f = 0$. ■

As will become evident from later examples, the functional F(u) is proportional to the total potential energy of the system under consideration. In such cases, Theorem 10.2.1 provides a rigorous proof of the **principle of minimum total potential energy** and allows the problem of integrating a differential equation, under specified boundary conditions, to be replaced by the problem of seeking a function that minimizes the functional of Eq. 10.2.2. The direct methods that will be discussed later allow approximate solutions of the later problem to be found by relatively simple means.

Methods of solving problems of continuum mechanics that, instead of integrating the differential equation under specified boundary conditions, involve minimization of the functional of Eq. 10.2.2 are referred to as **energy methods**, or **variational methods**.

Example 10.2.1

Consider the equation $Au \equiv -d^2u/dx^2 = 2$, subject to boundary conditions $u'(0) = 0$ and $u'(1) + u(1) = 0$, which has the solution $u = 3 - x^2$. The operator A is positive definite on

$$D_A = \{ u \in C^2(0, 1): u'(0) = u'(1) + u(1) = 0 \}$$

To see this, form

$$(Au, v) = \int_0^1 -u''v\,dx = \int_0^1 u'v'\,dx - u'v\Big|_0^1$$

$$= \int_0^1 u'v'\,dx + u(1)\,v(1) = (u, Av)$$

Thus, A is symmetric. Further, with $u = v$,

$$(Au, u) = \int_0^1 (u')^2\,dx + u^2(1) \geq 0$$

and $(Au, u) = 0$ implies $u' = 0$ for $0 \leq x \leq 1$ and $u(1) = 0$. Thus, $u(x) = 0$, $0 \leq x \leq 1$, and A is positive definite.

By the minimum functional theorem, the function $3 - x^2$ gives a minimum value to the functional

$$F(u) = \int_0^1 [\,(u')^2(x) - 4u\,]\,dx + u^2(1) \tag{10.2.3}$$

in the class of functions that satisfy the given boundary conditions. ■

The Space of Functions With Finite Energy

In Section 10.1, it was observed that if A is a positive bounded below operator, defined on a dense subspace D_A of a Hilbert space H, then a new energy scalar product $[u, v]_A = (Au, v)$ can be defined on D_A, with the property that convergence in the energy norm implies convergence in the original norm of H; i.e., for a sequence $\{u_k\}$ in D_A,

$$\lim_{n,m\to\infty} \| u_n - u_m \|_A = 0 \text{ implies } \lim_{n,m\to\infty} \| u_n - u_m \| = 0 \tag{10.2.4}$$

Hence, a Cauchy sequence $\{u_n\}$ in $\|\bullet\|_A$ is also a Cauchy sequence in the H-norm. Since H is complete, there is a u_0 in H such that $\lim_{n\to\infty} u_n = u_0$. While u_0 may not be in D_A, all such limit functions may be added to D_A to obtain a larger space of functions.

Definition 10.2.1. The space H_A of **generalized functions** associated with a positive bounded below operator A consists of those elements u in the Hilbert space H for which there exists a sequence of elements $\{ u_n \}$ in D_A such that

$$\lim_{n,m \to \infty} \| u_n - u_m \|_A = 0 \text{ and } \lim_{n \to \infty} \| u_n - u \| = 0 \tag{10.2.5}$$

i.e.,

$$H_A \equiv \{ u \text{ in } H: \text{ there is a sequence } \{ u_n \} \text{ in } D_A \text{ that is Cauchy in } \| \bullet \|_A \text{ and converges in the norm of H to u } \} \tag{10.2.6}$$

The elements of H_A are said to be **functions of finite energy.** ■

Example 10.2.2

To understand how the space H_A of generalized functions is constructed, consider the string problem with static load,

$$Au \equiv -u_{xx} = f_n(x) \tag{10.2.7}$$

for $0 \leq x \leq 1$, with boundary conditions

$$u(0) = u(1) = 0$$

as shown in Fig. 10.2.1, where n = 1, 2, ... Consider a point force at $x = \xi$, $0 < \xi < 1$, as the limiting case of continuous forces $f_n(x)$ that vanish in (0, 1) if $| x - \xi | > 1/2n$, but for which the total intensity is

$$\int_0^1 f_n(x)\, dx = 1$$

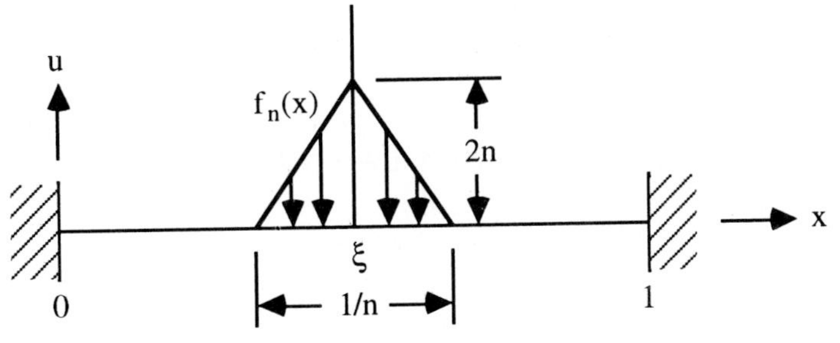

Figure 10.2.1 String with Distributed Load $f_n(x)$

For each $f_n(x)$, let the solution of Eq. 10.2.7 be denoted as $u_n(x)$. Then $u_n(x)$ is in $C^2(0, 1)$ and satisfies the boundary conditions, so $u_n(x)$ is in D_A for each n = 1, 2, ... However,

$$\lim_{n\to\infty} u_n(x) = u_0(x) = \begin{cases} (\xi - 1)\,x, & 0 \le x \le \xi \\ \xi x - \xi, & \xi \le x \le 1 \end{cases} \tag{10.2.8}$$

and $u_0(x)$ does not belong to D_A. On the other hand, $u_0(x)$ belongs to H_A. The energy norm of $u_0(x)$ is

$$\| u_0 \|_A^2 = \int_0^1 (u')^2\, dx = \int_0^\xi (\xi - x)^2\, dx + \int_\xi^1 \xi^2\, dx = \xi - \xi^2$$

Thus, $\| u_0 \|_A = (\xi - \xi^2)^{1/2}$, which is finite. ■

Theorem 10.2.2. Let u and v be in H_A for a positive bounded below operator A; i.e., there are Cauchy sequences $\{u_n\}$ and $\{v_n\}$ in H_A, u_n and v_n are in D_A, and $\lim_{n\to\infty} \| u - u_n \| = \lim_{n\to\infty} \| v - v_n \| = 0$. Then

$$\| u \|_A = \lim_{n\to\infty} \| u_n \|_A \tag{10.2.9}$$

$$[u, v]_A = \lim_{n\to\infty} [u_n, v_n]_A \tag{10.2.10}$$

are finite and define a norm and the associated scalar product on H_A. ■

To prove this result, note first that for any x, y, and z in D_A,

$$\| x + y \|_A \le \| x \|_A + \| y \|_A$$

since $\| \bullet \|_A$ is a norm on D_A. With $z = x + y$, this is

$$\| z \|_A - \| y \|_A \le \| z - y \|_A$$

Interchanging z and y, this shows that

$$\big|\, \| z \|_A - \| y \|_A \,\big| \le \| z - y \|_A \tag{10.2.11}$$

Thus,

$$\big|\, \| u_n \|_A - \| u_m \|_A \,\big| \le \| u_n - u_m \|_A$$

Hence, $\{ \| u_n \|_A \}$ is a Cauchy sequence of real numbers, so it converges to a finite real number; i.e.,

$$\lim_{n\to\infty} \| u_n \|_A \equiv \| u \|_A < \infty$$

Next note that

$$\begin{aligned} \left| [\, u_n, v_n \,]_A - [\, u_m, v_m \,]_A \right| &= \left| [\, u_n, v_n - v_m \,]_A + [\, u_n - u_m, v_m \,]_A \right| \\ &\le \left| [\, u_n, v_n - v_m \,]_A \right| + \left| [\, u_n - u_m, v_m \,]_A \right| \\ &\le \| u_n \|_A \| v_n - v_m \|_A + \| v_m \|_A \| u_n - u_m \|_A \end{aligned}$$

Since $\{ \| u_n \|_A \}$ and $\{ \| v_n \|_A \}$ converge to finite values, this shows that $\{ [\, u_n, v_n \,]_A \}$ is a Cauchy sequence of real numbers, so it is convergent to a finite real number; i.e.,

$$\lim_{n\to\infty} [\, u_n, v_n \,]_A \equiv [\, u, v \,]_A < \infty$$

Since $[\, \bullet, \bullet \,]_A$ is a scalar product on D_A and the limit in Eq. 10.2.10 is finite,

$$[\, u, v \,]_A = \lim_{n\to\infty} [\, u_n, v_n \,]_A = \lim_{n\to\infty} [\, v_n, u_n \,]_A = [\, v, u \,]_A$$

and

$$\begin{aligned} [\, \alpha u, v \,]_A &= \lim_{n\to\infty} [\, \alpha u_n, v_n \,]_A = \lim_{n\to\infty} \alpha [\, u_n, v_n \,]_A \\ &= \alpha \lim_{n\to\infty} [\, u_n, v_n \,]_A = \alpha [\, u, v \,]_A \end{aligned}$$

Since $[\, u_n, u_n \,]_A \ge 0$,

$$[\, u, u \,]_A = \lim_{n\to\infty} [\, u_n, u_n \,]_A \ge 0$$

Finally, if

$$[\, u, u \,]_A = \lim_{n\to\infty} [\, u_n, u_n \,]_A = 0$$

then, since $[\, u_n, u_n \,]_A \ge \gamma \| u_n \|^2$,

$$\lim_{n\to\infty} \| u_n \|^2 = 0$$

and $u = 0$.

To see that the norm $\| \bullet \|_A$ is generated by the scalar product $[\, \bullet, \bullet \,]_A$, note that

$$\begin{aligned} \| u \|_A^2 &= \lim_{n\to\infty} \| u_n \|_A^2 \\ &= \lim_{n\to\infty} [\, u_n, u_n \,]_A \\ &= [\, u, u \,]_A \end{aligned}$$

■

These results permit the proof of an important theorem of modern mechanics and applied mathematics.

Theorem 10.2.3. For a positive bounded below operator A, H_A with the norm and scalar product of Eqs. 10.2.9 and 10.2.10 is a Hilbert space. That is, H_A is complete. ■

To prove this result, first let $\{ u_n \}$ in D_A be a Cauchy sequence in the energy norm. Then $w_k = u - u_k$, where $\lim_{n\to\infty} \| u - u_n \| = 0$, is in H_A. For fixed k, the sequence $\{ w_{k_n} \} = \{ u_n - u_k \}$ is in D_A and is Cauchy, since

$$\| w_{k_n} - w_{k_m} \|_A = \| u_n - u_m \|_A$$

Thus, by Eq. 10.2.9,

$$\begin{aligned} \| u - u_k \|_A &= \| w_k \|_A \\ &= \lim_{n\to\infty} \| w_{k_n} \|_A \\ &= \lim_{n\to\infty} \| u_n - u_k \|_A \end{aligned}$$

Since $\{ u_n \}$ is Cauchy in the norm $\| \bullet \|_A$,

$$\lim_{k\to\infty} \| u - u_k \|_A = \lim_{k\to\infty} \lim_{n\to\infty} \| u_n - u_k \|_A = 0 \tag{10.2.12}$$

This shows that for any u in H_A, there is a u_k in D_A that is arbitrarily close to u in the energy norm. This result is of significant value in its own right.

To complete the proof of Theorem 10.2.3, let $\{ u_n \}$ be a Cauchy sequence in H_A. Thus, for any $\varepsilon > 0$, there is an integer N_1 such that

$$\| u_n - u_m \|_A \le \frac{\varepsilon}{2} \tag{10.2.13}$$

for n, m > N_1. As a result of Eq. 10.2.12, for each u_n in H_A, there is a ϕ_n in D_A such that

$$\| u_n - \phi_n \|_A \le \frac{\varepsilon}{4}$$

Thus,

$$\| \phi_n - \phi_m - (u_n - u_m) \|_A \le \| \phi_n - u_n \|_A + \| \phi_m - u_m \|_A \le \frac{\varepsilon}{2}$$

and by Eq. 10.2.11,

$$\| \phi_n - \phi_m \|_A - \| u_n - u_m \|_A \le \| \phi_n - \phi_m - (u_n - u_m) \|_A \le \frac{\varepsilon}{2}$$

For n, m > N_1, this result and Eq. 10.2.13 yield

$$\| \phi_n - \phi_m \|_A \leq \varepsilon$$

so $\{ \phi_n \}$ in D_A is a Cauchy sequence in the energy norm. There is thus a u in H_A such that, by Eq. 10.2.12,

$$\lim_{n\to\infty} \| u - \phi_n \|_A = 0$$

Thus, there is an integer N_2 such that

$$\| u - \phi_n \|_A < \frac{\varepsilon}{2}$$

if n > N_2. Finally,

$$\begin{aligned} \| u_n - u \|_A &= \| u_n - \phi_n + (\phi_n - u) \|_A \\ &\leq \| u_n - \phi_n \|_A + \| \phi_n - u \|_A \\ &\leq \frac{\varepsilon}{4} + \frac{\varepsilon}{2} < \varepsilon \end{aligned}$$

for n > max(N_1, N_2). Therefore,

$$\lim_{n\to\infty} \| u_n - u \|_A = 0$$

and H_A is complete in the energy norm. Taken with Theorem 10.2.2, this shows that H_A is a Hilbert space. ■

Example 10.2.3

Consider again the operator $A = - d^2/dx^2$ of Example 10.1.1. The energy space H_A of this operator is the space of functions on (0, 1) that are equal to zero at x = 0 and x = 1 and whose first derivatives are in $L_2(0, 1)$. Note that the first derivative of the solution $u_0(x)$ in Eq. 10.2.8 indeed belongs to $L_2(0, 1)$. This result is a special case of an **embedding theorem** [18, 22] that guarantees roughly that if A is a differential operator of order 2m, then H_A consists of functions that have continuous derivatives of order m – 1, derivatives of order m in H, and satisfy the boundary conditions.

To prove this, first let u be an element of H_A. Then there exists a sequence $\{ u_n \}$ in D_A such that $\lim_{n,m\to\infty} \| u_n - u_m \|_A = 0$ and $\lim_{n\to\infty} \| u_n - u \|_A = 0$; i.e.,

$$\lim_{n,m\to\infty} \int_0^1 [u_n'(x) - u_m'(x)]^2 \, dx = 0$$

Since $L_2(0, 1)$ is complete, there exists a w in $L_2(0, 1)$ such that $\lim_{n\to\infty} \| u_n' - w \| = 0$.

Define $u(x) = \int_0^x w(t)\,dt$. Then, $u' = w$ is in $L_2(0, 1)$ and it remains to show that $u(0) = u(1) = 0$. Since u_n is in D_A, it follows that $u_n(0) = u_n(1) = 0$ and

$$u_n(x) = u_n(0) + \int_0^x u_n'(t)\,dt = \int_0^x u_n'(t)\,dt$$

To see that $u_n(x) = \int_0^x u_n'(t)\,dt$ converges uniformly to $\int_0^x w(t)\,dt$ on the interval $[0, 1]$, note that

$$\left[\int_0^x u_n'(t)\,dt - \int_0^x w(t)\,dt\right]^2 = \left[\int_0^x [\,u_n'(t) - w(t)\,]\,dt\right]^2$$

$$\leq x\int_0^x [\,u_n'(t) - w(t)\,]^2\,dt$$

$$\leq \int_0^1 [\,u_n'(t) - w(t)\,]^2\,dt$$

$$= \| u_n' - w \|^2$$

Thus, $u(x) = \int_0^x w(t)\,dt$, for all x in $[0, 1]$, giving $u(0) = 0$. Similarly, using the formula

$$u_n(x) = u_n(1) - \int_x^1 u_n'(t)\,dt = -\int_x^1 u_n'(t)\,dt$$

$u(x) = -\int_x^1 w(t)\,dt$ and $u(1) = 0$.

Conversely, let u be a function with u' in $L_2(0, 1)$ and $u(0) = u(1) = 0$. To show that u is in the space H_A, a sequence $\{ u_n \}$ in D_A must be found such that $\lim_{n,m\to\infty} \| u_n - u_m \|_A = 0$ and $\lim_{n\to\infty} \| u_n - u \| = 0$. Since u' is in $L_2(0, 1)$, it can be expanded in a cosine series,

$$u'(x) = \sum_{k=0}^{\infty} a_k \cos k\pi x = \sum_{k=1}^{\infty} a_k \cos k\pi x$$

where

$$a_0 = \int_0^1 u'(x)\,dx = u(1) - u(0) = 0$$

Integrating this series term by term and taking into account $u(0) = 0$,

$$u(x) = \sum_{k=1}^{\infty} b_k \sin k\pi x$$

where

$$b_k = \frac{a_k}{k\pi}$$

Consider the sequence $\{ u_n \}$, where $u_n(x) = \sum_{k=1}^{n} b_k \sin k\pi x$. Clearly $\{ u_n \}$ is in D_A and $\lim_{n\to\infty} \| u_n - u \| = 0$. To show that $\lim_{n,m\to\infty} \| u_n - u_m \|_A = 0$, assume without loss of generality that $n > m$. Then $u_n(x) - u_m(x) = \sum_{k=m+1}^{n} b_k \sin k\pi x$ and

$$\| u_n - u_m \|_A^2 = \int_0^1 \left[\sum_{k=m+1}^{n} a_k \cos k\pi x \right]^2 dx = \frac{1}{2} \sum_{k=m+1}^{n} a_k^2$$

which approaches zero as $m, n \to \infty$. Thus, u is in H_A. ■

Generalized Solutions of Operator Equations

For a positive bounded below operator A on a domain D_A, the energy functional $F(u) = (Au, u) - 2(u, f)$ can be extended to all of the energy space H_A, as $F(u) = \| u \|_A^2 - 2(u, f)$. The following important existence theorem will now be proved.

Theorem 10.2.4. In the energy space H_A of a positive bounded below operator A, for each f in H, there exists one and only one element for which the energy functional attains a minimum. ■

To prove this theorem, consider the linear functional $\ell(u) = (u, f)$, defined on H_A. Then $| \ell(u) | = | (u, f) | \le \| u \| \, \| f \| \le (\| f \| / \gamma) \| u \|_A$, which implies that ℓ is a continuous linear functional on H_A. By the Riesz representation theorem, Theorem 9.1.3, there exists a unique element u_0 in H_A such that $(u, f) = [u, u_0]_A$, for all u in H_A. The energy functional F can now be written as

$$F(u) = \| u \|_A^2 - 2[u, u_0]_A = \| u - u_0 \|_A^2 - \| u_0 \|_A^2$$

from which it follows that the minimum of F is attained at u_0, proving the theorem. ■

Definition 10.2.2. For a positive bounded below operator A and a function f in H, the element u_0 in H_A that yields a minimum for the functional F is called the **generalized solution** of the equation Au = f. If u_0 is in D_A, it is called an **ordinary solution.** ■

EXERCISES 10.2

1. Show that the operator

$$Au = -\frac{d^2u}{dx^2}$$

with domain

$$D_A = \{ u \text{ in } C^2(0, 1): u'(0) = u'(1) + u(1) = 0 \}$$

is positive definite.

2. Consider the problem of integrating the equation

$$-\frac{du^2}{dx^2} = 2$$

subject to boundary conditions u(0) = u(1) = 0. Determine the functional whose minimum provides the solution to the above equation and determine the domain of this functional; i.e., the space of functions for which it is defined. The solution of the differential equation is $u_0 = x(1-x)$. Compute the value of the functional for $u_0 = x(1-x)$ and $u_1 = x(2-x)$. Is the result a contradiction of the minimum functional theorem? Why?

10.3 CALCULUS OF VARIATIONS

The minimum functional theorem of Section 10.2 motivates a brief introduction to the calculus of variations. Only necessary conditions for minimization of an integral functional are treated here, which is adequate for the purposes of this text.

Variations of Functionals

Definition 10.3.1. Let H be a Hilbert space (normally L_2) and T a functional (possibly nonlinear) that is defined on a subset D_T (not necessarily a linear subspace) of H. Let u be in D_T and η be in a linear subspace $\bar{D}_T$ of H, such that u + αη is in D_T for α sufficiently small. If the limit

$$\delta T(u; \eta) \equiv \lim_{\alpha \to 0} \frac{1}{\alpha} [T(u + \alpha\eta) - T(u)]$$

$$= \frac{d}{d\alpha} T(u + \alpha\eta) \Big|_{\alpha=0} \tag{10.3.1}$$

exists, it is called the **first variation** of T at u in the direction η. If this limit exists for every η in $\bar{D}_T$, T is said to be a **differentiable functional** at u. ■

Example 10.3.1

Let f(x, u, u') be a continuous, real valued function for x in [a, b], $-\infty < u < +\infty$ and $-\infty < u' < +\infty$, and let the partial derivatives f_u and $f_{u'}$ be continuous. Consider the integral functional

$$T(u) = \int_a^b f[\, x, u(x), u'(x)\,]\, dx \tag{10.3.2}$$

whose domain is $D_T = \{\, u \text{ in } C^1(a, b)\colon u(a) = d,\ u(b) = e \,\}$ and let $\bar{D}_T = \{\, \eta \text{ in } C^1(a, b)\colon \eta(a) = \eta(b) = 0 \,\}$. The first variation of T takes the form

$$\begin{aligned} \delta T(u; \eta) &= \frac{d}{d\alpha} T(\, u + \alpha\eta \,)\Big|_{\alpha=0} \\ &= \frac{d}{d\alpha} \int_a^b f(\, x, u + \alpha\eta, u' + \alpha\eta' \,)\, dx \Big|_{\alpha=0} \\ &= \int_a^b [\, f_u(x, u, u')\, \eta + f_{u'}(x, u, u')\, \eta' \,]\, dx \end{aligned} \tag{10.3.3}$$

If u(x) is such that $f_{u'}(x, u, u')$ is continuously differentiable, Eq. 10.3.3 can be integrated by parts to obtain

$$\delta T(u; \eta) = \int_a^b \left[\, f_u - \frac{d}{dx} f_{u'} \,\right] \eta\, dx \tag{10.3.4}$$

■

Example 10.3.2

Consider a region Ω in R^m, with a boundary Γ that consists of a finite number of piecewise smooth $(m-1)$-dimensional surfaces. Let the real valued function $f(x, u, z_1, z_2, \ldots, z_m)$ be twice continuously differentiable in all its arguments; i.e., x, u, and z_i. Consider the functional

$$T(u) = \int_\Omega f\left(\, x, u, \frac{\partial u}{\partial x_1}, \frac{\partial u}{\partial x_2}, \ldots, \frac{\partial u}{\partial x_m} \right) dx \tag{10.3.5}$$

defined on $D_T = \{\, u \text{ in } C^1(\Omega)\colon u(x) = g(x) \text{ on } \Gamma \,\}$ and let $\bar{D}_T = \{\, \eta \text{ in } C^1(\Omega)\colon \eta(x) = 0 \text{ on } \Gamma \,\}$. The first variation $\delta T(u; \eta)$ is then

$$\delta T(u;\eta) = \frac{d}{d\alpha} T(u+\alpha\eta)\Big|_{\alpha=0}$$

$$= \frac{d}{d\alpha}\int_\Omega f(x, u+\alpha\eta, u_1+\alpha\eta_1, \ldots, u_m+\alpha\eta_m)\, dx\Big|_{\alpha=0}$$

$$= \int_\Omega \left[\frac{\partial f}{\partial u}\eta + \sum_{k=1}^{m} \frac{\partial f}{\partial u_k}\eta_k \right] dx \tag{10.3.6}$$

where $u_k = \frac{\partial u}{\partial x_k}$ and $\eta_k = \frac{\partial \eta}{\partial x_k}$. Assume now that $\frac{\partial}{\partial x_k}\frac{\partial f}{\partial x_k}$ exists and is in $L_2(a, b)$. Integration by parts yields

$$\delta T(u;\eta) = \int_\Omega \left[\frac{\partial f}{\partial u} - \sum_{k=1}^{m} \frac{\partial}{\partial x_k}\left(\frac{\partial f}{\partial u_k}\right) \right] \eta\, dx \tag{10.3.7}$$

■

Minimization of Functionals

Given that a functional has a first variation, quantitative criteria can be defined for its minimization. The focus here is on necessary conditions for extrema. Presume $\hat{u}$ in D_T is such that

$$T(\hat{u}) \le T(u) \tag{10.3.8}$$

for all u in D_T. Then, $\hat{u}$ is said to minimize T over D_T. If Eq. 10.3.8 holds for all u in D_T that satisfy $\| u - \hat{u} \| \le \delta$, for some $\delta > 0$, T is said to have a **relative minimum** at $\hat{u}$. A change in the sense of the inequality of Eq. 10.3.8 leads to maximization. Results will be stated here only for minimization.

Employing the notation and conventions of Definition 10.3.1, observe that for any fixed η in $\bar{D}_T$ and for any α sufficiently small, if T has a relative minimum at $\hat{u}$, then

$$T(\hat{u}) = \min_\alpha T(\hat{u}+\alpha\eta) = T(\hat{u}+\alpha\eta)\Big|_{\alpha=0}$$

That is, for $\hat{u}$ and η fixed, the real valued function $T(\hat{u}+\alpha\eta)$ of the real parameter α is a minimum at $\alpha = 0$. If the functional has a first variation, then $T(\hat{u}+\alpha\eta)$ is a differentiable function of α and a necessary condition for a minimum of T at $\hat{u}$ is

$$\delta T(\hat{u};\eta) = \frac{d}{d\alpha} T(\hat{u}+\alpha\eta)\Big|_{\alpha=0} = 0 \tag{10.3.9}$$

for all η in $\bar{D}_T$. For cases such as Examples 10.3.1 and 10.3.2, Eqs. 10.3.4 and 10.3.7 are of the form,

$$\delta T(\hat{u}; \eta) = \int_{\Omega} E(\hat{u}, x)\, \eta(x)\, d\Omega = 0 \tag{10.3.10}$$

for all η in $\bar{D}_T$.

If the subspace $\bar{D}_T$ is dense in the underlying space H, then Eq. 10.3.10 is equivalent to requiring that the function $E(\hat{u}(x), x)$ be orthogonal to a complete set of functions in H; i.e., $\bar{D}_T$ contains a basis for H. Thus,

$$E(\hat{u}, x) = 0 \tag{10.3.11}$$

is a necessary condition for minimization of the functional T at $\hat{u}$. The function spaces $\bar{D}_T$ treated herein are large enough to contain the space $C_0^\infty(a, b)$ of functions in $C^\infty(a, b)$ that, with all their derivatives, vanish at a and b, which is itself dense in $L_2(a, b)$ [22], so they are dense in $L_2(a, b)$ and the condition of Eq. 10.3.11 holds. ■

Necessary Conditions for Minimization of Functionals

Theorem 10.3.1. The results of Examples 10.3.1 and 10.3.2 yield the following **Euler–Lagrange equations** as necessary conditions for minimization of the associated functionals:

(a) From Eq. 10.3.2,

$$\frac{\partial f}{\partial u} - \frac{d}{dx}\left(\frac{\partial f}{\partial u'}\right) = 0, \quad a \le x \le b \tag{10.3.12}$$

for minimization of the functional

$$T(u) = \int_a^b f(x, u, u')\, dx \tag{10.3.13}$$

(b) From Eq. 10.3.5,

$$\frac{\partial f}{\partial u} - \sum_{k=1}^{m} \frac{\partial}{\partial x_k}\left(\frac{\partial f}{\partial u_k}\right) = 0, \quad x \text{ in } \Omega \tag{10.3.14}$$

for minimization of the functional

$$T(u) = \int_{\Omega} f(x, u, u_1, u_2, \ldots, u_m)\, d\Omega \tag{10.3.15}$$

■

The integration by parts in Eqs. 10.3.3 and 10.3.6 resulted in no terms from the boundary conditions, because of the nature of D_T and $\bar{D}_T$ in those problems. In many practical problems, however, boundary terms do arise.

Example 10.3.3

Consider minimization of the functional of Eq. 10.3.2,

$$T(u) = \int_a^b f(x, u, u')\, dx$$

but with $D_T = \bar{D}_T = C^2(a, b)$. Then

$$\delta T(u; \eta) = \int_a^b (f_u \eta + f_{u'} \eta')\, dx$$

Integration by parts yields

$$\delta T(u; \eta) = \int_a^b \left[f_u - \frac{d}{dx} f_{u'} \right] \eta\, dx + f_{u'}(x, \hat{u}(x), \hat{u}'(x))\, \eta(x) \Big|_a^b = 0 \tag{10.3.16}$$

for all η in $\bar{D}_T$. The function η can be selected arbitrarily at the end points of the interval and in the interior of the interval, so both the integral and the boundary term must vanish. Thus, in addition to the necessary condition of Eq. 10.3.12, the **transversality condition**

$$f_{u'}(x, \hat{u}(x), \hat{u}'(x))\, \eta(x) \Big|_a^b = 0 \tag{10.3.17}$$

must hold for all η in $\bar{D}_T$. Since $\bar{D}_T = C^2(a, b)$, $\eta(a)$ and $\eta(b)$ are arbitrary. This means that the boundary conditions

$$\begin{aligned} f_{u'}(a, \hat{u}(a), \hat{u}'(a)) &= 0 \\ f_{u'}(b, \hat{u}(b), \hat{u}'(b)) &= 0 \end{aligned} \tag{10.3.18}$$

are also necessary conditions for a minimizing function $\hat{u}$. ■

To illustrate methods of the **calculus of variations**, it is instructive to briefly study special cases and applications of the fundamental problem of minimizing

$$T(u) = \int_a^b f(x, u, u')\, dx$$

subject to a variety of boundary conditions.

Example 10.3.4

The shortest path between points (a, u^0) and (b, u^1) in the x-u plane is to be found. As shown in Fig. 10.3.1, the particular path chosen between the points has a length associated with it. The problem is to choose the curve u(x), $a \le x \le b$, that has the shortest length. For a smooth curve u(x), the length is given by the functional

$$T(u) = \int_a^b \left[1 + \left(\frac{du}{dx}\right)^2 \right]^{\frac{1}{2}} dx \qquad (10.3.19)$$

Similarly, given points (a, u^0) and (b, u^1) in a vertical plane that do not lie on the same vertical line, a curve $u = \hat{u}(x)$, $a \le x \le b$, joining them is to be found so that a particle that moves without friction will start at rest from point (a, u^0) and traverse the curve to (b, u^1) in the shortest possible time. Candidate curves are shown in Fig. 10.3.2. This is called the **Brachistochrone problem.**

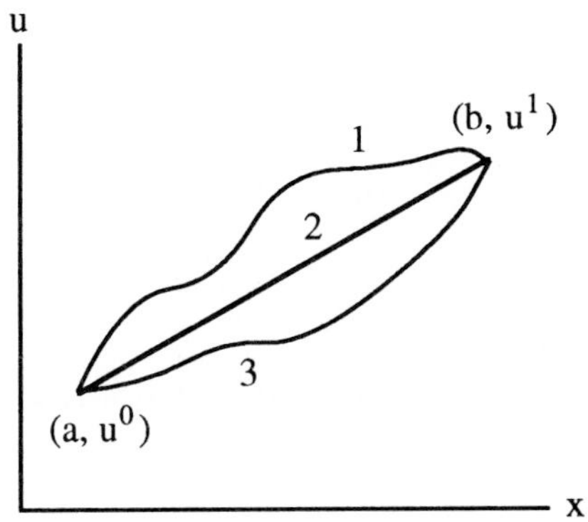

Figure 10.3.1 Shortest Path

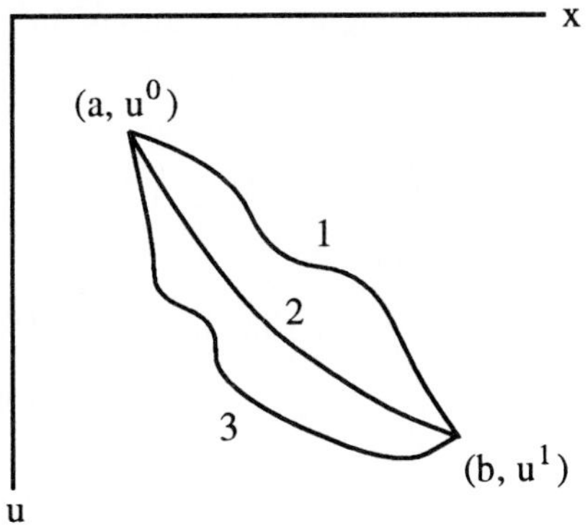

Figure 10.3.2 Curve for Minimum Time

Let m be the mass of the particle and g be the acceleration due to gravity. Since the particle starts at rest at (a, u^0) and there is no friction, the kinetic energy at any time is equal to the loss of potential energy; i.e.,

$$\frac{1}{2} m v^2 = mg\,(u - u^0) \qquad (10.3.20)$$

where v is velocity,

$$v = \left[\left(\frac{dx}{dt}\right)^2 + \left(\frac{du}{dt}\right)^2 \right]^{\frac{1}{2}} = \left[1 + \left(\frac{du}{dx}\right)^2 \right]^{\frac{1}{2}} \frac{dx}{dt} \qquad (10.3.21)$$

and t is time. Solving Eq. 10.3.20 for v, substituting the result into Eq. 10.3.21, and solving for dt yields

$$dt = \frac{\left[1 + \left(\frac{du}{dx}\right)^2 \right]^{1/2} dx}{[\, 2g\,(u - u^0)\,]^{1/2}}$$

The total time T required for the particle to move from (a, u^0) to (b, u^1) is thus

$$T = T(u) = \int_a^b \frac{\left[1 + \left(\frac{du}{dx}\right)^2\right]^{1/2}}{[2g(u - u^0)]^{1/2}}\, dx \tag{10.3.22}$$

This notation makes it clear that T depends on the curve traversed by the particle. The Brachistochrone problem, therefore, is reduced to finding a curve $u = \hat{u}(x)$, $a \leq x \leq b$, that passes through the given end points and makes T as small as possible. ■

In many problems, the form of the function $f(x, u, u')$ permits simplification of the Euler–Lagrange equations. In any case, Eq. 10.3.12 may be expanded, using the chain rule of differentiation and the notation

$$f_u = \frac{\partial f}{\partial u}, \quad f_{u'} = \frac{\partial f}{\partial u'}, \quad f_{xu'} = \frac{\partial^2 f}{\partial x\, \partial u'}, \quad f_{u'u} = \frac{\partial^2 f}{\partial u'\, \partial u}$$

to obtain

$$f_u - f_{u'x} - u' f_{u'u} - u'' f_{u'u'} = 0 \tag{10.3.23}$$

This is simply a second-order differential equation for $u(x)$. Several special cases are now considered.

Case 1. f does not depend on u'; i.e.,

$$f = f(x, u) \tag{10.3.24}$$

Equation 10.3.23, in this case, is

$$f_u(x, u) = 0 \tag{10.3.25}$$

This is simply an algebraic equation in x and u. Since there are no constants of integration, it will not generally be possible to pass the resulting curve through particular end points. This means that a solution to such a problem generally will not exist.

Example 10.3.5

As a concrete example of Case 1, find u in $C^2(0, 1)$ to minimize

$$\int_0^1 u^2\, dx$$

with

$$u(0) = 0, \quad u(1) = 1$$

The condition of Eq. 10.3.25 is

$$2u = 0$$

It is impossible to satisfy the condition u(1) = 1, so the problem has no solution.

To get an idea of what has gone wrong, note that since $u^2(x) \geq 0$ for each x,

$$\int_0^1 u^2(x)\, dx \geq 0$$

for any function u(x) on $0 \leq x \leq 1$. It is clear that if there were a function that minimized $\int_0^1 u^2\, dx$, then the minimum value of the integral would be nonnegative. Even though no minimum exists, consider the following family of functions in $C^2(0, 1)$:

$$u_n(x) = x^n$$

These functions all satisfy the end conditions and

$$T(u_n) = \int_0^1 x^{2n}\, dx = \frac{1}{2n+1}$$

It is possible to choose n large enough so that $\int_0^1 u_n^2\, dx$ is as close as desired to zero. However, the limit of $u_n(x)$ as n approaches infinity is the function

$$u_\infty(x) = \begin{cases} 0, & x < 1 \\ 1, & x = 1 \end{cases}$$

which is not even continuous, much less in $C^2(0, 1)$. Graphs of this sequence of functions are shown in Fig. 10.3.3.

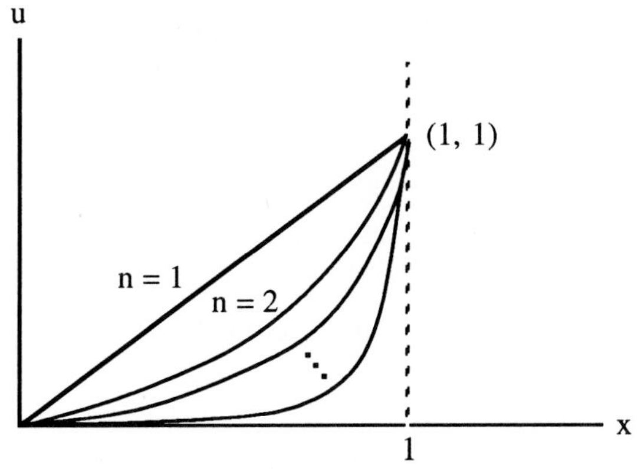

Figure 10.3.3 Minimizing Sequence

In this illustration, a solution of the problem exists in the class of piecewise continuous functions, but not in the class of twice continuously differentiable functions. This problem, therefore, serves as a warning that not all innocent-looking calculus of variations problems have solutions. ■

Case 2. f depends only on u'; i.e.,

$$f = f(u') \tag{10.3.26}$$

Equation 10.3.23 is, in this case,

$$f_{u'u'} u'' = 0 \tag{10.3.27}$$

Example 10.3.6

As a concrete example of Case 2, find the shortest curve in the x-u plane that passes through points (0, 0) and (1, 1). The function f from Eq. 10.3.19 is

$$f = [\, 1 + (u')^2 \,]^{1/2}$$

The form of the Euler–Lagrange equation in this case is, from Eq. 10.3.27,

$$-[\, 1 + (u')^2 \,]^{-3/2} u'' = 0$$

Since $(u')^2 \geq 0$, $[\, 1 + (u')^2 \,] \neq 0$ and $u'(x)$ is required to be continuous, so $[\, 1 + (u')^2 \,] \neq \infty$. Therefore, it is required that

$$u''(x) = 0$$

or

$$u(x) = ax + b$$

where a and b are constants. This implies that the shortest path between two points in a plane is a straight line, which is not surprising.

The end conditions yield

$$u(0) = b = 0$$

and

$$u(1) = a = 1$$

Therefore, the solution of the problem is

$$u(x) = x$$ ■

Case 3. f depends only on x and u'; i.e.,

$$f = f(x, u') \tag{10.3.28}$$

Equation 10.3.12 is, in this case,

$$\frac{d}{dx} f_{u'}(x, u') = 0$$

or

$$f_{u'}(x, u') = c \tag{10.3.29}$$

where c is an arbitrary constant.

Case 4. f depends only on u and u'; i.e.,

$$f = f(u, u') \tag{10.3.30}$$

Equation 10.3.23 is, in this case,

$$f_u - f_{u'u} u' - f_{u'u'} u'' = 0$$

Multiplying by u' yields

$$u' f_u - (u')^2 f_{u'u} - u'u'' f_{u'u'} = 0$$

This is just

$$\frac{d}{dx} (f - u' f_{u'}) = 0$$

so

$$f - u' f_{u'} = c \tag{10.3.31}$$

where c is an arbitrary constant.

Example 10.3.7

As a concrete example of Case 4, consider the Brachistochrone problem of Example 10.3.4. The function f of Eq. 10.3.22 is

$$f = \left[\frac{1 + (u')^2}{2gu} \right]^{\frac{1}{2}}$$

Equation 10.3.31 applies in this case and yields

$$\left[\frac{1 + (u')^2}{2gu} \right]^{\frac{1}{2}} - \frac{(u')^2}{(2gu)^{1/2} [1 + (u')^2]^{1/2}} = c$$

This reduces to

$$1 = \{ 2gu [1 + (u')^2] \}^{1/2} c$$

or

$$u [1 + (u')^2] = c_1$$

where c_1 is a new constant. The solution of this differential equation is a family of **cycloids** in parametric form; i.e.,

$$x = c_2 + \frac{c_1}{2} (s - \sin s)$$
$$u = \frac{c_1}{2} (1 - \cos s) \qquad (10.3.32)$$

The constants c_1 and c_2 are to be determined so that the cycloid passes through the given points. ■

Example 10.3.8

As an example that has more engineering significance, it is instructive to view H_A as a space of candidate generalized solutions of a boundary-value problem and see what sort of generalized solution is actually generated by Theorem 10.2.4. The differential equation and boundary conditions for displacement of the static, laterally loaded string of Fig. 10.3.4 are

$$- u'' = - f$$
$$u(0) = u(\ell) = 0 \qquad (10.3.33)$$

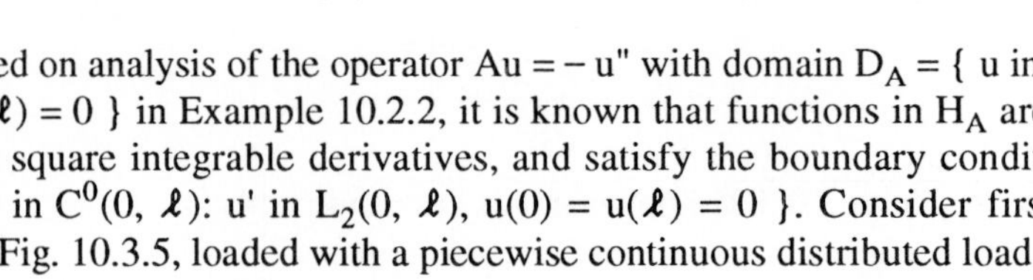

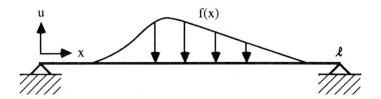

Figure 10.3.4 Static String

Based on analysis of the operator $Au = - u''$ with domain $D_A = \{ u \text{ in } C^2(0, \ell): u(0) = u(\ell) = 0 \}$ in Example 10.2.2, it is known that functions in H_A are continuous, have square integrable derivatives, and satisfy the boundary conditions; i.e., $H_A = \{ u \text{ in } C^0(0, \ell): u' \text{ in } L_2(0, \ell), u(0) = u(\ell) = 0 \}$. Consider first a string shown in Fig. 10.3.5, loaded with a piecewise continuous distributed load

$$f(x) = \begin{cases} 0, & 0 \le x \le a \\ 1, & a \le x \le \ell \end{cases}$$

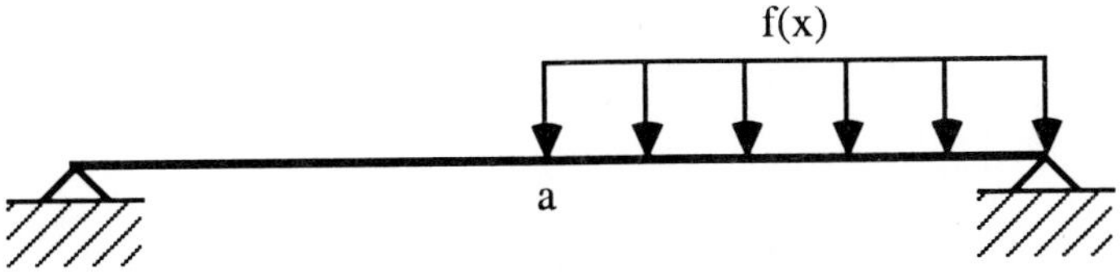

Figure 10.3.5 String With Piecewise Continuous Load

The energy functional $F(x) = \| u \|_A^2 - 2(f, u)$, in this problem, is

$$F(u) = \int_0^{\ell} [(u')^2 + 2fu] \, dx$$

Ideas of the calculus of variations may now be applied, recognizing that u' may be discontinuous at x = a, to obtain a function u in H_A that minimizes F(u). Setting the first variation of F(u) equal to zero,

$$\delta F(u; \eta) = \int_0^{\ell} [2u'\eta' + 2f\eta] \, dx = 0$$

for all η in H_A. Integration by parts from 0 to a and from a to ℓ yields, noting that u' may be discontinuous at x = a,

$$\int_0^{\ell} [-2u'' + 2f] \eta \, dx + 2u'\eta \Big|_{a+0}^{a-0} = 0$$

for arbitrary η in H_A. Thus,

$$-u'' + f = 0$$

except at x = a. Since η is continuous at x = a,

$$u'(a-0) - u'(a+0) = 0$$

so u'(x) is continuous at x = a, even though u" may not be defined there.

Solving the differential equation of Eq. 10.3.33,

$$u(x) = \begin{cases} \alpha_1 x, & 0 \le x \le a \\ \frac{1}{2}(x^2 - \ell^2) + \alpha_2 (x - \ell), & a \le x \le \ell \end{cases}$$

where α_1 and α_2 are constants of integration. Setting $u(a-0) = u(a+0)$ and $u'(a-0) = u'(a+0)$,

$$\alpha_1 = -\frac{(a^2 - \ell^2)}{2\ell}$$

$$\alpha_2 = -\frac{(a + \ell)^2}{2\ell}$$

Thus, the generalized solution is in $D^2(0, \ell) \cap C^1(0, \ell) \subset H_A$ and is the physically meaningful solution of this problem.

As a second loading, consider the static string with a point load applied, as shown in Fig. 10.3.6. This application allows consideration of less regular candidate solutions.

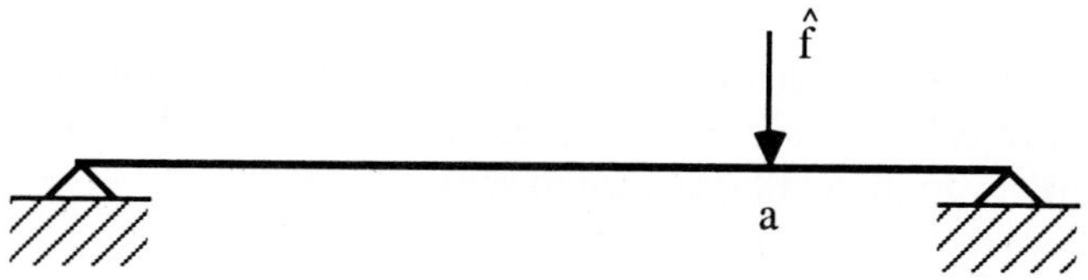

Figure 10.3.6 String With Point Load

In order to write F(u), first form a distributed load f(x) around x = a, with resultant $\hat{f}$. Then form the integral $\int_0^\ell$ fu dx and take the limit as the interval over which the load is distributed approaches zero, to obtain

$$\int_0^\ell fu\, dx = \hat{f}\, u(a)$$

Thus, the functional F(u) is

$$F(u) = \int_0^\ell (u')^2\, dx + 2\hat{f}\, u(a)$$

The first variation of F(u) is

$$\delta F(u; \eta) = \int_0^\ell 2u'\eta'\, dx + 2\hat{f}\, \eta(a) = 0$$

for all η in H_A. Integrating by parts, noting that u' may be discontinuous at x = a,

$$-\int_0^\ell 2u''\eta\, dx + 2u'\eta \Big|_{a+0}^{a-0} + 2\hat{f}\, \eta(a)$$

$$= \int_0^\ell 2u''\eta\, dx + 2\,[\, u'(a-0) - u'(a+0) + \hat{f}\,]\, \eta(a) = 0$$

since η is continuous at x = a. The integrand and boundary terms must vanish independently, so

$$u'' = 0 \tag{10.3.34}$$

except at x = a, and

$$u'(a-0) - u'(a+0) + \hat{f} = 0 \tag{10.3.35}$$

The solution of the differential equation of Eq. 10.3.34 is

$$u(x) = \begin{cases} \alpha_1 x, & 0 \leq x \leq a \\ \alpha_2(\ell - x), & a \leq x \leq \ell \end{cases}$$

Continuity of u at x = a implies

$$\alpha_1 a = \alpha_2(\ell - a)$$

so

$$u(x) = \begin{cases} \alpha_2 \dfrac{\ell - a}{a} x, & 0 \leq x \leq a \\ \alpha_2(\ell - x), & a \leq x \leq \ell \end{cases}$$

Substitution of u'(a − 0) and u'(a + 0) from this equation into Eq. 10.3.35,

$$\alpha_2\left(\frac{\ell - a}{a}\right) + \alpha_2 + \hat{f} = 0$$

or

$$\alpha_2 = -\frac{\hat{f}a}{\ell}$$

Thus, the generalized solution is

$$u(x) = -\frac{\hat{f}a}{\ell}\begin{cases} \dfrac{\ell - a}{a} x, & 0 \leq x \leq a \\ (\ell - x), & a \leq x \leq \ell \end{cases}$$

For this problem, the generalized solution is in $D^1(0, \ell) \cap C^0(0, \ell) \subset H_A$, which is physically meaningful for this problem. In fact, the condition of Eq. 10.3.35 for slope discontinuity at x = a, derived here from the minimum functional theorem, is identical to Eq. 6.3.13, which was derived on purely physical grounds. This result reinforces the applicability and usefulness of Theorem 10.2.4. ■

EXERCISES 10.3

1. Construct the first variation of the following functionals:

(a) $$T_1(u) = \int_0^1 [\,1 + (u')^2\,]^{1/2}\,dx$$

$$D_{T_1} = \{\, u \text{ in } C^2(0,1):\ u(0) = a,\ u(1) = b \,\}$$

$$\bar{D}_{T_1} = \{\, \eta \text{ in } C^2(0,1):\ \eta(0) = \eta(1) = 0 \,\}$$

(b) $$T_2(u) = \int_\Omega \left[\sum_{k=1}^{m} \left(\frac{\partial u}{\partial x_k} \right)^2 - 2fu \right] d\Omega$$

$$D_{T_2} = \bar{D}_{T_2} = \{\, u \text{ in } C^2(\Omega):\ u = 0 \text{ on } \Gamma \,\}$$

2. Extend the results given in Eq. 10.3.4 for a functional involving higher-order derivatives; i.e., find the first variation of the following functionals:

(a) $$T_1(u) = \int_0^1 f(x, u, u', u'')\,dx$$

$$D_{T_1} = \bar{D}_{T_1} = \{\, u \text{ in } C^4(0,1):\ u(0) = u'(0) = u(1) = u'(1) = 0 \,\}$$

(b) $$T_2(u) = \int_0^1 f(\,x, u, u', u'', \ldots, u^{(n)}\,)\,dx$$

$$D_{T_2} = \bar{D}_{T_2} = \{\, u \text{ in } C^{2n}(0,1):\ u(0) = \ldots = u^{(n-1)}(0)$$

$$= u(1) = \ldots = u^{(n-1)}(1) = 0 \,\}$$

3. Extend the result of Eq. 10.3.7 to the functional

$$T(u) = \int_\Omega f(\,\mathbf{x}, u, u_1, u_2, u_{11}, u_{12}, u_{22}\,)\,d\Omega$$

where Ω is a domain in R^2 and

$$D_T = \bar{D}_T = \left\{\, u \text{ in } C^4(\Omega):\ u(\mathbf{x}) = \frac{\partial u}{\partial n}(\mathbf{x}) = 0 \text{ in } \Gamma \,\right\}$$

4. Let $u(x) = [\,u_1(x), u_2(x), \ldots, u_k(x)\,]^T$, where each $u_i(x)$ is in $D = \bar{D} = \{\, u_i$ in $C^2(0, 1)$: $u_i(0) = u_i(1) = 0 \,\}$. Find the first variation of

$$T(u) = \int_0^1 f(\,x, u_1, \ldots, u_k, u_1', \ldots, u_k'\,)\,dx$$

5. State necessary conditions for minima of the functionals T_1 and T_2 of Exercise 1, but with

$$D_{T_1} = \{ u \text{ in } C^2(0, 1): u(0) = a \}$$

$$\bar{D}_{T_1} = \{ \eta \text{ in } C^2(0, 1): \eta(0) = 0 \}$$

and

$$D_{T_2} = \bar{D}_{T_2} = C^2(\Omega)$$

6. Obtain necessary conditions for the functional of Exercise 2(a), but with

$$D_{T_1} = \bar{D}_{T_1} = \{ u \text{ in } C^4(0, 1): u(0) = u(1) = 0 \}$$

7. Verify that the cycloids of Eq. 10.3.32 satisfy the Euler–Lagrange equations for the Brachistochrone problem.

10.4 NATURAL BOUNDARY CONDITIONS

In using the variational formulation of boundary-value problems given by Theorems 10.2.1 and 10.2.4, certain conclusions concerning candidate minimizing functions may be drawn from the calculus of variations.

Definition 10.4.1. Boundary conditions that must be satisfied by the function that minimizes $F(u) = \| u \|_A^2 - 2 (u, f)$ are called **natural boundary conditions**, and the remaining conditions specified in the definition of D_A are called **principal boundary conditions.** ■

Definition 10.4.1 can also be stated in terms of the energy space H_A, as follows [21]: Boundary conditions that must be satisfied by functions in D_A, but not necessarily by functions in H_A, are known as natural boundary conditions for the given operator.

Example 10.4.1

Consider the operator equation of Example 10.2.1, $Au \equiv - d^2u/dx^2 = 2$, with domain $D_A = \{ u \text{ in } C^2(0, 1): u'(0) = u'(1) + u(1) = 0 \}$. The energy functional, in this case, is

$$F(u) = (Au, u) - 2 (u, 2) = \int_0^1 [(u')^2 - 4u] \, dx + u^2(1)$$

Theorem 10.2.1 implies that if $\hat{u}$ minimizes F(u), over all functions in D_A, then it will be the solution of $Au = 2$, u in D_A. Consider minimizing F(u), over all u in $C^2(0, 1)$; i.e., disregarding boundary conditions. The first variation of F is

$$\begin{aligned}\delta F(u;\eta) &= \frac{d}{d\alpha} F(u+\alpha\eta)\Big|_{\alpha=0} \\ &= \frac{d}{d\alpha}\int_0^1 [(u' + \alpha\eta')^2 - 4(u + \alpha\eta)]\,dx \\ &\qquad + (u(1) + \alpha\,\eta(1))^2\Big|_{\alpha=0} \\ &= \int_0^1 [2u'\eta' - 4\eta]\,dx + 2\,u(1)\,\eta(1) \\ &= 2\int_0^1 [-u'' - 2]\,\eta\,dx + 2\,(u'(1) + u(1))\,\eta(1) \\ &\qquad - 2\,u'(0)\,\eta(0)\end{aligned} \tag{10.4.1}$$

In order that $\delta F(u;\eta) = 0$, for all η in $C^2(0, 1)$, the coefficients of η in the integrand, of $\eta(1)$, and of $\eta(0)$ must all be zero. This is,

$$\begin{aligned} -u'' &= 2, \quad 0 < x < 1 \\ u'(0) &= 0, \quad u'(1) + u(1) = 0 \end{aligned} \tag{10.4.2}$$

By Definition 10.4.1, these are natural boundary conditions for the problem. ■

In order to determine whether a given boundary condition is natural, the following rule [21] may be applied to a positive bounded below operator of order 2m: The natural boundary conditions for this class of operators contain derivatives of u of order m and higher. The principal boundary conditions contain derivatives of u up to only order $m - 1$. Thus, for the Laplace operator ($m = 1$), the condition $u|_\Gamma = 0$ is principal and the condition $(\partial u/\partial n + \sigma u)|_\Gamma = 0$ is natural. For the biharmonic operator ($m = 2$),

$$\nabla^4 u = \frac{\partial^4 u}{\partial x^4} + 2\frac{\partial^4 u}{\partial x^2 \partial y^2} + \frac{\partial^4 u}{\partial y^4}$$

which occurs in the theories of elasticity and thin plates [23, 24], the conditions for rigid fixing of the edge, $u|_\Gamma = 0$ and $(\partial u/\partial n)|_\Gamma = 0$, are both principal. In the conditions for a simply supported edge, $u|_\Gamma = 0$ and $\dfrac{\partial^2 u}{\partial n^2} + \nu\left(\dfrac{\partial^2 u}{\partial \tau^2} + \dfrac{1}{\rho}\dfrac{\partial u}{\partial n}\right) = 0$ (ρ is the radius of curvature of the edge), the first is principal and the second is natural.

A procedure that can be used to rigorously verify whether given boundary conditions are natural is the following:

(1) Transform the expression $(Au, v) = [u, v]_A$, where u and v in D_A; i.e., u and v satisfy the complete set of boundary conditions, to a symmetric form in u and v, using integration by parts and the boundary conditions.
(2) Substitute the symmetric expression for $[u, u]_A$ into F(u).
(3) Derive necessary conditions of the calculus of variations for minimization of F(u), without regard for boundary conditions.
(4) Those boundary conditions that are necessarily satisfied by the minimizing function are natural.

This method was used in Eq. 10.4.1 to show that the conditions $u'(0) = 0$ and $u'(1) + u(1) = 0$ are natural for the operator $Au \equiv - d^2u/dx^2$, $0 < x < 1$.

Example 10.4.2

As a further example, consider the equation

$$Au \equiv -\nabla^2 u = f \tag{10.4.3}$$

subject to boundary conditions

$$\frac{\partial u}{\partial n} + \sigma u = 0 \tag{10.4.4}$$

where $\sigma > 0$ on the boundary Γ of Ω. For u in D_A,

$$[u, v]_A = (-\nabla^2 u, v) = -\int_\Omega v \nabla^2 u \, d\Omega$$

Integration by parts reduces this to

$$[u, v]_A = \int_\Omega \nabla u^T \nabla v \, d\Omega - \int_\Gamma v \frac{\partial u}{\partial n} \, dS$$

Using the boundary condition, this reduces to the symmetric form

$$[u, v]_A = \int_\Omega \nabla u^T \nabla v \, d\Omega + \int_\Gamma \sigma u v \, dS \tag{10.4.5}$$

Thus,

$$[u, u]_A = \int_\Omega (\nabla u)^2 \, d\Omega + \int_\Gamma \sigma u^2 \, dS \tag{10.4.6}$$

and

$$\begin{aligned} F(u) &= [u, u]_A - 2 (f, u) \\ &= \int_\Omega [(\nabla u)^2 - 2fu] \, d\Omega + \int_\Gamma \sigma u^2 \, dS \end{aligned} \tag{10.4.7}$$

Taking the first variation of F,

$$\delta F(u;\eta) = 2\int_{\Omega}[\nabla u^T \nabla\eta - 2f\eta]\,d\Omega + \int_{\Gamma} 2\sigma u\eta\,dS$$

and integrating by parts yields

$$\delta F(u;\eta) = -2\int_{\Omega}\nabla^2 u\,\eta\,d\Omega + 2\int_{\Gamma}\frac{\partial u}{\partial n}\eta\,dS$$

$$-\int_{\Omega}2f\eta\,d\Omega + \int_{\Gamma}2\sigma u\eta\,dS$$

$$= -2\int_{\Omega}(\nabla^2 u - f)\,\eta\,d\Omega + 2\int_{\Gamma}\left(\frac{\partial u}{\partial n} + \sigma u\right)\eta\,dS$$

A necessary condition for u_0 to minimize F(u) is that $\delta F(u_0;\eta) = 0$, with no constraints imposed on the boundary values of u. Thus, it follows that

$$\left.\frac{\partial u_0}{\partial n} + \sigma u_0\right|_{\Gamma} = 0$$

from which it can be concluded that this is a natural boundary condition. ■

EXERCISES 10.4

1. Determine, by the procedure presented in this section, which of the following boundary conditions for the operator

$$Au = \frac{d^2}{dx^2}\left[a(x)\frac{d^2u}{dx^2}\right]$$

are natural:

$$a(0)\,u''(0) - u'(0) = 0$$

$$u(0) = 0$$

$$u''(1) = u(1) = 0$$

where $a(x) > 0$, $0 \le x \le 1$.

10.5 NONHOMOGENEOUS BOUNDARY CONDITIONS

Up to now, in discussing the equation

$$Au = f \tag{10.5.1}$$

it has been assumed that the operator A is defined for functions that satisfy homogeneous boundary conditions. In practice, nonhomogeneous boundary conditions often occur, so it is of interest to explain how the variational method can be extended for the case of nonhomogeneous boundary conditions. Equation 10.5.1 is to be solved, with the boundary conditions

$$G_1 u|_\Gamma = g_1,\ G_2 u|_\Gamma = g_2, \ldots,\ G_r u|_\Gamma = g_r \tag{10.5.2}$$

where $G_1, G_2, \ldots, G_r$ are linear operators and $g_1, g_2, \ldots, g_r$ are given functions. The number of boundary conditions is determined by the order k of the differential operator of Eq. 10.5.1 and by whether u(x) is a scalar or a vector function.

This problem is solved, under the following hypothesis [21]: There exists a function $\psi(x)$ that, together with its derivatives of order up to $k-1$, is continuous in $\bar{\Omega}$ and that has derivatives of order k that are piecewise continuous in $\bar{\Omega}$ and satisfy the boundary conditions of the problem; i.e.,

$$G_1 \psi|_\Gamma = g_1,\ G_2 \psi|_\Gamma = g_2, \ldots,\ G_r \psi|_\Gamma = g_r \tag{10.5.3}$$

Defining $u - \psi = v$ and substituting $u = \psi + v$ into Eq. 10.5.1, the new unknown function v satisfies the equation

$$Av = f_1(x) \tag{10.5.4}$$

where $f_1(x) = f(x) - A\psi$. Similarly, from Eq. 10.5.2, v satisfies the homogeneous boundary conditions

$$G_1 v|_\Gamma = 0,\ G_2 v|_\Gamma = 0, \ldots,\ G_r v|_\Gamma = 0 \tag{10.5.5}$$

Let the operator A be positive definite for the set of functions that satisfy Eq. 10.5.5. According to the minimum functional theorem, the solution of Eq. 10.5.4, subject to boundary conditions of Eq. 10.5.5, is equivalent to finding the function v that satisfies Eq. 10.5.5 and minimizes the functional

$$F(v) = (Av, v) - 2(v, f_1) \tag{10.5.6}$$

Replacing v by $u - \psi$ and f_1 by $f - A\psi$, the functional to be minimized by v, equivalently by u, is

$$\begin{aligned} F(v) &= (Au - A\psi) - 2(u - \psi, f - A\psi) \\ &= (Au, u) - 2(u, f) + (u, A\psi) - (Au, \psi) \\ &\qquad + 2(\psi, f) - (A\psi, \psi) \end{aligned} \tag{10.5.7}$$

Since ψ does not satisfy homogeneous boundary conditions $(u, A\psi) \neq (Au, \psi)$, but it is often possible to establish a formula for the operator A such that

$$(u, A\psi) - (Au, \psi) = \int_\Omega (uA\psi - \psi Au)\, d\Omega = \int_\Gamma r(u, \psi)\, dS \tag{10.5.8}$$

where the expression $r(u, \psi)$ depends on the form of the operator A. Using boundary conditions of Eqs. 10.5.2 and 10.5.3, it is often possible to represent $r(u, \psi)$ in the form $r(u, \psi) = N(u) + M$, where N(u) depends only on u and the functions $g_1, g_2, \ldots, g_r$ and M does not depend on u, but may depend on ψ. In such a case, F(v) of Eq. 10.5.7 is reduced to the form

$$F(v) = (Au, u) - 2(u, f) + \int_\Gamma N(u)\, dS + \left[2(\psi, f) - (A\psi, \psi) + \int_\Gamma M\, dS \right]$$

Since the terms in square brackets do not depend on u or v, minimization of the functional F(v) is thus equivalent to finding u to minimize

$$\Phi(u) = (Au, u) - 2(u, f) + \int_\Gamma N(u)\, dS \tag{10.5.9}$$

over all functions that satisfy boundary conditions of Eq. 10.5.2. Note that, in Eq. 10.5.9 (Au, u) is not in a symmetric form; i.e., an integration by parts has not been carried out.

The functional of Eq. 10.5.9 can be constructed without knowing the function ψ. However, in order that the minimum of this functional has meaning, such a function must exist. Otherwise, the variational problem may have no solution. In many classes of applications, it has been proved that a function ψ that satisfies Eq. 10.5.2 exists if the surface Γ is sufficiently smooth and the functions $g_1, g_2, \ldots, g_r$ that enter into Eq. 10.5.2 are differentiable a sufficient number of times in any direction tangential to the surface Γ [22].

It may happen that some of Eq. 10.5.5 will be natural for the functional F(v). The corresponding nonhomogeneous equations in Eq. 10.5.2 will then be natural for the functional $\Phi(u)$, the minimum of which can be sought over the class of functions that satisfy Eq. 10.5.2, except perhaps the natural boundary conditions.

Example 10.5.1

Consider the problem of integrating the Laplace equation

$$-\nabla^2 u = 0 \tag{10.5.10}$$

subject to the boundary conditions

$$u\big|_\Gamma = g(\mathbf{x}) \tag{10.5.11}$$

In this case, f = 0 and the second term in Eq. 10.5.9 vanishes. In order to find N(u), construct the expression

$$\int_\Omega (uA\psi - \psi Au)\, d\Omega = \int_\Omega (\psi \nabla^2 u - u \nabla^2 \psi)\, d\Omega$$

Integrating by parts,

$$\int_\Omega (\psi \nabla^2 u - u \nabla^2 \psi)\, d\Omega = \int_\Gamma \left(\psi \frac{\partial u}{\partial n} - u \frac{\partial \psi}{\partial n} \right) dS$$

where n is the external normal to Γ. Thus,

$$r(u, \psi) = \psi \frac{\partial u}{\partial n} - u \frac{\partial \psi}{\partial n}$$

By virtue of Eq. 10.5.11, which must be satisfied by both u and ψ, $r(u, \psi) = g\,(\partial u/\partial n) - g\,(\partial \psi/\partial n)$. Hence, in this case,

$$N(u) = g \frac{\partial u}{\partial n}$$

$$M = g \frac{\partial \psi}{\partial n}$$

This gives

$$\Phi(u) = -\int_\Omega u \nabla^2 u \, d\Omega + \int_\Gamma g \frac{\partial u}{\partial n} dS \qquad (10.5.12)$$

The first integral on the right can be simplified to

$$\begin{aligned} -\int_\Omega u \nabla^2 u \, d\Omega &= \int_\Omega (\nabla u)^2 \, d\Omega - \int_\Gamma u \frac{\partial u}{\partial n} dS \\ &= \int_\Omega (\nabla u)^2 \, d\Omega - \int_\Gamma g \frac{\partial u}{\partial n} dS \end{aligned} \qquad (10.5.13)$$

Substituting this into Eq. 10.5.12, the boundary integral vanishes. Solving Eqs. 10.5.10 and 10.5.11 thus reduces to finding the function u that minimizes

$$\Phi(u) = \int_\Omega (\nabla u)^2 \, d\Omega$$

over functions in $C^1(\Omega)$ that satisfy the principal boundary condition of Eq. 10.5.11. ■

EXERCISES 10.5

1. Convert the following boundary-value problems with nonhomogeneous boundary conditions to problems with homogeneous boundary conditions:

(a) $$-\frac{d}{dx}\left[(1+x^2)\frac{du}{dx}\right] + u = x, \quad 0 < x < 1$$

$$u(0) = 2$$

$$\frac{du}{dx}(1) = 1$$

(b) $$\frac{d^2}{dx^2}\left[x^2\frac{d^2u}{dx^2}\right] + (1+\sin x)u = e^x, \quad 0 < x < 1$$

$$u(0) = u(1) = 1$$

$$\frac{du}{dx}(0) + \frac{d^2u}{dx^2}(0) = 2$$

$$\frac{du}{dx}(1) = 1$$

(c) $$\nabla^2(f(\mathbf{x})\nabla^2 u) = 0$$

$$\left.\begin{aligned} u &= x_1 + x_2 \\ \nabla^2 u &= 0 \end{aligned}\right\} \quad \text{on } \Gamma$$

2. Find functionals whose minimization is equivalent to solving the boundary-value problems of Exercise 1.

3. Derive a functional that is minimized by the solution of

$$-u'' + u = f(x), \quad 0 < x < 1$$

$$u(0) = 0$$

$$u(1) = 1$$

4. Derive a functional that is minimized by the solution of

$$u'''' + u = f(x), \quad 0 < x < 1$$

$$u(0) = u'(0) = 0$$

$$u(1) = 1$$

$$u''(1) = 0$$

10.6 EIGENVALUE PROBLEMS

Completeness of Eigenfunctions in Energy

Properties of eigenvalues of positive bounded below operators and properties of the corresponding eigenfunctions were discussed in Section 9.2. Many of these properties carry over to the space of functions that have finite energy for the operator.

Theorem 10.6.1. If the eigenfunctions $\{ \phi_i \}$ of a positive bounded below operator A are complete in the L_2 sense, then they are also complete in energy. ■

If f has finite energy and if $[f, \phi_i]_A = 0$, $i = 1, 2, \ldots$, then it follows that $(A\phi_i, f) = \lambda_i (\phi_i, f) = 0$, $i = 1, 2, \ldots$. Since $\lambda_i > 0$, $(\phi_i, f) = 0$, $i = 1, 2, \ldots$, and since the ϕ_i are complete in the L_2 sense, $f = 0$. ■

Theorem 10.6.1 permits construction of the solution of the operator equation $Au = f$, where A is positive bounded below and has L_2 normalized eigenfunctions $\{ \phi_i \}$ that are complete in the L_2 sense. This follows by constructing an energy complete orthonormal sequence $\{ \psi_n \}$, as follows. First note that

$$\| \phi_n \|_A^2 = (A\phi_n, \phi_n) = \lambda_n (\phi_n, \phi_n) = \lambda_n$$

Normalizing the sequence $\{ \phi_i \}$ in the energy norm yields the energy complete orthonormal sequence

$$\psi_n = \frac{\phi_n}{\sqrt{\lambda_n}}$$

These observations lead to the following results.

Theorem 10.6.2. The solution u_0 of $Au = f$ can be represented by its energy Fourier series

$$u_0 = \sum_{n=1}^{\infty} [u_0, \psi_n] \psi_n = \sum_{n=1}^{\infty} [f, \psi_n] \psi_n = \sum_{n=1}^{\infty} \frac{(f, \phi_n)}{\lambda_n} \phi_n$$

This series converges in energy and in the L_2 sense. Furthermore, for

$$u_n = \sum_{i=1}^{n} \frac{(f, \phi_i) \phi_i}{\lambda_i}$$

$$Au_n = \sum_{i=1}^{n} \frac{1}{\lambda_i} (f, \phi_i) A\phi_i = \sum_{i=1}^{n} (f, \phi_i) \phi_i$$

so $\lim_{n \to \infty} \| Au_n - f \| = 0$. ■

If A is a symmetric **bounded below operator**; i.e., $(Au, u) \geq k (u, u)$, k possibly negative or zero, the result of Theorem 10.6.2 can still be shown to be valid [18, 22]. In particular, this shows that if ϕ_0 is an eigenfunction of A corresponding to a zero eigenvalue $\lambda_0 = 0$, then $Au = f$ has a solution only if $(f, \phi_0) = 0$. Another way of saying the same thing, which is a special case of the **theorem of the alternative**, is that f must be orthogonal to all solutions of $Au = 0$.

Minimization Principles for Eigenvalue Problems

If an operator A is positive bounded below, then there exists a positive constant γ such that $(Au, u) \geq \gamma \| u \|^2$, for all u in D_A, from which it follows that

$$\frac{(Au, u)}{(u, u)} \geq \gamma \tag{10.6.1}$$

for all $u \neq 0$ in D_A. This observation leads to the following theorem.

Theorem 10.6.3. Let A be a symmetric operator with domain D_A that is a dense subspace of L_2, such that there exists a greatest lower bound $d > -\infty$ of the functional

$$R(u) \equiv \frac{(Au, u)}{(u, u)} \tag{10.6.2}$$

which is called the **Rayleigh quotient**. If there exists a function $u_0 \neq 0$ from D_A such that

$$R(u_0) = \frac{(Au_0, u_0)}{(u_0, u_0)} = d \tag{10.6.3}$$

then d is the smallest eigenvalue of the operator A and u_0 is the corresponding eigenfunction. ■

To prove this theorem, let η be an arbitrary function from D_A and let t be an arbitrary real number. Since for any t, $u_0 + t\eta$ is in D_A and the function

$$\Phi(t) = \frac{(A(u_0 + t\eta), u_0 + t\eta)}{(u_0 + t\eta, u_0 + t\eta)}$$

$$= \frac{t^2 (A\eta, \eta) + 2t (Au_0, \eta) + (Au_0, u_0)}{t^2 (\eta, \eta) + 2t (u_0, \eta) + (u_0, u_0)}$$

has a minimum at $t = 0$, $\Phi'(0) = 0$. Evaluating $\Phi'(0)$ yields

$$(u_0, u_0) (Au_0, \eta) - (Au_0, u_0) (u_0, \eta) = 0$$

or, using Eq. 10.6.3,

$$(Au_0 - du_0, \eta) = 0 \tag{10.6.4}$$

for all η in D_A. Since D_A is dense in L_2, it follows that

$$Au_0 - du_0 = 0$$

i.e., d and u_0 are an eigenvalue and an associated eigenfunction of the operator A.

It follows that d is the smallest eigenvalue, since if λ_1 and $u_1 \neq 0$ are another eigenvalue and its corresponding eigenfunction of the operator A, then

$$\lambda_1 = \frac{(Au_1, u_1)}{(u_1, u_1)} \geq \min_{u \text{ in } D_A} \frac{(Au, u)}{(u, u)} = d$$

■

Theorem 10.6.3 reduces the problem of finding the smallest eigenvalue of an operator A that is only a bounded below operator; i.e., $(Au, u) \geq \beta (u, u)$ for some $\beta > -\infty$, but possibly $\beta < 0$, to the variational problem of finding the function that minimizes the Rayleigh quotient

$$R(u) \equiv \frac{(Au, u)}{(u, u)} \tag{10.6.5}$$

Consider next another formulation for this minimization problem, which is frequently more convenient. Letting $\psi = u/\| u \|$, $\| \psi \| = 1$, and

$$\frac{(Au, u)}{(u, u)} = (A\psi, \psi) \tag{10.6.6}$$

the variational problem can be formulated as follows: Find u in D_A to minimize the functional

$$(Au, u) \tag{10.6.7}$$

subject to the condition

$$(u, u) = 1 \tag{10.6.8}$$

A method may now be stated for determining subsequent eigenvalues.

Theorem 10.6.4. Let $\lambda_1 \leq \lambda_2 \leq \ldots \leq \lambda_n$ be the first n eigenvalues of an operator A that is bounded below and let $u_1, u_2, \ldots, u_n$ be corresponding orthonormal eigenfunctions. Let there exist a function $u = u_{n+1} \neq 0$ that minimizes the functional of Eq. 10.6.5, under the supplementary conditions

$$(u, u_1) = 0,\ (u, u_2) = 0, \ldots, (u, u_n) = 0 \tag{10.6.9}$$

Then u_{n+1} is an eigenfunction of the operator A that corresponds to the eigenvalue

$$\lambda_{n+1} = \frac{(Au_{n+1}, u_{n+1})}{(u_{n+1}, u_{n+1})} \tag{10.6.10}$$

This is the next larger eigenvalue than λ_n. ■

To prove this theorem, let ζ be an arbitrary function from D_A and define

$$\eta = \zeta - \sum_{k=1}^{n} (\zeta, u_k)\, u_k$$

Then η satisfies Eq. 10.6.9. In fact,

$$(\eta, u_m) = (\zeta, u_m) - \sum_{k=1}^{n} (\zeta, u_k)(u_k, u_m), \quad m = 1, \ldots, n \tag{10.6.11}$$

Since the u_k are orthonormal,

$$(\eta, u_m) = (\zeta, u_m) - (\zeta, u_m) = 0, \quad m = 1, \ldots, n$$

Just as η, the product $t\eta$ also satisfies Eq. 10.6.9, where t is any real number. Thus, the sum $u_{n+1} + t\eta$ satisfies Eq. 10.6.9.

Since u_{n+1} minimizes the functional of Eq. 10.6.5, for all such η,

$$\frac{(A(u_{n+1} + t\eta), u_{n+1} + t\eta)}{(u_{n+1} + t\eta, u_{n+1} + t\eta)}$$

has a minimum at $t = 0$. By repeating the same argument as in the proof of Theorem 10.6.3,

$$(Au_{n+1} - \lambda_{n+1}u_{n+1}, \eta) = 0 \tag{10.6.12}$$

Note at this point, however, that η is not arbitrary.

Consider the quantity

$$(Au_{n+1} - \lambda_{n+1}u_{n+1}, \zeta)$$

From the definition of η, Eqs. 10.6.11 and 10.6.12 lead to

$$(Au_{n+1} - \lambda_{n+1}u_{n+1}, \zeta) = (Au_{n+1} - \lambda_{n+1}u_{n+1}, \eta) + \sum_{k=1}^{n} (u_k, \zeta)(Au_{n+1} - \lambda_{n+1}u_{n+1}, u_k)$$

$$= \sum_{k=1}^{n} (u_k, \zeta)(Au_{n+1} - \lambda_{n+1}u_{n+1}, u_k)$$

Furthermore,

$$(Au_{n+1} - \lambda_{n+1}u_{n+1}, u_k) = (Au_{n+1}, u_k) - \lambda_{n+1} (u_{n+1}, u_k)$$

The second term on the right of this equation vanishes, by virtue of Eq. 10.6.9. Since A is symmetric, the first term is

$$(Au_{n+1}, u_k) = (u_{n+1}, Au_k)$$

Since u_k is an eigenfunction of A, $Au_k = \lambda_k u_k$, so $(Au_{n+1}, u_k) = \lambda_k (u_{n+1}, u_k) = 0$. It follows that $(Au_{n+1} - \lambda_{n+1}u_{n+1}, \zeta) = 0$ and, since ζ is arbitrary,

$$Au_{n+1} - \lambda_{n+1}u_{n+1} = 0$$

Thus, λ_{n+1} is an eigenvalue of the operator and u_{n+1} is the corresponding eigenfunction.

It remains to prove that λ_{n+1} is the smallest eigenvalue that is greater than or equal to λ_n. Let $\bar{\lambda}$ be any eigenvalue of the operator A that is greater than λ_n, with $\bar{u}$ the corresponding eigenfunction. Since A is symmetric, $\bar{u}$ satisfies Eq. 10.6.9 and

$$\bar{\lambda} = \frac{(A\bar{u}, \bar{u})}{(\bar{u}, \bar{u})} \geq \lambda_{n+1}$$

since λ_{n+1} is the minimum of the functional of Eq. 10.6.5, subject to Eq. 10.6.9. ■

As in the case of the smallest eigenvalue, the determination of λ_{n+1} can be reduced to the variational problem of minimizing the functional of Eq. 10.6.7, subject to the conditions of Eqs. 10.6.8 and 10.6.9.

EXERCISES 10.6

1. Verify that Theorems 10.6.1 and 10.6.2 remain valid if the operator is merely bounded below and does not have a zero eigenvalue; i.e., there is a number β (perhaps negative) such that $(Au, u) \geq \beta \| u \|^2$.

2. Note that the numerator of the Rayleigh quotient is the square of the energy norm of u, which contains only m derivatives of u when the order of A is 2m. Use the results

of Exercise 1 of Section 10.1.1 to construct Rayleigh quotients for the following operators:

(a) $$A_1 u = -\frac{d}{dx}\left(a(x)\frac{du}{dx} \right)$$

$$D_{A_1} = \{ u \text{ in } C^2(0, \ell) : u(0) = u(\ell) = 0 \}$$

(b) $$A_2 u = \frac{d^2}{dx^2}\left[EI(x)\frac{d^2u}{dx^2} \right]$$

$$D_{A_2} = \{ u \text{ in } C^4(0, \ell) : u(0) = u'(0) = u(\ell) = u'(\ell) = 0 \}$$

(c) $$A_3 u = \nabla^2(b(\mathbf{x}) \nabla^2 u)$$

$$D_{A_3} = \left\{ u \text{ in } C^4(\Omega) : u = \frac{\partial u}{\partial n} = 0 \text{ on } \Gamma \right\}$$

3. Show that if $\bar{a}(x) \ge a(x)$, $0 \le x \le \ell$; $\overline{EI}(x) \ge EI(x)$, $0 \le x \le \ell$; and $\bar{b}(\mathbf{x}) \ge b(\mathbf{x})$, $\mathbf{x}$ in Ω, then the eigenvalues $\bar{\lambda}_i$ of operators $\bar{A}_1$, $\bar{A}_2$, and $\bar{A}_3$ are greater than the eigenvalues λ_i of operators A_1, A_2, and A_3 of Exercise 2.

4. Let the natural frequency ω of a body be determined by the positive definite operator equation $Au = \omega^2 u$, with u in D_A. Show that if the system is more severely constrained (i.e., has a smaller domain $\bar{D}_A \subset D_A$), then the natural frequencies will increase (or at least not decrease).

5. Let the symmetric operators A and B be defined on the same domain D_{AB}, which is a dense subspace of L_2, such that there exists a greatest lower bound $d > -\infty$ of the functional (generalized Rayleigh quotient)

$$R(u) \equiv \frac{(Au, u)}{(Bu, u)} = \frac{\| u \|_A^2}{\| u \|_B^2}$$

Show that if there exists a function $u_0 \ne 0$ from D_{AB} such that

$$R(u_0) = \frac{(Au_0, u_0)}{(Bu_0, u_0)} = d$$

then d is the smallest eigenvalue of the eigenvalue problem $Au = \lambda Bu$ and u_0 is the corresponding eigenfunction.

10.7 MINIMIZING SEQUENCES AND THE RITZ METHOD

Minimizing Sequences

Definition 10.7.1. Let $\Phi(u)$ be a functional that is bounded below on a domain D_Φ, so there is a **greatest lower bound** d; i.e., $d = \inf_{u \text{ in } D_\Phi} \Phi(u)$. A sequence $\{ u_n \}$ of functions in D_Φ is called a **minimizing sequence** if

$$\lim_{n\to\infty} \Phi(u_n) = d$$

■

Let A be a positive definite operator with domain in H_A. If the equation

$$Au = f \tag{10.7.1}$$

has a solution u_0 in D_A, then, as in the proof of Theorem 10.2.1, the functional

$$F(u) = (Au, u) - 2(u, f)$$

reduces to the form

$$F(u) = \| u - u_0 \|_A^2 - \| u_0 \|_A^2 \tag{10.7.2}$$

It is clear that $\inf_{u \text{ in } D_A} F(u) = \min_{u \text{ in } D_A} F(u) = -\| u_0 \|_A^2$. A minimizing sequence $\{ u_n \}$ for the functional F(u) is characterized by the equation

$$\lim_{n\to\infty} F(u_n) = -\| u_0 \|_A^2 \tag{10.7.3}$$

The following theorem is of great importance for the variational method.

Theorem 10.7.1. If Eq. 10.7.1 has a solution in D_A, then any minimizing sequence for the functional of Eq. 10.7.2 converges in energy to this solution. If A is positive bounded below, the sequence converges in the mean. ■

To prove this theorem, let $\{ u_n \}$ be a minimizing sequence for the functional of Eq. 10.7.2. From Eqs. 10.7.2 and 10.7.3, it follows that

$$\lim_{n\to\infty} F(u_n) = \lim_{n\to\infty} \{ \| u_n - u_0 \|_A^2 - \| u_0 \|_A^2 \} = -\| u_0 \|_A^2$$

Hence, $\lim_{n\to\infty} \| u_n - u_0 \|_A^2 = 0$; i.e., u_n converges to u in energy. If A is a positive bounded below operator, u_n converges to u in the L_2 norm. ■

Theorem 10.7.1 underlies many numerical methods. Thus, to solve Eq. 10.7.1, assuming a solution exists, it suffices to construct a minimizing sequence for the functional of Eq. 10.7.2.

A minimizing sequence can also be defined for $F(u) = [u, u]_A - 2(u, f)$, in the energy space H_A associated with the symmetric and positive definite operator A. Let the sequence $\{u_n\}$ in H_A be such that $\lim_{n\to\infty} F(u_n) = -\| u_0 \|_A^2$, where u_0 is in H_A and minimizes F(u) on H_A; i.e., it is the generalized solution of Au = f. The sequence $\{u_n\}$ is thus a minimizing sequence in H_A and Theorem 10.7.1 may be generalized, as follows.

Theorem 10.7.2. If a symmetric operator A is positive definite, then any minimizing sequence of F(u) in H_A converges in energy to the generalized solution u_0 of Eq. 10.7.1 in H_A, which is guaranteed to exist. If A is positive bounded below, then convergence is also in L_2. ■

The proof requires no change from that of Theorem 10.7.1, except to name the limit u_0. ■

Similarly, but less complete, the following theorem provides a basis for numerical methods of solving the eigenvalue problem

$$Au = \lambda u \tag{10.7.4}$$

Theorem 10.7.3. Let A be a positive definite operator and let $\{\phi_n\}$, with $\| \phi_n \| = 1$, be a minimizing sequence in H_A for the functional $\| u \|_A^2$. If a subsequence $\{\phi_{n_k}\}$ converges in the mean, then $\lambda_1 = \lim_{k\to\infty} \| \phi_{n_k} \|_A^2$ is the smallest eigenvalue of Eq. 10.7.4 and $u = \lim_{k\to\infty} \phi_{n_k}$ is the corresponding eigenfunction. ■

For proof of this theorem, see Ref. 18. ■

This result is not as powerful as Theorem 10.7.2, which guarantees convergence of the minimizing sequence. Sharper results for the eigenvalue problem require a knowledge of the method that is employed to construct the minimizing sequence.

The Ritz Method of Generating Minimizing Sequences

One of the basic difficulties with applications of minimizing sequences, often called **direct methods** for solving boundary-value problems, is the construction of a complete sequence of functions, called **coordinate functions**, that are used to generate minimizing sequences. In view of the importance of this question, key theoretical results that are of practical value in applications are first summarized. The first such result is a practical test for completeness of a sequence of coordinate functions.

Theorem 10.7.4. Let A be a positive bounded below operator and let $\{\phi_n\}$ be a sequence of functions from its domain D_A. If the sequence of functions $\{A\phi_n\}$ is complete in L_2, then $\{\phi_n\}$ is complete in D_A, hence also in H_A, in the sense of energy convergence. ■

To prove this theorem, let v(x) be an arbitrary function from D_A. Since the sequence $\{ A\phi_n \}$ is complete in L_2, for any $\varepsilon > 0$ it is possible to find a positive integer n and constants $\alpha_1, \alpha_2, \ldots, \alpha_n$ such that

$$\left\| Av - \sum_{k=1}^{n} \alpha_k A\phi_k \right\| = \left\| A\left[v - \sum_{k=1}^{n} \alpha_k \phi_k \right] \right\| < \frac{1}{2} \gamma \varepsilon \qquad (10.7.5)$$

where γ is the constant in $(Au, u) \geq \gamma^2 \| u \|^2$. To obtain an estimate for the quantity $\| v - \sum_{k=1}^{n} \alpha_k \phi_k \|_A$, let $v_n = \sum_{k=1}^{n} \alpha_k \phi_k$. Thus,

$$\| v - v_n \|_A^2 = (A(v - v_n), v - v_n)$$

Using the Schwartz inequality and Eq. 10.7.5,

$$\| v - v_n \|_A^2 \leq \| A(v - v_n) \| \, \| v - v_n \| < \frac{1}{2} \varepsilon \gamma \| v - v_n \| \qquad (10.7.6)$$

Substituting $\| v - v_n \| \leq (1/\gamma) \| v - v_n \|_A$ into Eq. 10.7.6 and dividing by $\| v - v_n \|_A$,

$$\| v - v_n \|_A < \frac{1}{2} \varepsilon \qquad (10.7.7)$$

so $\{ \phi_n \}$ is complete in D_A.

To prove $\{ \phi_n \}$ is complete in H_A, let u(x) be an arbitrary function with finite energy; i.e., u is in H_A. Since D_A is dense in H_A, there exists a function $\bar{v}$ in D_A such that $\| u - \bar{v} \|_A \leq \varepsilon/2$. Since $\bar{v}$ is in D_A, there is a sequence $\bar{v}_m = \sum_{i=1}^{m} \bar{\alpha}_i \phi_i$ such that for m large enough, Eq. 10.7.7 is satisfied; i.e., $\| \bar{v} - \bar{v}_m \|_A \leq \varepsilon/2$. By the triangle inequality,

$$\left\| u - \sum_{k=1}^{m} \bar{\alpha}_k \phi_k \right\|_A = \| u - \bar{v}_m \|_A \leq \| u - \bar{v} \|_A + \| \bar{v} - \bar{v}_m \|_A < \varepsilon$$

which was to be proved. ■

Example 10.7.1

Let Ω be the rectangle $0 \leq x \leq a$, $0 \leq y \leq b$ and let $Au = -\nabla^2 u$, with $u = 0$ on the boundary Γ. Choose as coordinate functions

$$\phi_{m,n}(x, y) = \sin\frac{m\pi x}{a}\sin\frac{n\pi y}{b}, \quad m, n = 1, 2, \ldots \tag{10.7.8}$$

Then

$$A\phi_{m,n} = -\nabla^2\phi_{m,n} = \pi^2\left(\frac{m^2}{a^2} + \frac{n^2}{b^2}\right)\sin\frac{m\pi x}{a}\sin\frac{n\pi y}{b}$$

From the theory of double Fourier series, it is known that the sequence of functions $\{ \sin m\pi x/a \sin n\pi y/b \}$ is complete in the sense of convergence in the mean. Since multiplication of each function by a nonzero constant does not destroy completeness of the sequence, the sequence $\{ A\phi_{n,m} \}$is complete in $L_2(\Omega)$. It follows from Theorem 10.7.4 that the functions in Eq. 10.7.8 are complete in H_A.

This implies the following. Let u(x, y) be any function that is continuous and continuously differentiable in the rectangle $0 \le x \le a$, $0 \le y \le b$ and vanishes on its boundary. For any $\varepsilon > 0$, it is possible to find positive integers m and n and constants $\alpha_{k\ell}$, $k = 1, 2, \ldots, m$, $\ell = 1, 2, \ldots, n$, such that

$$\left\| u - \sum_{k=1}^{m}\sum_{\ell=1}^{n}\alpha_{k\ell}\sin\frac{k\pi x}{a}\sin\frac{\ell\pi y}{b}\right\|_A^2$$

$$= \int_0^a\int_0^b\left\{\left(\frac{\partial u}{\partial x} - \frac{\pi}{a}\sum_{k=1}^{m}\sum_{\ell=1}^{n}k\alpha_{k\ell}\cos\frac{k\pi x}{a}\sin\frac{\ell\pi y}{b}\right)^2\right.$$

$$\left. + \left(\frac{\partial u}{\partial y} - \frac{\pi}{b}\sum_{k=1}^{m}\sum_{\ell=1}^{n}\ell\alpha_{k\ell}\sin\frac{k\pi x}{a}\cos\frac{\ell\pi y}{b}\right)^2\right\} dx\, dy < \varepsilon^2$$

■

EXERCISES 10.7

1. Prove that the sequence

$$(x - a)^m (b - x)^m x^k, \quad k = 0, 1, 2, \ldots$$

is complete in the energy norm of the operator

$$(-1)^m \frac{d^{2m}u}{dx^{2m}}$$

on the interval [a, b], under the boundary conditions

$$u^{(k)}(a) = u^{(k)}(b) = 0, \quad k = 0, 1, \ldots, m-1$$

given that the polynomials $\phi_i(x) = x^i$ are complete in $L_2(a, b)$.

10.8 RITZ METHOD FOR EQUILIBRIUM PROBLEMS

The solution of the equation

$$Au = f \tag{10.8.1}$$

where A is a positive definite operator, reduces to finding a function in the domain D_A of A that minimizes the functional

$$F(u) = (Au, u) - 2(u, f) \tag{10.8.2}$$

Selection of Coordinate Functions

An approximate solution to this problem may be represented as a linear combination of a sequence of functions

$$\phi_1(x), \phi_2(x), \ldots, \phi_n(x), \ldots \tag{10.8.3}$$

called **coordinate functions,** which belong to the domain D_A of the operator. In order to yield good results, this sequence should be subject to two conditions:

(1) The sequence is in D_A and is complete in energy.
(2) For any n, the functions $\phi_1(x), \phi_2(x), \ldots, \phi_n(x)$ are linearly independent.

A linear combination of the first n coordinate functions

$$u_n(x) = \sum_{j=1}^{n} a_j \phi_j(x) \tag{10.8.4}$$

with numerical coefficients a_j is sought to approximate the solution of Eq. 10.8.1. Since the $\phi_j(x)$ are known, substituting $u_n(x)$ for u(x) in the energy functional of Eq. 10.8.2 makes $F(u_n)$ a function of the variables $a_1, a_2, \ldots, a_n$; i.e.,

$$\begin{aligned} F(u_n) &= \left(\sum_{j=1}^{n} a_j A\phi_j, \sum_{k=1}^{n} a_k\phi_k \right) - 2\left(\sum_{j=1}^{n} a_j\phi_j, f \right) \\ &= \sum_{j,k=1}^{n} (A\phi_j, \phi_k)\, a_j a_k - 2\sum_{j=1}^{n} (\phi_j, f)\, a_j \end{aligned} \tag{10.8.5}$$

Ritz Approximate Solutions

To create an approximation of the function u that minimizes F(u), the coefficients a_j are selected so that the function of Eq. 10.8.5 is a minimum. This requires that the a_i satisfy the conditions

$$\frac{\partial F(u_n)}{\partial a_i} = 0, \quad i = 1, 2, \ldots, n \tag{10.8.6}$$

Since the operator A is positive definite and the ϕ_i are linearly independent, if $\mathbf{a} = [\, a_1, \ldots, a_n \,]^T \neq \mathbf{0}$, $\sum_{i=1}^{n} a_i\phi_i \neq 0$ and

$$\sum_{j,k=1}^{n} (\, A\phi_j, \phi_k \,)\, a_j a_k = \left(A\sum_{i=1}^{n} a_i\phi_i, \sum_{i=1}^{n} a_i\phi_i \right) > 0$$

Thus, the quadratic form of Eq. 10.8.5 is positive definite, and the coefficients $a_1, a_2, \ldots, a_n$ that satisfy Eq. 10.8.6 yield a minimum of $F(u_n)$.

In order to evaluate the derivatives on the left side of Eq. 10.8.6, the function $F(u_n)$ can be rewritten from Eq. 10.8.5 as

$$\begin{aligned} F(u_n) = {} & (\, A\phi_i, \phi_i \,)\, a_i^2 + \sum_{k\neq i} (\, A\phi_i, \phi_k \,)\, a_i a_k \\ & + \sum_{j\neq i} (\, A\phi_j, \phi_i \,)\, a_j a_i - 2\, (\, f, \phi_i \,)\, a_i + C \end{aligned} \tag{10.8.7}$$

where i is a typical index, $1 \leq i \leq n$, and C denotes a group of terms that do not contain a_i. Using symmetry of A and combining terms, Eq. 10.8.7 may be rewritten as

$$F(u_n) = (\, A\phi_i, \phi_i \,)\, a_i^2 + 2\sum_{k\neq i} (\, A\phi_i, \phi_k \,)\, a_i a_k - 2\, (\, f, \phi_i \,)\, a_i + C$$

Now Eq. 10.8.6 is just

$$\frac{\partial F(u_n)}{\partial a_i} = \sum_{k=1}^{n} 2\, (\, A\phi_i, \phi_k \,)\, a_k - 2\, (\, f, \phi_i \,) = 0$$

or

$$\sum_{k=1}^{n} [\, \phi_i, \phi_k \,]_A\, a_k = (\, f, \phi_i \,), \quad i = 1, 2, \ldots, n \tag{10.8.8}$$

Writing this system in detail yields the **Ritz equations**,

$$\begin{aligned} [\, \phi_1, \phi_1 \,]_A\, a_1 + [\, \phi_1, \phi_2 \,]_A\, a_2 + \ldots + [\, \phi_1, \phi_n \,]_A\, a_n &= (\, f, \phi_1 \,) \\ [\, \phi_2, \phi_1 \,]_A\, a_1 + [\, \phi_2, \phi_2 \,]_A\, a_2 + \ldots + [\, \phi_2, \phi_n \,]_A\, a_n &= (\, f, \phi_2 \,) \\ &\vdots \\ [\, \phi_n, \phi_1 \,]_A\, a_1 + [\, \phi_n, \phi_2 \,]_A\, a_2 + \ldots + [\, \phi_n, \phi_n \,]_A\, a_n &= (\, f, \phi_n \,) \end{aligned} \tag{10.8.9}$$

The determinant of the coefficient matrix of this system of equations is

$$\begin{vmatrix} [\phi_1, \phi_1]_A & [\phi_1, \phi_2]_A & \cdots & [\phi_1, \phi_n]_A \\ [\phi_2, \phi_1]_A & [\phi_2, \phi_2]_A & \cdots & [\phi_2, \phi_n]_A \\ \vdots & \vdots & & \vdots \\ [\phi_n, \phi_1]_A & [\phi_n, \phi_2]_A & \cdots & [\phi_n, \phi_n]_A \end{vmatrix} \neq 0 \qquad (10.8.10)$$

which is the **Gram determinant** of the linearly independent functions $\phi_1(x), \phi_2(x), \ldots, \phi_n(x)$ with respect to the A-scalar product, so by Theorem 2.4.3 it is different from zero. It follows that the system of Ritz equations of Eq. 10.8.8 or 10.8.9 has a unique solution for the a_i, provided the operator A is positive definite. By finding the coefficients $a_1, a_2, \ldots, a_n$ and substituting them in Eq. 10.8.4, the function $u_n(x)$ is obtained. It is called the **Ritz approximate solution** of Eq. 10.8.1.

Theorem 10.8.1. If A is positive bounded below and if the coordinate functions used are complete in energy in D_A, the Ritz approximate solution u_n of the operator equation of Eq. 10.8.1 constitutes a minimizing sequence for the functional of Eq. 10.8.2. ■

For proof, denote the least value of the functional of Eq. 10.8.2 by d; i.e.,

$$d = \min_{u \text{ in } H_A} F(u) = -\| u_0 \|_A^2$$

where u_0 is the generalized solution of Eq. 10.8.1 in H_A. Note that $F(u_n) \geq d$; i.e., the Ritz method gives an upper bound for the value of the functional F. This follows from the fact that $u_n(x)$ is in D_A, since it is a linear combination of the functions $\phi_1(x), \phi_2(x), \ldots, \phi_n(x)$ that are in D_A, and the quantity d is the greatest lower bound of the functional F(u). For any $\varepsilon > 0$, by definition of the greatest lower bound and of H_A, there exists a function v(x) in D_A such that $d \leq F(v) < d + \varepsilon/2$.

Since $F(v) = [v, v]_A - (f, v)$ and $f = Au_0$,

$$F(v) = [v, v]_A - [v, u_0]_A$$
$$= [v - u_0, v - u_0]_A - [u_0, u_0]_A$$

Thus, for any $v_n = \alpha_1\phi_1 + \alpha_2\phi_2 + \ldots + \alpha_n\phi_n$,

$$F(v_n) - F(v) = \| v_n - u_0 \|_A^2 - \| v - u_0 \|_A^2$$
$$= [\, \| v_n - u_0 \|_A + \| v - u_0 \|_A \,]$$
$$\times [\, \| v_n - u_0 \|_A - \| v - u_0 \|_A \,]$$

By the triangle inequality,

$$\| v_n - u_0 \|_A - \| v - u_0 \|_A \le \| v_n - v \|_A$$

so

$$F(v_n) - F(v) \le [\, \| v_n - u_0 \|_A + \| v - u_0 \|_A \,] \, \| v_n - v \|_A$$

Since the sequence $\{ \phi_n \}$ is complete in energy, for any $\varepsilon > 0$, it is possible to choose an integer n and constants $\alpha_1, \alpha_2, \ldots, \alpha_n$ such that

$$\| v_n - v \|_A < \frac{\varepsilon}{k}$$

The value of k will be chosen later. Since $\| v_n \|_A - \| v \|_A \le \| v_n - v \|_A$,

$$\| v_n \|_A < \| v \|_A + \frac{\varepsilon}{k}$$

and

$$F(v_n) - F(v) < \left[2 \| v \|_A + 2 \| u_0 \|_A + \frac{\varepsilon}{k} \right] \frac{\varepsilon}{k}$$

By selecting k such that

$$\frac{1}{k} \left[2 \| v \|_A + 2 \| u_0 \|_A + \frac{\varepsilon}{k} \right] < \frac{1}{2}$$

the inequality $F(v_n) - F(v) < \varepsilon/2$ is obtained. Hence, it follows that

$$d \le F(v_n) \le F(v) + \frac{\varepsilon}{2} < d + \varepsilon$$

Let u_n be the function constructed by the Ritz method. Then

$$d \le F(u_n) \le F(v_n)$$

or

$$d \le F(u_n) \le d + \varepsilon$$

Letting $\varepsilon \to 0$, $F(u_n) \to d$; i.e., the sequence $\{ u_n \}$ is a minimizing sequence. ■

From the theorem just proved and from Theorem 10.7.1, it follows that Ritz approximate solutions converge to the solution of Eq. 10.8.1. To secure high accuracy, it may be necessary to take a large number of coordinate functions. This leads to the necessity of solving Eq. 10.8.8 with a large number of equations and unknowns.

Consider next the question: To what extent are the stipulations of completeness and linear independence that have been imposed on the coordinate functions necessary? The first stipulation is not entirely necessary. In fact, it is not necessary that every function from D_A be approximated in the energy sense by a linear combination of the coordinate functions selected. It is sufficient that the solution permits such an approximation. In the proof of Theorem 10.8.1, it is possible to take for v an appropriate linear combination of coordinate functions and the whole proof holds true. If it is known beforehand that the solution sought belongs to a class of functions that is more restricted than D_A, then it suffices that the sequence of coordinate functions be complete in this class. For example, if it is known that the function sought is even in a certain independent variable, then the coordinate functions may be selected as even functions of that variable. In this case, it suffices that the system of coordinate functions selected be complete with respect to even functions from D_A.

Consider next what would happen if the completeness condition were satisfied, but the coordinate functions were linearly dependent. Suppose, for simplicity, that there is only one function in Eq. 10.8.3 that is linearly dependent on a finite number of other functions of this system, and that this function is different from zero. In this case it is possible, by substituting linear combinations of other coordinate functions for a certain coordinate function, to eliminate a dependent function [say, $\phi_1(x)$] and that the functions $\phi_2(x)$, $\phi_3(x)$, . . . , are linearly independent. Equation 10.8.4 can then be written in the form

$$u_n(x) = \sum_{i=2}^{n} a_j \phi_j(x)$$

and the remainder of the analysis follows without alteration. In particular, the coefficients $a_2, a_3, \ldots, a_n$ are uniquely defined. Hence, it follows that the coefficients $a_1, a_2, \ldots, a_n$ are also defined, but not uniquely, since the coefficient a_1 can be specified arbitrarily. This means that Eq. 10.8.8 can be solved, although not uniquely, and any solution leads to the same function $u_n(x)$. Thus, the presence of one or several coordinate functions that are linearly dependent on the others does not alter the rule for constructing a Ritz approximate solution, although in this case the solution of the Ritz equations of Eq. 10.8.9 is more complex, because the coefficient matrix is singular.

In practical calculations, cases are encountered in which the coordinate functions are "quasi-linearly dependent"; i.e., the determinant of Eq. 10.8.10 is nearly zero, while the coordinate functions are technically linearly independent. In such a case, the relative error in the evaluation of the numerical values of the coefficients $a_1, a_2, \ldots, a_n$ fluctuates markedly with the degree of accuracy of the coefficients and of the independent terms in Eq. 10.8.8, but these fluctuations have little effect on the function $u_n(x)$. Thus, the presence of linearly dependent or quasi-linearly dependent functions does not preclude the use of the Ritz method. It should, however, be borne in mind that it complicates the nature and solution of the Ritz equations.

Finally, it is important to note that for the Ritz method, it is only required that the coordinate functions $\{ \phi_i \}$ be in H_A. Hence, they need not satisfy natural boundary conditions.

Theorem 10.8.2. If the coordinate functions used are complete in H_A and the operator A is positive bounded below, the Ritz approximate solution of the operator equation $Au = f$, on H_A, constitutes a minimizing sequence for $F(u) = [u, u]_A - 2(u, f)$. ■

The proof given in the foregoing remains unchanged.

EXERCISES 10.8

1. Show that if an operator A is positive definite and ϕ_i, $i = 1, \ldots, n$, are linearly independent in D_A, then the matrix $[(A\phi_i, \phi_j)] = [[\phi_i, \phi_j]_A]$ is positive definite.

2. Given the result of Exercise 1, show that Eq. 10.8.8 has a unique solution $\bar{a}_1, \ldots, \bar{a}_k$ and that $F(\bar{u}_n) \le F(u_n)$, where u_n is given by Eq. 10.8.4 and $\bar{u}_n = \sum_{i=1}^{n} \bar{a}_i \phi_i(x)$.

3. Let $\{\phi_i\}$ be orthonormal eigenfunctions of a positive definite operator A. Simplify and solve the Ritz equations for the operator equation $Au = f$.

10.9 RITZ METHOD FOR EIGENVALUE PROBLEMS

Let A be a positive bounded below operator. Clearly,

$$d = \inf_{u \in D_A, u \neq 0} \frac{(Au, u)}{(u, u)} \tag{10.9.1}$$

is the lowest eigenvalue of operator A, if there exists a function $u_0 \neq 0$ in D_A such that

$$d = \frac{(Au_0, u_0)}{(u_0, u_0)}$$

Assuming that such a function exists, the determination of the lowest eigenvalue of operator A reduces to finding the greatest lower bound of the functional of Eq. 10.9.1. Equivalently, this is the determination of the greatest lower bound

$$\inf_{u \in D_A} (Au, u) \tag{10.9.2}$$

under the supplementary condition

$$(u, u) = 1 \tag{10.9.3}$$

This problem can also be solved by the Ritz method.

Selection of Coordinate Functions

Consider a sequence of coordinate functions $\{ \phi_i(x) \}$ in D_A, subject to the requirements:

(1) The sequence of coordinate functions is complete in energy.
(2) For any n, the functions $\phi_1(x), \phi_2(x), \ldots, \phi_n(x)$, are linearly independent.

The goal is to approximate an eigenfunction as

$$u_n(x) = \sum_{k=1}^{n} a_k \phi_k(x)$$

where a_k are coefficients to be selected. Motivated by Eqs. 10.9.1 and 10.9.2, these coefficients are to be chosen so that u_n satisfies Eq. 10.9.3 and such that the quantity (Au_n, u_n) is a minimum. This requires the minimization of

$$(Au_n, u_n) = \sum_{k,i=1}^{n} (A\phi_k, \phi_i) a_k a_i \tag{10.9.4}$$

subject to the condition

$$(u_n, u_n) = \sum_{k,i=1}^{n} (\phi_k, \phi_i) a_k a_i = 1 \tag{10.9.5}$$

Ritz Approximate Solutions

In order to solve this constrained minimization problem, Lagrange multipliers are introduced to construct the function $\Phi = (Au_n, u_n) - \lambda [(u_n, u_n) - 1]$, where λ is a multiplier. As a necessary condition [8] for minimization of the function of Eq. 10.9.4, subject to the constraint of Eq. 10.9.5, the partial derivatives of Φ with respect to the a_i must be zero. This leads to the system of equations

$$\sum_{k=1}^{n} a_k [(A\phi_k, \phi_i) - \lambda (\phi_k, \phi_i)] = 0, \quad i = 1, 2, \ldots, n \tag{10.9.6}$$

Equation 10.9.6 is linear and homogeneous in the unknowns a_i, which cannot all be zero. It follows that the determinant of the coefficient matrix of Eq. 10.9.6 must be zero. This gives the following characteristic equation for λ:

$$\begin{vmatrix} (A\phi_1, \phi_1) - \lambda (\phi_1, \phi_1) & (A\phi_2, \phi_1) - \lambda (\phi_2, \phi_1) & \cdots & (A\phi_n, \phi_1) - \lambda (\phi_n, \phi_1) \\ (A\phi_1, \phi_2) - \lambda (\phi_1, \phi_2) & (A\phi_2, \phi_2) - \lambda (\phi_2, \phi_2) & \cdots & (A\phi_n, \phi_2) - \lambda (\phi_n, \phi_2) \\ \vdots & \vdots & & \vdots \\ (A\phi_1, \phi_n) - \lambda (\phi_1, \phi_n) & (A\phi_2, \phi_n) - \lambda (\phi_2, \phi_n) & \cdots & (A\phi_n, \phi_n) - \lambda (\phi_n, \phi_n) \end{vmatrix} = 0 \tag{10.9.7}$$

If the sequence $\{ \phi_n \}$ is orthonormal [i.e., $(\phi_i, \phi_j) = \delta_{ij}$], then Eq. 10.9.7 simplifies to

$$\begin{vmatrix} (A\phi_1, \phi_1) - \lambda & (A\phi_2, \phi_1) & \cdots & (A\phi_n, \phi_1) \\ (A\phi_1, \phi_2) & (A\phi_2, \phi_2) - \lambda & \cdots & (A\phi_n, \phi_2) \\ \vdots & \vdots & & \vdots \\ (A\phi_1, \phi_n) & (A\phi_2, \phi_n) & \cdots & (A\phi_n, \phi_n) - \lambda \end{vmatrix} = 0 \qquad (10.9.8)$$

As noted previously, the coordinate functions $\phi_1, \phi_2, \ldots, \phi_n$ are to be selected as linearly independent. Equation 10.9.7 will then be of nth degree, since the coefficient of $(-1)^n \lambda^n$ is the Gram determinant of the functions $\phi_1, \phi_2, \ldots, \phi_n$ relative to the A-scalar product. Hence, it follows that Eq. 10.9.7 has precisely n roots. Let λ_0 be any one of these roots. Substituting it in Eq. 10.9.6, the system has a nontrivial solution $a_k^{(0)}$, $k = 1, 2, \ldots, n$, that also satisfies Eq. 10.9.5.

Substituting $\lambda = \lambda_0$ and $a_k = a_k^{(0)}$ in Eq. 10.9.6,

$$\sum_{k=1}^{n} a_k^{(0)} (A\phi_k, \phi_i) = \lambda_0 \sum_{k=1}^{n} a_k^{(0)} (\phi_k, \phi_i), \quad i = 1, 2, \ldots, n \qquad (10.9.9)$$

Multiplying by $a_i^{(0)}$ and summing over i,

$$\sum_{k,i=1}^{n} (A\phi_k, \phi_i)\, a_k^{(0)} a_i^{(0)} = \lambda_0 \sum_{k,i=1}^{n} (\phi_k, \phi_i)\, a_k^{(0)} a_i^{(0)}$$

By virtue of Eq. 10.9.5, the right side of the above equation equals λ_0 and its left side equals $(Au_n^{(0)}, u_n^{(0)})$, where

$$u_n^{(0)} = \sum_{k=1}^{n} a_k^{(0)} \phi_k$$

Thus,

$$\lambda_0 = (Au_n^{(0)}, u_n^{(0)}) \qquad (10.9.10)$$

Equation 10.9.6 shows that the roots of Eq. 10.9.7 are real, if A is symmetric. Further, one of the functions $u_n^{(0)}$ minimizes the function of Eq. 10.9.4, subject to the constraint of Eq. 10.9.5. Equation 10.9.10 shows that this minimum equals the least of the roots of Eq. 10.9.7. The minimum thus constructed is denoted $\lambda_n^{(0)}$. It does not increase with increasing n and it is bounded below by d. Thus, $\lambda_n^{(0)}$ tends to a limit as $n \to \infty$, which is greater than or equal to d.

Theorem 10.9.1. The sequence $\lambda_n^{(0)}$ generated by the Ritz method for the positive bounded below operator equation $Au = \lambda u$ converges monotonically to d of Eq. 10.9.1; i.e., $\lim_{n \to \infty} \lambda_n^{(0)} = d$, $\lambda_n^{(0)} \geq \lambda_{n+k}^{(0)} \geq d$ for all n and $k \geq 1$. ■

To prove this, note first that for any $\varepsilon > 0$ there exists, by the definition of greatest lower bound, a function $\bar{u}$ in D_A such that $(\bar{u}, \bar{u}) = 1$ and

$$d \leq (A\bar{u}, \bar{u}) < d + \varepsilon$$

Equivalently,

$$\sqrt{d} \leq \| \bar{u} \|_A < \sqrt{d+\varepsilon} \tag{10.9.11}$$

Since the sequence $\{ \phi_n(x) \}$ is complete in energy, it is possible to find a function $\bar{u}_N$ of the form

$$\bar{u}_N = \sum_{k=1}^{N} b_k \phi_k$$

such that $\| \bar{u} - \bar{u}_N \|_A < \sqrt{\varepsilon}$. Using the triangle inequality, $\| \bar{u}_N \|_A - \| \bar{u} \|_A \leq \| \bar{u} - \bar{u}_N \|_A < \sqrt{\varepsilon}$. This and Eq. 10.9.11 yield

$$\| \bar{u}_N \|_A \leq \| \bar{u} \|_A + \sqrt{\varepsilon} < \sqrt{d+\varepsilon} + \sqrt{\varepsilon}$$

Squaring both sides of this inequality,

$$\| \bar{u}_N \|_A^2 = (A\bar{u}_N, \bar{u}_N) < (\sqrt{d+\varepsilon} + \sqrt{\varepsilon})^2 \tag{10.9.12}$$

From Eq. 10.9.1, it follows that $\| \bar{u}_N - \bar{u} \| \leq (1/\sqrt{d}) \| \bar{u}_N - \bar{u} \|_A < \sqrt{\varepsilon/d}$. Hence, $\| \bar{u}_N \| \geq \| \bar{u} \| - \sqrt{\varepsilon/d} = 1 - \sqrt{\varepsilon/d}$. This result and Eqs. 10.9.1 and 10.9.12 yield

$$d \leq \frac{(A\bar{u}_N, \bar{u}_N)}{(\bar{u}_N, \bar{u}_N)} = \frac{\| \bar{u}_N \|_A^2}{\| \bar{u}_N \|^2} < \frac{(\sqrt{d+\varepsilon} + \sqrt{\varepsilon})^2}{\left(1 - \sqrt{\frac{\varepsilon}{d}}\right)^2} = d + \eta$$

where η tends to zero with ε. Further, $\lambda_N^{(0)}$ is the minimum of

$$\frac{(Au_N, u_N)}{(u_N, u_N)}$$

over all a_k, where $u_N = \sum_{k=1}^{N} a_k\phi_k$. Thus,

$$d \le \lambda_N^{(0)} \le \frac{(A\bar{u}_N, \bar{u}_N)}{(u_N, u_N)} < d + \eta \tag{10.9.13}$$

If $n \ge N$, then $\lambda_n^{(0)} \le \lambda_N^{(0)}$ and $d \le \lambda_n^{(0)} < d + \eta$. Equation 10.9.13 shows that

$$\lim_{n\to\infty} \lambda_n^{(0)} = d \tag{10.9.14}$$

To complete the proof, note that as N increases, the space spanned by the coordinate functions chosen to minimize the Rayleigh quotient is enlarged. This yields the monotonicity property $\lambda_n^{(0)} \ge \lambda_{n+k}^{(0)}$, for $k \ge 1$. ■

Consider now the determination of the larger eigenvalues. To obtain an approximate value of the second eigenvalue, u_n is sought to minimize the scalar product of Eq. 10.9.4, subject to the constraints

$$\begin{aligned} (u_n, u_n) &= 1 \\ (u_n^{(0)}, u_n) &= \sum_{k,m=1}^{n} (\phi_k, \phi_m)\, a_k^{(0)} a_m = 0 \end{aligned} \tag{10.9.15}$$

where $u_n^{(0)} = \sum_{k=1}^{n} a_k^{(0)}\phi_k$ is the approximate value of the first normalized eigenfunction of operator A.

Using Lagrange multipliers λ and μ associated with the constraints of Eq. 10.9.15, the expression

$$\Psi = (Au_n, u_n) - \lambda[(u_n, u_n) - 1] - 2\mu(u_n, u_n^{(0)})$$

is constructed, and its partial derivatives with respect to the a_i are set to zero [8]. This leads to the system

$$\sum_{k=1}^{n} \{a_k[(A\phi_k, \phi_i) - \lambda(\phi_k, \phi_i)] - \mu(\phi_k, \phi_i)\, a_k^{(0)}\} = 0, \quad i = 1, \ldots, n \tag{10.9.16}$$

Multiplying both sides of this equation by $a_i^{(0)}$ and summing over i yields

$$\sum_{k,i=1}^{n} a_k a_i^{(0)} [(A\phi_k, \phi_i) - \lambda (\phi_k, \phi_i)] - \mu \sum_{k,i=1}^{n} a_k^{(0)} a_i^{(0)} (\phi_k, \phi_i) = 0 \tag{10.9.17}$$

The second summation is simply $(u_n^{(0)}, u_n^{(0)}) = 1$. Changing the order of the first summation,

$$\sum_{i=1}^{n} a_i \sum_{k=1}^{n} a_k^{(0)} [(A\phi_i, \phi_k) - \lambda (\phi_i, \phi_k)] \tag{10.9.18}$$

The inner sum may be written as

$$\sum_{k=1}^{n} a_k^{(0)} [(\phi_k, A\phi_i) - \lambda (\phi_k, \phi_i)]$$

or, since A is symmetric,

$$\sum_{k=1}^{n} a_k^{(0)} [(A\phi_k, \phi_i) - \lambda (\phi_k, \phi_i)]$$

By virtue of Eq. 10.9.9, with $a_k^{(0)}$ corresponding to $\lambda = \lambda_n^{(0)}$, this is

$$\begin{aligned} (\lambda_n^{(0)} - \lambda) \sum_{k=1}^{n} a_k^{(0)} (\phi_k, \phi_i) &= (\lambda_n^{(0)} - \lambda) \left(\sum_{k=1}^{n} a_k^{(0)} \phi_k, \phi_i \right) \\ &= (\lambda_n^{(0)} - \lambda) (u_n^{(0)}, \phi_i) \\ &= (\lambda_n^{(0)} - \lambda) (\phi_i, u_n^{(0)}) \end{aligned}$$

Equation 10.9.18 now becomes

$$(\lambda_n^{(0)} - \lambda) \sum_{i=1}^{n} a_i (\phi_i, u_n^{(0)}) = (\lambda_n^{(0)} - \lambda) (u_n, u_n^{(0)}) = 0$$

where the last equality follows from Eq. 10.9.15. From Eq. 10.9.17, $\mu = 0$, so Eq. 10.9.16 is identical to Eq. 10.9.6. Hence, the minimum sought is a root of Eq. 10.9.7. Similarly, it may be shown that all roots of Eq. 10.9.7 are approximations for larger eigenvalues of $Au = \lambda u$.

The Ritz method can also be developed for finding the minimum of the ratio $\frac{\| u \|_A^2}{\| u \|^2}$, over all $u \neq 0$ in H_A; i.e., ignoring natural boundary conditions. The coordinate functions are selected as before, with the requirement that they satisfy only principal boundary conditions. As before,

$$u_n(x) = \sum_{k=1}^{n} a_k \phi_k(x)$$

where it is required that

$$\| u_n \|_A^2 = \sum_{k,i=1}^{n} a_k a_i [\, \phi_k, \phi_i \,]_A$$

should be a minimum, subject to the condition that

$$\| u_n \|^2 = \sum_{k,i=1}^{n} a_k a_i (\, \phi_k, \phi_i \,) = 1$$

The necessary conditions for a constrained minimum yield

$$\begin{vmatrix} [\phi_1, \phi_1]_A - \lambda(\phi_1, \phi_1) & [\phi_2, \phi_1]_A - \lambda(\phi_2, \phi_1) & \cdots & [\phi_n, \phi_1]_A - \lambda(\phi_n, \phi_1) \\ [\phi_1, \phi_2]_A - \lambda(\phi_1, \phi_2) & [\phi_2, \phi_2]_A - \lambda(\phi_2, \phi_2) & \cdots & [\phi_n, \phi_2]_A - \lambda(\phi_n, \phi_2) \\ \vdots & \vdots & & \vdots \\ [\phi_1, \phi_n]_A - \lambda(\phi_1, \phi_n) & [\phi_2, \phi_n]_A - \lambda(\phi_2, \phi_n) & \cdots & [\phi_n, \phi_n]_A - \lambda(\phi_n, \phi_n) \end{vmatrix} = 0 \tag{10.9.19}$$

Note that Eq. 10.9.19 is equivalent to Eq. 10.9.7, if the coordinate functions are in D_A.

As before, the roots of Eq. 10.9.19 are approximate values of the eigenvalues of the operator and, as $n \to \infty$, the smallest root of Eq. 10.9.19 tends to the smallest eigenvalue of the operator.

Finally, consider briefly the **generalized eigenvalue problem**

$$Au = \lambda Bu \tag{10.9.20}$$

Assume that both operators A and B are positive bounded below and that D_A is a subset of D_B. For any function in D_A it is possible to define the energy in two ways, by associating it either with operator A or with operator B. These two forms of energy are called the **energy of operator A** or the **energy of operator B**.

The details of applying the Ritz method to Eq. 10.9.20 are left as an exercise. Let the sequence of coordinate functions $\{ \phi_i \}$ be in D_A, be linearly independent, and be complete in the energy of operator A. Equations 10.9.6 and 10.9.7 are then replaced by the Ritz equations

$$\sum_{k=1}^{n} a_k [(A\phi_k, \phi_i) - \lambda (B\phi_k, \phi_i)] = 0, \quad i = 1, 2, \ldots, n \qquad (10.9.21)$$

and the characteristic equation

$$\begin{vmatrix} [A\phi_1, \phi_1] - \lambda (B\phi_1, \phi_1) & [A\phi_2, \phi_1] - \lambda (B\phi_2, \phi_1) & \cdots & [A\phi_n, \phi_1] - \lambda (B\phi_n, \phi_1) \\ [A\phi_1, \phi_2] - \lambda (B\phi_1, \phi_2) & [A\phi_2, \phi_2] - \lambda (B\phi_2, \phi_2) & \cdots & [A\phi_n, \phi_2] - \lambda (B\phi_n, \phi_2) \\ \vdots & \vdots & & \vdots \\ [A\phi_1, \phi_n] - \lambda (B\phi_1, \phi_n) & [A\phi_2, \phi_n] - \lambda (B\phi_2, \phi_n) & \cdots & [A\phi_n, \phi_n] - \lambda (B\phi_n, \phi_n) \end{vmatrix} = 0$$

(10.9.22)

The condition that the ϕ_i be in D_A can be relaxed by merely requiring that the coordinate functions be in H_A. It can be proved that they also possess finite energy for operator B. Instead of Eqs. 10.9.21 and 10.9.22, the Ritz equations for the eigenvalue problem $Au = \lambda Bu$ are

$$\sum_{k=1}^{n} a_k \{ [\phi_k, \phi_i]_A - \lambda [\phi_k, \phi_i]_B \} = 0, \quad i = 1, 2, \ldots, n \qquad (10.9.23)$$

and the characteristic equation is

$$\begin{vmatrix} [\phi_1, \phi_1]_A - \lambda [\phi_1, \phi_1]_B & [\phi_2, \phi_1]_A - \lambda [\phi_2, \phi_1]_B & \cdots & [\phi_n, \phi_1]_A - \lambda [\phi_n, \phi_1]_B \\ [\phi_1, \phi_2]_A - \lambda [\phi_1, \phi_2]_B & [\phi_2, \phi_2]_A - \lambda [\phi_2, \phi_2]_B & \cdots & [\phi_n, \phi_2]_A - \lambda [\phi_n, \phi_2]_B \\ \vdots & \vdots & & \vdots \\ [\phi_1, \phi_n]_A - \lambda [\phi_1, \phi_n]_B & [\phi_2, \phi_n]_A - \lambda [\phi_2, \phi_n]_B & \cdots & [\phi_n, \phi_n]_A - \lambda [\phi_n, \phi_n]_B \end{vmatrix} = 0$$

(10.9.24)

EXERCISES 10.9

1. Use the result of Exercise 5 of Section 10.6 to derive the Ritz equations of Eqs. 10.9.21 and 10.9.23 for the generalized eigenvalue problem $Au = \lambda Bu$.

2. Write integral formulas for the coefficients in the Ritz equations for the eigenvalue problems

$$A_i u = \lambda\, g_i(x)\, u, \quad g_i(x) \geq g_0 > 0, \quad 0 \leq x \leq \ell$$

for the operators A_1, A_2, and A_3 of Exercise 1 of Section 10.1.

10.10 OTHER VARIATIONAL METHODS

The variational methods discussed thus far in this chapter are based on an equivalence between minimization of an energy functional and solving the linear operator equations $Au = f$ and $Au = \lambda Bu$. There are a number of other variational methods for approximating solutions of operator equations, but they are less restrictive in the hypotheses required in order that they are applicable. In particular, if a residual, or error term $R = Au - f$, associated with the operator equation $Au = f$ is formed, then the operator equation is satisfied when $R = 0$. This idea, and others that tend to cause the equation to be satisfied, form the basis for many variational methods.

Two commonly employed variational methods that are based on these ideas are summarized in this section. The presentation here is limited to formulation and application of these methods and does not treat their convergence properties, which are developed in Refs. 18 and 21.

The Method of Least Squares

Consider first the linear operator equation

$$Au = f \tag{10.10.1}$$

for u in D_A. The operator A is not necessarily symmetric or positive definite. It is clear that a function in D_A is the solution of Eq. 10.10.1 if and only if the norm of the **residual** or **error function** $R = Au - f$ is zero. This is equivalent to stating that the solution of Eq. 10.1.1 minimizes

$$\| R \|^2 = \| Au - f \|^2 \tag{10.10.2}$$

Selecting a sequence $\{ \phi_k(x) \}$ in D_A, an approximate solution of the form

$$u_n = \sum_{k=1}^{n} a_k \phi_k(x) \tag{10.10.3}$$

is constructed by choosing the a_k to minimize the function of Eq. 10.10.2. This is called the **least square method** of solution of Eq. 10.10.1. As for the Ritz method, the coordinate functions selected should be complete, in order that it is reasonable to expect that they can provide a good approximation of the solution. Substituting Eq. 10.10.3 into the functional of Eq. 10.10.2, the following function is obtained, which depends only on the undetermined coefficients:

$$\begin{aligned} \| Au_n - f \|^2 &= \left\| \sum_{k=1}^{n} a_k A\phi_k - f \right\|^2 \\ &= \left(\sum_{k=1}^{n} a_k A\phi_k - f, \sum_{j=1}^{n} a_j A\phi_j - f \right) \end{aligned} \tag{10.10.4}$$

In order for this function to be a minimum, it is necessary that its derivative with respect to each of the coefficients a_i be zero. This yields

$$\sum_{j=1}^{n} a_j \, (A\phi_i, A\phi_j) = (f, A\phi_i), \quad i = 1, 2, \ldots, n \qquad (10.10.5)$$

This set of linear equations can now be solved for the coefficients a_j to construct an approximate solution of Eq. 10.10.3. This approximation is called the **least square approximate solution** of Eq. 10.10.1.

It is shown in Ref. 21 that

(a) The least square approximation converges more slowly than the Ritz approximation, when the Ritz method applies.
(b) Au_n approaches f in the L_2 sense, which is not necessarily the case in the Ritz method.

While the least square method is not as efficient as the Ritz method, when the Ritz method applies, the least square method can be used for a much broader class of problems. For example, consider the nonlinear operator equation

$$Bu = f \qquad (10.10.6)$$

where u is in the linear space D_B. In this case, the least square method can still be applied, using the approximation

$$u_n = \sum_{k=1}^{n} a_k \phi_k \qquad (10.10.7)$$

but the equation

$$\frac{\partial}{\partial a_i} \| Bu_n - f \|^2 = \frac{\partial}{\partial a_i} \left\| B\left(\sum_{k=1}^{n} a_k \phi_k \right) - f \right\|^2 = 0, \quad i = 1, 2, \ldots, n \qquad (10.10.8)$$

is nonlinear in the unknown coefficients a_k. This creates the rather difficult task of solving a system of nonlinear algebraic equations. Formidable as this problem is, it is still solvable by approximate methods of numerical analysis, such as the Newton method for solution of nonlinear algebraic equations [8].

The Galerkin Method

Consider again the linear operator equation

$$Au = f \qquad (10.10.9)$$

for u in D_A, where the operator A is linear, but it may not be symmetric or positive definite. A function u in D_A is sought such that the residual

$$R = Au - f \tag{10.10.10}$$

is zero.

If a sequence $\{ \phi_k(x) \}$ in D_A is complete in L_2, then $R = 0$ if and only if

$$(\phi_i, R) = 0, \quad i = 1, 2, \ldots \tag{10.10.11}$$

To use this fact, consider an approximate solution of the form

$$u_n = \sum_{k=1}^{n} a_k \phi_k \tag{10.10.12}$$

Substituting this approximation into Eqs. 10.10.10 and 10.10.11 yields the linear equations

$$\left(\phi_i, \sum_{k=1}^{n} a_k A\phi_k - f \right) = 0, \quad i = 1, 2, \ldots, n$$

in the a_k, or

$$\sum_{k=1}^{n} a_k (A\phi_k, \phi_i) = (f, \phi_i), \quad i = 1, 2, \ldots, n \tag{10.10.13}$$

These equations serve to determine the coefficients a_k in the approximation. This process is called the **Galerkin method** of finding an approximate solution.

Note that if the operator A is symmetric and positive definite, Eq. 10.10.13 is precisely the Ritz equations of Eq. 10.8.8. Thus, the Galerkin method reduces to the Ritz method, when the Ritz method applies. The Galerkin method, however, is applicable to a much broader class of problems than is the Ritz method.

Likewise, for the eigenvalue problem $Au = \lambda Bu$, the residual

$$\bar{R} = Au - \lambda Bu \tag{10.10.14}$$

is to be zero. With the approximation of Eq. 10.10.12 and the conditions of Eq. 10.10.11,

$$\sum_{k=1}^{n} \{ (A\phi_k, \phi_i) - \lambda (B\phi_k, \phi_i) \} a_k = 0, \quad i = 1, 2, \ldots, n \tag{10.10.15}$$

as conditions that determine the unknown coefficients a_k and the eigenvalue λ. It is clear that the determinant of the coefficient matrix in Eq. 10.10.15 must be zero. This provides an approximate eigenvalue and the associated approximate eigenvector.

Note that Eq. 10.10.15 is the Ritz equations of Eq. 10.9.21 when the operators A and B are positive definite and symmetric. Thus, the Galerkin method for the eigenvalue problem also reduces to the Ritz method, when the Ritz method applies.

Note finally that if an operator B is nonlinear, with a linear domain D_B, then the Galerkin method applies and results in the following nonlinear equations for the coefficients a_k:

$$\left(B\left(\sum_{k=1}^{n} a_k \phi_k \right) - f,\ \phi_i \right) = 0, \quad i = 1, 2, \ldots, n \tag{10.10.16}$$

as an approximate solution in D_B of the operator equation

$$Bu = f \tag{10.10.17}$$

The formidable problem of solving these nonlinear algebraic equations for the unknown coefficients a_k remains.

EXERCISES 10.10

1. Consider the boundary-value problem

$$-\frac{d}{dx}\left(x^2 \frac{du}{dx} \right) + u = 2, \quad 0 < x < 1$$

$$u(0) = u(1) = 1$$

   (a) Formulate this problem as a linear operator equation.

   (b) Select a set of coordinate functions that can be used in construction of an approximate solution.

   (c) Using the first two coordinate functions selected in (b), find an approximate solution using the Galerkin method.

2. Consider the operator equation

$$Au = 4\frac{d^4u}{dx^4} - \frac{d}{dx}\left(x^2 \frac{du}{dx} \right) + e^x u = \sin x$$

$$D_A = \{ u \text{ in } C^4(0, 1):\ u(0) = u'(0) = u(1) = u''(1) = 0 \}$$

   (a) Determine rigorously which boundary conditions are principal.

   (b) Describe the energy space H_A of the operator.

   (c) Using the Galerkin method, find a two-term approximate solution for the following coordinate functions in H_A:

$$\phi_1(x) = x^2 (x - 1),\ \phi_2(x) = x^3 (x - 1)^2, \ldots,\ \phi_n(x) = x^{n+1} (x - 1)^n$$

3. Derive equations for the least square solution of

$$-u'' + u' + u = f(x), \quad 0 < x < \ell$$

$$u(0) = u(\ell) = 0$$

11

APPLICATIONS OF VARIATIONAL METHODS

In order to illustrate the variational theory developed in Chapter 10, a number of applications are studied in this chapter. The examples, which are of interest in themselves, illustrate the complexity encountered in general problems that precludes closed-form solutions. Concrete applications of physical significance also show that the power of variational methods rests on careful verification of hypotheses that underlie the methods.

11.1 BOUNDARY-VALUE PROBLEMS FOR ORDINARY DIFFERENTIAL EQUATIONS

Many problems in mechanics reduce to finding the solution of an ordinary differential equation that satisfies homogeneous boundary conditions. The Ritz method of Chapter 10 can be applied to such problems, if it can be established that the associated operator is symmetric and positive definite, or even better positive bounded below.

Theoretical Results

Consider first the ordinary differential equation

$$Au = -\frac{d}{dx}\left(p(x)\,\frac{du}{dx} \right) + q(x)\,u = f(x) \tag{11.1.1}$$

A solution is sought in $C^1(a, b)$ that satisfies the boundary conditions

$$\begin{aligned} \alpha\, u'(a) - \beta\, u(a) &= 0 \\ \gamma\, u'(b) + \delta\, u(b) &= 0 \end{aligned} \tag{11.1.2}$$

where α, β, γ, and δ are constants. The following assumptions are made regarding the functions $p(x)$ and $q(x)$ and constants α, β, γ, and δ:

(a) $p(x)$, $p'(x)$, and $q(x)$ are continuous in $a \le x \le b$.
(b) $p(x) \ge 0$, $q(x) \ge 0$.
(c) The function $p(x)$ can vanish at a number of points in $a \le x \le b$, but the integral

$$F = \int_a^b \frac{dx}{p(x)}$$

has finite value. In particular, this requirement is satisfied if $p(x) \geq k > 0$ in $a \leq x \leq b$.

(d) The constants α, β, γ, and δ are nonnegative and neither of the pairs (α, β) or (γ, δ) vanishes.

The domain D_A of the operator A defined in Eq. 11.1.1 is the set of functions in $C^2(a, b)$ that satisfy boundary conditions of Eq. 11.1.2; i.e.,

$$D_A = \{\, u(x) \in C^2(a, b):\ \alpha\, u'(a) - \beta\, u(a) = 0,\ \gamma\, u'(b) + \delta\, u(b) = 0 \,\} \tag{11.1.3}$$

To prove that the operator A is symmetric, form the scalar product (Au, v), for functions u and v in D_A,

$$(\, Au, v\,) = -\int_a^b v(x) \frac{d}{dx}\left(p(x) \frac{d\,u(x)}{dx} \right) dx + \int_a^b q(x)\, u(x)\, v(x)\, dx$$

Integrating the first integral by parts and using Eq. 11.1.2,

$$(\, Au, v\,) = \int_a^b \left(p(x) \frac{d\,u(x)}{dx} \frac{d\,v(x)}{dx} + q(x)\, u(x)\, v(x) \right) dx$$
$$+ \frac{\beta}{\alpha}\, p(a)\, u(a)\, v(a) + \frac{\delta}{\gamma}\, p(b)\, u(b)\, v(b) \tag{11.1.4}$$

if $\alpha \neq 0$ and $\gamma \neq 0$. If $\alpha = 0$ or $\gamma = 0$, then $u(a) = 0$ or $u(b) = 0$ and the corresponding terms on the right of Eq. 11.1.4 can be omitted. The expression on the right of Eq. 11.1.4 is symmetric in u(x) and v(x); hence (Au, v) = (u, Av) and the operator A is symmetric.

Putting v(x) = u(x) in Eq. 11.1.4,

$$(\, Au, u\,) = \int_a^b \left[\, p(x) \left(\frac{d\,u(x)}{dx} \right)^2 + q(x)\, u^2(x) \,\right] dx$$
$$+ \frac{\beta}{\alpha}\, p(a)\, u^2(a) + \frac{\delta}{\gamma}\, p(b)\, u^2(b) \tag{11.1.5}$$

Theorem 11.1.1. If p(a) and p(b) are positive, then the operator A of Eq. 11.1.1 with domain D_A of Eq. 11.1.3 is positive bounded below. ■

To prove this result, suppose first that $\beta > 0$ and $\alpha \neq 0 \neq \gamma$. On the right of Eq. 11.1.5, the second term under the integral and the last term may be deleted, to obtain the inequality

$$(Au, u) \geq \int_a^b p\left(\frac{du}{dx}\right)^2 dx + \frac{\beta}{\alpha}\, p(a)\, u^2(a)$$

$$\geq B\left[\int_a^b pu'^2\, dx + u^2(a)\right] \tag{11.1.6}$$

where B is the lesser of $(\beta/\alpha)\,p(a)$ and 1. Since

$$u(x) = u(a) + \int_a^x u'(t)\, dt$$

and $(a+b)^2 \leq (a+b)^2 + (a-b)^2 = 2(a^2+b^2)$,

$$u^2(x) \leq 2\,u^2(a) + 2\left[\int_a^x u'(t)\, dt\right]^2$$

Further,

$$\left(\int_a^x u'(t)\, dt\right)^2 = \left(\int_a^x \frac{1}{\sqrt{p(t)}}\sqrt{p(t)}\, u'(t)\, dt\right)^2$$

and by the Schwartz inequality,

$$\left(\int_a^x u'(t)\, dt\right)^2 \leq \int_a^x \frac{dt}{p(t)} \int_a^x p(t)\, u'^2(t)\, dt$$

Increasing the upper limit of integration on the right only strengthens the inequality, so

$$\left(\int_a^x u'(t)\, dt\right)^2 \leq \int_a^b \frac{dt}{p(t)} \int_a^b p(t)\, u'^2(t)\, dt = F\int_a^b p(t)\, u'^2(t)\, dt$$

and consequently

$$u^2(x) \leq 2F\int_a^b p(t)\, u'^2(t)\, dt + 2\,u^2(a)$$

Integrating this equation with respect to x in the range from a to b,

$$\| u \|^2 \leq C\left[\int_a^b p(t)\, u'^2(t)\, dt + u^2(a)\right] \tag{11.1.7}$$

where C is the larger of $2F(b-a)$ and $2(b-a)$. The inequalities of Eqs. 11.1.6 and 11.1.7 yield

$$(Au, u) \geq \frac{B}{C} \| u \|^2 \tag{11.1.8}$$

i.e., the operator A is positive bounded below.

It was assumed that $\alpha \neq 0$ and $\gamma \neq 0$. If either α or γ vanishes, then the statement about the operator A being positive bounded below is still true, the reasoning becoming even simpler. For example, let $\alpha = 0$. The first of the boundary conditions of Eq. 11.1.2 assumes the form $u(a) = 0$. In place of Eqs. 11.1.6 and 11.1.7, the simple inequalities

$$(Au, u) \geq \int_a^b p(x)\, u'^2(x)\, dx$$

and

$$\| u \|^2 \leq 2F (b - a) \int_a^b p(x)\, u'^2(x)\, dx$$

are obtained. Equation 11.1.8 follows, if B/C is replaced by $1/[\, 2F (b - a)\,]$. ■

By virtue of the minimum functional theorem, the problem of solving Eq. 11.1.1 subject to the boundary conditions of Eq. 11.1.2 reduces, if $\alpha \neq 0$ and $\gamma \neq 0$, to finding the function u(x) that minimizes the functional

$$\begin{aligned} F(u) &= (Au, u) - 2 (f, u) \\ &= \frac{\beta}{\alpha} p(a)\, u^2(a) + \frac{\delta}{\gamma} p(b)\, u^2(b) \\ &\quad + \int_a^b [\, p(x)\, u'(x) + q(x)\, u^2(x) - 2 f(x)\, u(x)\,]\, dx \end{aligned} \tag{11.1.9}$$

If $\alpha \neq 0$ or $\gamma \neq 0$, then the corresponding boundary conditions of Eq. 11.1.2 are natural. If $\alpha = 0$ or $\gamma = 0$, then the corresponding boundary conditions are principal.

Consider next an ordinary differential equation of arbitrary even order, of the form

$$Au = \sum_{k=0}^{m} (-1)^k \frac{d^k}{dx^k} \left[p_k(x) \frac{d^k u}{dx^k} \right] = f(x) \tag{11.1.10}$$

and confine attention to the simplest of boundary conditions; namely,

$$\begin{aligned} u(a) = u'(a) = \dots &= u^{(m-1)}(a) \\ &= u(b) = u'(b) = \dots = u^{(m-1)}(b) = 0 \end{aligned} \tag{11.1.11}$$

It is assumed that the coefficients $p_k(x)$, $k = 0, 1, 2, \dots, m - 1$, are nonnegative and that $p_m(x)$ is strictly positive. Integrating by parts and using Eq. 11.1.11,

$$(Au, u) = \sum_{k=0}^{m} \int_a^b p_k(x) \left(\frac{d^k u}{dx^k} \right)^2 dx$$

$$\geq \int_a^b p_m(x) \left(\frac{d^m u}{dx^m} \right)^2 dx \geq p_0 \| u^{(m)} \|^2 \qquad (11.1.12)$$

where

$$p_0 \equiv \min_{x \in [a, b]} p_m(x) > 0$$

Since $u(a) = 0$, $u(x) = \int_a^x u'(\xi)\, d\xi$. The Schwartz inequality yields

$$u^2(x) = \left(\int_a^x 1 \times u'(\xi)\, d\xi \right)^2 \leq (x - a) \int_a^x u'^2(\xi)\, d\xi$$

$$\leq (x - a) \int_a^b u'^2(\xi)\, d\xi = (x - a) \| u' \|^2$$

Integrating both sides of this inequality over (a, b),

$$\| u \|^2 \leq \frac{(b - a)^2}{2} \| u' \|^2$$

or

$$\| u \| \leq \frac{b - a}{\sqrt{2}} \| u' \|$$

$$\| u' \| \leq \frac{b - a}{\sqrt{2}} \| u'' \|$$

$$\vdots$$

$$\| u^{(m-1)} \| \leq \frac{b - a}{\sqrt{2}} \| u^{(m)} \|$$

Thus, $\| u^{(m)} \| \geq \left(\frac{\sqrt{2}}{b - a} \right)^m \| u \|$. Substituting this result into Eq. 11.1.12,

$$(Au, u) \geq \gamma^2 \| u \|^2$$

where $\gamma = \sqrt{p_0} \left(\frac{\sqrt{2}}{b - a} \right)^m$. This proves the following result.

Theorem 11.1.2. Under the conditions of Eq. 11.1.11, the operator of Eq. 11.1.10 is positive bounded below. ■

Bending of a Beam of Variable Cross Section on an Elastic Foundation

The equation for bending of a **beam on an elastic foundation** [25] has the form

$$Au = \frac{d^2}{dx^2}\left[EI(x) \frac{d^2u}{dx^2} \right] + Ku = f(x) \tag{11.1.13}$$

where u(x) is the deflection of the beam and I(x) is the moment of inertia of the cross section, which may vary with x. For simplicity, assume that nowhere does the beam cross-sectional area degenerate to zero, so that the moment of inertia I(x) never vanishes. In Eq. 11.1.13, E is Young's modulus of the beam material, K is the distributed spring constant for the elastic foundation, and f(x) is the applied normal loading. The length of the beam is ℓ. If the ends of the beam are clamped, then the following boundary conditions must be satisfied:

$$\begin{aligned} u(0) &= u(\ell) = 0 \\ u'(0) &= u'(\ell) = 0 \end{aligned} \tag{11.1.14}$$

Equations 11.1.13 and 11.1.14 are special cases of Eqs. 11.1.10 and 11.1.11. It is concluded immediately that the operator defined in Eq. 11.1.13 and the boundary conditions of Eq. 11.1.14 is positive bounded below and that the problem of bending of a beam with clamped ends that rests on an elastic foundation is equivalent to the problem of minimizing the functional

$$F(u) = (Au, u) - 2 (u, f)$$

over all functions u(x) that satisfy the boundary conditions of Eq. 11.1.14. Integrating by parts,

$$F(u) = \int_a^b [EIu''^2 + Ku^2 - 2fu]\, dx \tag{11.1.15}$$

The solution of the problem of finding the minimum of this functional can be obtained by the **Ritz method**, for which it is necessary to choose a sequence of coordinate functions $\{ \phi_n(x) \}$ that is complete in energy. Since the boundary conditions of Eq. 11.1.14 are principal, the coordinate functions must satisfy them; e.g.,

$$\phi_k(x) = (\ell - x)^2 x^{k+1}, \quad k = 1, 2, \ldots \tag{11.1.16}$$

Theorem 10.7.4 can be used to show that these coordinate functions are complete in energy (see Exercise 10.7.1).

An approximate solution of this problem is thus obtained by putting

$$u_n(x) = \sum_{k=1}^{n} a_k \phi_k(x) = (\ell - x)^2 \sum_{k=1}^{n} a_k x^{k+1} \tag{11.1.17}$$

and determining the coefficients from the Ritz equations of Eq. 10.8.8, which may be written as

$$\sum_{k=1}^{n} a_k A_{ik} = b_i, \quad i = 1, 2, \ldots, n \tag{11.1.18}$$

where

$$b_k = (f, \phi_k) = \int_0^{\ell} f(x)(\ell - x)^2 x^{k+1} \, dx$$

The coefficients A_{ik} can be represented as

$$A_{ik} = (A\phi_i, \phi_k)$$

$$= \int_0^{\ell} \phi_k A\phi_i \, dx = \int_0^{\ell} \phi_k \left[\frac{d^2}{dx^2}\left(EI \frac{d^2\phi_i}{dx^2} \right) + K\phi_i \right] dx$$

or

$$A_{ik} = [\phi_i, \phi_k]_A = \int_0^{\ell} \left[EI \frac{d^2\phi_i}{dx^2} \frac{d^2\phi_k}{dx^2} + K\phi_i\phi_k \right] dx$$

Eigenvalues of a Second-order Ordinary Differential Equation

To illustrate use of the Ritz method for an eigenvalue problem, consider the equation

$$Au \equiv -\frac{d}{dx}\left[\sqrt{(1+x)} \frac{du}{dx} \right] = \lambda u \tag{11.1.19}$$

with the boundary conditions

$$u(0) = u(1) = 0 \tag{11.1.20}$$

In seeking a solution by the Ritz method, coordinate functions may be chosen as

$$\phi_k(x) = (1 - x) x^k, \quad k = 1, 2, \ldots$$

which satisfy the principal boundary conditions of Eq. 11.1.20. Consider the first three coordinate functions,

$$\phi_1(x) = (1 - x)x, \quad \phi_2(x) = (1 - x)x^2, \quad \phi_3(x) = (1 - x)x^3$$

Applying the method of Section 10.9, Eq. 10.9.7 yields the characteristic equation [21]

$$\begin{vmatrix} 0.4048 - \dfrac{\lambda}{30} & 0.2161 - \dfrac{\lambda}{60} & 0.1350 - \dfrac{\lambda}{105} \\ 0.2161 - \dfrac{\lambda}{60} & 0.5108 - \dfrac{\lambda}{105} & 0.6754 - \dfrac{\lambda}{168} \\ 0.1350 - \dfrac{\lambda}{105} & 0.6754 - \dfrac{\lambda}{168} & 1.0238 - \dfrac{\lambda}{262} \end{vmatrix} = 0$$

The crudest one-term approximation to the first eigenvalue is obtained by equating the first diagonal minor to zero; i.e., $0.4048 - \lambda/30 = 0$. This corresponds to using only one coordinate function $\phi_1(x)$ in the Ritz method, yielding the approximation $\lambda_1^{(1)} = 12.14$. The superscript denotes the number of terms used in the approximation and the subscript denotes the number of the eigenvalue.

The two-term approximation of the first eigenvalue is obtained by equating the first 2×2 minor to zero; i.e.,

$$\begin{vmatrix} 0.4048 - \dfrac{\lambda}{30} & 0.2161 - \dfrac{\lambda}{60} \\ 0.2161 - \dfrac{\lambda}{60} & 0.5108 - \dfrac{\lambda}{105} \end{vmatrix} = 0$$

The smallest root of this equation is $\lambda_1^{(2)} = 12.12$, which gives a more precise approximation to the smallest eigenvalue.

Solving the three-term approximation by Newton's method yields

$$\lambda_1^{(3)} = 12.12$$

which should be quite accurate.

Stability of a Column

Consider a column of variable cross section that is compressed by longitudinal forces of magnitude P applied to its ends. The differential equation describing the deformed axis of the column [25] is

$$E \frac{d^2}{dx^2}\left[I(x) \frac{d^2u}{dx^2} \right] = P\left(-\frac{d^2u}{dx^2} \right) \tag{11.1.21}$$

where E is Young's modulus for the material and I(x) is the centroidal moment of inertia of the cross section. In addition, boundary conditions that define the nature of the supports at the ends of the column must be specified. Consideration here is limited to the following two types of supports:

(1) Both ends of the column are clamped. Assuming that the column in an unbent state occupies the segment $[0, \ell]$ of the x-axis, this is

$$u(0) = u(\ell) = 0, \quad u'(0) = u'(\ell) = 0 \tag{11.1.22}$$

(2) Both ends of the column are pinned. In this case,

$$u(0) = u(\ell) = 0, \quad u''(0) = u''(\ell) = 0 \tag{11.1.23}$$

The investigation of other types of boundary conditions does not present any essential difficulties.

Equations 11.1.21 and 10.9.20 are of the same type, P playing the part of the parameter λ. In this case,

$$Au = E\frac{d^2}{dx^2}\left[I(x)\frac{d^2u}{dx^2}\right]$$

$$Bu = -\frac{d^2u}{dx^2}$$

both of which are positive definite operators, with the boundary conditions of Eqs. 11.1.22 or 11.1.23.

It is easy to show by integration by parts, with the boundary conditions of Eqs. 11.1.22 or 11.1.23, that

$$\| u \|_A^2 = (Au, u) = E\int_0^\ell I(x)\left[\frac{d^2u(x)}{dx^2}\right]^2 dx$$

$$\| u \|_B^2 = (Bu, u) = \int_0^\ell \left[\frac{du(x)}{dx}\right]^2 dx$$

The problem of **stability of a column** reduces to the problem of finding the eigenvalues of Eq. 11.1.21, under boundary conditions of Eq. 11.1.22 or 11.1.23. For the clamped case of Eq. 11.1.22, the lowest critical load is given by

$$P_1 = \min_u \frac{\| u \|_A^2}{\| u \|_B^2} = E\min_u \frac{\int_0^\ell I(x)\, u''^2(x)\, dx}{\int_0^\ell u'^2(x)\, dx} \tag{11.1.24}$$

the minimum being sought in the class of functions that satisfy the boundary conditions of Eq. 11.1.22.

The case of pinned ends of Eq. 11.1.23 can also be investigated by the Ritz method. The lowest critical load is determined by Eq. 11.1.24, but the minimum can be sought in a wider class of functions; namely, those satisfying only the principal boundary conditions $u(0) = u(\ell) = 0$, the conditions $u''(0) = u''(\ell) = 0$ of Eq. 11.1.23 being natural. In general, widening of the class of functions decreases the minimum. Hence, it follows that, for pinned ends, the smallest critical load is lower than for clamped ends.

Vibration of a Beam of Variable Cross Section

The equation for **free vibration of a beam** of variable cross section [25] is

$$E \frac{\partial^2}{\partial x^2}\left[I(x) \frac{\partial^2 u}{\partial x^2} \right] + \rho\, S(x) \frac{\partial^2 u}{\partial t^2} = 0 \qquad (11.1.25)$$

where the x-axis is directed along the axis of the beam, u(x, t) is its transverse displacement, I(x) and S(x) are the geometrical moment of inertia and the area of the cross section, and E and ρ are Young's modulus and the density of the beam material. Suppose that one end of the beam is clamped and the other end is free; i.e., the beam is cantilevered. Denote the length of the beam by ℓ and place the coordinate origin at its clamped left end. The boundary conditions can then be written as

$$u\big|_{x=0} = 0, \quad \frac{\partial u}{\partial x}\bigg|_{x=0} = 0 \qquad (11.1.26)$$

$$\frac{\partial^2 u}{\partial x^2}\bigg|_{x=\ell} = 0, \quad \frac{\partial^3 u}{\partial x^3}\bigg|_{x=\ell} = 0 \qquad (11.1.27)$$

For harmonic motion; i.e., free vibration, a solution is sought in the form

$$u(x, t) = v(x) \sin(\sqrt{\lambda}\, t + \alpha)$$

Equations 11.1.25, 11.1.26, and 11.1.27 transform to the following eigenvalue problem:

$$Au \equiv E \frac{d^2}{dx^2}\left[I(x) \frac{d^2 v}{dx^2} \right] = \rho\lambda\, S(x)\, v \equiv \lambda Bu \qquad (11.1.28)$$

$$v(0) = 0, \quad v'(0) = 0 \qquad (11.1.29)$$

$$v''(\ell) = 0, \quad v'''(\ell) = 0 \qquad (11.1.30)$$

The smallest eigenvalue λ_1 of this problem equals the minimum of the integral

$$\int_0^{\ell} E\, v(x) \frac{d^2}{dx^2}\left[I(x) \frac{d^2 v(x)}{dx^2} \right] dx = E \int_0^{\ell} I(x) \left(\frac{d^2 v(x)}{dx^2} \right)^2 dx \qquad (11.1.31)$$

over the set of functions v(x) that satisfy boundary conditions of Eqs. 11.1.29 and 11.1.30 and the supplementary condition

$$\int_0^{\ell} \rho\, S(x)\, v^2(x)\, dx = 1 \qquad (11.1.32)$$

In finding the minimum of the integral of Eq. 11.1.31 by the Ritz method, coordinate functions are selected as the eigenfunctions of the operator d^4v/dx^4, under boundary conditions of Eqs. 11.1.29 and 11.1.30. These functions form a complete orthonormal system and have the form

$$\phi_k(\xi) = \frac{\sin \alpha_k\xi}{\sin \frac{\alpha_k}{2}} + \frac{\cosh \alpha_k\xi}{\cosh \frac{\alpha_k}{2}}, \quad k = 2m - 1$$

$$\phi_k(\xi) = \frac{\cos \alpha_k\xi}{\cos \frac{\alpha_k}{2}} + \frac{\sinh \alpha_k\xi}{\sinh \frac{\alpha_k}{2}}, \quad k = 2m$$

where the α_k are the roots of the equation $\cos \alpha \cosh \alpha = 1$ and $\xi = (2x - \ell)/2\ell$. Putting

$$v(x) = \sum_{k=1}^{n} a_k \psi_k(x)$$

where

$$\psi_k(x) = \phi_k\left(\frac{2x - \ell}{2\ell}\right)$$

the characteristic equation is

$$\begin{vmatrix} A_{11} - \lambda B_{11} & A_{12} - \lambda B_{12} & \dots & A_{1n} - \lambda B_{1n} \\ A_{21} - \lambda B_{21} & A_{22} - \lambda B_{22} & \dots & A_{2n} - \lambda B_{2n} \\ \vdots & \vdots & & \vdots \\ A_{n1} - \lambda B_{n1} & A_{n2} - \lambda B_{n2} & \dots & A_{nn} - \lambda B_{nn} \end{vmatrix} = 0 \tag{11.1.33}$$

where

$$\begin{aligned} A_{ik} &= E\int_0^{\ell} I(x)\, \psi_i''(x)\, \psi_k''(x)\, dx \\ B_{ik} &= \int_0^{\ell} \rho\, S(x)\, \psi_i(x)\, \psi_k(x)\, dx \end{aligned} \tag{11.1.34}$$

As a specific example, consider a tube with linearly varying diameter [21]. An axial section of this tube is shown in Fig. 11.1.1. In this case,

$$I(x) = \frac{\pi}{4}(y^4 - a^4)$$

$$S(x) = \pi(y^2 - a^2)$$

where

$$y = R - \frac{R - r}{\ell} x = R - \left(\xi + \frac{1}{2} \right) (R - r)$$

The coefficients A_{ik} and B_{ik} of Eq. 11.1.34 are thus

$$A_{ik} = \frac{\pi E}{4\ell^3} \left\{ \left[\frac{1}{16} (R + r)^4 - a^4 \right] I_{ik}^{(0)} - \frac{1}{2} (R + r)(R - r) I_{ik}^{(1)} + \frac{3}{2} (R + r)^2 (R - r)^2 I_{ik}^{(2)} - 2 (R + r)(R - r)^3 I_{ik}^{(3)} + (R - r)^4 I_{ik}^{(4)} \right\}$$

$$B_{ik} = \pi \rho \ell \left\{ \left[\frac{(R + r)^2}{4} - a^2 \right] \tilde{I}_{ik}^{(0)} - \frac{1}{8} (R^2 - r^2) \tilde{I}_{ik}^{(1)} + (R - r)^2 \tilde{I}_{ik}^{(2)} \right\}$$

where $I_{ik}^{(n)}$ and $\tilde{I}_{ik}^{(n)}$ are the integrals

$$I_{ik}^{(n)} = \int_{-1/2}^{1/2} \xi^n \phi_i''(\xi) \phi_k''(\xi) d\xi, \quad n = 0, 1, 2, 3, 4$$

$$\tilde{I}_{ik}^{(n)} = \int_{-1/2}^{1/2} \xi^n \phi_i(\xi) \phi_k(\xi) d\xi, \quad n = 0, 1, 2$$

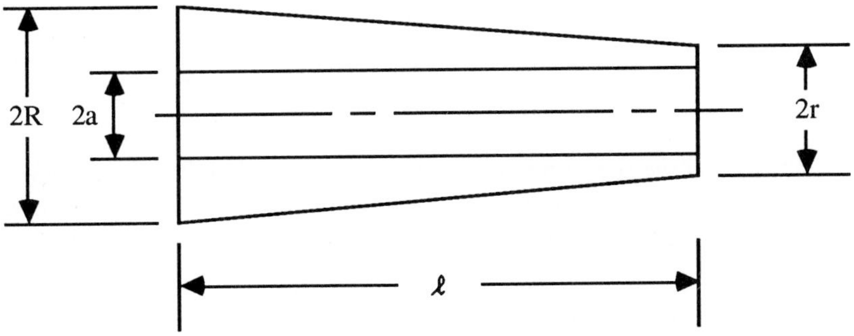

Figure 11.1.1 Tapered Beam

For numerical calculations, the following dimensions are adopted for the tube: $\ell = 4000$ mm, $R = 400$ mm, $r = 200$ mm, $a = 100$ mm.

Putting n = 1, 2, and 3 in

$$v = \sum_{k=1}^{n} a_k \phi_k$$

and evaluating terms in Eq. 11.1.33, detailed calculations [21] yield the following equations for λ:

$$(1)\quad 0.8526 - 0.1917 \times 10^9 \lambda \frac{\rho}{E} = 0$$

$$(2)\quad \begin{vmatrix} 0.8526 - 0.1917 \times 10^9 \lambda \frac{\rho}{E} & -1.5341 + 0.0632 \times 10^9 \lambda \frac{\rho}{E} \\ -1.5341 + 0.0632 \times 10^9 \lambda \frac{\rho}{E} & 39.4262 + 0.3526 \times 10^9 \lambda \frac{\rho}{E} \end{vmatrix} = 0$$

$$(3)\quad \begin{vmatrix} 0.8526 - 0.1917 \times 10^9 \lambda \frac{\rho}{E} & -1.5341 + 0.0632 \times 10^9 \lambda \frac{\rho}{E} & 1.1077 + 0.0037 \times 10^9 \lambda \frac{\rho}{E} \\ -1.5341 - 0.0632 \times 10^9 \lambda \frac{\rho}{E} & 39.4262 + 0.3526 \times 10^9 \lambda \frac{\rho}{E} & -26.614 + 0.0865 \times 10^9 \lambda \frac{\rho}{E} \\ 1.1077 - 0.0037 \times 10^9 \lambda \frac{\rho}{E} & -26.614 + 0.0865 \times 10^9 \lambda \frac{\rho}{E} & 156.59 + 0.3168 \times 10^9 \lambda \frac{\rho}{E} \end{vmatrix} = 0$$

the smallest roots of which are

$$\lambda_1^{(1)} = 0.4446 \times 10^{-8} \frac{E}{\rho}$$

$$\lambda_1^{(2)} = 0.4226 \times 10^{-8} \frac{E}{\rho}$$

$$\lambda_1^{(3)} = 0.4171 \times 10^{-8} \frac{E}{\rho}$$

Note that the successive values of λ_1 converge rapidly. The relative error in passing from $\lambda_1{}^{(1)}$ to $\lambda_1{}^{(3)}$ is about 5% and in passing from $\lambda_1{}^{(2)}$ to $\lambda_1{}^{(3)}$, it is about 1.3%. For the second eigenvalues, the approximate values are

$$\lambda_2^{(2)} = 11.628 \times 10^{-8} \frac{E}{\rho}$$

$$\lambda_2^{(3)} = 11.074 \times 10^{-8} \frac{E}{\rho}$$

and the relative error in passing from $\lambda_2{}^{(2)}$ to $\lambda_2{}^{(3)}$ is approximately 5%.

11.2 SECOND-ORDER PARTIAL DIFFERENTIAL EQUATIONS

Boundary-value Problems for Poisson and Laplace Equations

Consider the **Poisson equation**

$$-\nabla^2 u = f(\mathbf{x}) \tag{11.2.1}$$

where $\mathbf{x}$ is a point in a bounded m-dimensional region Ω, $m = 2$ or 3. Consider first the **Dirichlet problem**; i.e., find the solution of Eq. 11.2.1 that is continuous in the closed physical domain $\bar{\Omega} = \Omega + \Gamma$ and satisfies the boundary condition

$$u\Big|_\Gamma = 0 \tag{11.2.2}$$

As the domain of the Laplace operator $Au = -\nabla^2 u$, take the linear space of functions D_A that satisfy the following conditions:

(1) They are continuous, together with their first and second derivatives in the closed domain $\bar{\Omega} = \Omega + \Gamma$.
(2) They vanish on Γ.

For the linear space D_A, it was shown in Section 9.2 that the operator $-\nabla^2$ is positive definite.

It follows that the Dirichlet problem of Eqs. 11.2.1 and 11.2.2 is equivalent to the problem of minimizing the functional

$$F(u) = (-\nabla^2 u, u) - 2(u, f)$$

Integrating by parts and using the boundary condition of Eq. 11.2.2, this is

$$F(u) = \int_\Omega \{ (\nabla u)^2 - 2uf \} \, d\Omega \tag{11.2.3}$$

As shown in Section 6.3, the problem of the deflection of a membrane that is fixed at its edge, under the influence of a normal load, reduces to Eqs. 11.2.1 and 11.2.2, where $f(\mathbf{x})$ is proportional to the loading. The functional of Eq. 11.2.3 is proportional to the potential energy of the deformed membrane. Thus, the minimum functional theorem is just the **principle of minimum total potential energy.**

A similar result is obtained if, in place of Eq. 11.2.2, the boundary condition for the **mixed problem,**

$$\left[\frac{\partial u}{\partial n} + \sigma(\mathbf{x})\, u \right]_\Gamma = 0 \tag{11.2.4}$$

is applied, where $\sigma(\mathbf{x})$ is a nonnegative continuous function that is not identically zero. For this problem, the domain of the Laplace operator is denoted as the linear space M_σ of functions that satisfy the boundary equation of Eq. 11.2.4 and the same conditions of continuity and differentiability as functions in the linear space D_A. To show that, with domain M_σ, the operator $-\nabla^2 u$ is also positive definite, note that

$$(-\nabla^2 u, u) = -\int_\Gamma u \frac{\partial u}{\partial n} dS + \int_\Omega (\nabla u)^2 d\Omega$$

$$= \int_\Gamma \sigma u^2 dS + \int_\Omega (\nabla u)^2 d\Omega \geq 0$$

If $(-\nabla^2 u, u) = 0$, then it is necessary that

$$\int_\Gamma \sigma u^2 dS = \int_\Omega (\nabla u)^2 d\Omega = 0$$

From the second equation, it follows that $u = c$, a constant. Substituting this in the first equation,

$$c^2 \int_\Gamma \sigma \, dS = 0$$

Since the function $\sigma(\mathbf{x})$ is continuous, nonnegative, and not identically zero, it follows that $\int_\Gamma \sigma \, dS > 0$, so $c = 0$. Thus, $u(\mathbf{x}) = 0$, for all $\mathbf{x}$ in Ω, so the operator $Au = -\nabla^2 u$ with domain M_σ is positive definite.

The problem of integrating the Poisson equation of Eq. 11.2.1, subject to the boundary condition of Eq. 11.2.4, can now be replaced by the problem of minimizing the functional

$$F(u) = (-\nabla^2 u, u) - 2 (u, f)$$

$$= \int_\Omega \{ (\nabla u)^2 - 2uf \} d\Omega + \int_\Gamma \sigma u^2 dS \qquad (11.2.5)$$

over the linear space M_σ. Actually, the boundary condition of Eq. 11.2.4 is natural, so it can be ignored in minimization of the functional of Eq. 11.2.5.

Finally, a solution of the **Neumann problem** for Eq. 11.2.1 is sought that is continuous and continuously differentiable in $\bar{\Omega}$ and that satisfies the free edge boundary condition

$$\left. \frac{\partial u}{\partial n} \right|_\Gamma = 0 \qquad (11.2.6)$$

Define a linear space M_0 that consists of functions that satisfy Eq. 11.2.6 and the same conditions of continuity and differentiability as the functions of the linear space D_A. However, the operator $-\nabla^2 u$ will not be positive definite on this linear space. In fact,

$$(-\nabla^2 u, u) = \int_\Omega (\nabla u)^2 \, d\Omega \geq 0 \tag{11.2.7}$$

From the equality $(-\nabla^2 u, u) = 0$, it does not follow that $u = 0$. Indeed, the function $u = 1$ clearly belongs to the linear space M_0 and $(-\nabla^2 u, u) = 0$.

To resolve this problem with positive definiteness of the Laplace operator on M_0, the theorem of the alternative [18, 22] is used (see discussion in Section 3.1 for matrices and discussion following Theorem 10.6.2). The Laplace operator on M_0 is symmetric, so it is required that

$$\int_\Omega f\bar{u} \, d\Omega = 0 \tag{11.2.8}$$

for all solutions $\bar{u}$ of

$$-\nabla^2 \bar{u} = 0, \quad \text{in } \Omega$$

$$\frac{\partial \bar{u}}{\partial n} = 0, \quad \text{on } \Gamma$$

Note that

$$\bar{u}(\mathbf{x}) = c \neq 0$$

is a solution of this problem. Substituting this result into Eq. 11.2.8, it follows that

$$\int_\Omega f \, d\Omega = 0 \tag{11.2.9}$$

That is, the Neumann problem of Eqs. 11.2.1 and 11.2.6 is solvable only if the applied force f satisfies Eq. 11.2.9. Further, if this problem is solvable, it has an infinite number of solutions, all differing from one another by a constant. To obtain a unique solution $u(\mathbf{x})$, it is required that

$$\int_\Omega u(\mathbf{x}) \, d\Omega = 0 \tag{11.2.10}$$

Considering the Neumann problem for Eq. 11.2.1, the domain of the Laplace operator is defined as the linear space M_0^* of functions that

(1) are continuous, together with their first and second derivatives in $\bar{\Omega}$.
(2) satisfy Eq. 11.2.6.
(3) satisfy Eq. 11.2.10.

It will next be proved that the operator $-\nabla^2 u$ is positive definite for all functions in the linear space M_0^*. For any u in M_0^*, Eq. 11.2.7 holds. If $(-\nabla^2 u, u) = 0$, then from Eq. 11.2.7, u = c. But Eq. 11.2.10 implies c = u = 0. Thus, the operator is positive definite.

This Neumann problem can thus be replaced by the variational problem of finding the function in M_0^* that minimizes

$$F(u) = (-\nabla^2 u, u) - 2(u, f)$$

or, by virtue of the boundary condition of Eq. 11.2.6,

$$F(u) = \int_\Omega \{ (\nabla u)^2 - 2fu \} \, d\Omega \tag{11.2.11}$$

The boundary condition of Eq. 11.2.6 is natural, so the functional of Eq. 11.2.11 can be minimized without regard for boundary conditions.

It is often required to solve the Laplace equation with nonhomogeneous boundary conditions. It was shown in Section 10.5 that integration of the Laplace equation in the domain Ω, under the boundary condition of the **Dirichlet problem**

$$u \big|_\Gamma = g(\mathbf{x}) \tag{11.2.12}$$

reduces to finding the minimum of the functional

$$\int_\Omega (\nabla u)^2 \, d\Omega \tag{11.2.13}$$

over the set of functions that satisfy Eq. 11.2.12.

If the boundary condition has the form of the **Neumann problem**

$$\frac{\partial u}{\partial n}\bigg|_\Gamma = h(\mathbf{x}) \tag{11.2.14}$$

then the problem reduces to finding the minimum of the functional

$$\int_\Omega (\nabla u)^2 \, d\Omega - 2\int_\Gamma uh \, dS \tag{11.2.15}$$

without regard to boundary conditions.

Note that in the case of boundary conditions of the **mixed problem**; i.e.,

$$\left[\frac{\partial u}{\partial n} + \sigma(\mathbf{x})\, u \right]_\Gamma = h_1(\mathbf{x}) \tag{11.2.16}$$

the energy functional has the form

$$F(u) = \int_\Omega (\nabla u)^2 \, d\Omega + \int_\Gamma (\sigma u^2 - 2uh_1) \, dS \tag{11.2.17}$$

The positive definite character of the operator $-\nabla^2 u$ has been established here for each of the domains D_A, M_σ, M_0, and M_0^*. However, in order to be able to use the Ritz method with full confidence, it is important to establish that the operators are positive bounded below. It was shown in Section 9.2 that the Laplace operator is positive bounded below over the set D_A. It is shown in Ref. 21 that the Laplace operator is also positive bounded below over the domains M_σ and M_0.

Torsion of a Rod of Rectangular Cross Section

The problem of **torsion of a rectangular rod** is considered in some detail, because a relatively simple exact solution is known [26], to which approximations can be compared. This problem reduces [23, 26] to integration of the Poisson equation

$$-\nabla^2 u = 1 \tag{11.2.18}$$

in the rectangle $-a \le x_1 \le a$, $-b \le x_2 \le b$, under the boundary conditions

$$u(\pm a, x_2) = u(x_1, \pm b) = 0 \tag{11.2.19}$$

From symmetry, it is clear that the function u is even in both x_1 and x_2. A sequence of polynomials that possess this property and vanish on the edges of the rectangle (i.e., on the straight lines $x_1 = \pm a$, $x_2 = \pm b$) has the form

$$(x_1^2 - a^2)(x_2^2 - b^2)(a_1 + a_2 x_1^2 + a_3 x_2^2 + \dots) \tag{11.2.20}$$

Restricting attention to three terms, the approximate solution will be of the form

$$u \approx u_3 = (x_1^2 - a^2)(x_2^2 - b^2)(a_1 + a_2 x_1^2 + a_3 x_2^2)$$

Performing the necessary calculations [21], the Ritz equations for the unknowns a_1, a_2, and a_3 are

$$\frac{128}{45} a^3 b^3 (a^2 + b^2) a_1 + \frac{128}{45} a^5 b^3 \left[\frac{a^2}{7} + \frac{b^2}{5} \right] a_2 + \frac{128}{45} a^3 b^5 \left[\frac{a^2}{5} + \frac{b^2}{7} \right] a_3 = \frac{16 a^3 b^3}{9}$$

$$\frac{128}{45} a^5 b^3 \left[\frac{a^2}{7} + \frac{b^2}{5} \right] a_1 + \frac{128}{45 \times 7} a^5 b^5 \left[\frac{11}{5} b^2 + \frac{1}{3} a^2 \right] a_2 + \frac{128}{45 \times 35} a^5 b^5 (a^2 + b^2) a_3 = \frac{16 a^5 b^3}{45}$$

$$\frac{128}{45}a^3b^3\left[\frac{a^2}{5}+\frac{b^2}{7}\right]a_1+\frac{128}{45\times 35}a^5b^5(a^2+b^2)a_2$$

$$+\frac{128}{45\times 7}\left[\frac{11}{5}a^2+\frac{1}{3}b^2\right]a_3=\frac{16a^3b^5}{45}$$

The solution of these equations is

$$a_1=\frac{35(9a^4+130a^2b^2+9b^4)}{16(45a^6+509a^4b^2+509a^2b^4+45b^6)}$$

$$a_2=\frac{105(9a^2+b^2)}{16(45a^6+509a^4b^2+509a^2b^4+45b^6)} \tag{11.2.21}$$

$$a_3=\frac{105(a^2+9b^2)}{16(45a^6+509a^4b^2+509a^2b^4+45b^6)}$$

If only one coefficient a_1 were used; i.e., if

$$u\approx u_1=a_1(x_1^2-a^2)(x_2^2-b^2)$$

then

$$a_1=\frac{5}{8(a^2+b^2)}$$

A similar calculation can be performed to obtain a two-term approximation.

A second sequence of coordinate functions that are even and satisfy the boundary conditions is

$$\phi_{ij}(x_1,x_2)=\cos\frac{i\pi x_1}{2a}\cos\frac{j\pi x_2}{2b} \tag{11.2.22}$$

where i and j are odd integers. These functions are energy orthogonal, since

$$-\int_{-b}^{b}\int_{-a}^{a}\phi_{mn}\nabla^2\phi_{rs}\,dx_1\,dx_2$$

$$=\frac{\pi^2}{4}\left[\frac{r^2}{a^2}+\frac{s^2}{b^2}\right]\left[\int_{-a}^{a}\cos\frac{m\pi x_1}{2a}\cos\frac{r\pi x_1}{2a}\,dx_1\right]$$

$$\times\left[\int_{-b}^{b}\cos\frac{n\pi x_2}{2b}\cos\frac{s\pi x_2}{2b}\,dx_2\right]$$

$$=0,\quad \text{if } m\neq r \text{ or } n\neq s$$

Since the coordinate functions are orthogonal, the Ritz equations are decoupled and easily solved. Direct calculation [21] yields the following one-, three-, and six-term approximations:

$$u_1(x_1, x_2) = \frac{64a^2b^2}{\pi^4 (a^2 + b^2)} \cos\frac{\pi x_1}{2a} \cos\frac{\pi x_2}{2b}$$

$$u_3(x_1, x_2) = \frac{64a^2b^2}{\pi^4}\left[\frac{1}{a^2 + b^2} \cos\frac{\pi x_1}{2a} \cos\frac{\pi x_2}{2b} + \frac{1}{3(9a^2 + b^2)} \cos\frac{\pi x_1}{2a} \cos\frac{3\pi x_2}{2b} + \frac{1}{3(a^2 + 9b^2)} \cos\frac{3\pi x_1}{2a} \cos\frac{\pi x_2}{2b} \right]$$

$$u_6(x_1, x_2) = \frac{64a^2b^2}{\pi^4}\left[\frac{1}{a^2 + b^2} \cos\frac{\pi x_1}{2a} \cos\frac{\pi x_2}{2b} + \frac{1}{3(9a^2 + b^2)} \cos\frac{\pi x_1}{2a} \cos\frac{3\pi x_2}{2b} + \frac{1}{3(a^2 + 9b^2)} \cos\frac{3\pi x_1}{2a} \cos\frac{\pi x_2}{2b} + \frac{1}{5(25a^2 + b^2)} \cos\frac{\pi x_1}{2a} \cos\frac{5\pi x_2}{2b} + \frac{1}{81(a^2 + b^2)} \cos\frac{3\pi x_1}{2a} \cos\frac{3\pi x_2}{2b} + \frac{1}{5(a^2 + 25b^2)} \cos\frac{5\pi x_1}{2a} \cos\frac{\pi x_2}{2b} \right] \tag{11.2.23}$$

With the aid of these approximate solutions, two mechanical characteristics of the torsional problem can be calculated (namely, the torque and the maximum tangential stress) as functions of α. These values for the approximate solutions can be obtained and compared with known exact solutions. From Ref. 26, the torque required to create a twist of α rad. per unit length of the rod is

$$T = 4G\alpha \int_{-b}^{b} \int_{-a}^{a} u \, dx_1 \, dx_2 \tag{11.2.24}$$

which can be represented in the form

$$T \equiv G\alpha\, (2a)^3\, (2b)\, k_1(\gamma) \tag{11.2.25}$$

where $\gamma = b/a$ and $k_1(\gamma)$ is a nondimensional factor. For $b > a$, the maximum tangential stress in the rectangular section occurs halfway along the side [26]. The greatest tangential stress τ [26] is

$$\tau = -2G\alpha \frac{\partial u}{\partial n} \equiv 2G\alpha a\, k(\gamma) \tag{11.2.26}$$

where $k(\gamma)$ is a nondimensional factor.

For numerical comparison with known solutions, denote the nondimensional approximations obtained above with the polynomial and Fourier coordinate functions as $(k_p^{(i)}, k_{1_p}^{(i)})$ and $(k_f^{(i)}, k_{1_f}^{(i)})$, respectively. For the Fourier approximation, numerical evaluation of the integral in Eq. 11.2.24, using the notation of Eq. 11.2.25, yields the approximations

$$k_{1_f}^{(1)}(\gamma) = \frac{256}{\pi^6}\left[\frac{\gamma^2}{1+\gamma^2}\right]$$

$$k_{1_f}^{(3)}(\gamma) = \frac{256\gamma^2}{\pi^6}\left[\frac{1}{1+\gamma^2} + \frac{1}{9(9+\gamma^2)} + \frac{1}{9(1+9\gamma^2)}\right]$$

$$k_{1_f}^{(6)}(\gamma) = \frac{256\gamma^2}{\pi^6}\left[\frac{1}{1+\gamma^2} + \frac{1}{9(9+\gamma^2)} + \frac{1}{9(1+9\gamma^2)} + \frac{1}{25(25+\gamma^2)} + \frac{1}{25(1+25\gamma^2)}\right]$$

Likewise, presuming $b \geq a$, the maximum shear stress occurs at $(a, 0)$. Evaluating Eq. 11.2.26 at this point, where $\partial/\partial n = \partial/\partial x_1$, yields the approximations

$$k_f^{(1)}(\gamma) = \frac{32}{\pi^3}\left[\frac{\gamma^2}{1+\gamma^2}\right]$$

$$k_f^{(3)}(\gamma) = \frac{32\gamma^2}{\pi^3}\left[\frac{1}{1+\gamma^2} - \frac{1}{3(\gamma^2+9)} + \frac{1}{1+9\gamma^2}\right]$$

$$k_f^{(6)}(\gamma) = \frac{32\gamma^2}{\pi^3}\left[\frac{1}{1+\gamma^2} - \frac{1}{3(\gamma^2+9)} + \frac{1}{1+9\gamma^2} + \frac{1}{5(\gamma^2+25)} + \frac{1}{1+25\gamma^2}\right]$$

Tables 11.2.1 and 11.2.2 provide a comparison of the exact values [26] of the quantities k_1 and k with their approximate values.

Table 11.2.1 Fourier Approximate Values of k_1

γ	k_1	$k_{1_f}^{(1)}$	$k_{1_f}^{(3)}$	$k_{1_f}^{(6)}$
1	0.1406	0.133	0.139	0.1401
2	0.229	0.213	0.225	0.228
3	0.263	0.240	0.258	0.261
4	0.281	0.251	0.273	0.278
5	0.291	0.256	0.281	0.287
∞	0.333	0.266	0.299	0.311

Table 11.2.2 Fourier Approximate Values of k

γ	k	$k_f^{(1)}$	$k_f^{(3)}$	$k_f^{(6)}$
1	0.675	0.516	0.585	0.613
2	0.930	0.826	0.831	0.870
3	0.985	0.929	0.870	0.931
4	0.997	0.971	0.865	0.954
5	0.999	0.992	0.854	0.961
∞	1.000	1.032	0.803	1.012

Tables 11.2.1 and 11.2.2 show that the approximate solutions give fairly good values of the torque, but poorer values for the maximum tangential stress. This applies particularly to $u_3(x_1, x_2)$. This is caused by the fact that the precision of the evaluation of the torque k_1 depends on the rapidity of convergence of the series in the mean, whereas the precision of the evaluation of k depends on the rapidity of uniform convergence of the series of derivatives. It is seen that the series converges in the mean fairly rapidly, but that the convergence of derivatives of this series is slow.

For comparison, values of $k_{1_p}^{(1)}(\gamma)$, $k_{1_p}^{(3)}(\gamma)$, $k_p^{(1)}(\gamma)$, and $k_p^{(3)}(\gamma)$, evaluated from the polynomial approximations, are given in Tables 11.2.3 and 11.2.4.

Table 11.2.3
Polynomial Approximate Values of k_1

γ	k_1	$k_{1_p}^{(1)}$	$k_{1_p}^{(3)}$
1	0.1406	0.139	0.1404
2	0.229	0.222	0.228
3	0.263	0.250	0.263
4	0.281	0.261	0.279
5	0.291	0.267	0.290
∞	0.333	0.278	0.311

Table 11.2.4
Polynomial Approximate Values of k

γ	k	$k_p^{(1)}$	$k_p^{(3)}$
1	0.675	0.625	0.703
2	0.930	1.000	0.951
3	0.985	1.125	0.982
4	0.997	1.176	0.969
5	0.999	1.202	0.951
∞	1.000	1.250	0.875

Comparison of Tables 11.2.1 through 11.2.4 shows that, for this application, an approximation of the same order with polynomials gives better results than with trigonometric functions for torque, and worse approximations for the maximum tangential stress. However, this circumstance may not be valid for approximations with greater numbers of terms.

11.3 HIGHER-ORDER EQUATIONS AND SYSTEMS OF PARTIAL DIFFERENTIAL EQUATIONS

The power of variational methods becomes even more clear for problems that are described by higher-order partial differential equations and systems of second-order partial differential equations. The basic operator properties of equations that govern such systems are the same as for the simpler ordinary and second-order partial differential operator equations studied in the preceding sections. More important, the Ritz method offers a practical method of solving such problems, with the aid of the modern high-speed digital computer. The examples presented in this section are intended to illustrate the power and generality of the Ritz method.

Bending of Thin Plates

The equation for **bending of a thin elastic plate** [24] may be written as

$$\nabla^4 u = \frac{f(x_1, x_2)}{D} \tag{11.3.1}$$

or in more detailed form,

$$\nabla^4 u \equiv \frac{\partial^4 u}{\partial x_1^4} + 2\frac{\partial^4 u}{\partial x_1^4 \partial x_2^4} + \frac{\partial^4 u}{\partial x_2^4} = \frac{f(x_1, x_2)}{D}$$

Here, $u(x_1, x_2)$ is the lateral deflection of the plate, $f(x_1, x_2)$ is the intensity of the normal load on the plate, and

$$D = \frac{Eh^3}{12(1 - \sigma^2)}$$

where E and $\sigma < 1$ are Young's modulus and Poisson's ratio for the material from which the plate is made and h is its thickness. The region covered by the plate in the (x_1, x_2) plane is denoted by Ω, with boundary Γ. Depending on the manner of support of the plate edge, the following boundary conditions are most frequently encountered [24]:

(a) The edge of the plate is clamped;

$$\begin{aligned} u\big|_\Gamma &= 0 \\ \frac{\partial u}{\partial n}\bigg|_\Gamma &= 0 \end{aligned} \qquad (11.3.2)$$

(b) The edge of the plate is simply supported;

$$\begin{aligned} u\big|_\Gamma &= 0 \\ \frac{\partial^2 u}{\partial n^2} + \sigma\left(\frac{\partial^2 u}{\partial \tau^2} + \frac{1}{\rho}\frac{\partial u}{\partial n}\right)\bigg|_\Gamma &= 0 \end{aligned} \qquad (11.3.3)$$

In Eqs. 11.3.2 and 11.3.3, n and τ are the exterior normal and tangent to Γ, respectively, and ρ is the radius of curvature of the boundary Γ. Different parts of the edge Γ may be fixed differently. Accordingly, Γ can be split into several parts, on each of which either Eq. 11.3.2 or Eq. 11.3.3 holds.

If at least one of Eq. 11.3.2 or 11.3.3 holds on each part of the boundary Γ, the energy functional is

$$\begin{aligned} F(u) &= (Au, u) - 2\left(u, \frac{f}{D}\right) \\ &= \iint_\Omega \left[\left(\frac{\partial^2 u}{\partial x_1^2}\right)^2 + 2\sigma\frac{\partial^2 u}{\partial x_1^2}\frac{\partial^2 u}{\partial x_2^2} + \left(\frac{\partial^2 u}{\partial x_2^2}\right)^2\right. \\ &\quad \left. + 2(1 - \sigma)\left(\frac{\partial^2 u}{\partial x_1 \partial x_2}\right)^2 - \frac{2f}{D}u\right] d\Omega \end{aligned} \qquad (11.3.4)$$

If the entire boundary of the plate is clamped, the functional of Eq. 11.3.4 can be represented in either of the following simpler forms:

$$F(u) = (Au, u) - 2\left(u, \frac{f}{D}\right)$$

$$= \iint_{\Omega}\left[(\nabla^2 u)^2 - \frac{2f}{D} u \right] d\Omega$$

$$= \iint_{\Omega}\left[\left(\frac{\partial^2 u}{\partial x_1^2}\right)^2 + 2\left(\frac{\partial^2 u}{\partial x_1 \partial x_2}\right)^2 + \left(\frac{\partial^2 u}{\partial x_2^2}\right)^2 - \frac{2f}{D} u \right] d\Omega \quad (11.3.5)$$

Clearly, for the case of a clamped boundary, $(Au, u) \geq 0$. If $(Au, u) = 0$, all the second derivatives of u are zero in Ω. In particular, $-\nabla^2 u = 0$. Since $u = 0$ on Γ, this implies $u = 0$ in Ω. Thus, the biharmonic operator with clamped boundary conditions is positive definite. While not shown for other boundary conditions, the same result holds. In fact, the operator is positive bounded below [18].

The foregoing results show that the problem of bending of a plate is equivalent to minimizing the energy functional over the set of functions in the class H_A that satisfy Eq. 11.3.2 on the clamped part of the boundary and the condition $u|_{\Gamma} = 0$ on the simply supported part of the boundary. The remaining boundary condition is natural and need not be satisfied by coordinate functions.

If the plate has variable thickness h(x), then Eq. 11.3.1 is replaced by

$$\frac{\partial^2}{\partial x_1^2}\left(h^3 \frac{\partial^2 u}{\partial x_1^2} \right) + \sigma \frac{\partial^2}{\partial x_2^2}\left(h^3 \frac{\partial^2 u}{\partial x_1^2} \right) + \sigma \frac{\partial^2}{\partial x_1^2}\left(h^3 \frac{\partial^2 u}{\partial x_2^2} \right)$$
$$+ \frac{\partial^2}{\partial x_2^2}\left(h^3 \frac{\partial^2 u}{\partial x_2^2} \right) + 2(1 - \sigma)\frac{\partial^2}{\partial x_1 \partial x_2}\left(h^3 \frac{\partial^2 u}{\partial x_1 \partial x_2} \right) = \frac{f}{D'} \quad (11.3.6)$$

where

$$D' = \frac{E}{12(1 - \sigma^2)}$$

The boundary conditions on the clamped or simply supported segments of the boundary have the same form as for a plate of constant thickness. If the thickness h of the plate is nonzero, the operator of Eq. 11.3.6 is positive definite on the set of functions that satisfy Eqs. 11.3.2 and 11.3.3 on the clamped and simply supported parts of the boundary, respectively. The corresponding variational problem consists of minimizing the integral

$$F(u) = \iint_\Omega \left\{ h^3 \left[\left(\frac{\partial^2 u}{\partial x_1^2} \right)^2 + 2\sigma \frac{\partial^2 u}{\partial x_1^2} \frac{\partial^2 u}{\partial x_2^2} + \left(\frac{\partial^2 u}{\partial x_2^2} \right)^2 \right. \right.$$

$$\left. \left. + 2(1 - \sigma) \left(\frac{\partial^2 u}{\partial x_1 \partial x_2} \right)^2 \right] - \frac{2f}{D'} u \right\} d\Omega \qquad (11.3.7)$$

subject to the conditions of Eq. 11.3.2 on the clamped part of the boundary and the condition $u|_\Gamma = 0$ on the simply supported part of the boundary.

If the thickness of the plate is constant, the frequency of **vibration of a thin elastic plate** is proportional to the eigenvalues of the **biharmonic operator**; i.e.,

$$\nabla^4 u \equiv \frac{\partial^4 u}{\partial x_1^4} + 2 \frac{\partial^4 u}{\partial x_1^2 \partial x_2^2} + \frac{\partial^4 u}{\partial x_2^4} = \lambda h u$$

subject to the boundary conditions of Eq. 11.3.2 or 11.3.3. For the general case, the boundary of the plate is split into segments Γ_1 and Γ_2, on which Eqs. 11.3.2 and 11.3.3, respectively, are satisfied.

The smallest eigenvalue of the biharmonic operator is the minimum value of the functional

$$(Au, u) = \iint_\Omega \left\{ \left(\frac{\partial^2 u}{\partial x_1^2} \right)^2 + 2\sigma \frac{\partial^2 u}{\partial x_1^2} \frac{\partial^2 u}{\partial x_2^2} + \left(\frac{\partial^2 u}{\partial x_2^2} \right)^2 \right.$$

$$\left. + 2(1 - \sigma) \left(\frac{\partial^2 u}{\partial x_1 \partial x_2} \right)^2 \right\} d\Omega \qquad (11.3.8)$$

the minimum taken over the set of functions that satisfy

$$(Bu, u) = \iint_\Omega h u^2 \, d\Omega = 1 \qquad (11.3.9)$$

and the boundary conditions

$$u \Big|_{\Gamma_1 + \Gamma_2} = 0$$
$$\frac{\partial u}{\partial n} \Big|_{\Gamma_1} = 0 \qquad (11.3.10)$$

If the entire boundary of the plate is clamped, the smallest eigenvalue of the biharmonic operator is equal to the minimum of the functional

$$\iint_{\Omega} (\nabla^2 u)^2 \, dx_1 \, dx_2$$

where the function u(**x**) satisfies Eq. 11.3.9 and the boundary conditions $u|_\Gamma = 0$ and $(\partial u/\partial n)|_\Gamma = 0$.

The nth eigenvalue λ_n of the biharmonic operator is related to the nth natural frequency ω_n of the plate by the expression [24]

$$\lambda_n = \frac{\gamma \omega_n^2}{D}$$

where γ is the density of the plate and D is the resistance of the plate to bending.

If the thickness h of the plate varies, the equation for natural vibration of the plate takes the form

$$\frac{\partial^2}{\partial x_1^2}\left(h^3 \frac{\partial^2 u}{\partial x_1^2} \right) + \sigma \frac{\partial^2}{\partial x_2^2}\left(h^3 \frac{\partial^2 u}{\partial x_1^2} \right) + \sigma \frac{\partial^2}{\partial x_1^2}\left(h^3 \frac{\partial^2 u}{\partial x_2^2} \right)$$
$$+ \frac{\partial^2}{\partial x_2^2}\left(h^3 \frac{\partial^2 u}{\partial x_2^2} \right) + 2(1-\sigma)\frac{\partial^2}{\partial x_1 \partial x_2}\left(h^3 \frac{\partial^2 u}{\partial x_1 \partial x_2} \right) = \lambda h u \qquad (11.3.11)$$

where the parameter λ is related to the natural frequency ω by the expression [24]

$$\lambda = \frac{12(1-\sigma^2)\gamma}{E}\omega^2$$

The smallest eigenvalue λ_1 is the minimum of the functional

$$\iint_{\Omega} h^3 \left[\left(\frac{\partial^2 u}{\partial x_1^2} \right)^2 + 2\sigma \frac{\partial^2 u}{\partial x_1^2}\frac{\partial^2 u}{\partial x_2^2} + \left(\frac{\partial^2 u}{\partial x_2^2} \right)^2 \right.$$
$$\left. + 2(1-\sigma)\left(\frac{\partial^2 u}{\partial x_1 \partial x_2} \right)^2 \right] d\Omega \qquad (11.3.12)$$

subject to the conditions of Eqs. 11.3.9 and 11.3.10.

Bending of a Clamped Rectangular Plate

Lateral displacement u of a plate satisfies the biharmonic equation

$$\nabla^4 u = \frac{f}{D} = p \qquad (11.3.13)$$

where f is the load intensity and D is the rigidity of the plate, subject to the boundary conditions

$$u\Big|_{\Gamma} = 0 \qquad \frac{\partial u}{\partial n}\Big|_{\Gamma} = 0 \tag{11.3.14}$$

The lengths of the sides of the plate are 2a and 2b, where the coordinate axes are parallel to the sides of the plate and the origin of coordinates is at the center of the plate.

As was shown in Eq. 11.3.5, this problem reduces to finding the minimum of the functional

$$F(u) = \int_{-a}^{a}\int_{-b}^{b} \{ (\nabla^2 u)^2 - 2pu \} \, dx_2 \, dx_1 \tag{11.3.15}$$

over the space of functions that satisfy Eq. 11.3.14. This problem can be solved by the Ritz method. In order to simplify calculations, suppose that the load is distributed uniformly; i.e., p = constant.

As coordinate functions, consider polynomials of the form

$$(x_1^2 - a^2)^2 (x_2^2 - b^2)^2 (a_1 + a_2 x_1^2 + a_3 x_2^2 + \dots) \tag{11.3.16}$$

Odd powers of x_1 and x_2 are omitted because the solution $u(x_1, x_2)$ must be symmetric about the coordinate axes.

Restricting attention to three terms in Eq. 11.3.16,

$$\phi_1 = (x_1^2 - a^2)^2 (x_2^2 - b^2)^2, \quad \phi_2 = x_1^2 \phi_1, \quad \phi_3 = x_2^2 \phi_1$$

the three-term Ritz approximation is

$$u \approx u_3 = a_1\phi_1 + a_2\phi_2 + a_3\phi_3$$

The Ritz equations in this case have the form

$$\sum_{k=1}^{3} (\nabla^2\phi_k, \nabla^2\phi_m) \, a_k = p(1, \phi_m), \quad m = 1, 2, 3 \tag{11.3.17}$$

Performing the necessary calculations, Eq. 11.3.17 becomes [21]

$$\left(\gamma^2 + \frac{1}{\gamma^2} + \frac{4}{7}\right) a_1 + \left(\frac{1}{7} + \frac{1}{11\gamma^4}\right) b^2 a_2 + \left(\frac{1}{7} + \frac{\gamma^4}{11}\right) a^2 a_3 = \frac{7}{128a^2b^2} p$$

$$\left(\frac{\gamma^2}{7} + \frac{1}{11\gamma^2}\right) a_1 + \left(\frac{3}{7} + \frac{4}{143\gamma^4} + \frac{4}{77\gamma^2}\right) b^2 a_2 + \frac{1}{77}(1 + \gamma^4) a^2 a_3 = \frac{1}{128a^2b^2} p$$

$$\left(\frac{1}{7\gamma^2} + \frac{\gamma^2}{11}\right) a_1 + \frac{1}{77}\left(1 + \frac{1}{\gamma^4}\right) b^2 a_2 + \left(\frac{3}{7} + \frac{3\gamma^4}{143} + \frac{4\gamma^2}{77}\right) a^2 a_3 = \frac{1}{128a^2b^2} p$$

where $\gamma = b/a$. If the plate is square, then $\gamma = 1$ and these equations assume the simpler form

$$\frac{18}{7} a_1 + \frac{18}{77} a^2 a_2 + \frac{18}{77} a^2 a_3 = \frac{7}{128a^4} p$$

$$\frac{18}{77} a_1 + \frac{502}{1001} a^2 a_2 + \frac{2}{77} a^2 a_3 = \frac{1}{128a^4} p$$

$$\frac{18}{77} a_1 + \frac{2}{77} a^2 a_2 + \frac{502}{1001} a^2 a_3 = \frac{1}{128a^4} p$$

Solving this system,

$$u \approx u_3 = \frac{p}{a^4}(x_1^2 - a^2)^2 (x_2^2 - a^2)^2 \left(0.02067 + 0.0038 \frac{x_1^2 + x_2^2}{a^2}\right)$$

The approximate deflection at the center of the plate is thus

$$u_3 \bigg|_{x_1=0, x_2=0} = 0.02067 pa^4$$

If the approximation for u were restricted to one term; i.e.,

$$u \approx u_1 = a_1\phi_1 = a_1 (x_1^2 - a^2)^2 (x_2^2 - b^2)^2$$

then the Ritz equations reduce to the single equation

$$\left(\gamma^2 + \frac{1}{\gamma^2} + \frac{4}{7} \right) a_1 = \frac{7}{128a^2b^2}\, p$$

so

$$a_1 = \frac{49}{128\,(7a^4 + 7b^4 + 4a^2b^2)}\, p$$

which gives the value

$$u_1 \Big|_{x_1=0, x_2=0} = \frac{49a^4b^4}{128\,(7a^4 + 7b^4 + 4a^2b^2)}$$

for the deflection at the center of the plate. In the case of a square plate (a = b),

$$u_1 \Big|_{x_1=0, x_2=0} = 0.02127pa^4$$

Note that u_1 and u_3 give similar values at the center of the plate, suggesting that u_3 is a good approximation of the solution.

Computer Solution of a Clamped Plate Problem

To illustrate the effect of the number of coordinate functions used in solving problems with the Ritz technique, the clamped plate with both constant thickness and variable thickness is solved for displacement due to a uniformly distributed lateral load f and for the fundamental natural frequency and the associated eigenfunction. In these problems, static deflection is governed by the differential equation of Eq. 11.3.6 and the boundary conditions of Eq. 11.3.10. For vibration, the deflection satisfies Eq. 11.3.11 and the same boundary conditions. Since all boundary conditions in Eq. 11.3.10 are principal, the coordinate functions must satisfy them.

Since the problem is symmetric, the following even trigonometric coordinate functions are selected:

$$\begin{aligned} u_M &= \sum_{m=1}^{M} \sum_{n=1}^{M} a_{mn} \cos \frac{(2m-1)\pi x_1}{a} \cos \frac{\pi x_1}{a} \cos \frac{(2n-1)\pi x_2}{b} \cos \frac{\pi x_2}{b} \\ &= \frac{1}{4} \sum_{m=1}^{M} \sum_{n=1}^{M} a_{mn} \left(\cos \frac{2m\pi x_1}{a} + \cos \frac{(2m-2)\pi x_1}{a} \right) \\ &\qquad \times \left(\cos \frac{2n\pi x_2}{b} + \cos \frac{(2n-2)\pi x_2}{b} \right) \end{aligned} \qquad (11.3.18)$$

The Ritz equations for this formulation are

$$
\begin{aligned}
&\frac{1}{2}\sum_{m=1}^{M}\sum_{n=1}^{M} a_{mn}\int_{-b}^{b}\int_{-a}^{a} D\Bigg\{\Bigg[\Bigg[\left\{\left(\frac{2m\pi}{a}\right)^2\cos\frac{2m\pi x_1}{a}+\left(\frac{(2m-2)\pi}{a}\right)^2\cos\frac{(2m-2)\pi x_1}{a}\right\}\\
&\quad\times\left\{\cos\frac{2n\pi x_2}{b}+\cos\frac{(2n-2)\pi x_2}{b}\right\}+\nu\left\{\cos\frac{2m\pi x_1}{a}+\cos\frac{(2m-2)\pi x_1}{a}\right\}\\
&\quad\times\left\{\left(\frac{2m\pi}{b}\right)^2\cos\frac{2m\pi x_2}{b}+\left(\frac{(2m-2)\pi}{b}\right)^2\cos\frac{(2m-2)\pi x_2}{b}\right\}\\
&\quad+\left\{\left(\frac{2k\pi}{a}\right)^2\cos\frac{2k\pi x_1}{a}+\left(\frac{(2k-2)\pi}{a}\right)^2\cos\frac{(2k-2)\pi x_1}{a}\right\}\\
&\quad\times\left\{\cos\frac{2\ell\pi x_2}{b}+\cos\frac{(2\ell-2)\pi x_2}{b}\right\}\Bigg]+\Bigg[\left(\cos\frac{2m\pi x_1}{a}+\cos\frac{(2m-2)\pi x_1}{a}\right)\\
&\quad\times\left\{\left(\frac{2m\pi}{b}\right)^2\cos\frac{2m\pi x_2}{b}+\left(\frac{(2m-2)\pi}{b}\right)^2\cos\frac{(2m-2)\pi x_2}{b}\right\}\\
&\quad+\nu\left\{\left(\frac{2m\pi}{a}\right)^2\cos\frac{2m\pi x_1}{a}+\left(\frac{(2m-2)\pi}{a}\right)^2\cos\frac{(2m-2)\pi x_1}{a}\right\}\\
&\quad\times\left(\cos\frac{2m\pi x_2}{b}+\cos\frac{(2m-2)\pi x_2}{b}\right)\left(\cos\frac{2k\pi x_1}{a}+\cos\frac{(2k-2)\pi x_1}{a}\right)\\
&\quad\times\left\{\left(\frac{2\ell\pi}{b}\right)^2\cos\frac{2\ell\pi x_2}{b}+\left(\frac{(2\ell-2)\pi}{b}\right)^2\cos\frac{(2\ell-2)\pi x_2}{b}\right\}\\
&\quad+2(1-\nu)\left(\frac{2m\pi}{a}\sin\frac{2m\pi x_1}{a}+\frac{(2m-2)\pi}{a}\sin\frac{(2m-2)\pi x_1}{a}\right)\\
&\quad\times\left(\frac{2m\pi}{b}\sin\frac{2m\pi x_2}{b}+\frac{(2m-2)\pi}{b}\sin\frac{(2m-2)\pi x_2}{b}\right)\\
&\quad\times\left(\frac{2k\pi}{a}\sin\frac{2k\pi x_1}{a}+\frac{(2k-2)\pi}{a}\sin\frac{(2k-2)\pi x_1}{a}\right)\\
&\quad\times\left(\frac{2\ell\pi}{b}\sin\frac{2\ell\pi x_2}{b}+\frac{(2\ell-2)\pi}{b}\sin\frac{(2\ell-2)\pi x_2}{b}\right)\Bigg]\Bigg\}\,dx_1\,dx_2\\
&=\frac{1}{2}\int_{-b}^{b}\int_{-a}^{a} f\left(\cos\frac{2k\pi x_1}{a}+\cos\frac{(2k-2)\pi x_2}{a}\right)\left(\cos\frac{2\ell\pi x_2}{b}+\cos\frac{(2\ell-2)\pi x_2}{b}\right)dx_1\,dx_2,\\
&\qquad k=1,2,\ldots,M,\quad \ell=1,2,\ldots,M \qquad (11.3.19)
\end{aligned}
$$

These equations were programmed and solved for a square plate (b = a), with 9 and 16 coordinate functions; i.e., M = 3 and M = 4, respectively. Results for deflection and bending moment M_{x_1} at the points of interest shown in Fig. 11.3.1 and for the smallest eigenvalue for a square plate of constant thickness are given in Table 11.3.1. The Ritz eigenvalue equations were also solved for the eigenvalue $\zeta = \rho\omega^2$, where ρ is the mass density of material and ω is the natural frequency, in rad/sec. Results are presented in Table 11.3.1. Note that the polynomial 3-term approximation for deflection was 0.0206 (fa^4/D), whereas the 16-term approximation here gives 0.0201 (fa^4/D). Note also that while accurate displacement and eigenvalues are obtained, since bending moment involves two derivatives of displacement [26] (see also Eq. 11.4.11), it is less accurate.

Table 11.3.1 Results for Clamped Plate (Constant Thickness)

	M_{x_1} @ A $\times 4fa^2$	M_{x_1} @ B $\times 4fa^2$	u @ A $\times \frac{16fa^4}{D}$	$\zeta = \rho\omega^2 \times \frac{1}{4a^2}\sqrt{\frac{D}{\rho h}}$
M = 3	0.025092	– 0.038922	0.0012626	36.106
M = 4	0.021518	– 0.041473	0.0012586	36.044
Solution [26]	0.0231	– 0.0513	0.00126	35.98

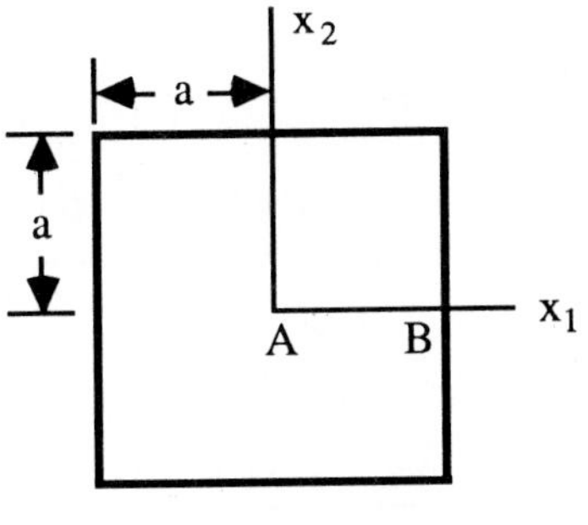

Figure 11.3.1 Points of Interest on Plate

The same problem was solved for a clamped plate of variable thickness, with

$$h(x) = 2\left(1 - \frac{|x_1|}{a}\right)\left(1 - \frac{|x_2|}{a}\right)h_0$$

The same coordinate functions, with 4, 9, and 16 terms (corresponding to M = 2, 3, and 4, respectively) were used to solve this problem. Numerical results are given in Table 11.3.2.

Table 11.3.2 Results for Clamped Plate (Varying Thickness)

	M_{x_1} @ A $\times 4fa^2$	M_{x_1} @ B $\times 4fa^2$	u @ A $\times \dfrac{16fa^4}{D^*}$	$\zeta = \rho\omega^2$ $\times \dfrac{1}{a^2}\sqrt{\dfrac{D^*}{\rho h_0}}$
M = 2	0.029754	– 0.0062549	0.00060060	40.706
M = 3	0.063371	– 0.0085592	0.00065468	39.376
M = 4	0.040864	– 0.010112	0.00067205	38.774

$$* \ D = \frac{Eh_0^3}{12(1-\nu)}$$

11.4 SOLUTION OF BEAM AND PLATE PROBLEMS BY THE GALERKIN METHOD

To illustrate use of the Galerkin method, problems of the type treated in Section 11.3 for beams and plates are solved numerically using the Galerkin method.

Static Analysis of Beams

The governing equations of a beam may be written in the second-order form

$$A\begin{bmatrix} u \\ M \end{bmatrix} \equiv \begin{bmatrix} -\dfrac{d^2M}{dx^2} \\ -\dfrac{d^2u}{dx^2} - \dfrac{M}{EI} \end{bmatrix} = \begin{bmatrix} f(x) \\ 0 \end{bmatrix} \tag{11.4.1}$$

or in the more usual fourth-order form

$$\frac{d^2}{dx^2}\left(EI\,\frac{d^2u}{dx^2} \right) = f(x) \tag{11.4.2}$$

where u is deflection, M is moment, E is Young's modulus, I is moment of inertia of the beam cross section, and f(x) is distributed load, taken as constant in the examples treated here.

The second-order form of Eq. 11.4.1 has the advantage that both displacement u and moment M are determined with comparable accuracy. This feature allows direct computation of bending stress in the beam, without having to compute derivatives of an approximate numerical solution u(x), as was done in the case of plates in Tables 11.3.1 and 11.3.2.

As has been shown, under the usual boundary conditions, the fourth-order operator of Eq. 11.4.2 is symmetric and positive definite. For the second-order operator of Eq. 11.4.1, under the usual boundary conditions,

$$\begin{aligned}\int_{-\ell/2}^{\ell/2} \begin{bmatrix} \bar{u} \\ \bar{M} \end{bmatrix}^T A \begin{bmatrix} u \\ M \end{bmatrix} dx &= \int_{-\ell/2}^{\ell/2} \left(-\bar{u}M'' - \bar{M}u'' - \frac{\bar{M}M}{EI} \right) dx \\ &= \int_{-\ell/2}^{\ell/2} \left(-u\bar{M}'' - M\bar{u}'' - \frac{M\bar{M}}{EI} \right) dx \\ &= \int_{-\ell/2}^{\ell/2} \begin{bmatrix} u \\ M \end{bmatrix}^T A \begin{bmatrix} \bar{u} \\ \bar{M} \end{bmatrix} dx\end{aligned}$$

so the operator A is symmetric. However,

$$\int_{-\ell/2}^{\ell/2} \begin{bmatrix} u \\ M \end{bmatrix}^T A \begin{bmatrix} \bar{u} \\ \bar{M} \end{bmatrix} dx = \int_{-\ell/2}^{\ell/2} \left[2u'M' - \frac{M^2}{EI} \right] dx$$

may be made negative by choosing admissible u' and M' that are orthogonal, so the first term vanishes and a negative integral results. Thus, the second-order operator A is not positive definite and the Ritz method cannot be used directly for this formulation.

Simply Supported Beam

Two simply supported beam problems are solved using the Galerkin method. One has a uniform cross section and the other has a varying cross section, with area

$$a(x) = 2a_0\left(1 - \frac{|x|}{\ell} \right), \quad -\frac{\ell}{2} < x < \frac{\ell}{2}$$

and moment of inertia $I(x) = \alpha[a(x)]^2$, which is symmetric with respect to the center of the span. The boundary conditions are

$$u\left(\pm\frac{\ell}{2}\right) = M\left(\pm\frac{\ell}{2}\right) = 0 \tag{11.4.3}$$

Under symmetric loading, the solution will be symmetric, so cosine functions are selected as coordinate functions. This leads to the approximate solutions

$$u = \sum_{n=1}^{N} a_n \cos \frac{(2n-1)\pi x}{\ell}$$
$$M = \sum_{m=1}^{M} b_m \cos \frac{(2m-1)\pi x}{\ell} \tag{11.4.4}$$

Substituting these sums into the Galerkin equations of Eq. 10.10.13 and performing the necessary calculations,

$$\int_{-\ell/2}^{\ell/2} \begin{bmatrix} \bar{u} \\ \bar{M} \end{bmatrix}^T A \begin{bmatrix} u \\ M \end{bmatrix} dx$$
$$= \sum_{m=1}^{M} b_m \left(\frac{(2m-1)\pi}{\ell} \right)^2$$
$$\times \int_{-\ell/2}^{\ell/2} \left[\cos \frac{(2m-1)\pi x}{\ell} \cos \frac{(2n-1)\pi x}{\ell} \right] dx$$
$$= \int_{-\ell/2}^{\ell/2} f \left[\cos \frac{(2n-1)\pi x}{\ell} \right] dx = 0, \quad n = 1, 2, \ldots, N \tag{11.4.5}$$

where $[\,\bar{u}, \bar{M}\,]^T = \left[\cos \frac{(2n-1)\pi x}{\ell}, 0 \right]^T$, and

$$\int_{-\ell/2}^{\ell/2} \begin{bmatrix} \bar{u} \\ \bar{M} \end{bmatrix}^T A \begin{bmatrix} u \\ M \end{bmatrix} dx$$
$$= \sum_{n=1}^{N} a_n \left(\frac{(2n-1)\pi}{\ell} \right)^2$$
$$\times \int_{-\ell/2}^{\ell/2} \cos \frac{(2n-1)\pi x}{\ell} \cos \frac{(2m-1)\pi x}{\ell} dx$$
$$- \sum_{j=1}^{M} b_j \int_{-\ell/2}^{\ell/2} \left[\frac{1}{EI} \cos \frac{(2j-1)\pi x}{\ell} \cos \frac{(2m-1)\pi x}{\ell} \right] dx = 0,$$
$$m = 1, 2, \ldots, M \tag{11.4.6}$$

where $[\,\bar{u},\ \bar{M}\,]^T = \left[\ 0,\ \cos\dfrac{(2m-1)\pi x}{\ell}\ \right]^T$.

Numerical results for the uniform beam are presented in the first two columns of Table 11.4.1, for six and ten coordinate functions. Similar results for a beam with variable cross section are given in Table 11.4.2. As expected, results for deflections are very accurate. While there is some error in the calculation of bending moment, it is much smaller than would be experienced if numerically determined displacements were differentiated twice.

Table 11.4.1
Results for Simply Supported Beam
(Uniform Cross Section, Second-order Method)

EI_0 = bending rigidity of beam
a_0 = cross-sectional area of beam
ρ = mass density per unit volume

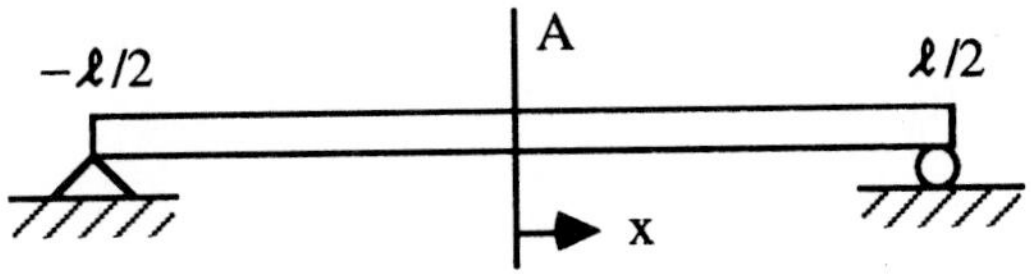

	Moment @ A $\times f\ell^2$	u @ A $\times \dfrac{f\ell^4}{EI_0}$	$\zeta = \rho\omega^2$ $\times \dfrac{1}{\ell^2}\sqrt{\dfrac{EI_0}{\rho a_0}}$
N = 3 M = 3	0.12526	0.013021	9.86960
N = 5 M = 5	0.12506	0.013021	9.86960
EXACT	0.12500	0.013021	9.86960

Table 11.4.2
Results for Simply Supported Beam
(Varying Cross Section, Second-order Method)

$$a(x) = 2a_0\left(1 - \frac{|x|}{\ell}\right)$$

$$I_0 = \alpha a_0^2$$

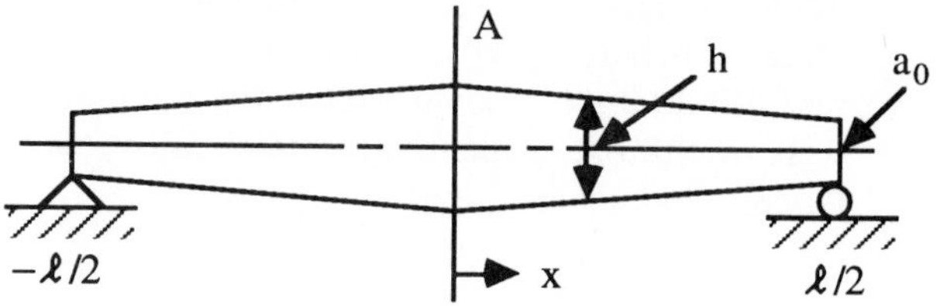

	Moment @ A $\times f\ell^2$	u @ A $\times \dfrac{f\ell^4}{EI_0}$	$\zeta = \rho\omega^2$ $\times \dfrac{1}{\ell^2}\sqrt{\dfrac{EI_0}{\rho a_0}}$
N = 3, M = 3	0.12526	0.0046279	12.598
N = 5, M = 5	0.12506	0.0046247	12.599

Clamped Beam

Clamped beam problems are now solved, using both the second- and fourth-order formulations. The boundary conditions for this case are

$$u\left(\pm\frac{\ell}{2}\right) = \frac{du}{dx}\left(\pm\frac{\ell}{2}\right) = 0 \tag{11.4.7}$$

The following coordinate function approximation, which satisfies these principal boundary conditions, is selected:

$$\begin{aligned} u &= \sum_{n=1}^{N} a_n \cos\frac{(2n-1)\pi x}{\ell}\cos\frac{\pi x}{\ell} \\ &= \sum_{n=1}^{N} a_n\left[\cos\frac{2n\pi x}{\ell} + \cos\frac{(2n-2)\pi x}{\ell}\right] \\ M &= \sum_{m=1}^{M} b_m \cos\frac{(m-1)\pi x}{\ell} \end{aligned} \tag{11.4.8}$$

where u and M are represented in symmetric form.

Substituting the above into the Galerkin equations of Eq. 10.10.13 and carrying out the necessary computation yield

$$\int_{-\ell/2}^{\ell/2} \begin{bmatrix} \bar{u} \\ \bar{M} \end{bmatrix}^T A \begin{bmatrix} u \\ M \end{bmatrix} dx$$

$$= \sum_{m=0}^{M} b_m \int_{-\ell/2}^{\ell/2} \left[\left(\frac{(m-1)\pi}{\ell} \right)^2 \cos \frac{2n\pi x}{\ell} + \left(\frac{(m-1)\pi}{\ell} \right)^2 \cos \frac{(2n-2)\pi x}{\ell} \right] \cos \frac{(m-1)\pi x}{\ell} dx$$

$$= \int_{-\ell/2}^{\ell/2} f \left[\cos \frac{2n\pi x}{\ell} + \cos \frac{(2n-2)\pi x}{\ell} \right] dx, \quad n = 1, \ldots, N \tag{11.4.9}$$

where $[\bar{u}, \bar{M}]^T = [\cos 2n\pi x/\ell + \cos\{(2n-2)\pi x/\ell\}, 0]^T$, and

$$\int_{-\ell/2}^{\ell/2} \begin{bmatrix} \bar{u} \\ \bar{M} \end{bmatrix}^T A \begin{bmatrix} u \\ M \end{bmatrix} dx$$

$$= \sum_{n=1}^{N} a_n \int_{-\ell/2}^{\ell/2} \left[\left(\frac{2n\pi}{\ell} \right)^2 \cos \frac{2n\pi x}{\ell} + \left(\frac{(2n-2)\pi}{\ell} \right)^2 \cos \frac{(2n-2)\pi x}{\ell} \right] \cos \frac{(m-1)\pi x}{\ell} dx$$

$$- \sum_{j=1}^{M} b_j \int_{-\ell/2}^{\ell/2} \frac{2}{EI} \cos \frac{(j-1)\pi x}{\ell} \cos \frac{(m-1)\pi x}{\ell} dx = 0,$$

$$m = 1, \ldots, M \tag{11.4.10}$$

where $[\bar{u}, \bar{M}]^T = [0, \cos\{(m-1)\pi x/\ell\}]^T$.

Inserting the first of Eq. 11.4.8 into the Galerkin equations of Eq. 10.10.13 for the fourth-order operator and performing the necessary calculations yield

$$\sum_{n=0}^{N} a_n \int_{-\ell/2}^{\ell/2} \frac{1}{2} EI \left\{ \left(\frac{2n\pi}{\ell} \right)^2 \cos \frac{2n\pi x}{\ell} + \left(\frac{(2n-2)\pi}{\ell} \right)^2 \cos \frac{(2n-2)\pi x}{\ell} \right\}$$

$$\times \left\{ \left(\frac{2k\pi}{\ell} \right)^2 \cos \frac{2k\pi x}{\ell} \left(\frac{(2k-2)\pi}{\ell} \right)^2 \cos \frac{(2k-2)\pi x}{\ell} \right\} dx$$

$$= \int_{-\ell/2}^{\ell/2} f \left[\cos \frac{2k\pi x}{\ell} + \cos \frac{(2k-2)\pi x}{\ell} \right] dx,$$

$$k = 1, \ldots, N \tag{11.4.11}$$

The second-order approximate solution of Eqs. 11.4.9 and 11.4.10 and the fourth-order approximate solution of Eq. 11.4.11 were solved numerically for uniform and tapered clamped beams. Results for varying numbers of coordinate functions are given in Tables 11.4.3 and 11.4.4. Note that displacement results are accurate with both second- and fourth-order formulations, slightly more accurate with the fourth-order formulation. Bending moments, in contrast, are much more accurate with the second-order formulation. This is due partly to error induced in differentiating numerically computed displacements to obtain bending moments in the fourth-order formulation.

Table 11.4.3
Results for Clamped Beam (Uniform Cross Section)

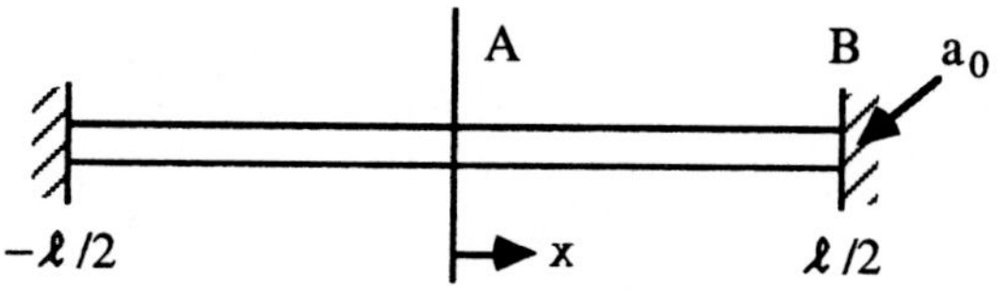

		Moment @ A $\times f\ell^2$	Moment @ B $\times f\ell^2$	u @ A $\times \frac{f\ell^4}{EI_0}$	$\zeta = \rho\omega^2$ $\times \frac{1}{\ell^2}\sqrt{\frac{EI_0}{\rho a_0}}$
2nd Order	N = 3, M = 4	0.041676	− 0.083179	0.0026213	22.374
	N = 5, M = 6	0.041668	− 0.083404	0.0026131	22.373
4th Order	N = 5	0.042485	− 0.074147	0.0026023	22.383
	N = 10	0.041185	− 0.078765	0.0026023	22.374
Exact		0.041667	− 0.083333	0.0026042	22.373

Table 11.4.4
Results for Clamped Beam (Varying Cross Section)

$$a(x) = 2a_0\left(1 - \frac{|x|}{\ell}\right)$$

$$I_0 = \alpha a_0^2$$

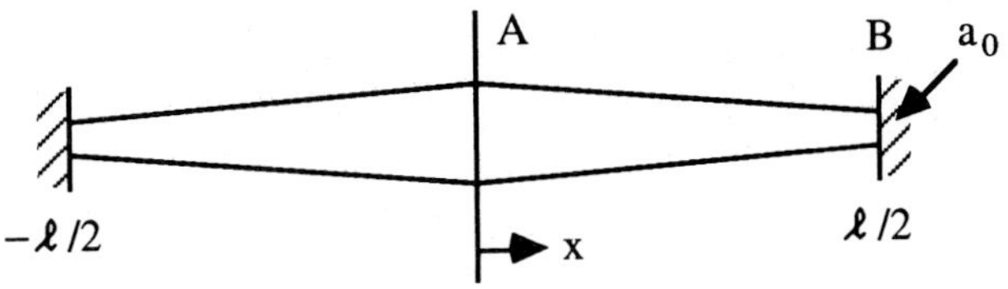

		Moment @ A $\times f\ell^2$	Moment @ B $\times f\ell^2$	u @ A $\times \dfrac{f\ell^4}{EI_0}$	$\zeta = \rho\omega^2 \times \dfrac{1}{\ell^2}\sqrt{\dfrac{EI_0}{\rho a_0}}$
2nd Order	N = 3, M = 4	0.056862	− 0.067993	0.0013586	23.410
2nd Order	N = 5, M = 6	0.056855	− 0.068216	0.0013476	23.413
4th Order	N = 5	0.061550	− 0.053045	0.0013252	23.487
4th Order	N = 10	0.056509	− 0.060293	0.0013320	23.428

Static Analysis of Plates

The governing equations for bending of a plate [24] can be written in the second-order form

$$A\begin{bmatrix} M_{x_1} \\ M_{x_2} \\ M_{x_1x_2} \\ u \end{bmatrix} \equiv \begin{bmatrix} -\dfrac{\partial^2 u}{\partial x_1^2} - \dfrac{12}{Eh^3}M_{x_1} + \nu\dfrac{12}{Eh^3}M_{x_2} \\ -\dfrac{\partial^2 u}{\partial x_2^2} - \dfrac{12}{Eh^3}M_{x_2} + \nu\dfrac{12}{Eh^3}M_{x_1} \\ 2\dfrac{\partial^2 u}{\partial x_1 \partial x_2} - \dfrac{24(1+\nu)}{Eh^3}M_{x_1x_2} \\ -\dfrac{\partial^2 M_{x_1}}{\partial x_1^2} - \dfrac{\partial^2 M_{x_2}}{\partial x_2^2} + 2\dfrac{\partial^2 M_{x_1x_2}}{\partial x_1 \partial x_2} \end{bmatrix} = \begin{bmatrix} 0 \\ 0 \\ 0 \\ f(x_1, x_2) \end{bmatrix} \qquad (11.4.12)$$

where M_{x_1} and M_{x_2} are bending moments on cross sections normal to the x_1 and x_2 axes, $M_{x_1x_2}$ is a torsional moment, u is deflection, h is thickness of the plate, ν is Poisson's ratio, E is Young's modulus, $D = Eh^3/[12(1-\nu)]$, and f is the distributed applied load. Equation 11.4.12 applies to plates with varying thickness. The applied load f is taken as uniformly distributed in the examples treated here.

Simply Supported Plate

Simply supported plates with uniform and varying thickness are solved using the second-order equations of Eq. 11.4.12. The boundary conditions are

$$\begin{aligned} u(\pm a, x_2) &= u(x_1, \pm b) = 0 \\ M_{x_1}(\pm a, x_2) &= M_{x_2}(x_1, \pm b) = 0 \end{aligned} \tag{11.4.13}$$

The solution is approximated by the following coordinate function expansions:

$$\begin{aligned} M_{x_1} &= \sum_{m=1}^{M}\sum_{n=1}^{N} a_{mn} \cos\frac{(2m-1)\pi x_1}{2a} \cos\frac{(2n-1)\pi x_2}{2b} \\ M_{x_2} &= \sum_{m=1}^{M}\sum_{n=1}^{N} b_{mn} \cos\frac{(2m-1)\pi x_1}{2a} \cos\frac{(2n-1)\pi x_2}{2b} \\ M_{x_1x_2} &= \sum_{m=1}^{M}\sum_{n=1}^{N} c_{mn} \sin\frac{m\pi x_1}{2a} \sin\frac{n\pi x_2}{2b} \\ u &= \sum_{m=1}^{M}\sum_{n=1}^{N} d_{mn} \cos\frac{(2m-1)\pi x_1}{2a} \cos\frac{(2n-1)\pi x_2}{2b} \end{aligned} \tag{11.4.14}$$

where the coordinate functions are chosen so that M_{x_1}, M_{x_2}, and u satisfy the principal boundary conditions. Note that since the moment $M_{x_1x_2}$ is antisymmetric, its coordinate functions are selected in antisymmetric form.

Substituting Eq. 11.4.14 into the Galerkin equations of Eq. 10.10.13 and carrying out necessary computations,

$$\int_\Omega [\bar{M}_{x_1}, \bar{M}_{x_2}, \bar{M}_{x_1x_2}, \bar{u}]\, A\, [M_{x_1}, M_{x_2}, M_{x_1x_2}, u]^T d\Omega$$

$$= \int_\Omega \sum_{m=1}^{M}\sum_{n=1}^{M} \left\{ d_{mn}\left(\frac{(2m-1)\pi}{2a}\right)^2 - \frac{12}{Eh^3} a_{mn} + \nu \frac{12}{Eh^3} b_{mn} \right\}$$

$$\times \cos\frac{(2m-1)\pi x_1}{2a} \cos\frac{(2n-1)\pi x_2}{2b} \cos\frac{(2\alpha-1)\pi x_1}{2a} \cos\frac{(2\beta-1)\pi x_2}{2b}\, d\Omega = 0,$$

$$\alpha, \beta = 1, \ldots, M$$

where $[\bar{M}_{x_1}, \bar{M}_{x_2}, \bar{M}_{x_1x_2}, \bar{u}]^T = \left[\cos\frac{(2\alpha-1)\pi x_1}{2a} \cos\frac{(2\beta-1)\pi x_2}{2b}, 0, 0, 0\right]^T$,

$$\int_\Omega [\bar{M}_{x_1}, \bar{M}_{x_2}, \bar{M}_{x_1x_2}, \bar{u}]\, A\, [M_{x_1}, M_{x_2}, M_{x_1x_2}, u]^T \, d\Omega$$

$$= \int_\Omega \sum_{m=1}^{M}\sum_{n=1}^{M} \left\{ d_{mn}\left(\frac{(2n-1)\pi}{2b}\right)^2 - \frac{12}{Eh^3} b_{mn} + \nu \frac{12}{Eh^3} a_{mn} \right\}$$

$$\times \cos\frac{(2m-1)\pi x_1}{2a} \cos\frac{(2n-1)\pi x_2}{2b} \cos\frac{(2\alpha-1)\pi x_1}{2a} \cos\frac{(2\beta-1)\pi x_2}{2b}\, d\Omega = 0,$$

$$\alpha, \beta = 1, \ldots, M$$

where $[\bar{M}_{x_1}, \bar{M}_{x_2}, \bar{M}_{x_1x_2}, \bar{u}]^T = \left[\, 0, \cos\frac{(2\alpha-1)\pi x_1}{2a} \cos\frac{(2\beta-1)\pi x_2}{2b}, 0, 0 \,\right]^T,$

$$\int_\Omega [\bar{M}_{x_1}, \bar{M}_{x_2}, \bar{M}_{x_1x_2}, \bar{u}]\, A\, [M_{x_1}, M_{x_2}, M_{x_1x_2}, u]^T \, d\Omega$$

$$= \int_\Omega 2 \sum_{m=1}^{M}\sum_{n=1}^{M} \left\{ d_{mn} \frac{(2m-1)\pi}{2a}\frac{(2n-1)\pi}{2b} - c_{mn}\frac{24(1+\nu)}{Eh^3} \right\}$$

$$\times \sin\frac{m\pi x_1}{2a} \sin\frac{n\pi x_2}{2b} \sin\frac{\alpha\pi x_1}{2a} \sin\frac{\beta\pi x_2}{2b}\, d\Omega = 0, \quad \alpha, \beta = 1, \ldots, M$$

where $[\bar{M}_{x_1}, \bar{M}_{x_2}, \bar{M}_{x_1x_2}, \bar{u}]^T = \left[\, 0, 0, \sin\frac{\alpha\pi x_1}{2a} \sin\frac{\beta\pi x_2}{2b}, 0 \,\right]^T$, and

$$\int_\Omega [\bar{M}_{x_1}, \bar{M}_{x_2}, \bar{M}_{x_1x_2}, \bar{u}]\, A\, [M_{x_1}, M_{x_2}, M_{x_1x_2}, u]^T \, d\Omega$$

$$= \int_\Omega \sum_{m=1}^{M}\sum_{n=1}^{M} \left\{ \left[a_{mn}\left(\frac{(2m-1)\pi}{2a}\right)^2 + b_{mn}\left(\frac{(2n-1)\pi}{2b}\right)^2 \right]\right.$$

$$\times \cos\frac{(2m-1)\pi x_1}{2a} \cos\frac{(2n-1)\pi x_2}{2b} \cos\frac{(2\alpha-1)\pi x_1}{2a} \cos\frac{(2\beta-1)\pi x_2}{2b}$$

$$\left. + 2c_{mn}\left(\frac{m\pi}{2a}\frac{n\pi}{2b}\right) \cos\frac{m\pi x_1}{2a} \cos\frac{n\pi x_2}{2b} \cos\frac{(2\alpha-1)\pi x_1}{2a} \cos\frac{(2\beta-1)\pi x_2}{2b} \right\} d\Omega$$

$$= \int_\Omega f \cos\frac{(2\alpha-1)\pi x_1}{2a} \cos\frac{(2\beta-1)\pi x_2}{2b}\, d\Omega, \quad \alpha, \beta = 1, \ldots, M \tag{11.4.15}$$

where $[\bar{M}_{x_1}, \bar{M}_{x_2}, \bar{M}_{x_1x_2}, \bar{u}]^T = \left[\, 0, 0, 0, \cos\frac{(2\alpha-1)\pi x_1}{2a} \cos\frac{(2\beta-1)\pi x_2}{2b} \,\right]^T.$

Numerical solutions for both uniform and variable thickness simply supported square plates are presented in Tables 11.4.5 and 11.4.6, where points of interest are defined in Fig. 11.4.1. Note that the accuracy of both displacement and bending moments is good for the uniform plate and only $M_{x_1x_2}$ at point C shows significant error (or slower convergence) for the variable thickness plate.

Table 11.4.5
Results for Simply Supported Plate
(Uniform Thickness, Second-order Method)

	M_{x_1} @ A $\times 4fa^2$	$M_{x_1x_2}$ @ C $\times 4fa^2$	u @ A $\times 16fa^4D$	$\zeta = \rho\omega^2 \times \frac{1}{4a^2}\sqrt{\frac{D}{\rho h_0}}$
M = 2	0.046895	0.033244	0.0040547	19.739
M = 3	0.048235	0.032865	0.0040639	19.739
M = 4	0.047709	0.032271	0.0040619	19.739
EXACT [26]	0.0479	0.0325	0.00406	19.739

Table 11.4.6
Results for Simply Supported Plate
(Varying Thickness, Second-order Method)

$$h = 2h_0\left(1 - \frac{|x_1|}{2a}\right)\left(1 - \frac{|x_2|}{2a}\right)$$

	M_{x_1} @ A $\times 4fa^2$	$M_{x_1x_2}$ @ C $\times 4fa^2$	u @ A $\times 16fa^4D$	$\zeta = \rho\omega^2 \times \frac{1}{4a^2}\sqrt{\frac{D}{\rho h_0}}$
M = 2	0.075359	0.0054289	0.0019412	22.923
M = 3	0.079413	0.0039519	0.0019674	22.942
M = 4	0.080911	0.0040777	0.0019570	22.955

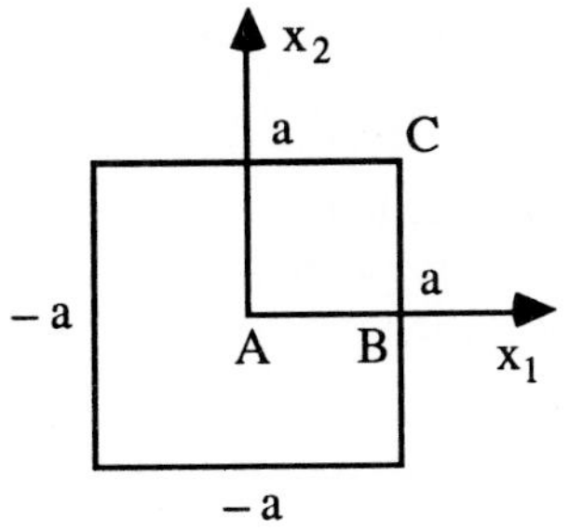

Figure 11.4.1 Points of Interest on Plate

Clamped Plate

Clamped plates with uniform and varying thickness are solved, using both the second- and fourth-order equations. The boundary conditions are

$$
\begin{aligned}
&u(\pm a, x_2) = u(x_1, \pm b) = 0 \\
&\frac{\partial u}{\partial x_1}(\pm a, x_2) = \frac{\partial u}{\partial x_2}(x_1, \pm b) = 0
\end{aligned}
\tag{11.4.16}
$$

The solution is approximated by coordinate functions of the form

$$
\begin{aligned}
M_{x_1} &= \sum_{m=0}^{M}\sum_{n=0}^{M} a_{mn} \cos\frac{2m\pi x_1}{a} \cos\frac{2n\pi x_2}{b} \\
M_{x_2} &= \sum_{m=0}^{M}\sum_{n=0}^{M} b_{mn} \cos\frac{2m\pi x_1}{a} \cos\frac{2n\pi x_2}{b} \\
M_{x_1x_2} &= \sum_{m=1}^{M}\sum_{n=1}^{M} c_{mn} \sin\frac{4m\pi x_1}{a} \sin\frac{4n\pi x_2}{b} \\
u &= \sum_{m=1}^{M}\sum_{n=1}^{M} d_{mn} \cos\frac{2(2m-1)\pi x_1}{a} \cos\frac{2\pi x_1}{a} \cos\frac{2(2n-1)\pi x_2}{b} \cos\frac{2\pi x_2}{b} \\
&= \frac{1}{4}\sum_{m=1}^{M}\sum_{n=1}^{M} d_{mn}\left[\cos\frac{4m\pi x_1}{a} + \cos\frac{2(2m-2)\pi x_1}{a}\right] \\
&\qquad \times \left[\cos\frac{4n\pi x_2}{b} + \cos\frac{2(2n-2)\pi x_2}{b}\right]
\end{aligned}
\tag{11.4.17}
$$

where u in Eq. 11.4.17 satisfies the boundary conditions. The moments M_{x_1} and M_{x_2} are symmetric with respect to the x_1 and x_2 axes, so they are approximated by cosine functions. The moment $M_{x_1x_2}$, on the other hand, is antisymmetric and must be zero along the boundary, so its coordinate functions are selected with these properties.

Substitution of these approximations into the Galerkin equations of Eq. 10.10.13 and performing necessary calculations yields the following equations:

$$\int_\Omega [\bar{M}_{x_1}, \bar{M}_{x_2}, \bar{M}_{x_1x_2}, \bar{u}]\, A\, [M_{x_1}, M_{x_2}, M_{x_1x_2}, u]^T \, d\Omega$$

$$= \int_{-b}^{b}\int_{-a}^{a}\left(\sum_{m=1}^{M}\sum_{n=1}^{M} d_{mn}\left[\left(\frac{2m\pi}{a}\right)^2 \cos\frac{4m\pi x_1}{a} + \left(\frac{(2m-2)\pi}{a}\right)^2 \cos\frac{2(2m-2)\pi x_1}{a} \right] \right.$$

$$\times \left\{ \cos\frac{4n\pi x_2}{b} + \cos\frac{2(2n-2)\pi x_2}{b} \right\} - \frac{12}{Eh^3}\sum_{m=0}^{M}\sum_{n=0}^{M} a_{mn}\cos\frac{2m\pi x_1}{a}\cos\frac{2n\pi x_2}{b}$$

$$\left. + \nu\frac{12}{Eh^3}\sum_{m=0}^{M}\sum_{n=0}^{M} b_{mn}\cos\frac{2m\pi x_1}{a}\cos\frac{2n\pi x_2}{b} \right)\cos\frac{2\alpha\pi x_1}{a}\cos\frac{2\beta\pi x_2}{b}\, dx_1\, dx_2 = 0,$$

$$\alpha, \beta = 0, 1, \ldots, M \qquad (11.4.18)$$

where $[\bar{M}_{x_1}, \bar{M}_{x_2}, \bar{M}_{x_1x_2}, \bar{u}]^T = \left[\cos\frac{2\alpha\pi x_1}{a}\cos\frac{2\beta\pi x_2}{b}, 0, 0, 0 \right]^T$,

$$\int_\Omega [\bar{M}_{x_1}, \bar{M}_{x_2}, \bar{M}_{x_1x_2}, \bar{u}]\, A\, [M_{x_1}, M_{x_2}, M_{x_1x_2}, u]^T \, d\Omega$$

$$= \int_{-b}^{b}\int_{-a}^{a}\left(\sum_{m=1}^{M}\sum_{n=1}^{M} d_{mn}\left[\left(\frac{2n\pi}{b}\right)^2 \cos\frac{4n\pi x_2}{b} + \left(\frac{(2n-2)\pi}{b}\right)^2 \cos\frac{2(2n-2)\pi x_2}{b} \right] \right.$$

$$\times \left\{ \cos\frac{4m\pi x_1}{a} + \cos\frac{2(2m-2)\pi x_1}{a} \right\} - \frac{12}{Eh^3}\sum_{m=0}^{M}\sum_{n=0}^{M} b_{mn}\cos\frac{2m\pi x_1}{a}\cos\frac{2n\pi x_2}{b}$$

$$\left. + \nu\frac{12}{Eh^3}\sum_{m=0}^{M}\sum_{n=0}^{M} a_{mn}\cos\frac{2m\pi x_1}{a}\cos\frac{2n\pi x_2}{b} \right)\cos\frac{2\alpha\pi x_1}{a}\cos\frac{2\beta\pi x_2}{b}\, dx_1\, dx_2 = 0,$$

$$\alpha, \beta = 1, 2, \ldots, M \qquad (11.4.19)$$

where $[\bar{M}_{x_1}, \bar{M}_{x_2}, \bar{M}_{x_1x_2}, \bar{u}]^T = \left[0, \cos\frac{2\alpha\pi x_1}{a}\cos\frac{2\beta\pi x_2}{b}, 0, 0 \right]^T$,

$$\int_\Omega [\bar{M}_{x_1}, \bar{M}_{x_2}, \bar{M}_{x_1x_2}, \bar{u}]\, A\, [M_{x_1}, M_{x_2}, M_{x_1x_2}, u]^T d\Omega$$

$$= \int_{-b}^{b}\int_{-a}^{a} 2\sum_{m=1}^{M}\sum_{n=1}^{M}\left\{ d_{mn}\left[\left(\frac{2m\pi}{a}\right)\sin\frac{4m\pi x_1}{a} + \left(\frac{(2m-2)\pi}{a}\right)\sin\frac{2(2m-2)\pi x_1}{a}\right]\right.$$

$$\times\left[\left(\frac{2n\pi}{b}\right)\sin\frac{4n\pi x_2}{b} + \left(\frac{(2n-2)\pi}{b}\right)\sin\frac{2(2n-2)\pi x_2}{b}\right]$$

$$\left. - \frac{24(1+\nu)}{Eh^3}\sin\frac{4m\pi x_1}{a}\sin\frac{4n\pi x_2}{b}\right\}\sin\frac{4\alpha\pi x_1}{a}\sin\frac{4\beta\pi x_2}{b}\,dx_1\,dx_2 = 0,$$

$$\alpha, \beta = 1, 2, \ldots, M \qquad (11.4.20)$$

where $[\bar{M}_{x_1}, \bar{M}_{x_2}, \bar{M}_{x_1x_2}, \bar{u}]^T = \left[0, 0, \sin\frac{4\alpha\pi x_1}{a}\sin\frac{4\beta\pi x_2}{b}, 0\right]^T$, and

$$\int_\Omega [\bar{M}_{x_1}, \bar{M}_{x_2}, \bar{M}_{x_1x_2}, \bar{u}]\, A\, [M_{x_1}, M_{x_2}, M_{x_1x_2}, u]^T d\Omega$$

$$= \int_{-b}^{b}\int_{-a}^{a}\left\{\sum_{m=0}^{M}\sum_{n=0}^{M}\left[a_{mn}\left(\frac{2m\pi}{a}\right)^2 + b_{mn}\left(\frac{2n\pi}{b}\right)^2\right]\cos\frac{2m\pi x_1}{a}\cos\frac{2n\pi x_2}{b}\right.$$

$$\left. + 2\sum_{m=1}^{M}\sum_{n=1}^{M} c_{mn}\left(\frac{4m\pi}{a}\right)\left(\frac{4n\pi}{b}\right)\cos\frac{4m\pi x_1}{a}\cos\frac{4n\pi x_2}{b}\right\}$$

$$\times\cos\frac{2(2\alpha-1)\pi x_1}{a}\cos\frac{2\pi x_1}{a}\cos\frac{2(2\beta-1)\pi x_2}{b}\cos\frac{2\pi x_2}{b}\,dx_1\,dx_2$$

$$= \int_{-b}^{b}\int_{-a}^{a} f(x_1, x_2)\cos\frac{2(2\alpha-1)\pi x_1}{a}\cos\frac{2\pi x_1}{a}\cos\frac{2(2\beta-1)\pi x_2}{b}\cos\frac{2\pi x_2}{b}\,dx_1\,dx_2,$$

$$\alpha, \beta = 1, 2, \ldots, M \qquad (11.4.21)$$

where

$$[\bar{M}_{x_1}, \bar{M}_{x_2}, \bar{M}_{x_1x_2}, \bar{u}]^T$$

$$= \left[0, 0, 0, \cos\frac{2(2\alpha-1)\pi x_1}{a}\cos\frac{2\pi x_1}{a}\cos\frac{2(2\beta-1)\pi x_2}{b}\cos\frac{2\pi x_2}{b}\right]^T$$

Numerical results, obtained with both the second-order and the fourth-order equations for square plates, are presented in Tables 11.4.7 and 11.4.8, where points of interest are defined in Fig. 11.4.1. As in prior examples, the accuracy of displacement and bending moments is uniformly good with the second-order formulation. In contrast, bending moments calculated by differentiating approximate displacements in the fourth-order formulation show significant error for even the uniform thickness plate and large error for the variable thickness plate.

Table 11.4.7
Results for Clamped Plate (Uniform Thickness)

		Moment @ A $\times 4fa^2$	Moment @ B $\times 4fa^2$	u @ A $\times 16fa^4D$	$\zeta = \rho\omega^2 \times \frac{1}{4a^2}\sqrt{\frac{D}{\rho h_0}}$
2nd Order	M = 2	0.022891	– 0.048742	0.0012075	36.047
	M = 3	0.022788	– 0.050480	0.0012958	35.990
	M = 4	0.022987	– 0.050932	0.0012438	35.974
4th Order	M = 3	0.025092	– 0.038922	0.0012626	36.106
	M = 4	0.021518	– 0.041473	0.0012586	36.044
EXACT [26]		0.0213	– 0.0513	0.00126	35.98

Table 11.4.8
Results for Clamped Plate (Varying Thickness)

$$h = 2h_0\left(1 - \frac{|x_1|}{2a}\right)\left(1 - \frac{|x_2|}{2a}\right)$$

		Moment @ A $\times 4fa^2$	Moment @ B $\times 4fa^2$	u @ A $\times 16fa^4D$	$\zeta = \rho\omega^2 \times \frac{1}{4a^2}\sqrt{\frac{D}{\rho h_0}}$
2nd Order	M = 2	0.040755	– 0.011680	0.00061224	38.058
	M = 3	0.041235	– 0.014495	0.00075683	38.141
	M = 4	0.043061	– 0.014452	0.00065104	38.175
4th Order	M = 2	0.029754	– 0.0062549	0.00060060	40.706
	M = 3	0.063371	– 0.0085592	0.00065468	39.376
	M = 4	0.040864	– 0.0101120	0.00067205	38.774

Vibration of Beams and Plates

The beam vibration eigenvalue problem can be written in second-order form as

$$A\begin{bmatrix} u \\ M \end{bmatrix} = \begin{bmatrix} -\dfrac{d^2M}{dx^2} \\ -\dfrac{d^2u}{dx^2} - \dfrac{M}{EI} \end{bmatrix} = \begin{bmatrix} \zeta a u \\ 0 \end{bmatrix} \tag{11.4.22}$$

where M is bending moment, or in fourth-order form as

$$\frac{d^2}{dx^2}\left[EI \frac{d^2u}{dx^2} \right] = \zeta a u \tag{11.4.23}$$

where $\zeta = \rho\omega^2$ and a is the cross-sectional area.

The equations for vibration of a plate can be written in second-order form as

$$A\begin{bmatrix} M_{x_1} \\ M_{x_2} \\ M_{x_1 x_2} \\ u \end{bmatrix} \equiv \begin{bmatrix} -\dfrac{\partial^2 u}{\partial x_1^2} - \dfrac{12}{Eh^3}M_{x_1} + \nu\dfrac{12}{Eh^3}M_{x_2} \\ -\dfrac{\partial^2 u}{\partial x_2^2} - \dfrac{12}{Eh^3}M_{x_2} + \nu\dfrac{12}{Eh^3}M_{x_1} \\ 2\dfrac{\partial^2 u}{\partial x_1 \partial x_2} - \dfrac{24(1+\nu)}{Eh^3}M_{x_1 x_2} \\ -\dfrac{\partial^2 M_{x_1}}{\partial x_1^2} - \dfrac{\partial^2 M_{x_2}}{\partial x_2^2} + 2\dfrac{\partial^2 M_{x_1 x_2}}{\partial x_1 \partial x_2} \end{bmatrix} = \begin{bmatrix} 0 \\ 0 \\ 0 \\ \zeta h u \end{bmatrix} \tag{11.4.24}$$

or in fourth-order form as

$$\frac{\partial^2}{\partial x_1^2}\left\{ D\left[\frac{\partial^2 u}{\partial x_1^2} + \nu \frac{\partial^2 u}{\partial x_2^2} \right] \right\} + \frac{\partial^2}{\partial x_2^2}\left\{ D\left[\frac{\partial^2 u}{\partial x_2^2} + \nu \frac{\partial^2 u}{\partial x_1^2} \right] \right\} + 2\frac{\partial^2}{\partial x_1 \partial x_2}\left\{ D(1-\nu)\frac{\partial^2 u}{\partial x_1 \partial x_2} \right\} = \zeta h u \tag{11.4.25}$$

where $\zeta = \rho\omega^2$ and h is the plate thickness.

The Galerkin method for eigenvalue analysis was implemented for simply supported and clamped beams and plates, with both uniform and variable cross sections and thicknesses. The same coordinate functions employed in static analysis were used to obtain numerical results that are presented in Tables 11.4.1 through 11.4.8. It is instructive to note that when the second-order equations are employed, the approximate eigenvalues do not necessarily decrease as more coordinate functions are employed; e.g., see Tables 11.4.2, 11.4.4, 11.4.6, and 11.4.8. This is due to the fact that the operators for second-order systems of equations are not positive definite. Note, however, that since the fourth-order operators are positive definite, the approximate eigenvalues always decrease as more coordinate functions are employed.

12

AN INTRODUCTION TO FINITE ELEMENT METHODS

As seen in Chapters 10 and 11, variational formulations of operator equations provide direct methods of constructing convergent numerical approximations to solutions of complex boundary-value problems. The Ritz method is particularly attractive for symmetric, positive bounded below operators that arise in nonhomogeneous equations and in eigenvalue problems. The principal limitation on the Ritz method is the construction of coordinate functions $\{ \phi_i(x) \}$ for use in constructing approximations of the form

$$u_n(x) = \sum_{i=1}^{n} a_i \phi_i(x)$$

In practical applications, it is sometimes difficult to find coordinate functions in H_A that are complete in energy and for which the Ritz equations are well posed. If the $\phi_i(x)$ are linearly dependent, or if the Gram determinant $| [\phi_i, \phi_j]_A |$ is nearly zero, practical computational difficulties arise. The finite element method for constructing coordinate functions is introduced in this chapter, for use in variational solution of operator equations that arise in a wide variety of applications. The basic idea is to subdivide the underlying physical domain Ω into subsets Ω_i, over which nonzero coordinate functions $\phi_i(x)$ are defined; i.e., $\phi_j(x) = 0$ on Ω_i, for all $j \neq i$. The result is a technique that systematically constructs coordinate functions that can be computer generated, even for irregularly shaped regions, and for which the resulting Ritz equations are computer generated and are well conditioned for numerical calculation. The development presented here is motivated by the elegant and more theoretically complete treatment of the subject by Strang and Fix [27]. The serious student of finite element methods is referred to Ref. 27 and related engineering books [6, 25, 28].

12.1 FINITE ELEMENTS FOR ONE-DIMENSIONAL BOUNDARY-VALUE PROBLEMS

As an introduction, several finite elements are formulated for the **Sturm–Liouville equation**; e.g., for deflection of a string. In this one-dimensional problem, the construction of finite elements is simple and natural. The equation

$$-\frac{d}{dx}\left(p(x) \frac{du}{dx} \right) + q(x)\,u = f(x) \tag{12.1.1}$$

for deflection of a string is first investigated. In order to illustrate the treatment of both principal and natural boundary conditions, the left end is fixed and the right end is free. Thus, at x = 0, there is a principal boundary condition

$$u(0) = 0 \tag{12.1.2}$$

As shown in Section 11.1, the operator

$$Au = -\frac{d}{dx}\left(p(x) \frac{du}{dx} \right) + q(x)\,u \tag{12.1.3}$$

on the domain

$$D_A = \{ u \text{ in } C^2(0, \pi):\ u(0) = u'(\pi) = 0 \} \tag{12.1.4}$$

is positive bounded below if p(x) and q(x) are positive in $(0, \pi)$. Therefore, from the minimum functional theorem, the solution of the operator equation

$$Au = f \tag{12.1.5}$$

is equivalent to finding a function in H_A that minimizes

$$F(u) = \int_0^{\pi} [\, p(u')^2 + qu^2 - 2uf \,]\, dx \tag{12.1.6}$$

Further, in Section 10.2, it was shown that the space H_A consists of functions that have derivatives in $L_2(0, \pi)$. Thus, functions of the class $D^1(0, \pi) \cap C^0(0, \pi)$ are considered here.

To construct a **finite element subspace** S^h of $D^1(0, \pi) \cap C^0(0, \pi)$, the interval $(0, \pi)$ is divided into segments of length $h = \pi/n$, on each of which the coordinate functions $\phi_j^h(x)$ are polynomials. Some degree of continuity is imposed at the boundaries between segments, but no more than is required in order that the following hold:

(1) The functions $\phi_j^h(x)$ are admissible in the variational principle.
(2) Quantities of physical interest, usually displacements, stresses, or moments, can be recovered from the approximate solution $u^h(x)$.

Linear Elements

In the string example, the admissible space is H_A, whose members are continuous. This rules out piecewise constant functions. Therefore, the simplest choice for S^h is the space of functions that are linear over each subinterval $[\,(j-1)\,h, jh\,]$, continuous at the nodes $x = jh$, and zero at $x = 0$. The derivative of such a function is piecewise constant and obviously has finite energy, so S^h is a subspace of H_A. These coordinate functions are called **linear finite elements**, the subintervals are called **finite elements** (or simply **elements**), and the points jh are called **nodes**.

For $j = 1, \ldots, n$, let $\phi_j^h(x)$ be the function that equals one at the node $x = jh$ and is zero at all the other nodes, as shown in Fig. 12.1.1. These functions constitute a basis for the subspace S^h, since every member of S^h can be written in the form

$$u^h(x) = \sum_{j=1}^{n} a_j \phi_j^h(x) \tag{12.1.7}$$

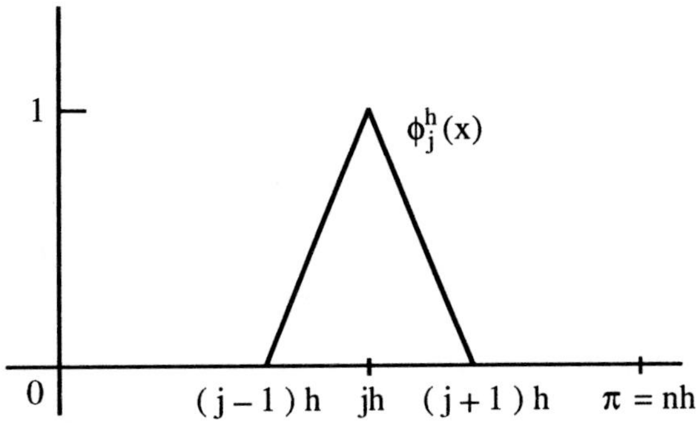

Figure 12.1.1 Piecewise Linear Coordinate Functions

Note that the coefficient a_j is the value of $u^h(x)$ at the jth node, $x = jh$; i.e., $u^h(jh) = a_j$. Since the constants a_j are **nodal values** of the function, they have direct physical significance. They are the Ritz approximations to the displacement of the string at the nodes. Note also that the coordinate functions $\phi_j^h(x)$ form a **local basis**, since each function $\phi_j^h(x)$ is zero, except in an interval of length 2h. In fact, $\phi_{j-1}^h(x)$ is orthogonal to $\phi_{j+1}^h(x)$, since wherever one of these functions can be nonzero, the other vanishes. Even though the basis is not orthogonal, only coordinate functions on adjacent elements have nonzero scalar products.

Considering $p = q = 1$ in Eqs. 12.1.1 and 12.1.6,

$$F(u^h) = \int_0^{\pi} [(u^{h\prime})^2 + (u^h)^2 - 2fu^h] \, dx \tag{12.1.8}$$

With $u^h = \sum_{j=1}^{n} a_j \phi_j^h(x)$, this integral is a function of the coefficients $a_1, \ldots, a_n$ and it can be independently computed over each element. On the jth element on $[(j-1)h, jh]$, the function $u^h(x)$ varies linearly from a_{j-1} to a_j and $u^{h\prime}(x) = (a_j - a_{j-1})/h$. Therefore,

$$\int_{(j-1)h}^{jh} (u^{h\,\prime})^2 \, dx = \frac{(a_j - a_{j-1})^2}{h} \tag{12.1.9}$$

A similar computation yields

$$\int_{(j-1)h}^{jh} (u^h)^2 \, dx = \frac{h}{3}(a_j^2 + a_j a_{j-1} + a_{j-1}^2) \tag{12.1.10}$$

For the entire interval $[0, \pi]$, summing from Eqs. 12.1.9 and 12.1.10,

$$\int_0^{\pi} [(u^{h\,\prime})^2 + (u^h)^2] \, dx$$

$$= \sum_{j=1}^{N} \left[\frac{(a_j - a_{j-1})^2}{h} + \frac{h(a_j + a_j a_{j-1} + a_{j-1}^2)}{3} \right] \tag{12.1.11}$$

It would be better to have this result as a quadratic form $\mathbf{a}^T\mathbf{K}\mathbf{a}$, since the positive semidefinite or positive definite **global stiffness matrix K** is needed. The expression $F(u^h)$ is quadratic in the vector parameter $\mathbf{a} = [\,\mathbf{a}_1, \ldots, \mathbf{a}_n\,]^T$, of the form

$$F(u^h) = \mathbf{a}^T\mathbf{K}\mathbf{a} - 2\mathbf{F}^T\mathbf{a} \tag{12.1.12}$$

The minimum of such an expression is thus determined by the matrix equation

$$\mathbf{K}\mathbf{a} = \mathbf{F} \tag{12.1.13}$$

which is a necessary and sufficient condition for minimizing $F(u^h)$ of Eq. 12.1.12 if **K** is positive definite. Therefore, all that is needed is the matrix **K** and the vector **F**.

The contribution to **K** from each element; i.e., each subinterval of the string, can be found by returning to Eq. 12.1.9 and calculating the right side of the matrix form

$$\int_{(j-1)h}^{jh} (u^{h\,\prime})^2 \, dx = [a_{j-1} \;\; a_j] \frac{1}{h} \begin{bmatrix} 1 & -1 \\ -1 & 1 \end{bmatrix} \begin{bmatrix} a_{j-1} \\ a_j \end{bmatrix}$$

$$\equiv [a_{j-1} \;\; a_j] \, \mathbf{k}_1 \begin{bmatrix} a_{j-1} \\ a_j \end{bmatrix} \tag{12.1.14}$$

where $\mathbf{k}_1$ is called an **element stiffness matrix**. Evaluation of $\mathbf{k}_1$ needs to be done only once. Similarly, the calculation of the term in Eq. 12.1.10 over a single element is carried out only once, yielding the **element mass matrix** $\mathbf{k}_0$, defined by

$$\int_{(j-1)h}^{jh} (u^h)^2 \, dx = [a_{j-1} \;\; a_j] \frac{h}{6} \begin{bmatrix} 2 & 1 \\ 1 & 2 \end{bmatrix} \begin{bmatrix} a_{j-1} \\ a_j \end{bmatrix}$$

$$\equiv [a_{j-1} \;\; a_j] \, \mathbf{k}_0 \begin{bmatrix} a_{j-1} \\ a_j \end{bmatrix} \tag{12.1.15}$$

The summation of Eqs. 12.1.14 and 12.1.15 over elements $j = 1, \ldots, n$ defines the assembled **global stiffness matrix K**; i.e., the element matrices are added into proper positions in the global stiffness matrix.

The matrix associated with $\int_0^{\pi} (u^{h\,\prime})^2\, dx$, using the fact that $a_0 = 0$, is

$$\mathbf{K}_1 = \frac{1}{h}\begin{bmatrix} 1 & \mathbf{0} & \\ & & \\ \mathbf{0} & \mathbf{0} & \end{bmatrix} + \frac{1}{h}\begin{bmatrix} 1 & -1 & & \\ -1 & 1 & \mathbf{0} & \\ & & & \\ & \mathbf{0} & \mathbf{0} & \\ & & & \end{bmatrix} + \ldots + \frac{1}{h}\begin{bmatrix} & & & \\ \mathbf{0} & \mathbf{0} & & \\ & & & \\ \mathbf{0} & & 1 & -1 \\ & & -1 & 1 \end{bmatrix}$$

$$= \frac{1}{h}\begin{bmatrix} 2 & -1 & 0 & 0 & \cdot \\ -1 & 2 & -1 & \cdot & 0 \\ 0 & -1 & \cdot & -1 & 0 \\ 0 & \cdot & -1 & 2 & -1 \\ \cdot & 0 & 0 & -1 & 1 \end{bmatrix} \tag{12.1.16}$$

The relationship between the matrix $\mathbf{K}_1$ and the integral of Eq.12.1.9 is

$$\int_0^{\pi} (u^{h\,\prime})^2\, dx = \mathbf{a}^T\mathbf{K}_1\mathbf{a} \tag{12.1.17}$$

The integral of $(u^h)^2$ is given by $\mathbf{a}^T\mathbf{K}_0\mathbf{a}$, where the matrix $\mathbf{K}_0$ is formed by the same assembling process; i.e.,

$$\mathbf{K}_0 = \frac{h}{6}\begin{bmatrix} 4 & 1 & 0 & 0 & \cdot \\ 1 & 4 & 1 & \cdot & 0 \\ 0 & 1 & \cdot & 1 & 0 \\ 0 & \cdot & 1 & 4 & 1 \\ \cdot & 0 & 0 & 1 & 2 \end{bmatrix} \tag{12.1.18}$$

The global stiffness matrix is thus obtained as $\mathbf{K} = \mathbf{K}_1 + \mathbf{K}_0$.

Since the coefficients $p(x)$ and $q(x)$ in Eq. 12.1.6 depend on x, the integrals of $p(x)(u^{h\,\prime})^2$ and $q(x)(u^h)^2$ are computed numerically on each element. The assembled results are stored as stiffness matrices, whose quadratic forms are $\mathbf{a}^T\mathbf{K}_1\mathbf{a}$ and $\mathbf{a}^T\mathbf{K}_0\mathbf{a}$.

It remains to compute the term

$$\int_0^{\pi} f u^h\, dx = \sum_{j=1}^{n} a_j \int_0^{\pi} f\phi_j^h\, dx \equiv \mathbf{F}^T\mathbf{a} \tag{12.1.19}$$

where

$$F_j = \int_0^{\pi} f\phi_j^h \, dx \tag{12.1.20}$$

These terms are computed by integrating over one element at a time.

To summarize, writing $\mathbf{K} = \mathbf{K}_1 + \mathbf{K}_0$, the computations thus yield

$$F(u^h) = F\left(\sum_{j=1}^{n} a_j\phi_j^h\right) = \mathbf{a}^T\mathbf{K}\mathbf{a} - 2\mathbf{F}^T\mathbf{a} \tag{12.1.21}$$

This is the expression that is to be minimized in the Ritz method. The minimizing vector is thus determined by the matrix equation

$$\mathbf{K}\mathbf{a} = \mathbf{F} \tag{12.1.22}$$

which is referred to as the **finite element equation.** If the element length h is small, Eq. 12.1.22 will be a large system of equations. If the operator A of Eq. 12.1.3 is positive definite, with boundary conditions accounted for, the matrix **K** is guaranteed to be positive definite and therefore invertable, since for $p(x) > 0$,

$$\mathbf{a}^T\mathbf{K}\mathbf{a} = \int_0^{\pi}\left[p(x)\left(\sum_{j=1}^{n} a_j\,\phi_j^{h\,\prime}(x)\right)^2 + q(x)\left(\sum_{j=1}^{n} a_j\,\phi_j^h(x)\right)^2\right]dx \geq 0 \tag{12.1.23}$$

can be zero only if $\sum_{j=1}^{n} a_j\phi_j^{h\,\prime}(x)$ is identically zero and this happens only if $a_j = 0$, $j = 1, 2, \ldots, n$.

An extension of the foregoing development involves the introduction of finite elements that are more refined than piecewise linear functions. A general function u(x) in $C^2(0, \pi)$ is better approximated by a higher-degree polynomial. Therefore, it is natural to construct coordinate functions in spaces S^h that are composed of polynomials of higher degree.

Quadratic Elements

As a first refinement, let S^h consist of all piecewise quadratic functions that are continuous at the nodes $x = jh$ and satisfy $u^h(0) = 0$, called **quadratic elements**. The first objective is to determine the dimension of S^h (the number of free parameters a_j) and define a basis for S^h. Note that continuity imposes only one constraint on the quadratic function at each node, so two coefficients in the polynomial remain arbitrary. Therefore, the dimension of S^h is twice the number of quadratic functions, or 2n.

A basis for S^h can be constructed by introducing the midpoints $x = (j - 1/2)h$ of the elements as nodes. There are then 2n nodes, since $x = 0$ is excluded and $nh = \pi$. Denote them by z_j, j = 1, . . . , 2n. For each node, there is a continuous piecewise quadratic function that equals one at z_j and zero at z_i, for all $i \neq j$; i.e.,

$$\phi_i(z_j) = \delta_{ij} \tag{12.1.24}$$

These coordinate functions are of the three kinds shown in Fig. 12.1.2. Note that each is continuous and therefore in H_A.

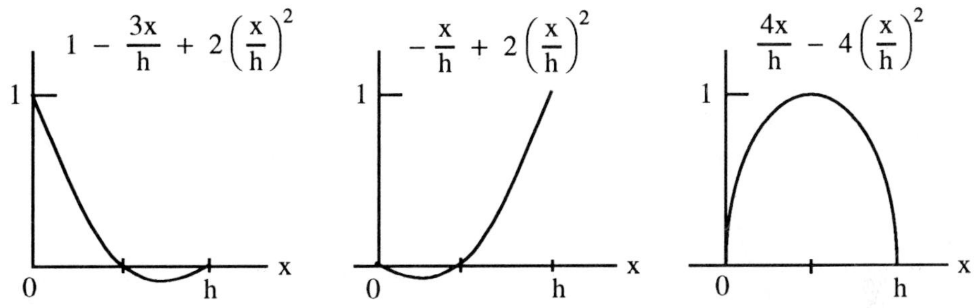

Figure 12.1.2 Coordinate Functions for Piecewise Quadratic Elements

The quadratic approximate displacement on $0 \le x \le h$ thus equals a_0 at x = 0, $a_{1/2}$ at the midpoint x = h/2, and a_1 at x = h; i.e.,

$$u^h(x) = a_0\left[1 - 3\left(\frac{x}{h}\right) + 2\left(\frac{x^2}{h^2}\right)\right] + a_{1/2}\left[4\left(\frac{x}{h}\right) - 4\left(\frac{x^2}{h^2}\right)\right]$$

$$+ a_1\left[-\left(\frac{x}{h}\right) + 2\left(\frac{x^2}{h^2}\right)\right] \tag{12.1.25}$$

The element stiffness matrix $\mathbf{k}_1$ is computed by integrating $(u^{h\prime}(x))^2$ and writing the result as $[\,a_0\ a_{1/2}\ a_1\,]\,\mathbf{k}_1\,[\,a_0\ a_{1/2}\ a_1\,]^T$. Notice that $\mathbf{k}_1$ is a 3×3 matrix, since only three of the parameters appear in any given interval. More specifically,

$$\mathbf{k}_1 = \frac{1}{3h}\begin{bmatrix} 7 & -8 & 1 \\ -8 & 16 & -8 \\ 1 & -8 & 7 \end{bmatrix} \tag{12.1.26}$$

Note that $\mathbf{k}_1$ is singular, since applied to the nonzero vector $[\,1, 1, 1\,]^T$, it yields the zero vector. This vector (i.e., $a_0 = 1$, $a_{1/2} = 1$, $a_1 = 1$) corresponds to a quadratic function u(x) that is constant (i.e., u(x) = 1), so its derivative is zero. The matrix $\mathbf{k}_0$, in contrast, is nonsingular.

Cubic Elements

There is a **cubic element** that is better in almost every respect than the quadratic element just discussed. It is constructed by imposing continuity not only on the function u(x), but also on its first derivative. There is a **double node** at each point x = jh. Instead of being determined by its values at four distinct points such as 0, h/3, 2h/3, and h, the cubic polynomial is now determined by its values and the values of its first derivatives at two end points; i.e., by u_0, u_0', u_1, and u_1'. Both u_1 and u_1' are shared by the cubic in the next subinterval, thus assuring continuity of both u(x) and u'(x). These coordinate functions are shown in Fig. 12.1.3. These functions have a double zero at the ends $(j \pm 1)h$; i.e., the function and its first derivative are both zero. They are called **Hermite cubics**.

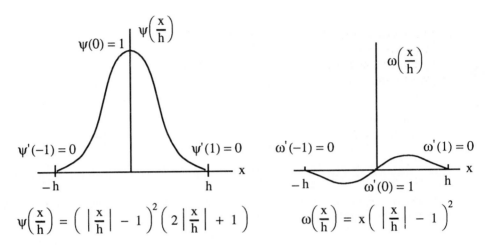

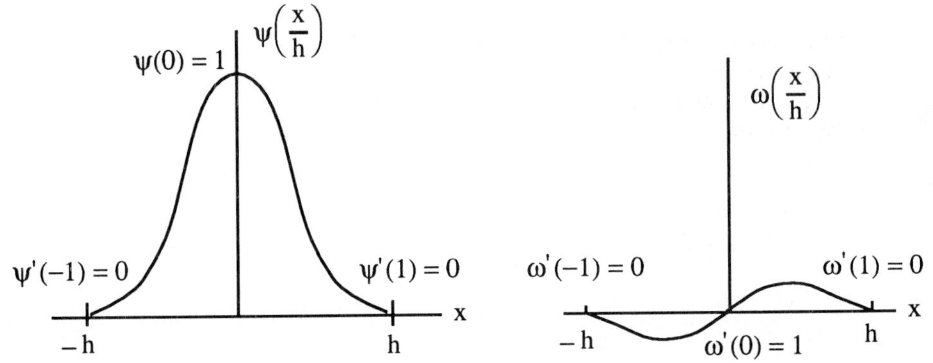

$$\psi\left(\frac{x}{h}\right) = \left(\left|\frac{x}{h}\right| - 1\right)^2\left(2\left|\frac{x}{h}\right| + 1\right) \qquad \omega\left(\frac{x}{h}\right) = x\left(\left|\frac{x}{h}\right| - 1\right)^2$$

Figure 12.1.3 Hermite Cubics

The cubic polynomial on $0 \le x \le h$ that takes on the values u_0, u_0', u_1, and u_1' is

$$\begin{aligned} u^h(x) &= u_0\psi\left(\frac{x}{h}\right) + u_0'\omega\left(\frac{x}{h}\right) + u_1\psi\left(\frac{x-h}{h}\right) + u_1'\omega\left(\frac{x-h}{h}\right) \\ &= u_0 + u_0'x + (3u_1 - 3u_0 - u_1'h - 2u_0'h)\frac{x^2}{h^2} \\ &\qquad + (2u_0 - 2u_1 + hu_1' + hu_0)\frac{x^3}{h^3} \\ &= q_0 + q_1x + q_2x^2 + q_3x^3 \end{aligned} \tag{12.1.27}$$

The 4×4 element matrices are computed as follows, where $\mathbf{a} = [u_0, u_0', u_1, u_1']^T$:

mass matrix $\mathbf{k}_0$:

$$\int_0^h (u^h)^2\,dx = \mathbf{a}^T\mathbf{k}_0\mathbf{a} \tag{12.1.28}$$

stiffness matrix $\mathbf{k}_1$:

$$\int_0^h (u^{h\,\prime})^2\, dx = \mathbf{a}^T\mathbf{k}_1\mathbf{a} \tag{12.1.29}$$

bending matrix $\mathbf{k}_2$:

$$\int_0^h (u^{h\,\prime\prime})^2\, dx = \mathbf{a}^T\mathbf{k}_2\mathbf{a} \tag{12.1.30}$$

Note that because $u^h(x)$ is in $D^2 \cap C^1$, it can be used in applications involving fourth-order operators, where the bending matrix $\mathbf{k}_2$ is needed.

One method of computing the $\mathbf{k}_i$ is to form a matrix $\mathbf{H}$ that connects the four nodal parameters in the vector $\mathbf{a}$ to the four coefficients $\mathbf{q} = [\, q_0, q_1, q_2, q_3\,]^T$ of the cubic polynomial in Eq. 12.1.27 for $u^h(x) = (\, q_0 + q_1 x + q_2 x^2 + q_3 x^3\,)$; i.e., $\mathbf{q} = \mathbf{Ha}$. From Eq. 12.1.27,

$$\mathbf{q} = \begin{bmatrix} q_0 \\ q_1 \\ q_2 \\ q_3 \end{bmatrix} = \begin{bmatrix} 1 & 0 & 0 & 0 \\ 0 & 1 & 0 & 0 \\ -\dfrac{3}{h^2} & -\dfrac{2}{h} & \dfrac{3}{h^2} & -\dfrac{1}{h} \\ \dfrac{2}{h^3} & \dfrac{1}{h^2} & -\dfrac{2}{h^3} & \dfrac{1}{h^2} \end{bmatrix} \begin{bmatrix} u_0 \\ u_0' \\ u_1 \\ u_1' \end{bmatrix} \equiv \mathbf{Ha} \tag{12.1.31}$$

The integration of $[\, u^h(x)\,]^2$ in Eq. 12.1.28 yields

$$\int_0^h (u^h)^2\, dx = [\, q_0 \;\; q_1 \;\; q_2 \;\; q_3\,] \begin{bmatrix} h & \dfrac{h^2}{2} & \dfrac{h^3}{3} & \dfrac{h^4}{4} \\ \dfrac{h^2}{2} & \dfrac{h^3}{3} & \dfrac{h^4}{4} & \dfrac{h^5}{5} \\ \dfrac{h^3}{3} & \dfrac{h^4}{4} & \dfrac{h^5}{5} & \dfrac{h^6}{6} \\ \dfrac{h^4}{4} & \dfrac{h^5}{5} & \dfrac{h^6}{6} & \dfrac{h^7}{7} \end{bmatrix} \begin{bmatrix} q_0 \\ q_1 \\ q_2 \\ q_3 \end{bmatrix} \equiv \mathbf{q}^T\mathbf{N}_0\mathbf{q} \tag{12.1.32}$$

From Eqs. 12.1.31 and 12.1.32,

$$\int_0^h (u^h)^2\, dx = \mathbf{q}^T\mathbf{N}_0\mathbf{q} = \mathbf{a}^T\mathbf{H}^T\mathbf{N}_0\mathbf{H}\mathbf{a}$$

so the element mass matrix is

$$\mathbf{k}_0 = \mathbf{H}^T\mathbf{N}_0\mathbf{H} \tag{12.1.33}$$

For the stiffness matrix, the only difference is that

$$\int_0^h (u^{h\,\prime})^2\,dx = \int_0^h (q_1 + 2q_2x + 3q_3x^2)^2\,dx$$

$$= \mathbf{q}^T \begin{bmatrix} 0 & 0 & 0 & 0 \\ 0 & h & h^2 & h^3 \\ 0 & h^2 & \frac{4h^3}{3} & \frac{3h^4}{2} \\ 0 & h^3 & \frac{3h^4}{2} & \frac{9h^5}{5} \end{bmatrix} \mathbf{q} \equiv \mathbf{q}^T\mathbf{N}_1\mathbf{q} \qquad (12.1.34)$$

so the element stiffness matrix is $\mathbf{k}_1 = \mathbf{H}^T\mathbf{N}_1\mathbf{H}$.

The results of these computations, and a similar one for $\mathbf{k}_2$, are the following matrices:

$$\mathbf{k}_0 = \frac{h}{420}\begin{bmatrix} 156 & 22h & 54 & -13h \\ 22h & 4h^2 & 13h & -3h^2 \\ 54 & 13h & 156 & -22h \\ -13h & -3h^2 & -22h & 4h^2 \end{bmatrix} \qquad (12.1.35)$$

$$\mathbf{k}_1 = \frac{1}{30h}\begin{bmatrix} 36 & 3h & -36 & 3h \\ 3h & 4h^2 & -3h & -h^2 \\ -36 & -3h & 36 & -3h \\ 3h & -h^2 & -3h & 4h^2 \end{bmatrix} \qquad (12.1.36)$$

$$\mathbf{k}_2 = \frac{1}{h^3}\begin{bmatrix} 12 & 6h & -12 & 6h \\ 6h & 4h^2 & -6h & 2h^2 \\ -12 & -6h & 12 & -6h \\ 6h & 2h^2 & -6h & 4h^2 \end{bmatrix} \qquad (12.1.37)$$

The matrix $\mathbf{k}_0$ is positive definite, but $\mathbf{k}_1$ has a zero eigenvalue that corresponds to the constant function $u^h(x) = 1$; i.e., to $\mathbf{a} = [\,1, 0, 1, 0\,]^T$. The matrix $\mathbf{k}_2$ is also singular, since for every linear $u^h(x)$, $u^{h\,\prime\prime}(x) = 0$.

A fourth-order **beam equation**

$$Au = (ru'')'' - (pu')' + qu = f \qquad (12.1.38)$$

with $r(x) \geq r_{min} > 0$, $p(x) \geq 0$ and $q(x) \geq 0$ can now be treated. With physically reasonable boundary conditions, the associated energy scalar product is

$$[\,u, v\,]_A = \int_0^{\pi} (\, ru''v'' + pu'v' + quv \,)\, dx \tag{12.1.39}$$

and the minimum functional theorem leads to $F(u) = [\,u, u\,]_A - 2\,(\,f, u\,)$. For the Ritz method to apply, the coordinate functions must have finite energy, which means they must be in $D^2 \cap C^1$. The Hermite cubics are therefore applicable.

If a beam is clamped at $x = 0$, the boundary conditions at $x = 0$ are

$$u(0) = u'(0) = 0 \tag{12.1.40}$$

which are principal. The candidate solution $u^h(x)$ must have a double zero at $x = 0$; i.e., it must satisfy Eq. 12.1.40. To see what the natural boundary conditions are at $x = \pi$, integrate $F(u)$ by parts and require that its first variation vanishes. The result is that, for every δu in the admissible space $D^2 \cap C^1$,

$$\delta F = \int_0^{\pi} [\,(\,ru''\,)'' - (\,pu'\,)' + qu - f\,]\,\delta u\, dx + ru''\delta u'\Big|_0^{\pi}$$

$$+ (\,pu' - (\,ru''\,)'\,)\,\delta u\Big|_0^{\pi} = 0 \tag{12.1.41}$$

Thus, the natural boundary conditions on $u(x)$ are those that physically correspond to a free end; i.e.,

$$\begin{aligned} u''(\pi) &= 0 \\ [\,pu' - (\,ru''\,)'\,](\pi) &= 0 \end{aligned} \tag{12.1.42}$$

These are the well-known zero moment and zero shear conditions at the free end of a beam [25].

Finite elements with Hermite cubic coordinate functions of Eq. 12.1.27 may be selected, which satisfy boundary conditions of Eq. 12.1.40, but not Eq. 12.1.42. Presuming $r(x)$, $p(x)$, and $q(x)$ are piecewise constant on the elements, the element matrices of Eqs. 12.1.35 through 12.1.37 may be applied in Eqs. 12.1.28 through 12.1.30. Thus,

$$\begin{aligned} F(u^h) &= \sum_{i=1}^{n} \int_{(i-1)h}^{ih} [\,r_i\,(u^{h''})^2 + p_i\,(u^{h'})^2 + q_i\,(u^h)^2 - 2u^h f\,]\, dx \\ &= \sum_{i=1}^{n} [\,\mathbf{a}^{iT} r_i \bar{\mathbf{k}}_2 \mathbf{a}^i + \mathbf{a}^{iT} p_i \bar{\mathbf{k}}_1 \mathbf{a}^i + \mathbf{a}^{iT} q_i \bar{\mathbf{k}}_0 \mathbf{a}^i - 2\mathbf{a}^{iT}\mathbf{F}^i\,] \\ &\equiv \mathbf{a}^T\mathbf{K}\mathbf{a} - 2\mathbf{a}^T\mathbf{F} \end{aligned} \tag{12.1.43}$$

and the Ritz equations are

$$\mathbf{K}\mathbf{a} = \mathbf{F} \tag{12.1.44}$$

12.2 FINITE ELEMENTS FOR TWO-DIMENSIONAL BOUNDARY-VALUE PROBLEMS

While development of the finite element method for elliptic partial differential equations in two independent variables is theoretically no more difficult than for problems with only one independent variable, the algebraic complexity is somewhat greater. For this reason, attention is restricted to a general description of the method and the reader is referred to the literature for detailed development of element formulas [25, 27]. In this section, some finite elements in the plane are defined. Development of such elements involves only elementary algebra, but the results are of great importance. The goal is to choose piecewise polynomials that are determined by a small set of nodal values, and yet have the desired degree of continuity.

There are many triangular and rectangular finite elements, and it is not clear whether it is more efficient to subdivide the region into triangles or rectangles. Triangles are better for approximating a curved boundary, but there are advantages to rectangles in the interior. There are fewer of them, and they permit very simple elements of high degree.

Consider first subdividing the region Ω into triangles, as shown in Fig. 12.2.1. The union of these triangles is a polygon Ω^h. The longest edge of the ith triangle will be denoted by h_i and $h = \max_i h_i$. Assume that no vertex $z_j = (x_1^j, x_2^j)$ of one triangle lies on the edge of another triangle.

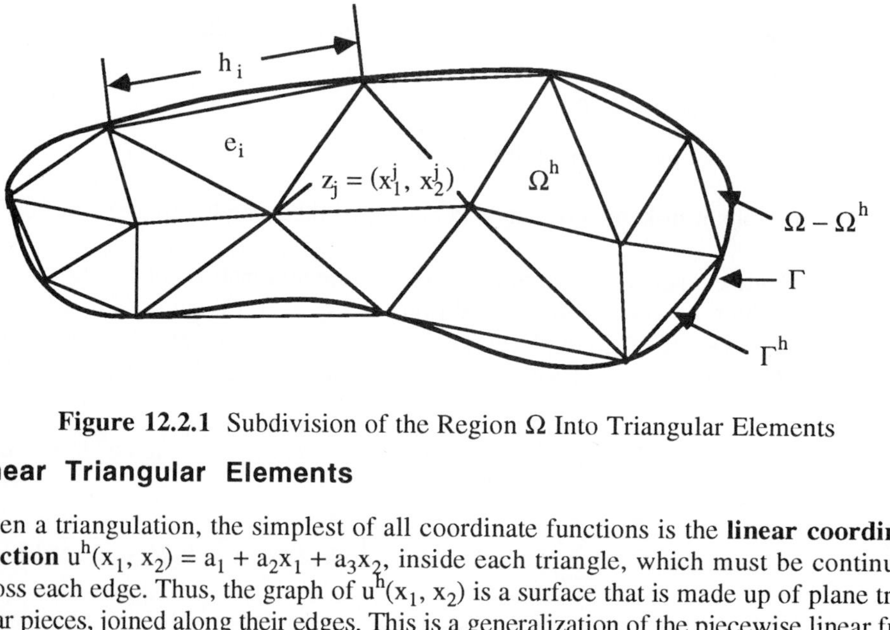

Figure 12.2.1 Subdivision of the Region Ω Into Triangular Elements

Linear Triangular Elements

Given a triangulation, the simplest of all coordinate functions is the **linear coordinate function** $u^h(x_1, x_2) = a_1 + a_2x_1 + a_3x_2$, inside each triangle, which must be continuous across each edge. Thus, the graph of $u^h(x_1, x_2)$ is a surface that is made up of plane triangular pieces, joined along their edges. This is a generalization of the piecewise linear functions in one dimension used in Section 12.1. The space S^h of these piecewise linear functions is a subspace of $D^1(\Omega^h) \cap C^0(\Omega^h)$, whose first derivatives are piecewise constant. As seen in Chapter 9, this class of functions is reasonable for solution of second-order equations. It is not regular enough, however, for fourth-order equations.

The simplicity of piecewise linear functions lies in the fact that within each triangle, the three coefficients of $u^h(x_1, x_2) = a_1 + a_2x_1 + a_3x_2$ are uniquely determined by the nodal values of $u^h(x_1, x_2)$ at the three vertices. Furthermore, along any edge $u^h(x_1, x_2)$ reduces to a linear function of the arc length variable along that edge and this function is determined by its values at the two end points of the edge. The value of $u^h(x_1, x_2)$ at the third vertex has no effect on the function along this edge. Therefore, continuity of $u^h(x_1, x_2)$ across the edge is assured by continuity at the vertices.

For a principal boundary condition [say, $u(x_1, x_2) = 0$ on Γ], the simplest space S^h is formed by requiring the functions to be zero on the polygonal boundary Γ^h shown in Fig. 12.2.1. The dimension of the space S^h; i.e., the number n of free parameters in the functions $u^h(x_1, x_2)$, is the number of unconstrained nodes. Let $\phi_j(x_1, x_2)$ be a coordinate function that equals 1 at the jth node and zero at all other nodes, as shown in Fig. 12.2.2. These **pyramid functions** $\phi_j(x_1, x_2)$ form a basis for the space S^h. An arbitrary $u^h(x_1, x_2)$ in S^h can be expressed as

$$u^h(x_1, x_2) = \sum_{j=1}^{n} a_j \phi_j(x_1, x_2) \tag{12.2.1}$$

The **nodal coordinate** a_j has the physical significance of displacement $u^h(x_1, x_2)$ at the jth node $z_j = (x_1^j, x_2^j)$.

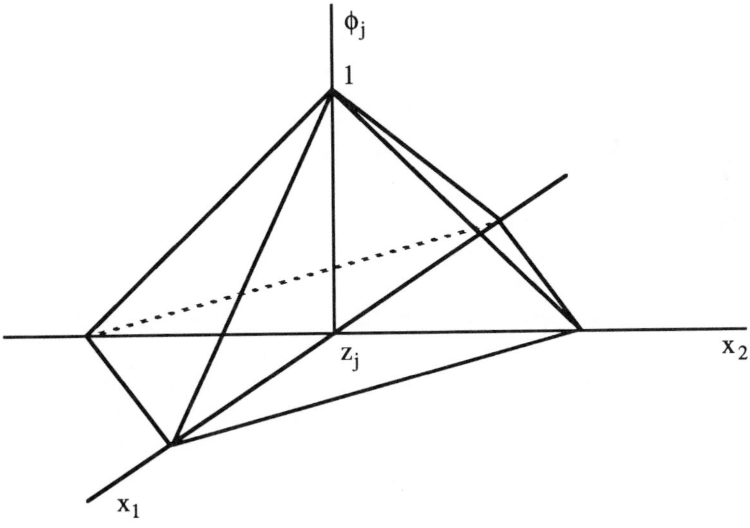

Figure 12.2.2 Pyramid Coordinate Function

The coordinates a_j are determined by minimizing the energy functional $F(u^h)$, which is quadratic in $a_1, \ldots, a_n$. The minimizing function $u^h(x_1, x_2) = \sum_{j=1}^{n} a_j \phi_j(x_1, x_2)$ is determined by the solution of a matrix equation of the form

$$\mathbf{Ka} = \mathbf{F}$$

The matrix **K** is well conditioned and sparse, since two nodal coordinates are coupled only if they belong to the same element. Furthermore, the coefficients $k_{ij} = [\ \phi_i, \phi_j\ ]_A$ and $F_j = (\ f, \phi_j\ )$ in the Ritz equation can be found by a systematic evaluation of scalar products from the triangular elements, one at a time. This means that the integrals are computed over each element, yielding a set of element stiffness matrices $\mathbf{k}_i$, involving only the nodes of the ith triangle of the finite element grid. The global stiffness matrix **K** that contains all such scalar products is then assembled from these pieces. This process is carried out just as in the one-dimensional case developed in Section 12.1.

Quadratic Triangular Elements

Consider next a more refined element. Rather than using a linear function within each triangle, $u^h(x_1, x_2)$ will be the **quadratic coordinate function**

$$u^h(x_1, x_2) = c_1 + c_2x_1 + c_3x_2 + c_4x_1^2 + c_5x_1x_2 + c_6x_2^2 \qquad (12.2.2)$$

In order to be in $D^1 \cap C^0$, $u^h(x_1, x_2)$ must be continuous across edges between adjacent triangles. To define a basis for this space, a set of continuous piecewise quadratic functions $\phi_j(x_1, x_2)$ must be found, such that every member of S^h has a unique expansion as

$$u^h(x_1, x_2) = \sum_{j=1}^{n} a_j\ \phi_j(x_1, x_2) \qquad (12.2.3)$$

To form such a basis, additional nodes may be placed at the midpoints of the edges, as in Fig. 12.2.3(a). With each node, whether it is a vertex or the midpoint of an edge, the coordinate function $\phi_j(x_1, x_2)$ is defined as 1 at that node and 0 at all others. This rule specifies $\phi_j(x_1, x_2)$ at six points in each triangle, the three vertices and three midpoints, thereby determining the six coefficients in Eq. 12.2.2.

The **piecewise quadratic coordinate function** that is determined by this rule, in each separate triangle, must be shown to be continuous across edges between triangles. Along an edge, $u^h(x_1, x_2)$ is quadratic in the boundary arc length variable and there are three nodes on the edge. Therefore, the quadratic polynomial is uniquely determined by the three nodal values that are shared by the triangles that meet at that edge. The nodal values elsewhere in the triangles have no effect on $u^h(x_1, x_2)$ along the edge, so continuity holds. For an edge that is on the outer boundary Γ^h with boundary condition $u = 0$, the quadratic is zero.

Any $u^h(x_1, x_2)$ in S^h can be expressed as $u^h(x_1, x_2) = \sum_{j=1}^{n} a_j\ \phi_j(x_1, x_2)$, where the nodal coordinate a_j is the value of $u^h(x_1, x_2)$ at the jth node. Therefore, these $\phi_j(x_1, x_2)$ form a basis for S^h and its dimension is the number of unconstrained nodes.

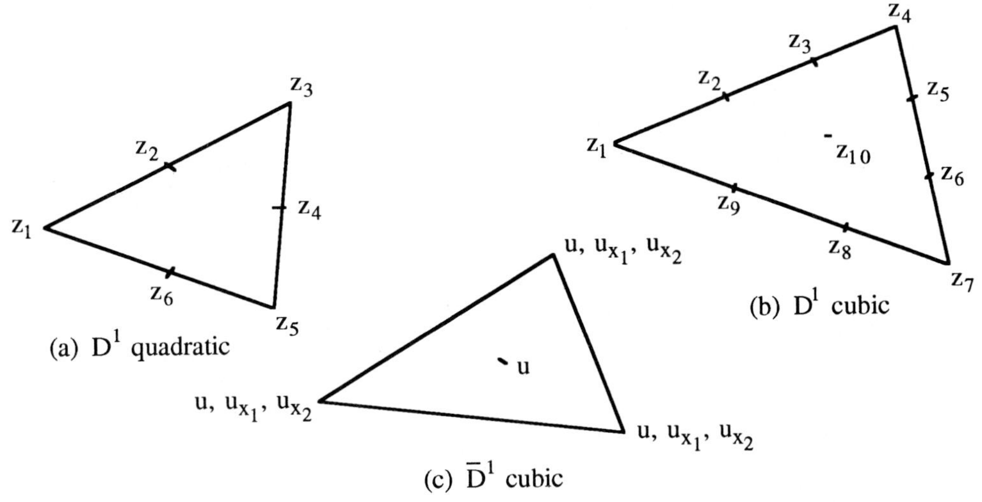

Figure 12.2.3 Nodal Placement for Quadratic and Cubic Elements

Cubic Triangular Elements

Using continuous **piecewise cubic** polynomials, a basis can be constructed in a similar way. A cubic in the variables x_1 and x_2 is determined by 10 coefficients, in particular by its values at the 10-node triangle of Fig. 12.2.3(b). The four nodes on each edge determine the **piecewise cubic coordinate function** uniquely on that edge, so continuity is assured.

In the cubic case of Fig. 12.2.3(c), the condition that the first derivatives of $u^h(x_1, x_2)$ should be continuous at every vertex can be added. This is clearly a subspace of the previous cubic case, in which only continuity of $u^h(x_1, x_2)$ was required. If m triangles meet at a given interior vertex, the continuity of $u^h_{x_1}(x_1, x_2)$ and $u^h_{x_2}(x_1, x_2)$ imposes new constraints on the coordinate functions, so the dimension is correspondingly reduced. To construct a basis for S^h, the midedge nodes are removed and a **triple node** is defined at each vertex, as shown in Fig. 12.2.3(c). The 10 coefficients in the cubic are determined by u, u_{x_1}, and u_{x_2} at each vertex, together with u at the centroid. A unique cubic is determined by these 10 nodal values. Along each edge, it is determined by a subset of four of the values of u(x) and its first derivative along the edge at each end, a combination that assures matching along the edge. The result is a useful set of cubic coordinate functions in $D^1(\Omega^h)$, denoted $\bar{D}^1$.

Piecewise polynomial finite elements in $\bar{D}^1$ are easy to describe in the interior of the domain, but care is required at the boundary. In the case of a principal boundary condition $u(x_1, x_2) = 0$ on Γ, there is first the constraint $u^h(x_1, x_2) = 0$ along all of Γ^h. Thus, the derivatives along both element boundary edges must be set equal to zero, leaving no parameters free at that vertex.

Quintic Triangular Elements

None of the spaces defined above can be applied to the biharmonic equation, since they are not subspaces of $D^2 \cap C^1$. In order to construct coordinate functions that belong to C^1, the normal derivative must be continuous across element boundaries. If piecewise polynomials are to be used, they must be of fifth degree; i.e., quintics.

A polynomial of fifth degree in x_1 and x_2 has 21 coefficients to be determined, of which 18 can come from the values of u, u_{x_1}, u_{x_2}, $u_{x_1x_1}$, $u_{x_1x_2}$, and $u_{x_2x_2}$ at the vertices. These second derivatives represent **bending moments** that are of physical interest in plates; e.g., Eq. 11.4.11, which will be continuous at the vertices and available as output from the finite element analysis process. Furthermore, with all second derivatives present, there is no difficulty in allowing an irregular finite element triangulation. The **piecewise quintic coordinate function** along the edge between two triangles is the same from both sides, since conditions are shared at each end of the edge on u and its first two tangential derivatives u_s and u_{ss}, which are computable from the set of six parameters at each vertex.

It remains to determine three more constraints in such a way that the normal derivative $u_n(x_1, x_2)$ is also continuous across edges. One technique is to add the value of u_n at the midpoint of each edge to the first of nodal parameters, as shown in Fig. 12.2.4(a). Then, since $u_n(x_1, x_2)$ is a fourth-degree polynomial in s along the edge, it is uniquely determined by this midpoint nodal value of u_n and the values of u_n and u_{ns} at each end. The normal n at the middle of an edge points into one triangle and out of the adjacent one, somewhat complicating the bookkeeping. This construction produces a complete $C^1 \cap D^2$ quintic.

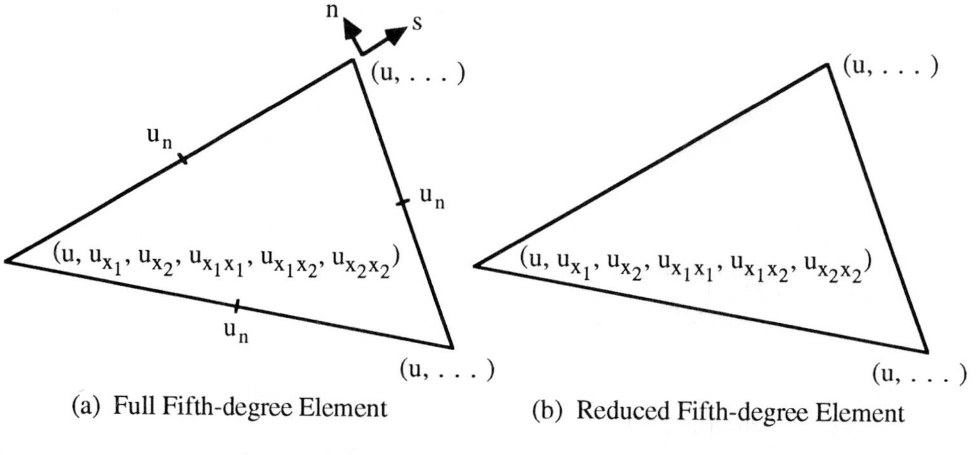

Figure 12.2.4 Two Fifth-degree Triangles

Another way of determining the three remaining constraints is to require that $u_n(x_1, x_2)$ reduces to a cubic along each edge; i.e., that the leading coefficient in what would otherwise be a fourth-degree polynomial vanishes. With this constraint, the nodal values of u_n and u_{ns} determine the cubic $u_n(x_1, x_2)$ along the edge, and the element is in $C^1 \cap D^2$. The complete quintic is lost, however, reducing the order of displacement error [27]. At the same time, the dimension of S^h is significantly reduced [see Fig. 12.2.4(b)].

Rectangular Elements

Consider next **rectangular elements**. A number of important problems in the plane are defined on a rectangle or on a union of rectangles. The boundary of a more general region cannot be described sufficiently well without using triangles, but it will often be possible to combine rectangular elements in the interior, with triangular elements near the boundary.

The simplest coordinate functions on rectangles are **piecewise bilinear functions**; i.e., $u^h(x_1, x_2) = c_1 + c_2x_1 + c_3x_2 + c_4x_1x_2$. The four coefficients are determined by the values of u^h at the vertices. There is thus a set of **piecewise bilinear coordinate functions**, in which $\phi_j(x_1, x_2)$ is 1 at node j and is 0 at the others. It is a product $\psi(x_1)\,\psi(x_2)$ of piecewise linear functions in one variable, so that the space S^h is a product of two simpler spaces. Such elements are called **bilinear rectangular elements**. They are continuous, so they can be used for second-order problems. Bilinear elements on rectangles can be merged with linear elements on triangles, since both are completely determined by the value of u^h at the nodes.

One other rectangular element is defined here, as an extension of the one-dimensional Hermite cubic element of Fig. 12.1.3. The basis in one dimension contains two kinds of functions, $\psi(x)$ and $\omega(x)$, which interpolate values of the function and its first derivative, respectively,

$$\psi_j = \begin{cases} 1, & \text{at node } x = x_j \\ 0, & \text{at other nodes} \end{cases} \qquad \omega_j = 0, \text{ at all nodes}$$

$$\frac{d\psi_j}{dx} = 0, \text{ at all nodes}, \qquad \frac{d\omega_j}{dx} = \begin{cases} 1, & \text{at node } x = x_j \\ 0, & \text{at other nodes} \end{cases}$$

The **Hermite bicubic space** is a product of two such cubic spaces. The four parameters at a typical node (x_1^j, x_2^j) lead to four corresponding coordinate functions,

$$\begin{aligned} \phi_1(x_1, x_2) &= \psi_j(x_1)\,\psi_j(x_2), & \phi_2(x_1, x_2) &= \psi_j(x_1)\,\omega_j(x_2) \\ \phi_3(x_1, x_2) &= \omega_j(x_1)\,\psi_j(x_2), & \phi_4(x_1, x_2) &= \omega_j(x_1)\,\omega_j(x_2) \end{aligned} \tag{12.2.4}$$

The degree of continuity of the **Hermite bicubic coordinate functions** follows immediately from the basis given in Eq. 12.2.4. Since $\psi(x_1)$ and $\omega(x_2)$ are both in C^1, all their products have this property. Therefore, this bicubic may be applied to fourth-order equations. The trial functions lie in D^2. Furthermore, even the cross derivatives $u_{x_1x_2}(x_1, x_2)$ are continuous.

The rather messy algebra and calculus that must be carried out to form the Ritz equations,

$$\sum_{j=1}^{n} \left(\iint_{\Omega} \phi_j A \phi_i \, d\Omega \right) a_i = \iint_{\Omega} f \phi_j \, d\Omega \tag{12.2.5}$$

for the operator equation Au = f are not developed here. The reader who wishes to go into such detail is referred to the extensive finite element literature [25, 27, 28].

12.3 FINITE ELEMENT SOLUTION OF A SYSTEM OF SECOND-ORDER PARTIAL DIFFERENTIAL EQUATIONS

Second-order Plate Boundary-value Problems

The governing equations for **bending of a plate** can be written, as in Eq. 11.4.11, as

$$\mathbf{Az} = \begin{bmatrix} -\dfrac{12}{Eh^3} z_1 + \dfrac{12\nu}{Eh^3} z_2 - \dfrac{\partial^2 z_4}{\partial x_1^2} \\[2ex] \dfrac{12\nu}{Eh^3} z_1 - \dfrac{12}{Eh^3} z_2 - \dfrac{\partial^2 z_4}{\partial x_2^2} \\[2ex] -\dfrac{24(1+\nu)}{Eh^3} z_3 + 2\dfrac{\partial^2 z_4}{\partial x_1 \partial x_2} \\[2ex] -\dfrac{\partial^2 z_1}{\partial x_1^2} - \dfrac{\partial^2 z_2}{\partial x_2^2} + 2\dfrac{\partial^2 z_3}{\partial x_1 \partial x_2} \end{bmatrix} = \begin{bmatrix} 0 \\ 0 \\ 0 \\ f \end{bmatrix} \tag{12.3.1}$$

where $z_1(x_1, x_2)$ and $z_2(x_1, x_2)$ are bending moments on cross sections normal to the x_1 and x_2 axes, respectively, $z_3(x_1, x_2)$ is the torsional moment on a cross section normal to either the x_1 or x_2 axes, $z_4(x_1, x_2)$ is displacement in the direction normal to the plate middle surface, ν is Poisson's ratio, h is the thickness of the plate, and $f(x_1, x_2)$ represents the distributed load.

For the rectangular plate of Fig. 12.3.1 that is simply supported along its edges, the boundary conditions are

$$z_1(\pm a, x_2) = z_2(x_1, \pm b) = z_4(\pm a, x_2) = z_4(x_1, \pm b) = 0 \tag{12.3.2}$$

On the other hand, if the plate is clamped along its edges, the boundary conditions are

$$\frac{\partial z_4}{\partial x_1}(\pm a, x_2) = \frac{\partial z_4}{\partial x_2}(x_1, \pm b) = z_4(\pm a, x_2) = z_4(x_1, \pm b) = 0 \tag{12.3.3}$$

It may be noted that having the slope equal to zero along the boundary implies that $z_3(x_1, x_2)$ is equal to zero along the boundary. The latter condition can be handled more easily and will be used to solve clamped plate problems.

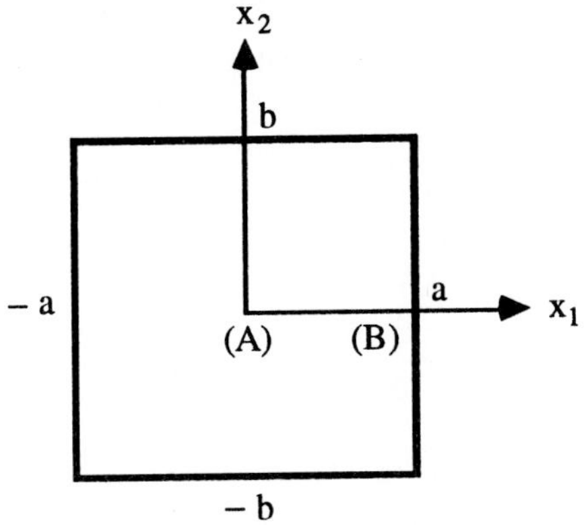

Figure 12.3.1 Rectangular Plate

Galerkin Approximate Solution

Here, the plate problem is solved by the finite element Galerkin method, for comparison with numerical results of Section 11.4. A stiffness matrix is formulated for the system of partial differential equations of Eq. 12.3.1. To formulate the finite element approximation, an element geometry must first be chosen. Considering the rectangular shape of the plate in Fig. 12.3.1, a rectangular element is chosen, with the eight nodes shown in Fig. 12.3.2. The equations contain four unknown variables $z_i(x_1, x_2)$, $i = 1, \ldots, 4$, which can be approximated by their values, Z_i^α, $i = 1, \ldots, 4$, $\alpha = 1, \ldots, 8$, at the nodes; i.e.,

$$
\begin{aligned}
z_i^h(x_1, x_2) &= \sum_{\alpha=1}^{8} Z_i^\alpha \phi_\alpha(x_1, x_2), \quad i = 1, \ldots, 4 \\
f_i^h(x_1, x_2) &= \sum_{\alpha=1}^{8} F_i^\alpha \phi_\alpha(x_1, x_2), \quad i = 1, \ldots, 4
\end{aligned}
\tag{12.3.4}
$$

where the $\phi_\alpha(x_1, x_2)$ are coordinate functions. Note that for computational convenience, the forcing functions $f_i(x_1, x_2)$ have been expanded in terms of the coordinate functions. A higher-order element that is developed in Ref. 28 is used. Details of the calculations are not presented here.

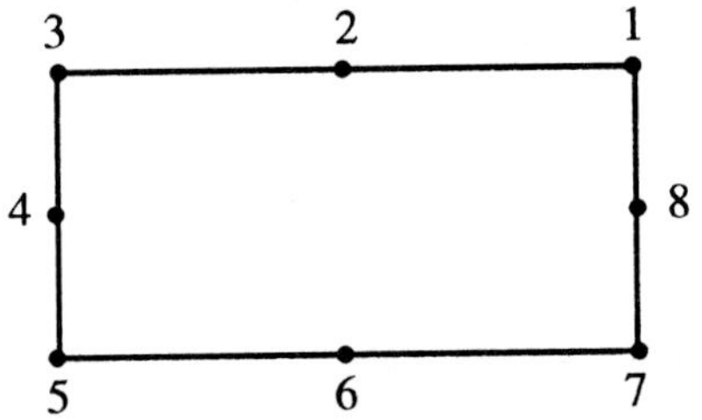

Figure 12.3.2 Element With Eight Nodes

Using the Galerkin technique, the differential equations may be reduced to a stiffness matrix and force vector on the ith element, given symbolically by

$$\iint_{\Omega_i} \left\{ \sum_{\alpha=1}^{8} \begin{bmatrix} -\dfrac{12}{Eh^3}\phi_\alpha\phi_\beta & \dfrac{12v}{Eh^3}\phi_\alpha\phi_\beta & 0 & -\dfrac{\partial^2\phi_\alpha}{\partial x_1^2}\phi_\beta \\ \dfrac{12v}{Eh^3}\phi_\alpha\phi_\beta & -\dfrac{12}{Eh^3}\phi_\alpha\phi_\beta & 0 & -\dfrac{\partial^2\phi_\alpha}{\partial x_2^2}\phi_\beta \\ 0 & 0 & -\dfrac{24\,(1+v)}{Eh^3}\phi_\alpha\phi_\beta & 2\dfrac{\partial^2\phi_\alpha}{\partial x_1\,\partial x_2}\phi_\beta \\ -\dfrac{\partial^2\phi_\alpha}{\partial x_1^2}\phi_\beta & -\dfrac{\partial^2\phi_\alpha}{\partial x_2^2}\phi_\beta & 2\dfrac{\partial^2\phi_\alpha}{\partial x_1\,\partial x_2}\phi_\beta & 0 \end{bmatrix}_i \begin{bmatrix} Z_1^\alpha \\ Z_2^\alpha \\ Z_3^\alpha \\ Z_4^\alpha \end{bmatrix} - \begin{bmatrix} F_1^\alpha\phi_\alpha\phi_\beta \\ F_2^\alpha\phi_\alpha\phi_\beta \\ F_3^\alpha\phi_\alpha\phi_\beta \\ F_4^\alpha\phi_\alpha\phi_\beta \end{bmatrix}_i \right\} dx_1\,dx_2 = 0, \quad \beta = 1, \ldots, 8$$

By performing integration by parts, the above relation yields

$$\iint_{\Omega_i} \left\{ \sum_{\alpha=1}^{8} \begin{bmatrix} -\dfrac{12}{Eh^3}\phi_\alpha\phi_\beta & \dfrac{12v}{Eh^3}\phi_\alpha\phi_\beta & 0 & \dfrac{\partial\phi_\alpha}{\partial x_1}\dfrac{\partial\phi_\beta}{\partial x_1} \\ \dfrac{12v}{Eh^3}\phi_\alpha\phi_\beta & -\dfrac{12}{Eh^3}\phi_\alpha\phi_\beta & 0 & \dfrac{\partial\phi_\alpha}{\partial x_2}\dfrac{\partial\phi_\beta}{\partial x_2} \\ 0 & 0 & -\dfrac{24\,(1+v)}{Eh^3}\phi_\alpha\phi_\beta & -\dfrac{\partial\phi_\alpha}{\partial x_1}\dfrac{\partial\phi_\beta}{\partial x_2} - \dfrac{\partial\phi_\alpha}{\partial x_2}\dfrac{\partial\phi_\beta}{\partial x_1} \\ \dfrac{\partial\phi_\alpha}{\partial x_1}\dfrac{\partial\phi_\beta}{\partial x_1} & \dfrac{\partial\phi_\alpha}{\partial x_2}\dfrac{\partial\phi_\beta}{\partial x_2} & -\dfrac{\partial\phi_\alpha}{\partial x_1}\dfrac{\partial\phi_\beta}{\partial x_2} - \dfrac{\partial\phi_\alpha}{\partial x_2}\dfrac{\partial\phi_\beta}{\partial x_1} & 0 \end{bmatrix}_i \begin{bmatrix} Z_1^\alpha \\ Z_2^\alpha \\ Z_3^\alpha \\ Z_4^\alpha \end{bmatrix} - \begin{bmatrix} F_1^\alpha\phi_\alpha\phi_\beta \\ F_2^\alpha\phi_\alpha\phi_\beta \\ F_3^\alpha\phi_\alpha\phi_\beta \\ F_4^\alpha\phi_\alpha\phi_\beta \end{bmatrix}_i \right\} dx_1\,dx_2 = 0, \quad \beta = 1, \ldots, 8 \tag{12.3.5}$$

where Ω_i is the subdomain corresponding to the ith element.

After integration, Eq. 12.3.5 becomes

$$\mathbf{k}_i\mathbf{z}_i = \mathbf{F}_i \tag{12.3.6}$$

where $\mathbf{k}_i$ is a stiffness matrix for the ith element. This integration over a subdomain may be conveniently carried out by introducing local coordinates and using a Gaussian quadrature formula. The details are explained in Ref. 28. Assembling Eq. 12.3.6 for each element yields a system of linear equations

$$\mathbf{Kz} = \mathbf{F} \tag{12.3.7}$$

that represents the finite element approximation of the system of plate equations of Eq. 12.3.1.

In order to compare with numerical results of Section 11.4, the following square plate problems are analyzed:

(1) simply supported boundary
(2) clamped boundary

For each boundary condition, the following thicknesses are studied:

(1) uniform thickness
(2) varying thickness $h(x_1, x_2) = 2\left(1 - \frac{|x_1|}{a}\right)\left(1 - \frac{|x_2|}{a}\right) h_0$

Numerical results for four test problems are presented in Tables 12.3.1 and 12.3.2. Note that the finite element method gives consistently good approximations, in comparison with results presented in Section 11.4.

Table 12.3.1
Simply Supported Square Plate (Finite Element Method)

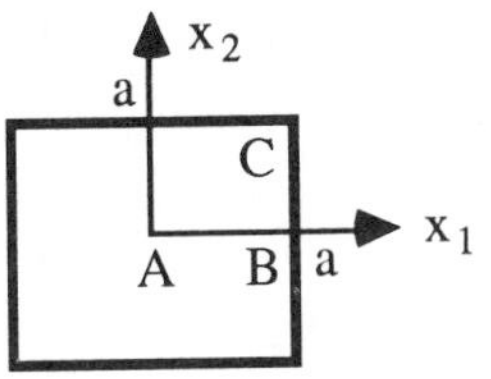

(a) Uniform thickness

	M_{x_1} @ A $\times 4fa^2$	$M_{x_1x_2}$ @ C $\times 4fa^2$	$z_4(u)$ @ A $\times 16fa^4$
4 elements	0.048206	0.034505	0.0040563
9 elements	0.047855	0.033768	0.0040612
Exact [26]	0.0479	0.0325	0.00406

(b) Varying thickness

	M_{x_1} @ A $\times 4fa^2$	$M_{x_1x_2}$ @ C $\times 4fa^2$	$z_4(u)$ @ A $\times 16fa^4$
4 elements	0.086030	0.0028338	0.0019514
9 elements	0.086501	0.0038296	0.0019556

Table 12.3.2
Clamped Square Plate (Finite Element Method)

(a) Uniform thickness

	M_{x_1} @ A $\times 4fa^2$	M_{x_1} @ B $\times 4fa^2$	$z_4(u)$ @ A $\times 16fa^4$
4 elements	0.021366	− 0.055412	0.0012677
9 elements	0.023162	− 0.053113	0.0012661
Exact [26]	0.0231	− 0.0513	0.00126

(b) Varying thickness

	M_{x_1} @ A $\times 4fa^2$	M_{x_1} @ B $\times 4fa^2$	$z_4(u)$ @ A $\times 16fa^4$
4 elements	0.045565	− 0.056682	0.0006928
9 elements	0.046390	− 0.055672	0.0006937

12.4 FINITE ELEMENT METHODS IN DYNAMICS

In the preceding subsections, only finite element methods for elliptic equations are discussed. Galerkin's method is flexible enough to apply also to initial-value problems, which are considered in this section with finite element approximations.

The Heat Equation

A natural setting in which to illustrate the finite element method for dynamics is the heat equation,

$$\frac{\partial u}{\partial t} - \frac{\partial^2 u}{\partial x^2} = h(x, t), \quad 0 < x < \pi, \quad t > 0 \tag{12.4.1}$$

This parabolic differential equation represents heat conduction in a rod, where u(x, t) is the temperature at point x and time $t > 0$ and h(x, t) is a heat-source term. Boundary conditions at $x = 0$ and at $x = \pi$ are

$$u(0, t) = \frac{\partial u}{\partial x}(\pi, t) = 0, \quad t > 0 \tag{12.4.2}$$

The first condition means physically that the temperature at the left end of the rod is held at 0. The second condition means that the right end of the rod is insulated, so there is no temperature gradient at $x = \pi$. To complete the statement of the problem, the initial temperature is

$$u(x, 0) = f(x), \quad 0 \le x \le \pi \tag{12.4.3}$$

This classical formulation of heat conduction is fraught with difficulties. For example, Eqs. 12.4.2 and 12.4.3 are contradictory if f(x) fails to vanish at $x = 0$, or if $\partial f(x)/\partial x$ does not vanish at $x = \pi$. In addition, h(x, t) could be a point source at x_0, which is singular, so Eq. 12.4.1 is not meaningful at that point. In each of these cases, the underlying physical problem still makes sense.

Since the equation is not symmetric, the Galerkin formulation is employed. At each time $t > 0$, it is desired that for arbitrary v(x) in $C^2(0, \pi)$,

$$\int_0^\pi (u_t - u_{xx} - h)\, v\, dx = 0 \tag{12.4.4}$$

In the steady state, with $u_t(x, t) = 0$ and $h(x, t) = h(x)$, this coincides with the earlier Galerkin formulation.

To achieve some symmetry, integrate the term $-u_{xx}v$ in Eq. 12.4.4 by parts. The boundary term $u_x v$ vanishes at $x = \pi$ and it is required that $v(0) = 0$. Equation 12.4.4 then becomes

$$\int_0^\pi (u_t v + u_x v_x - hv)\, dx = 0 \tag{12.4.5}$$

for all v(x) in $H_A(0, \pi)$.

This is the starting point for finite element approximation. Given an n-dimensional subspace S^h of $H_A(0, \pi)$, the goal of the Galerkin method is to find a function $u^h(x, t)$

that, for each $t > 0$, is in S^h of Section 12.1 and satisfies

$$\int_0^{\pi} (u_t^h v^h + u_x^h v_x^h - h v^h) \, dx = 0 \qquad (12.4.6)$$

for all $v^h(x)$ in S^h.

Notice that the time variable is still continuous; i.e., the Galerkin formulation is discrete in the space variables and yields a system of ordinary differential equations in time. It is these equations that must be solved numerically. To make this formulation useful, a basis $\{\phi_i(x)\}$ is chosen for the space S^h and the unknown solution is expanded as

$$u^h(x, t) = \sum_{j=1}^{n} q_j(t)\, \phi_j(x) \qquad (12.4.7)$$

The coefficients $q_j(t)$ are determined by the Galerkin method; i.e.,

$$\sum_{j=1}^{n} \int_0^{\pi} \left[\dot{q}_j \phi_j \phi_k + q_j \frac{\partial \phi_j}{\partial x} \frac{\partial \phi_k}{\partial x} - h \phi_k \right] dx = 0, \quad k = 1, \ldots, n \qquad (12.4.8)$$

Since every $v^h(x)$ is a combination of the coordinate functions $\phi_k(x)$, it is enough to apply the principle only to these functions. The result is a system of n ordinary differential equations for the n unknowns, $q_1(t), \ldots, q_n(t)$, and the initial condition $u(x, 0) = f(x)$ is still to be accounted for. Equation 12.4.8 may be rewritten as

$$\sum_{j=1}^{n} \left[\dot{q}_j(t) \int_0^{\pi} \phi_j \phi_k \, dx + q_j \int_0^{\pi} \phi_j' \phi_k' \, dx \right] = \int_0^{\pi} h(x, t)\, \phi_k(x)\, dx,$$
$$k = 1, \ldots, n \qquad (12.4.9)$$

Defining

$$m_{jk} = \int_0^{\pi} \phi_j \phi_k \, dx$$
$$k_{ij} = \int_0^{\pi} \phi_j' \phi_k' \, dx \qquad (12.4.10)$$
$$p_k(t) = \int_0^{\pi} h(x, t)\, \phi_k \, dx$$

Eq. 12.4.9 may be written in matrix form as

$$\mathbf{M}\dot{\mathbf{q}} + \mathbf{K}\mathbf{q} = \mathbf{p}(t) \qquad (12.4.11)$$

where **M** and **K** are $n \times n$ symmetric, positive definite matrices. This system of equations can be solved numerically by efficient numerical integration techniques for $\mathbf{q}(t) = [\, q_1(t), \ldots, q_n(t)\,]^T$. The result may be substituted into Eq. 12.4.7, to obtain the approximate solution of the transient heat problem.

Hyperbolic Equations

As an introduction to the finite element method for hyperbolic equations, consider the dynamics problem

$$\begin{aligned} u_{tt} + Au &= h(x, t) \\ u(x, 0) = f(x),\ \ u_t(x, 0) &= g(x) \end{aligned} \tag{12.4.12}$$

with x in $\Omega \subset R^n$ and u(x, t) in D_A, which is a subset of $C^{2m}(\Omega \times T)$ that satisfies homogeneous boundary conditions, A is a positive definite operator of order 2m for fixed t, and h(x, t) is the forcing function.

The Galerkin method begins by requiring that, for all v(x) in D_A and for all t,

$$(\, u_{tt}, v\,) + (\, Au, v\,) = (\, h, v\,) \tag{12.4.13}$$

Integrating by parts in the second term on the left,

$$(\, u_{tt}, v\,) + [\, u, v\,]_A = (\, h, v\,) \tag{12.4.14}$$

for all v(x) in H_A and $t > 0$. In the Galerkin approximation, u(x, t) and v(x) are replaced by $u^h(x, t) = \sum_{j=1}^{n} q_j(t)\, \phi_j(x)$ and $v^h(x)$ in S^h. The time-dependent coefficients $q_j(t)$, $j = 1, \ldots, n$, are determined by

$$\sum_{j=1}^{n} \ddot{q}_j\, (\, \phi_j, \phi_k\,) + \sum_{j=1}^{n} q_j\, [\, \phi_j, \phi_k\,]_A = (\, h, \phi_k\,),\ \ k = 1, \ldots, n \tag{12.4.15}$$

This is an ordinary differential equation in the time variable. In this case, the same mass and stiffness matrices as in the parabolic case appear, but the equation is of second order; i.e.,

$$\mathbf{M}\ddot{\mathbf{q}} + \mathbf{K}\mathbf{q} = \mathbf{p}(t) \tag{12.4.16}$$

Since the matrices **M** and **K** in Eq. 12.4.16 are positive definite, methods of matrix differential equations may be directly applied. The equations may be decoupled using a modal matrix transformation and the resulting simple second-order scalar differential equations may be solved in closed form, or with simple and effective numerical integration methods. This method is routinely applied in the field of structural mechanics, with good results.

REFERENCES

1. Strang, G., *Linear Algebra and Its Applications*, 2nd ed., Academic Press, New York, 1980.
2. Kolman, B., *Elementary Linear Algebra*, Macmillan, New York, 1970.
3. Davis, H. F., *Introduction to Vector Analysis*, 4th ed., Allyn & Bacon, Boston, 1979.
4. Pipes, L. A., *Matrix Methods for Engineering*, Prentice Hall, Englewood Cliffs, NJ, 1961.
5. Langhaar, H. L., *Energy Methods in Applied Mechanics*, Wiley, New York, 1962.
6. Przemieniecki, J. S., *Theory of Matrix Structural Analysis*, McGraw-Hill, New York, 1968.
7. Kreyszig, E., *Advanced Engineering Mathematics*, 5th ed., Wiley, New York, 1983.
8. Taylor, A. E., and Mann, W. R., *Advanced Calculus*, 3rd ed., Ginn & Company, Boston, 1983.
9. Hildebrand, F. B., *Advanced Calculus for Applications*, 2nd ed., Prentice Hall, Englewood Cliffs, NJ, 1976.
10. Hochstadt, H., *The Functions of Mathematical Physics*, Wiley-Interscience, New York, 1971.
11. Courant, R., and Hilbert, D., *Methods of Mathematical Physics*, Vol. I, Wiley-Interscience, New York, 1962.
12. Hildebrand, F. B., *Methods of Applied Mathematics*, Prentice Hall, Englewood Cliffs, NJ, 1952.
13. Lanczos, C., *Linear Differential Operators*, Van Nostrand, New York, 1961.
14. Royden, H. L., *Real Analysis*, Macmillan, New York, 1963.

15. Reed, M., and Simon, B., *Methods of Modern Mathematical Physics, Vol. I: Functional Analysis*, Academic Press, New York, 1972.

16. Synge, J. L., and Schild, A., *Tensor Calculus*, University of Toronto Press, Toronto, 1949.

17. Tychonov, A. N., and Samarski, A. A., *Partial Differential Equations of Mathematical Physics*, Vol. I, Holden-Day, San Francisco, 1964.

18. Mikhlin, S. G., *Mathematical Physics, An Advanced Course*, North-Holland, Amsterdam, 1970.

19. Garabedian, P. R., *Partial Differential Equations*, Wiley, New York, 1964.

20. Coddington, E. A., and Levington, N., *Theory of Ordinary Differential Equations*, McGraw-Hill, New York, 1955.

21. Mikhlin, S. G., *Variational Methods in Mathematical Physics*, Pergamon Press, Oxford, U.K., 1964.

22. Aubin, J. P., *Applied Functional Analysis*, Wiley, New York, 1979.

23. Sokolnikoff, I. S., *Mathematical Theory of Elasticity*, McGraw-Hill, New York, 1956.

24. Gould, P. L., *Analysis of Shells and Plates*, Springer-Verlag, New York, 1988.

25. Shames, I. H., and Dym, C. L., *Energy and Finite Element Methods in Structural Mechanics*, McGraw-Hill, New York, 1985.

26. Timoshenko, S., and Goodier, J. N., *Theory of Elasticity*, McGraw-Hill, New York, 1951.

27. Strang, G., and Fix, G. J., *An Analysis of the Finite Element Method*, Prentice Hall, Englewood Cliffs, NJ, 1973.

28. Zienkiewicz, O. C., *The Finite Element Method*, 3rd ed., McGraw-Hill, New York, 1977.

INDEX

A

B

C

D

E

F

G

H

N

O

P

Q

R

S

T

U

V

W

Y

Z